Table C Standard normal distribution (*continued*)

Z	0.00	0.01	0.02	0.03	0.04	0.05	0.06	0.07	0.08	0.09
0.0	0.5000	0.5040	0.5080	0.5120	0.5160	0.5199	0.5239	0.5279	0.5319	0.5359
0.1	0.5398	0.5438	0.5478	0.5517	0.5557	0.5596	0.5636	0.5675	0.5714	0.5753
0.2	0.5793	0.5832	0.5871	0.5910	0.5948	0.5987	0.6026	0.6064	0.6103	0.6141
0.3	0.6179	0.6217	0.6255	0.6293	0.6331	0.6368	0.6406	0.6443	0.6480	0.6517
0.4	0.6554	0.6591	0.6628	0.6664	0.6700	0.6736	0.6772	0.6808	0.6844	0.6879
0.5	0.6915	0.6950	0.6985	0.7019	0.7054	0.7088	0.7123	0.7157	0.7190	0.7224
0.6	0.7257	0.7291	0.7324	0.7357	0.7389	0.7422	0.7454	0.7486	0.7517	0.7549
0.7	0.7580	0.7611	0.7642	0.7673	0.7704	0.7734	0.7764	0.7794	0.7823	0.7852
0.8	0.7881	0.7910	0.7939	0.7967	0.7995	0.8023	0.8051	0.8078	0.8106	0.8133
0.9	0.8159	0.8186	0.8212	0.8238	0.8264	0.8289	0.8315	0.8340	0.8365	0.8389
1.0	0.8413	0.8438	0.8461	0.8485	0.8508	0.8531	0.8554	0.8577	0.8599	0.8621
1.1	0.8643	0.8665	0.8686	0.8708	0.8729	0.8749	0.8770	0.8790	0.8810	0.8830
1.2	0.8849	0.8869	0.8888	0.8907	0.8925	0.8944	0.8962	0.8980	0.8997	0.9015
1.3	0.9032	0.9049	0.9066	0.9082	0.9099	0.9115	0.9131	0.9147	0.9162	0.9177
1.4	0.9192	0.9207	0.9222	0.9236	0.9251	0.9265	0.9279	0.9292	0.9306	0.9319
1.5	0.9332	0.9345	0.9357	0.9370	0.9382	0.9394	0.9406	0.9418	0.9429	0.9441
1.6	0.9452	0.9463	0.9474	0.9484	0.9495	0.9505	0.9515	0.9525	0.9535	0.9545
1.7	0.9554	0.9564	0.9573	0.9582	0.9591	0.9599	0.9608	0.9616	0.9625	0.9633
1.8	0.9641	0.9649	0.9656	0.9664	0.9671	0.9678	0.9686	0.9693	0.9699	0.9706
1.9	0.9713	0.9719	0.9726	0.9732	0.9738	0.9744	0.9750	0.9756	0.9761	0.9767
2.0	0.9772	0.9778	0.9783	0.9788	0.9793	0.9798	0.9803	0.9808	0.9812	0.9817
2.1	0.9821	0.9826	0.9830	0.9834	0.9838	0.9842	0.9846	0.9850	0.9854	0.9857
2.2	0.9861	0.9864	0.9868	0.9871	0.9875	0.9878	0.9881	0.9884	0.9887	0.9890
2.3	0.9893	0.9896	0.9898	0.9901	0.9904	0.9906	0.9909	0.9911	0.9913	0.9916
2.4	0.9918	0.9920	0.9922	0.9925	0.9927	0.9929	0.9931	0.9932	0.9934	0.9936
2.5	0.9938	0.9940	0.9941	0.9943	0.9945	0.9946	0.9948	0.9949	0.9951	0.9952
2.6	0.9953	0.9955	0.9956	0.9957	0.9959	0.9960	0.9961	0.9962	0.9963	0.9964
2.7	0.9965	0.9966	0.9967	0.9968	0.9969	0.9970	0.9971	0.9972	0.9973	0.9974
2.8	0.9974	0.9975	0.9976	0.9977	0.9977	0.9978	0.9979	0.9979	0.9980	0.9981
2.9	0.9981	0.9982	0.9982	0.9983	0.9984	0.9984	0.9985	0.9985	0.9986	0.9986
3.0	0.9987	0.9987	0.9987	0.9988	0.9988	0.9989	0.9989	0.9989	0.9990	0.9990
3.1	0.9990	0.9991	0.9991	0.9991	0.9992	0.9992	0.9992	0.9992	0.9993	0.9993
3.2	0.9993	0.9993	0.9994	0.9994	0.9994	0.9994	0.9994	0.9995	0.9995	0.9995
3.3	0.9995	0.9995	0.9995	0.9996	0.9996	0.9996	0.9996	0.9996	0.9996	0.9997
3.4	0.9997	0.9997	0.9997	0.9997	0.9997	0.9997	0.9997	0.9997	0.9997	0.9998

The cover image for *Discovering Statistics* is taken from one of the bo[illegible]
"Trial of the Pyx: How Much Gold Is in Your Gold Coins?" (from C[illegible]
Distributions). The Case Study describes a procedure held annually in Lo[illegible]
Ages to ensure that the coins of the realm contain the proper amount of g[illegible]
pling distribution of the sample mean, the probability is calculated that th[illegible]
would have been caught and punished had he cheated the throne by deba[illegible]
in the coins.

CASE STUDY | **Trial of the Pyx: How Much Gold Is in Your Gold Coins?**

The kings of bygone England had a problem: How much gold should they put into their gold coins? After all, the very commerce of the kingdom depended on the purity of the currency. How did the lords of the realm ensure that the coins floating around the kingdom contained reliable amounts of gold?

From the year 1282, the Trial of the Pyx has been held annually in London to ensure that newly minted coins adhere to the standards of the realm. It is the responsibility of the presiding judge to ensure that the trial proceeds lawfully and to inform Her Majesty's Treasury of the verdict. Six members of the Company of Goldsmiths compose the jury, who are given two months to test the coins. It works like this: A ceremonial boxwood chest, called the Pyx, is brought forth, and a sample of 100 of the coins cast that year at the mint is put into it. The Pyx is then weighed. In times past, each gold coin, called a guinea, had an expected weight of 128 grams, so the total weight of the guineas in the Pyx was expected to be 12,800 grams.

If the weight of the coins in the Pyx was much less than 12,800 grams, the jury concluded that the Master of the Mint was cheating the crown by pocketing the excess gold, and he was severely punished. On the other hand, if the coins in the Pyx weighed much more than 12,800 grams, that wasn't good either, since it cut down on the profits produced by the kings' coin-minting monopoly.

By how much could the Master of the Mint debase the coinage before getting caught? We shall see in the chapter's Case Study, The Trial of the Pyx, which unfolds in Section 7.2.

Learn more as this Case Study unfolds in Chapter 7, pages 338.

CASE STUDY | **Trial of the Pyx: How Much Gold Is in Your Gold Coins?**

Medieval English kings and queens devised a procedure to ensure that the coins of the realm contained the proper amount of gold. A sample of 100 of the gold coins that were cast each year was placed in a ceremonial box called the Pyx. At the chosen time, the Company of Goldsmiths jury weighed the gold coins. The weight of the entire sample of coins was supposed to be 12,800 grams. If the weight was much less than 12,800 grams, the jury concluded that the Master of the Mint was cheating the crown by pocketing the excess gold, and he was severely punished. We assume that, for the Master of the Mint to keep his head, the *sum* of the weights of the gold coins had to lie between 12,768 and 12,832 grams, and therefore that the average weight $\bar{x}$ of the sample of the 100 coins had to lie between 127.68 grams and 128.32 grams.

Discovering Statistics

Brief Version

DANIEL T. LAROSE

Central Connecticut State University

W. H. FREEMAN AND COMPANY / NEW YORK

Senior Publisher:	Craig Bleyer
Publisher:	Ruth Baruth
Senior Acquisitions Editor:	Roland Cheyney
Development Editors:	Shona Burke, Tony Palermino
Executive Marketing Manager:	Jennifer Somerville
Market Development Manager:	Kirsten Watrud
Media Editor:	Laura Capuano
Assistant Media Editor:	Catriona Kaplan
Associate Editor:	Katrina Wilhelm
Editorial Assistant:	Lauren Kimmich
Photo Editor:	Cecilia Varas
Photo Researcher:	Patty Cateura
Cover and Text Designer:	Blake Logan
Senior Project Editor:	Mary Louise Byrd
Illustrations:	MPS Limited, A Macmillan Company
Production Manager:	Susan Wein
Composition:	MPS Limited, A Macmillan Company
Printing and Binding:	RR Donnelley

Cover photo: Roz Woodward/Photodisc/Getty Images

TI-83™ screen shots are used with permission of the publisher: ©1996, Texas Instruments Incorporated. TI-83™ Graphic Calculator is a registered trademark of Texas Instruments Incorporated. Minitab is a registered trademark of Minitab, Inc. Microsoft© and Windows© are registered trademarks of the Microsoft Corporation in the United States and other countries. Excel screen shots are reprinted with permission from the Microsoft Corporation.

Library of Congress Control Number: 2009940462

Paperback ISBN-13: 978-1-4292-4462-6
ISBN-10: 1-4292-4462-3

Instructor's Edition ISBN-13: 978-1-4292-4506-7
ISBN-10: 1-4292-4506-9

Printed in the United States of America

First printing

W. H. Freeman and Company
41 Madison Avenue
New York, NY 10010
Houndmills, Basingstoke RG21 6XS, England
www.whfreeman.com

Brief Contents

Handwritten annotations: Lecture 1 ½2 ½8 · lecture · Lecture 3 · Lecture 4 · Lecture 6 ½ 13 · Lecture 7. 126 pg. · Lecture 9 · Lecture 11

Contents

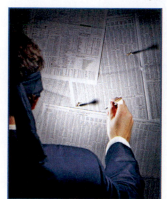

Chapter 3   Describing Data Numerically   77

Chapter 4   Describing the Relationship Between Two Variables   147

Our twenty-first-century world is flooded with data. Stock market returns and sports results snake across our TV screens in a nonstop stream. Grocery purchases are beep-beeped into data warehouses that enable the retailer to analyze the purchases and recommend individualized offers to their customers. Political candidates recite statistical facts and figures often massaged to support their positions on the issues. To develop a deeper sense of meaning and comprehension of data, students today need to study **statistics**: the art and science of collecting, analyzing, presenting, and interpreting data. *Discovering Statistics Brief Version* will help you develop the quantitative and analytical tools needed to understand statistics in today's data-saturated world.

Introductory Statistics Course

Discovering Statistics Brief Version is intended for an algebra-based, undergraduate, one-semester course in general introductory statistics for non-majors. The only prerequisite is intermediate algebra. Introductory statistics will prepare you to work with data in fields such as: psychology, business, nursing, education, and liberal arts, to name a few.
The text was written with the following goals in mind:

- Emphasize student motivation.
- Emphasize statistical literacy and develop statistical thinking.
- Use real data.
- Stress conceptual understanding in addition to knowledge of procedures.
- Foster active learning in the classroom.
- Use technology for developing conceptual understanding and analyzing data.
- Use assessments to improve and evaluate student learning.
- Demonstrate, through Case Studies, how newly acquired analytic tools may be applied to a familiar problem.

These goals follow the Guidelines for Assessment and Instruction in Statistics Education (GAISE) model, endorsed by the American Statistical Association.

Approach of *Discovering Statistics Brief Version*

Balanced analytical and computational coverage. The text integrates data interpretation and discovery-based methods with complete computational coverage of introductory statistics topics. Through unique and careful use of pedagogy, the text helps you develop your "statistical sense"—understanding the meaning behind the numbers. Equally, the text includes integrated and comprehensive computational coverage, including step-by-step solutions within examples. Select examples include screen shots and computer output from TI-83/84, Excel, and Minitab, with keystroke instructions located in the Step-by-Step Technology Guides at the end of sections.

Communication of results. *Discovering Statistics Brief Version* emphasizes how, in the real world, to explain statistical results to others who have never taken a statistics course. The text emphasizes the importance of keeping in mind how to interpret these results to non-specialists.

Emphasis on variability. The importance of variability in the introductory statistics curriculum cannot be overstated. Without a solid appreciation of how statistics may vary, there is little chance that you will be able to understand the crucial topic of sampling distributions.

Emphasis on sampling distributions and the Central Limit Theorem. All statistical inference in this text depends on sampling distributions. All test statistics, all confidence interval formulas depend on sampling distributions. It is vital that you have a grasp of sampling distributions and the Central Limit Theorem to completely understand classical statistical inference.

Use of powerful, current examples with real data. The terrorist attacks on September 11, 2001, the bird flu pandemic, and the popularity of online dating represent the variety of examples included in the text. Example and exercise topics reflect real-world problems and engage your interest in their solution. Real data (with sources cited) are frequently used to further demonstrate relevance of topics.

Features of *Discovering Statistics Brief Version*

Case Studies. A Case Study begins each chapter and is developed throughout the section examples, using the new set of tools that the section provides.

The Big Picture. Brief paragraphs look at "where we are coming from and where we are headed . . . ". (Chapter 2, page 32)

The Big Picture Where we are coming from, and where we are headed...

In Chapter 1, we became acquainted with the building blocks of data analysis, such as the concepts of population, sample, types of variables, and the importance of a random sample. In Chapter 2, we apply the adage "A picture is worth a thousand words." The human mind can assess information presented pictorially, better than it can through words and numbers alone. We can recognize the entire scope of the interrelationships within a picture and somehow formulate a synthesis. Psychologists sometimes call this innate pattern recognition ability the "gestalt" experience. Statistical graphics take advantage of these abilities to quickly summarize data.

Matched Objectives. Each section begins with a list of numbered objectives headed "By the end of this section, I will be able to . . .". The objective numbers are matched with the numbered topics within each section as well as with the end-of-section summary. (Chapter 9, pages 411, 416)

9.1 Introduction to Hypothesis Testing

Objectives: By the end of this section, I will be able to…

1. Construct the null hypothesis and the alternative hypothesis from the statement of the problem.

2. State the two types of errors made in hypothesis tests: the Type I error, made with probability α, and the Type II error, made with probability β.

1 Constructing the Hypotheses

Recall that inferential statistics refers to methods for learning about the unknown value of a population parameter by studying the information in a sample. The basic idea of hypothesis testing is the following:

Section 9.1 SUMMARY

1. Statistical hypothesis testing is a way of formalizing the decision-making process so that a decision can be rendered about the unknown value of the parameter. The status quo hypothesis that represents what has been tentatively assumed about the value of the parameter is called the null hypothesis and is denoted as H_0. The alternative hypothesis, or research hypothesis, denoted as H_a, represents an alternative conjecture about the value of the parameter.

2. When performing a hypothesis test, there are two ways of making a correct decision: to not reject H_0 when H_0 is true and to reject H_0 when H_0 is false. Also, there are two types of error: a Type I error is to reject H_0 when H_0 is true, and a Type II error is to not reject H_0 when H_0 is false. The probability of a Type I error is denoted as α (alpha). The probability of a Type II error is denoted as β (beta).

A Decision Is Not Proof

It is important to understand that the decision to reject or not reject H_0 does not prove anything. The decision represents whether or not there is sufficient evidence against the null hypothesis. This is our best judgment given the data available. You cannot claim to have *proven* anything about the value of a population parameter unless you elicit information from the entire population, which in most cases is not possible. ■

Developing Your Statistical Sense. This feature will empower you with useful perspectives that data analysts need to know. You will learn to think like a real-world statistical analyst. (Chapter 9, page 415)

What Does This Mean?

Feature boxes foster an intuitive approach and interpretation of results. Whenever a new formula or statistic is being introduced, the emphasis is on "What does this really mean?" Developing this understanding is just as important as getting the right answer, especially when the software can do the calculations. In the workplace, you may need to explain to your managers what the statistical results really mean. (Chapter 4, page 172)

Interpreting the Slope

What is the meaning of the slope coefficient b_1? In data analysis, we interpret the slope of the regression line as the *estimated change in y per unit increase in x.* In our temperature example, the units are degrees Fahrenheit, and the value of b_1 is 0.9865. Therefore, $b_1 = 0.9865$ means the following: "For each increase of 1°F in low temperature, the estimated high temperature increases by 0.9865°F." Also note that the slope is positive, which concurs with our assessment of the overall relationship between high and low temperatures. Further, is it a coincidence that both the slope and the correlation coefficient are positive? Not at all.

What If Scenarios. The scenarios help you focus on statistical thinking rather than rote computation. Because of the availability of powerful statistical computer packages, statistical analysis is easy to do badly. The wrong analysis is worse than useless. It can cost companies lots of money, may convince lawmakers to pass legislation affecting millions of people, can incorrectly determine effects of pharmaceuticals or environmental pollution, and can have many other serious ramifications. The What If scenarios are extensions of examples or exercises aimed at honing your critical-thinking skills. In What If exercises, the original problem set-up is altered in a specific but non-quantifiable way. You are then asked to think about how that change would percolate through the results, without recourse to calculations. The exercises as well as the scenarios are marked with the What If icon. (Chapter 3, page 86)

Give the Calculator a Rest

What If Scenarios offer you a chance to reflect on how changes in the initial conditions will percolate through the various aspects of a problem. The only requirement is to put your calculator down and think through the problem. No data are given. You are asked to find the answers by using your knowledge of what the statistics *represent.*

Consider Example 3.5 once again. Now imagine: *what if* there was a data entry error (a typo), and the subscription cost for the most expensive journal, the *Journal of Pure and Applied Algebra,* was not $3631 but actually some *unspecified* dollar amount greater than $3631. Describe how and why this change would have affected the following, if at all:

a. The mean subscription cost

b. The median subscription cost

c. The mode subscription cost

Solution

a. Consider Figure 3.3, a dotplot of the subscription costs, with the triangle indicating the mean cost of $1185.10. Recall that this represents the balance point of the data. As the cost of the most expensive journal increases (blue arrow), the point at which the data balance (that is, the mean) also moves somewhat to the right. Thus, the

Stepped Example Solutions. In selected examples, you are guided through the key steps needed to work the calculations and find the solution. (Chapter 9, page 436)

Solution

Step 1   **State the hypotheses.** In Example 9.8 our hypotheses were

$$H_0: \mu = 11 \quad \text{versus} \quad H_a: \mu > 11$$

where μ is the population mean number of times people use their debit cards per month.

Step 2   **Find Z_{crit} and state the rejection rule.** We have a right-tailed test and level of significance $\alpha = 0.01$, which, from Table 9.6, tell us that $Z_{crit} = 2.33$. Using Table 9.7, because we have a right-tailed test, the rejection rule will be "Reject H_0 if Z_{data} is greater then Z_{crit}"; that is, "Reject H_0 if Z_{data} is greater than 2.33" (see Figure 9.20).

Step 3   **Find Z_{data}.** We have a sample of $n = 36$ people who used their debit cards an average of $\bar{x} = 11.5$ times, and we assume $\sigma = 3$. Also, μ_0 is the hypothesized value of μ, in this case 11. Using these values, we get

$$Z_{data} = \frac{\bar{x} - \mu_0}{\sigma_{\bar{x}}} = \frac{\bar{x} - \mu_0}{\sigma/\sqrt{n}} = \frac{11.5 - 11.0}{3/\sqrt{36}} = 1.0$$

What Results Might We Expect? This feature, located in example solutions, challenges you to predict what the results of a particular problem will be. You are presented with a graphical view of the situation, and, before performing any calculations, you are asked to bring your intuition and common sense to bear on the problem and to state what results we might expect once we do the number crunching. (Chapter 7, page 329)

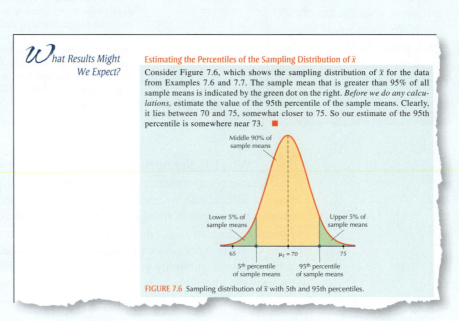

*W*hat Results Might We Expect?

Estimating the Percentiles of the Sampling Distribution of $\bar{x}$

Consider Figure 7.6, which shows the sampling distribution of $\bar{x}$ for the data from Examples 7.6 and 7.7. The sample mean that is greater than 95% of all sample means is indicated by the green dot on the right. *Before we do any calculations,* estimate the value of the 95th percentile of the sample means. Clearly, it lies between 70 and 75, somewhat closer to 75. So our estimate of the 95th percentile is somewhere near 73. ■

FIGURE 7.6 Sampling distribution of $\bar{x}$ with 5th and 95th percentiles.

Definitions and Formulas. Easily located in highlighted boxes, key definitions and formulas are important to understand when working examples and exercises. Important vocabulary and formulas are also listed (with page references) at the end of each chapter. (Chapter 1, page 6)

What Is Statistics?
Statistics is the *art* and *science* of

- collecting,
- analyzing,
- presenting, and
- interpreting data.

Exercises. *Discovering Statistics Brief Version* contains a rich and varied collection of section and chapter exercises:

- Clarifying the Concepts (conceptual)
- Practicing the Techniques (skill-based)
- Applying the Concepts (real-world applications)

At the end of each chapter, **Review Exercises** and a **Chapter Quiz** help to test your overall understanding of each chapter's concepts and to practice for exams. The answers to odd-numbered exercises and all chapter quiz exercises are located in the back of the book.

CHAPTER 5 *Quiz*

TRUE OR FALSE
1. True or false: An outcome is a collection of a series of events from the sample space of an experiment.
2. True or false: For any event A (even events like A: the moon is made of green cheese) the probability of A plus the probability of A^c always add up to 1.

FILL IN THE BLANK
3. The minimum value that a probability can take is _____ and the maximum value is _____.

7. For which type of sampling are consecutive draws independent?
8. For two events A and B, what do we call the event containing only those outcomes that belong to both A and B?

CALCULATIONS AND INTERPRETATIONS
9. The experiment is to roll a fair die twice. Find the following probabilities.
 a. Pair of 1's
 b. Pair of 2's

Step-by-Step Technology Guide. This feature covers TI-83/84 calculators, Excel, and Minitab, providing stepped keystroke instructions for working through selected examples in the text. Often, screen shots are provided showing the required steps. Screen shots of the results are often provided as well, either within the Step-by-Step Technology Guide or in the corresponding example. (Chapter 4, page 174)

STEP-BY-STEP TECHNOLOGY GUIDE: Correlation and Regression

We illustrate using Example 4.5, the temperature data (page 159).

TI-83/84
Step 1 Turn diagnostics on as follows. Press **2nd 0**. Scroll down and select **DiagnosticOn** (Figure 4.23). Press **ENTER** twice to turn diagnostics on.
Step 2 **Enter** the X (**Low Temp**) data in **L1**, and the Y (**High Temp**) data in **L2**.
Step 3 Press **STAT** and highlight **CALC**.
Step 4 Press **8** (not **4**) to choose **LinReg(a+bx)**, as shown in Figure 4.24.
Step 5 On the home screen, **LinReg(a+bx)** appears. Press **ENTER**.

Figure 4.23

Figure 4.24

MINITAB
Correlation Coefficient *r*
Step 1 Select **Stat > Basic Statistics > Correlation...**
Step 2 Select the variables you want to analyze and click **OK**.

EXCEL
Step 1 Enter the *x* variable in column **A** and the *y* variable in column **B**.
Step 2 Click on **Data > Data Analysis > Regression** and click **OK**.
Step 3 For **Input Y Range**, select cells **B1–B10**. For **Input X Range**, select cells **A1–A10**. Click **OK**.

Regression
Step 1 Enter the *x* variable variable in **C1** and the *y* variable in **C2**.
Step 2 Click on **Stat > Regression > Regression**.
Step 3 Select the *y* variable for the **Response Variable** and the *x* variable for the **Predictor Variable**. Click **OK**.

Try This in Class! In-class activities promote active and cooperative learning. Each chapter ends with a set of class-based activities, Try This in Class!, in which you can engage with other students during class time using manipulatives such as dice. (Chapter 9, page 474)

Try This in Class! In-class activities to enhance your understanding of statistics

TESTING FOR THE POPULATION MEAN OF THE SUM OF TWO FAIR DICE

Is the population mean μ of the sum of two fair dice equal to 7? The population standard deviation is $\sigma = 2.4152$. Each student will generate sample data to perform the following hypothesis test with level of significance $\alpha = 0.10$:

$$H_0: \mu = 7 \quad \text{versus} \quad H_a: \mu \neq 7$$

where μ represents the population mean sum of two fair dice. Assume that the distribution is normal. The instructor should prepare a number line on the board to display the collection of test statistic Z_{data} values. For $\alpha = 0.10$ and a two-tailed test, the value of Z_{crit} is 1.645. Thus, the instructor should draw lines separating the critical regions $Z_{data} < -1.645$ and $Z_{data} > 1.645$ from the noncritical region $-1.645 < Z_{data} < 1.645$.

1. Each student should toss two dice 10 times, recording the sum each time, and then calculate the mean sum $\bar{x}$ of this sample of size 10.

2. Using $n = 10$, $\sigma = 2.4152$, and $\mu_0 = 7$, each student should calculate his or her own value for the test statistic Z_{data} using his or her own value of $\bar{x}$ generated above.

3. Give each student a Post-It Note. Each student should record the following on the note: his or her name and the sample mean $\bar{x}$ and test statistic Z_{data} for his or her sample.

4. Each student should place the Post-It Note on the number line that the instructor prepared earlier.

5. Identify any Post-It Notes that indicate evidence against the null hypothesis that the population mean is 7. What is the value of Z_{data}? Of the sample mean $\bar{x}$? Do you agree that these sample means are unusual?

6. Approximately what percentage of sample means indicate evidence against $H_0: \mu_0 = 7$? Approximately what percentage of sample means indicate evidence consistent with $H_0: \mu_0 = 7$?

7. Overall, what is your personal conclusion about whether or not the population mean of the dice equals 7? What evidence can you point to in support of your conclusion?

Applets. Interactive statistical applets are located on your CD and the book companion Web site: **www.whfreeman.com/discostatbrief**. Applet icons in the text mark the related chapter material and exercises.

Caution notes. Signaled by the Caution icon, these warnings in the text help you to avoid common errors and misconceptions.

Supplements

The following electronic and print supplements are available with *Discovering Statistics Brief Version*.

courses.bfwpub.com/discostatbrief (access code required; available packaged with *Discovering Statistics Brief Version* or for purchase online). StatsPortal is the digital gateway to *Discovering Statistics Brief Version,* designed to enrich the course and enhance your study skills through a collection of Web-based tools. StatsPortal integrates a suite of diagnostic, assessment, tutorial, and enrichment features, enabling you to master statistics at your own pace. It is organized around the following learning components:

    Interactive eBook offers a complete and customizable online version of the text, fully integrated with all the media resources available with *Discovering Statistics Brief Version*. The eBook allows you to quickly search the text, highlight key areas, and add notes about what you're reading.

Resources organizes all the resources for *Discovering Statistics Brief Version* into one location for ease of use. The resources include the following:

- **StatTutor Tutorials** offer audio-multimedia tutorials tied directly to the textbook, including video, applets, and animations.
- **Stats@Work Simulations** put you in the role of a statistical consultant, helping you to better understand statistics interactively in the context of real-life scenarios. You are asked to interpret and analyze data presented in report form, as well as to interpret current events.
- **Statistical Applets** help you master key statistical concepts and work exercises from the text. There are 12 interactive applets.
- **CrunchIt! 2.0®**
- **EESEE Case Studies**
- **Data Sets** are available in ASCII, Excel, TI, Minitab, SPSS, and JMP formats.
- **Statistical Software Manuals** for TI-83/84, Excel, Minitab, and SPSS are provided.
- **WHFStat Add-ins for Excel**
- **Student Solutions Manual**

Assignments (for instructor use only) organizes assignments and grades through an easy-to-create assignment process providing access to questions from the Test Bank, Web Quizzes, and Exercises from *Discovering Statistics Brief Version*.

Online Study Center: www.whfreeman.com/osc/discostatbrief (access code required; available for purchase online) offers all the resources available in StatsPortal except the eBook and Assignments Center.

Companion Web Site: www.whfreeman.com/discostatbrief is an open-access Web site that includes statistical applets, data sets, and self-quizzes.

Interactive Student CD-ROM is included with every new copy of *Discovering Statistics Brief Version;* it contains access to data sets and applets (also found on the companion Web site).

Printed Student Solutions Manual offers detailed solutions for key exercises from each section of *Discovering Statistics Brief Version*. ISBN: 1-4292-4511-5

EESEE (Electronic Encyclopedia of Statistical Examples and Exercises) Case Studies. Developed by The Ohio State University Statistics Department, these electronic case studies provide a wide variety of timely, real examples with real data. EESEE case studies are available via an access code–protected Web site. Access codes are included with new copies of *Discovering Statistics Brief Version* (printed behind the CD in the back of the book), or subscriptions can be purchased online.

Acknowledgments

I would like to join W. H. Freeman and Company in thanking the reviewers who offered comments that assisted in the development and refinement of the *Brief Version*:

David E. Ching, *University of Hawai'i at Manoa*

David Gurney, *Southeastern Louisiana University*

Kristin M. Hartford, *Long Beach City College*

Jackie MacLaughlin, *Central Piedmont Community College*

Shai Neumann, *Brevard Community College*

Martha Tapia, *Berry College*

Additionally, I am grateful for the careful accuracy reviewing of manuscript by the students and advisers of the following mathematics and statistics clubs:

Actuarial Science Club, *Youngstown State University*
Adviser: G. Andy Chang, Professor, Mathematics and Statistics
Kappa Mu Epsilon Math Club, *Western Illinois University*
Adviser: Dr. Boris Petracovici
Math Club / Pi Mu Epsilon Illinois Zeta chapter, *Southern Illinois University Edwardsville*
Adviser: Adam G. Weyhaupt
Math Club, *Saint Mary's of Minnesota*
Adviser: Dr. Todd Iverson
Math Club, *University of South Florida*
Math/Computer Club, *Felician College*
Adviser: Dr. Elizabeth B. Uptegrove

Mathematics Club, *Arkansas Tech*
Adviser: Dr. Tom Limperis
Michigan Undergraduate Students of Statistics, Department of Statistics at *The University of Michigan*
Adviser: Dr. Brenda Gunderson and Co-Adviser: Dr. Kerby Shedden
Mu Sigma Rho at *Virginia Commonwealth University*
Adviser: Dr. Scott Street
Suffolk University Mathematics Society (SUMS)
Advisers: Dr. Xinxin Jiang and Dr. Adam Glesser
The Math and Stat Club, *Winona State University*
Advisers: Christopher J. Malone and April Kerby

I would also like to thank the many instructors from across the United States and Canada who offered comments on the full version of *Discovering Statistics,* upon which the *Brief Version* is based:

ARKANSAS George Bratton, *University of Central Arkansas*
ARIZONA Cheryl Ossenfort, *Coconino Community College*
CALIFORNIA Christine Cole, *Moorpark College*; Carol Curtis, *Fresno City College*; Kevin Fox, *Shasta College*; Kristin M. Hartford, *Long Beach City College*; Elizabeth Hamman, *Cypress College*; Sara Jones, *Santa Rosa Junior College*; Wendy Miao, *El Camino College*; Michael A. Nasab, *Long Beach City College*; Keith Oberlander, *Pasadena City College*; Greg Perkins, *Hartnell College*; Zika Perovic, *MiraCosta College*; Ladera Rosenburg, *Long Beach City College*; Sherman Sowby, *California State University, Fresno*; James Wan, *Long Beach City College*; Michael Zeitzew, *El Camino College*
CANADA Susan Chen, *Camosun College*; Shaun Fallat, *University of Regina*; Dorothy Levay, *Brock University*
COLORADO Holly Ashton, *Pikes Peak Community College*; Dean Barchers, *Red Rocks Community College*; Nels Grevstad, *Metropolitan State College of Denver*; Jay Schaffer, *University of Northern Colorado*
DELAWARE Derald E. Wentzien, *Wesley College*
FLORIDA Abraham Biggs, *Broward Community College*; Lisa M. Borzewski, *St. Petersburg College*; Janette H. Campbell, *Palm Beach Community College*; Zhao Chen, *Florida Gulf Coast University*; Lani Kempner, *Broward Community College*; Nancy Liu, *Miami Dade College*; Panagiotis Nikolopoulos, *Nova Southeastern University*; William Radulovich, *Florida Community College at Jacksonville*; Traci M. Reed, *St. Johns River Community College*; Pali Sen, *University of North Florida*; Jerry Shawver, *Florida Community College at Jacksonville*; Deanna Voehl, *Indian River State College*
GEORGIA Donna Brouillette, *Georgia Perimeter College*; Ayona Chatterjee, *University of West Georgia*; Wanda Eanes, *Macon State College*; Todd Hendricks, *Georgia Perimeter College*; Shahryar Heydari, *Piedmont College*; Barry J. Monk,

Macon State College; Chandler Pike, *University of Georgia*; Kim Robinson, *Clayton State University*; Howard L. Sanders, *Georgia Perimeter College*; Karen H. Smith, *University of West Georgia*; Martha Tapia, *Berry College*
HAWAII David Ching, *University of Hawai'i at Mānoa*; Eric Matsuoka, *Leeward Community College*
ILLINOIS Virginia Coil, *College of Lake County*; James Cicarelli, *Roosevelt University*; Faye Dang, *Joliet Junior College*; Linda Hoffman, *McKendree University*; Glenn Jablonski, *Triton College*; Julius Nadas, *Wilbur Wright College*; Stephen G. Zuro, *Joliet Junior College*
INDIANA Ewa Misiolek, *Saint Mary's College*
IOWA Russell Campbell, *University of Northern Iowa*
KANSAS Linda Herndon, *Benedictine College*; James Leininger, *MidAmerica Nazarene University*; Leesa Pohl, *Donnelly College*
KENTUCKY Brooke Buckley, *Northern Kentucky University;* Lloyd Jaisingh, *Morehead State University*; Christopher Schroeder, *Morehead State University*; Marlene Will, *Spalding University*
LOUISIANA Arun K. Agarwal, *Grambling State University*; David Busekist, *Southeastern Louisiana University*; Julien Doucet, *Louisiana State University at Alexandria*; Diane Fisher, *University of Louisiana at Lafayette*; David Gurney, *Southeastern Louisiana University*; Nabendu Pal, *University of Louisiana at Lafayette*; Victor S. Swaim, *Southeastern Louisiana University*
MARYLAND Cathy Hess, *Anne Arundel Community College*; Steven Hundert, *College of Southern Maryland*; Annette Noble, *University of Maryland Eastern Shore*; Steve Prehoda, *Frederick Community College*; Fary Sami, *Harford Community College*; Kim Sheppard, *Cecil College*
MASSACHUSETTS Mary Fowler, *Worcester State College*; LeRoy P. Hammerstrom, *Eastern Nazarene College*; Bonnie Wicklund, *Mount Wachusett Community College*

MICHIGAN Jennifer Borrello, *Grand Rapids Community College*; Lorraine Gregory, *Lake Superior State University*; Linda Reist, *Macomb Community College*; Kathy Zhong, *University of Detroit Mercy*
MINNESOTA Ken Grace, *Anoka-Ramsey Community College*; Mezbahur Rahman, *Minnesota State University, Mankato*
MISSOURI Kathy Carroll, *Drury University*; Christina Morian, *Lincoln University of Missouri*
MONTANA Debra Wiens, *Rocky Mountain College*
NEBRASKA Polly Amstutz, *University of Nebraska at Kearney*; Kathy Woitaszewski, *Central Community College*
NEW JERSEY Robert Thurston, *Rowan University*; Cathleen Zucco-Teveloff, *Rowan University*
NEW YORK David Bernklau, *Long Island University*; Jadwiga Domino, *Medaille College*; Reva Fish, *University at Buffalo*; Maryann Justinger, *Erie Community College*; Michael Kent, *Borough of Manhattan Community College*; William Price, *North Country Community College*; Sharon Testone, *Onondaga Community College*; Nicholas Zaino, *University of Rochester*
NORTH CAROLINA Emma B. Borynski, *Durham Technical Community College*; Ayesha Delpish, *Elon University*; Jackie MacLaughlin, *Central Piedmont Community College*; Jeanette Szwec, *Cape Fear Community College*; John Russell Taylor, *The University of North Carolina at Charlotte*; James Truesdell, *Chowan University*
OHIO G. Andy Chang, *Youngstown State University*; Don Davis, *Lakeland Community College*; Arjun Gupta, *Bowling Green State University*; William Huepenbecker, *BGSU Firelands*; Gaurab Mahapatra, *The University of Akron*; Mahbobeh Vezvaei, *Kent State University*
OKLAHOMA Mickle Duggan, *East Central University*; John Nichols, *Oklahoma Baptist University*; William Warde, *Oklahoma State University*
OREGON Jong Sung Kim, *Portland State University*; Carrie Kyser, *Clackamas Community College*

PENNSYLVANIA Linda M. Myers, *Harrisburg Area Community College*; Sandra Nypaver, *Mount Aloysius College*
SOUTH CAROLINA Diana J. Asmus, *Greenville Technical College*; Thomas Fitzkee, *Francis Marion University*; Erwin Walker, *Clemson University*
TENNESSEE Aniekan Ebiefung, *University of Tennessee at Chattanooga*; Frankie E. Harris, *Southwest Tennessee Community College*; Marc Loizeaux, *University of Tennessee at Chattanooga*; Mary Ella Poteat, *Northeast State Technical Community College*
TEXAS Ananda Bandulasiri, *Sam Houston State University*; Ferry Butar Butar, *Sam Houston State University*; Ola Disu, *Tarrant County College*; Emmett Elam, *Texas Tech University*; Maggie Foster, *Tarrant County College*; Grady Grizzle, *North Lake College*; Jada P. Hill, *Richland College*; Melinda Holt, *Sam Houston State University*; Jianguo Liu, *University of North Texas*; David D. Marshall, *Texas Woman's University*; Melissa Reeves, *East Texas Baptist University*; Ricardo Rodriguez, *Eastfield College*; Daniela Stoevska-Kojouharov, *Tarrant County College*; Jo Tucker, *Tarrant County College*
UTAH Kari Arnoldsen, *Snow College*; Joe Gallegos, *Salt Lake Community College*; Ruth Trygstad, *Salt Lake Community College*
VIRGINIA John Avioli, *Christopher Newport University*; Robert May, *Virginia Highlands Community College*; Mike Shirazi, *Germanna Community College*; Glenn Weber, *Christopher Newport University*; Ken Wissmann, *Shenandoah University*
WASHINGTON Margaret Balachowski, *Everett Community College*; Kelly Brooks, *Pierce College*; Abel Gage, *Skagit Valley College*; John Kellermeier, *Tacoma Community College*
WISCONSIN William K. Applebaugh, *University of Wisconsin – Eau Claire*; David M. Reineke, *University of Wisconsin – La Crosse*; Vicki Whitledge, *University of Wisconsin – Eau Claire*

Discovering Statistics Brief Version owes much to the untiring efforts of the team of professionals at W. H. Freeman and Company. I would like to thank Craig Bleyer, Mary Louise Byrd, Shona Burke, Kirsten Watrud, and Jennifer Somerville for contributing their talents to the creation of the book. A big thank you goes to Tony Palermino, whose development editing greatly improved the text, as well as to Pam Bruton, Penny Hull, Ann Cannon, Christina Morian, Sivanandan Balakumar, and Jackie Miller. Most especially, I would like to thank Ruth Baruth, Mathematics and Statistics Publisher, who recognized the need for a book like *Discovering Statistics* and helped make it a reality.

I also wish to thank Dr. Timothy Craine and Dr. Chun Jin, Chair and Assistant Chair of the Department of Mathematical Sciences at Central Connecticut State University, Dr. Dipak K. Dey, Chair of the Department of Statistics at the University of Connecticut, and Dr. John Judge, Chair of the Department of Mathematics at Westfield State College. Thanks to my daughter and statistician-in-training Chantal Danielle (21), for carrying on the love of statistics to the next generation, and to my twin children, Tristan Spring and Ravel Renaissance (10), for demonstrating that there is life beyond the computer screen. Above all, I extend my deepest gratitude to my darling wife of 24 years, Debra J. Larose, for her love, understanding, and true belief in the project.

Since his days of collecting baseball cards as a youngster and scrutinizing the statistics of his favorite players, Dan Larose has loved statistics. He also loved language and writing, so when Dan went to college he majored in French, then philosophy, and finally linguistics and computer science. This background in the liberal arts honed his writing ability. However, his love of statistics never left him, so he went on to earn an M.S. (1993) and a Ph.D. in statistics (1996) from the University of Connecticut. Today, Dan is Professor of Statistics in the Department of Mathematical Sciences at Central Connecticut State University.

At CCSU, Dan designed, developed, and now directs the world's first online Master of Science degree and Graduate Certificate program in data mining. He has published three books on data mining and one book on SAS programming. *Discovering Knowledge in Data: An Introduction to Data Mining* has been translated into French and Polish. He is the founder of DataMiningConsultant.com, and his consulting clients include *The Economist* magazine, the CIT Group, KPMG International, Sonalysts, Inc., Booz Allen and Hamilton, the Hospital for Special Care, and Microsoft. His consulting work includes a $750,000 Phase II grant from the Air Force Office of Research, Storage Efficient Data Mining of High Speed Data Streams. He is the Series Editor for the Wiley Series on Methods and Applications in Data Mining.

However, Dan's favorite work is imparting a love of statistics to a new generation, and he trusts that *Discovering Statistics Brief Version* will help to do so.

Dan lives in Tolland, Connecticut, with his wife and three children.

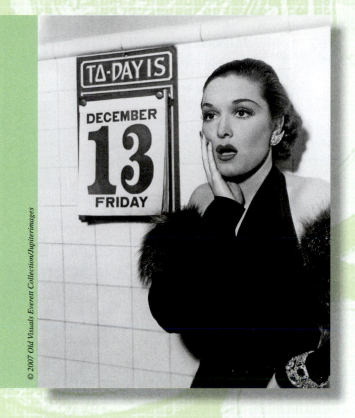

The Nature of Statistics

CASE STUDY

Does Friday the 13th Change Human Behavior?

Superstitions affect most of us. Some people will never walk under a ladder, while others will alter their path to avoid a black cat. Do you think that people change their behavior on Friday the 13th? Perhaps, suspecting that it may be unlucky, some people might elect to stay home and watch television rather than venture outdoors or drive on the highway.

But how does one go about researching such a question? How would *you* do it? In this chapter, we will learn about a British study that considered this question.

The Big Picture

Where we are coming from, and where we are headed...

We begin Chapter 1 by sharing some stories with you, data stories. Hidden within many data sets lie undiscovered items of real human interest. For example:

- Why did many of the survivors of Hurricane Katrina choose not to evacuate?
- How did you feel right after the events of September 11, 2001? Do you think that men and women experienced different emotions?

Our goal is to demonstrate that the field of statistics, often said to be too dry, actually provides us with the tools to understand and appreciate the emotions and experiences of the people behind the numbers.

In Section 1.2, we will learn about the building blocks of data analysis, including many basic but crucial concepts that the rest of the book depends on, such as *population, sample, variable, element,* and *observation*. In Section 1.3, we will investigate ways of gathering data, including random sampling, questionnaires, and statistical studies.

Where do we go from here? Once we have gathered the data, we will need to find ways to describe the characteristics of the data set. Humans, especially "right-brain people," understand graphical representations of information very well, and so in Chapter 2 we will investigate methods of *describing data using graphs and tables*. Later, in Chapter 3, we will look into *describing data numerically,* which is often characterized as a left-brain function. ■

1.1 Data Stories: The People Behind the Numbers

Objective: By the end of this section, I will be able to...

1. Realize that behind each data set lies a story about real people undergoing real-life experiences.

We begin *Discovering Statistics* by sharing some data stories. The field of statistics is often regarded as dry, too antiseptic and unfeeling. We hope to dispel this misunderstanding by sharing these data stories with you. We hope that these stories will kindle a response in you, be it sympathy or curiosity or concern, for behind every data set lies a story about the lives of real people. Individual people are speaking to us from behind the numbers.

Example 1.1   The reasons Katrina survivors did not evacuate

Hurricane Katrina was the costliest and one of the deadliest hurricanes in American history. Damages exceeded $50 billion and fatalities exceeded 1300, according to the National Oceanic and Atmospheric Administration. In September 2005, a survey was

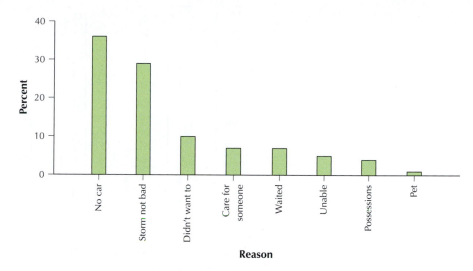

FIGURE 1.1 Bar graph of Katrina survivors' reasons for not evacuating.

conducted of a group of hurricane survivors who had later been moved to shelters in the Greater Houston area. The respondents who did not evacuate were asked what was their most important reason for not evacuating. Figure 1.1 provides a bar graph of the responses, with Table 1.1 supplying more detailed information. ■

Table 1.1 Katrina survivors' most important reasons for not evacuating

Reason	Percent
I did not have a car or a way to leave	36
I thought the storm and its aftermath would not be as bad as they were	29
I just didn't want to leave	10
I had to care for someone who was physically unable to leave	7
I waited too long	7
I was physically unable to leave	5
I worried that my possessions would be stolen or damaged if I left	4
I didn't want to leave my pet	1

Example 1.2   Were there gender differences in the emotions experienced immediately after September 11, 2001?

What was your strongest emotion in the days following the September 11, 2001, terrorist attacks in New York City and Washington, DC? Do you think that men and women felt the same emotions, or did the reactions of men and women differ? In an NBC News Terrorism Poll conducted the day after the tragic events, the following question was asked: "Which one of the following emotions do you feel the most strongly in response to these terrorist attacks: sadness, fear, anger, disbelief, vulnerability?" Figure 1.2 is called a *clustered bar graph* and shows the results. The dominant emotion felt by the men was anger, while the women tended to feel either sadness, anger, or disbelief. Note how the bar graph makes these findings—that there were indeed systematic differences in the emotions felt by men and women regarding the events of September 11, 2001— crystal clear. We will learn how to construct bar graphs in Chapter 2, "Describing Data Using Graphs and Tables." ■

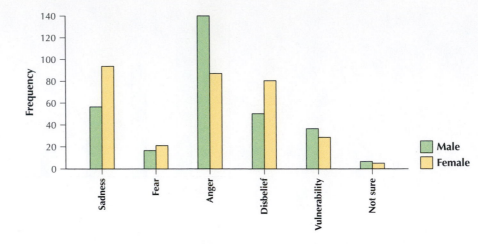

FIGURE 1.2   Clustered bar graph of strongest emotions felt regarding the September 11, 2001, attacks (by gender).

Example 1.3   The ballot that changed history

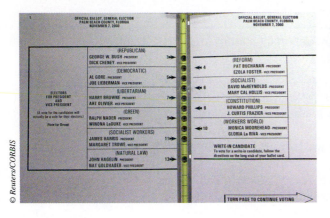

© Reuters/CORBIS

Look at the accompanying illustration. Do you find this 2000 presidential election ballot confusing? There is evidence that many Palm Beach County, Florida, residents did find the ballot confusing. According to a review of discarded ballots by the *Palm Beach Post,* confused voters marked more than one choice on the county's "butterfly ballot," thereby presumably costing Gore 6607 votes.[1] Bush officially won Florida by 537 votes—and with it the presidency.

There may also be evidence that many confused Palm Beach County voters chose Reform Party candidate Pat Buchanan by mistake as well. The scatterplot in Figure 1.3 shows, for each of Florida's counties, the number of votes for Buchanan versus the total number of votes. There is a clear trend of a linear (straight-line) relationship except for one glaring point that lies outside the trend, which is Palm Beach County. Do you think that this confusing ballot had an effect on the results of the election? We will learn how to construct and interpret scatterplots in Chapter 4, "Describing the Relationship Between Two Variables."

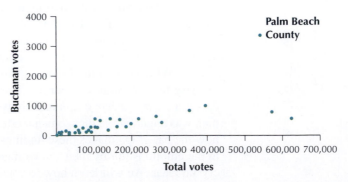

FIGURE 1.3   Scatterplot of Buchanan votes versus total votes in Florida counties, 2000.

Section 1.1 EXERCISES

Refer to Example 1.1 for Exercises 1–4.

1. If you had been in New Orleans during Hurricane Katrina, which of the reasons in Table 1.1 might have been most powerful for convincing you not to evacuate? Which of the reasons would not have played a part in convincing you not to evacuate?

2. Refer to Figure 1.1.
  a. What does the graph say was the most common reason why the Katrina survivors did not evacuate?
  b. What does Table 1.1 say was the most common reason?
  c. Which is more descriptive, the table or the figure?
  d. Why do you think the text in Figure 1.1 has been shortened?

3. If you were writing a news story that sought to display the Katrina survivors in the most sympathetic light, which reasons from Table 1.1 might you emphasize?

4. If you were writing a news story that sought to display the Katrina survivors in a less favorable light, which reasons from Table 1.1 might you emphasize?

Refer to Example 1.2 for Exercises 5–8.

5. What was your strongest emotion in the days following the September 11, 2001, terrorist attacks in New York City and Washington, DC?

6. Can you find the bar in Figure 1.2 that reflects the emotion you felt? Was this the most common emotion felt by others of your gender?

7. Do you think the emotions felt were different for men and women? If so, then what evidence from Figure 1.2 would you offer in support of such a view?

8. Suppose you did not believe that the emotions felt were different for men and women. What evidence from Figure 1.2 could be offered in support of that position?

Refer to Example 1.3 for Exercises 9–12.

9. Do you think that the Palm Beach County votes for Buchanan were unusually high? If so, what is it about Figure 1.3 that leads you to think so?

10. If the number of votes for Buchanan followed the pattern of the other counties in Florida, about how many votes would Buchanan have gotten in Palm Beach County? (*Hint:* Look at the vertical axis, and give a rough estimate.)

11. About how many votes did Buchanan actually receive in Palm Beach County?

12. Do you think that Figure 1.3 gives convincing evidence that what occurred in Palm Beach County was unusual?

1.2 An Introduction to Statistics

Objectives: **By the end of this section, I will be able to…**

1 Describe what statistics is.

2 State the meaning of descriptive statistics.

3 Understand that inferential statistics refers to learning about a population by studying a sample from that population.

1 What Is Statistics?

Do you believe in aliens? According to a recent survey, 54% of the men surveyed responded that they believed in aliens, and 33% of the women did (Figure 1.4). These numbers are examples of *statistics*. Think about these numbers. Here are some questions we could ask about this survey:

- How did the pollsters arrive at these figures?

- Are the figures accurate? Could they be inaccurate?

- Why do pollsters never ask me my opinion about aliens?

- This survey found that more men than women believed in aliens. But is this difference meaningful or just a product of random chance (like getting 5 heads in a row when tossing a coin)?

FIGURE 1.4 Graphs comparing percentages of men and women who believe in aliens. (© *USA Today*)

These are some of the types of questions we shall be investigating throughout this book.

Examples of Statistics

Many people, including the author, first became interested in statistics as children collecting baseball cards. The back of each card contains the player's statistics season by season. You can always tell diehard baseball fans by the number and quality of the statistics that they are able to recite about their favorite teams. Television networks routinely employ sports statisticians to collect and report statistics about sports figures. Table 1.2, for example, contains batting averages of the league-leading hitters from 2002 to 2006.

Table 1.2 Batting-average leaders, Major League Baseball, 2002–2006

Season	Player	Team	Batting average
2006	Joe Mauer	Minnesota Twins	.347
2005	Derrek Lee	Chicago Cubs	.340
2004	Ichiro Suzuki	Seattle Mariners	.372
2003	Albert Pujols	St. Louis Cardinals	.359
2002	Barry Bonds	San Francisco Giants	.370

The informal meaning of the term *statistic* refers to a number that describes a person, a group, or a set of items. (On page 12, we provide a more precise definition of statistic.) For example, Joe Mauer's batting average of .347 is a statistic, because it uses his batting performance for the 2006 season to describe his batting performance for his entire career. Apart from sports, most people become familiar with statistics through exposure to media reports or advertising, such as

- "Polls indicate a majority of Democrats support stem cell research."

- "The median home sales price in Connecticut has climbed in recent months to $250,000."

- "Three out of four dentists surveyed recommend sugarless gum for their patients who chew gum."

You may have noticed that the section title, "What Is Statistics?" refers to **statistics** in the singular. Why? Because the *field of statistics* involves much more than just collecting and reporting numerical facts. The field of statistics may be defined as follows.

> **What Is Statistics?**
> **Statistics** is the *art* and *science* of
> - collecting,
> - analyzing,
> - presenting, and
> - interpreting data.

A statistician, then, is not simply a sports analyst but any person trained in the art and science of statistics. You may be surprised at the inclusion of the word *art* in the definition of statistics. But there is no question that judgment, experience, and even a little intuition are indispensable tools for any statistician's portfolio.

For today's college student, the field of statistics is especially relevant and useful.

- For example, a *business major* may be interested in whether she should consider diversifying her portfolio to tech stocks, based on their price/earnings ratio.

- A *psychology major* may be interested in determining whether there are differences in therapeutic outcomes between traditional counseling methods and a new cognitive approach.

- An *education major* may be interested in whether listening to a Mozart sonata before taking an exam can significantly improve your grade.

The field of statistics can help solve each of these puzzles.

© 2007 Old Visuals Everett Collection/Jupiterimages

CASE STUDY | Does Friday the 13th Change Human Behavior?

How would researchers go about studying whether superstitions change the way people behave? What kind of evidence would support the hypothesis that Friday the 13th causes a change in human behavior? T. J. Scanlon and his coresearchers thought that if there were fewer vehicles on the road on Friday the 13th than on the previous Friday, this would be evidence that some people were playing it safe on Friday the 13th and staying off the roads.[2] Note that the researchers didn't simply argue about the validity of the Friday the 13th superstition. Such discussions are interesting but largely subjective. What they deemed important is the effect of such a superstition on human behavior and how to measure such an effect as a change in behavior.

Phase 1 **Data collection.**   The first phase of a statistical study, as in the definition of statistics, is to *collect* the data. The researchers obtained data kept by the British Department of Transport on the traffic flow through certain junctions of the M25 motorway in England.

Phase 2 **Data analysis.**   Next comes the analysis of the data. The authors compared the number of vehicles passing through certain junctions on the M25 motorway on Friday the 13th and the previous Friday during 1990, 1991, and 1992.

Table 1.3 Traffic through M25 junctions, 1990–1992

Friday the 6th	Friday the 13th	Difference
139,246	138,548	698
134,012	132,908	1104
137,055	136,018	1037
133,732	131,843	1889
123,552	121,641	1911
121,139	118,723	2416
128,293	125,532	2761
124,631	120,249	4382
124,609	122,770	1839
117,584	117,263	321

Table 1.3 shows that, in every instance, the number of vehicles passing through these junctions on Friday the 13th was less than on the preceding Friday. Now, let's

examine the data graphically. The clustered bar graph in Figure 1.5 illustrates the difference in the number of vehicles traveling on the M25 motorway on Friday the 6th (in green) and the subsequent Friday the 13th (in yellow) for 10 pairs (clusters) of dates. Note that, *in every instance,* the green bar is longer than its partner yellow bar. This indicates that the number of vehicles on the motorway *decreased* on Friday the 13th when compared with the previous Friday in every instance.

FIGURE 1.5
Clustered bar graph of motorway traffic, with Friday the 6th in green and Friday the 13th in yellow.

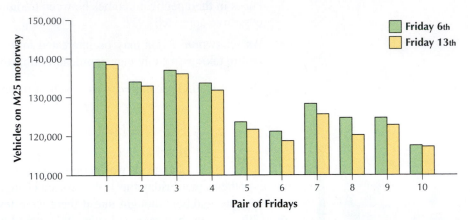

Phase 3 **Data presentation.**   The *presentation* of the results is important, and the researchers found a highly respectable journal, the *British Medical Journal,* in which to publish their findings. Other avenues for presentation are delivering a talk at a conference, writing up a report for one's supervisor, or presenting a class project.

What Do These Numbers Mean?

Phase 4 **Data interpretation.**   Finally, the last facet in our definition of statistics is *interpretation.* It is crucial for those who are performing a statistical study to make their results understandable to nonstatisticians. It is not sufficient for the statistician alone to understand the results. Rather, the statistician must communicate the results clearly, whether in writing or orally. In this case, the researchers chose the decrease in number of vehicles as the criterion on which to base support for their hypothesis that people changed their behavior on Friday the 13th. Their finding of a consistent decrease in traffic on Friday the 13th supports their hypothesis. ■

❷ Descriptive Statistics: The Building Blocks of Data Analysis

Every data set holds within it a story waiting to be told, as we saw in the Friday the 13th Case Study. To provide us with the tools to uncover these stories we need to learn some simple concepts, *the building blocks of data analysis.*

> **Descriptive statistics** refers to methods for summarizing and organizing the information in a data set.

In **descriptive statistics** we use numbers (such as counts and percents) and graphics to describe the data set, as a first step in data analysis. In Chapters 2 to 4, we will examine descriptive methods much more closely. First, we need to introduce a few terms. Suppose a data analyst for a health maintenance organization (HMO) is collecting data about the patients in a particular hospital, including the diagnosis, length of stay, gender, and total cost. The sources of the information (the patients) are called the **elements.** The patients' characteristics (for example, diagnosis, length of stay) are called the **variables.** Finally, the complete set of characteristics for a particular patient is called an **observation.**

> **Elements, Variables, and Observations**
>
> An **element** is a specific entity for which information is collected.
>
> A **variable** is a characteristic of an element, which can assume different values for different elements.
>
> An **observation** is the set of values of the variables for a given element.

When data are presented in tables and spreadsheets, it is typical practice to have the columns indicate the variables, and the rows to indicate the elements. So, for the hospital patients, the observation (specific values for the set of all the variables) for each element (patient) would appear as a row in the table.

Example 1.4 A remarkable student named Maria

Information was collected on four students from two area colleges and is presented in Table 1.4. What are the elements? What are the variables? Provide the observation for Maria.

Table 1.4 Data set of four elements and seven variables

Student	Age	Gender	Ethnicity	No. of children	Marital status	GPA	College
Jamal	19	Male	African American	0	Single	4.00	Western CC
Maria	25	Female	Hispanic	2	Married	3.95	Northern State Univ.
Chang	20	Female	Asian	0	Single	3.90	Northern State Univ.
Michael	47	Male	European American	3	Divorced	3.75	Western CC

Solution

The elements are the students Jamal, Maria, Chang, and Michael. The seven variables are

- age
- gender
- ethnicity
- number of children
- marital status
- GPA
- college

Suppose we are especially interested in Maria, a 25-year-old married Hispanic mother of two, who attends Northern State University and has a grade point average (GPA) of 3.95. Since the observation for Maria consists of the values for the variables in Maria's entire row, her observation is ■

Student	Age	Gender	Ethnicity	No. of children	Marital status	GPA	College
Maria	25	Female	Hispanic	2	Married	3.95	Northern State Univ.

Notice that we have variables that can take on various types of values, some of which are numbers and some of which are categories. For example, Maria is 25 years old, has two children, and has a GPA of 3.95, each of which is numeric. On the other hand, Maria is Hispanic, married, and enrolled at Northern State University, characteristics that do not have numeric values but instead are categories. This leads us to define two types of variables: **qualitative** and **quantitative.**

> A **qualitative variable** allows the elements to be classified according to some attribute or characteristic. A **quantitative variable** takes numeric values and allows arithmetic operations (such as addition and subtraction) to be performed on it meaningfully.

Qualitative variables are also called *categorical variables*, because they can be grouped into categories. Maria's qualitative variables include her gender, ethnicity, marital status, and college. In contrast, Maria's grade point average is an example of a *quantitative variable*. Other quantitative variables include age and number of children.

Quantitative variables can be classified as either **discrete** or **continuous.**

> A **discrete variable** can take either a finite or a countable number of values. Each value can be graphed as a separate point on a number line, with space between each point. A **continuous variable** can take infinitely many values, forming an interval on the number line, with no space between the points.

Example 1.5   Discrete or continuous?

In Table 1.4, determine whether the following variables are discrete or continuous: (a) number of children and (b) GPA.

Solution

a.   Since the number of children per student is finite, the variable "number of children" is discrete.
b.   Since GPA can take an infinite number of possible values, for example in the interval 0.0 to 4.0, the variable "GPA" is continuous.   ■

Example 1.6   Most active stocks, June 30, 2007

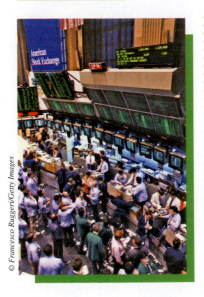

© Francesco Ruggeri/Getty Images

One of the most widespread applications of statistical analysis occurs in the business world. Managers examine patterns and trends in data, thereby hoping to increase profitability. Table 1.5 shows the five most active stocks on the New York Stock Exchange (NYSE), as reported by *USA Today* for June 30, 2007. What are the elements and the variables of this data set? Which variables are qualitative? Which are quantitative? Provide the observation for Pfizer, Inc.

Table 1.5 Most active stocks on NYSE, June 30, 2007

Stock	Symbol	Last	Volume	Change
Ford Motor Company	F	$ 9.42	50,308,258	−0.07
General Electric Company	GE	$38.28	44,692,276	+0.16
Pfizer, Inc.	PFE	$25.57	44,346,322	−0.06
General Growth Properties, Inc.	GGP	$52.95	40,926,559	−0.05
AT&T, Inc.	T	$41.50	37,579,025	+0.76

Solution

The *elements* are the five most active stocks traded on the NYSE on this day in 2007. The *variables* are as follows:

- Symbol: The Ticker symbol used to represent the stock.
- Last: The most recent trading price for the stock.
- Volume: How many shares of the stock were traded that day.
- Change: The change in share price (in dollars) between the opening price and the closing price that day.

The Ticker symbol, since it is a label for the stock, is qualitative. The other variables are quantitative. The observation for Pfizer, Inc., includes the Ticker symbol and the set of the day's stock data for that company. That stock has a Ticker symbol of PFE, its last share price was $25.57 per share, 44,346,322 shares of its stock were traded, and the price decreased by $0.06 per share. ■

Stock	Symbol	Last	Volume	Change
Pfizer, Inc.	PFE	$25.57	44,346,322	–0.06

Levels of Measurement

Data may be classified according to the following four *levels of measurement.*

- *Nominal data* consist of names, labels, or categories. There is no natural or obvious ordering of nominal data (such as high to low). Arithmetic cannot be carried out on nominal data.

- *Ordinal data* can be arranged in a particular order. However, no arithmetic can be performed on ordinal data.

- *Interval data* are similar to ordinal data, with the extra property that subtraction may be carried out on interval data. There is no *natural zero* for interval data.

- *Ratio data* are similar to interval data, with the extra property that division may be carried out on ratio data. There does exist a natural zero for ratio data.

Example 1.7   Levels of measurement

Identify which level of measurement is represented by the following data.
a.   Years covered in European History 101: 1066–1492
b.   Annual income of students in Statistics 101 class: $0–$15,000
c.   Course grades in English 101: A, B, C, D, F
d.   Student gender: male, female

Solution
a.   The years 1066 to 1492 represent interval data. There is no natural zero (no "year zero"; the calendar goes from 1 B.C. to A.D. 1). Also, division (1492/1066) does not make sense in terms of years, so that the data are not ratio data. However, subtraction does make sense, in that the course covers 1492 − 1066 = 426 years.
b.   Student income represents ratio data. Here division does make sense. That is, someone who made $4000 last year made twice as much as someone who made $2000 last year. Also, some students probably had no income last year, so that $0, the natural zero, also makes sense.
c.   Course grades represent ordinal data, since (a) they may be arranged in a particular order, and (b) arithmetic cannot be performed on them. The quantity A − B makes no sense.
d.   Student gender represents nominal data, since there is no natural or obvious way that the data may be ordered. Also, no arithmetic can be carried out on student gender. ■

3   Inferential Statistics: How Do We Get There from Here?

Descriptive methods of data analysis are widespread and quite informative. However, the modern field of statistics involves much more than simply summarizing a data set. For example, suppose a sociologist claims that one-third of American teenagers have been

© *Somos Images/CORBIS*

the targets of cyberbullying, that is, have received a threatening message or have had their emails or text messages forwarded without their consent, an embarrassing picture posted without permission, or rumors spread about them online. How should the sociologist go about collecting evidence to support her claim? One method would be to ask each and every person in the population of all American teenagers. In general, a **population** is the collection of *all* elements (persons, items, or data) of interest in a particular study.

However, to ask every teenager in America about his or her online experiences is a daunting task that is expensive, time-consuming, and, in the end, simply impossible. So, unfortunately, the population proportion of American teenagers who have been the targets of cyberbullying remains *unknown*. This proportion who have been targets of cyberbullying is one characteristic of the population of American teenagers. A characteristic of a population is called a **parameter.** The actual value of a population parameter is often unknown.

A **sample** is a subset of the population from which information is collected. For example, from a sample of 100 teenagers at a local mall, 18 said they had been the targets of cyberbullying. That is, the sample proportion of students who had been targets is $18/100 = 18\%$. This proportion is a characteristic of the sample and is called a **statistic.** The advantage here is that, since the sample is relatively small, the characteristics of the sample can be determined.

> **Populations, Parameters, Samples, and Statistics**
> A **population** is the collection of *all* elements (persons, items, or data) of interest in a particular study. A **parameter** is a characteristic of a population.
> A **sample** is a subset of the population from which information is collected. A **statistic** is a characteristic of a sample.

A sample is a subset of a population.

The U.S. Constitution requires that a census be conducted every 10 years. A **census** is the collection of data from every element in the population. As you can imagine, such a task is very difficult and very expensive. In fact, the Census Bureau estimates that the 2000 U.S. census "undercounted the actual U.S. population by over three million individuals."[3]

Because the population you are interested in may be too large to allow you to elicit information from every element, it is often best to gather data from a sample, a subset of that population. Also, time and money often constrain the researcher to choosing a sample rather than studying the entire population. Also, in some experiments, the resource is exhausted when testing is done, for example, in estimating the mean lifetime of light bulbs. Finally, it may be simply impossible to gather information from the entire population, such as when studying the quality of water in Lake Erie.

To estimate the proportion of all American teenagers who have been subjected to cyberbullying, we can use statistical inference. **Statistical inference** refers to learning about the characteristics of a population by studying those characteristics in a subset of the population (that is, in a *sample*). The Pew Internet and American

Life Project conducted a survey of 886 teenagers and found that 284 of them (32%) said they had been the targets of cyberbullying.[4] These 886 teenagers represent a sample, and their characteristics can be known. For example, we know that 284 of the 886 teenagers in the sample said they have been subjected to cyberbullying. At this point, the sociologist can make the *inference* that the proportion of *all* American teenagers who have been subjected to cyberbullying is 32%, because this is the proportion in the sample. In doing this, the sociologist is performing *statistical inference*.

> **Statistical inference** consists of methods for estimating and drawing conclusions about population characteristics based on the information contained in a subset (sample) of that population.

"Now wait just a minute," you might object. "How can you say that the proportion of *all* American teenagers who have been subjected to cyberbullying is 32% just because your *sample* proportion is 32%?" Actually, you have a point. We *are* generalizing. We are taking what we know about a portion of the whole (a sample) and using it to draw a conclusion about the whole (the population). But even though the true proportion of American teenagers who have been the targets of cyberbullying is probably not exactly 32%, it is most likely not very far from 32%. The 32% is an *estimate,* an approximation based on sample data. Later we will learn how we can get the estimate as close as we wish to the actual value just by taking a large enough sample.

Finally, we need to point out one further attribute of parameters and statistics. The value of a parameter, even though it is unknown, is a fixed constant. For example, the average age of all persons in your home state (population) at noon today is unknown, but it still exists, and it is a single number. On the other hand, the value of a statistic depends on the sample. For example, a sample of 100 people in your hometown may produce an average age of 31. The average age of a sample of 100 people in a neighboring town may be 32. Later, we will learn that this is because a statistic is a *random variable.*

Of course, to deliver a valid estimate, the sample needs to be *representative* of the population. The sample should not differ systematically in any major characteristic from the population. We will learn more about this in Section 1.3, when we study sampling methods. Table 1.6 summarizes the attributes of a population and a sample. Isn't it convenient that the characteristic associated with the population starts with the same letter, and the same is true for sample?

Table 1.6 Summary of attributes of population and sample

	Population	Sample
Thumbnail definition	All elements	Subset of population
Characteristic	Parameter	Statistic
Value	Usually unknown	Usually known
Status	Constant	Depends on sample

Example 1.8   Descriptive statistics or statistical inference?

State whether the following situations illustrate the use of descriptive statistics or statistical inference.
a. In Baltimore County, Maryland, the average amount spent per week on gasoline consumption in a sample of 500 commuters was $75. The county government infers that the average amount spent weekly by all Baltimore County commuters is $75.

b. A sample of 100 residents of Broward County, Florida, yielded 27 residents who work for the government at the local, state, or federal level. Thus, 27% of these 100 residents work for the government.

c. The average age of a sample of 200 residents of Garden City, New York, was 34 years old.

d. In a survey of 1000 citizens in the Seattle, Washington, metropolitan area, 570 said they would pay higher prices in order to reduce greenhouse emissions. City planners conclude that 57% of all Seattle citizens would do so.

Solution

a. Statistical inference. A sample was taken, and a sample statistic ($75 per week) was calculated. Then, the County government used this statistic to make the *statistical inference* that this was the average amount spent by all Baltimore County commuters.

b. Descriptive statistics. Though a sample was taken, there was no attempt to make an inference from this sample of 100 workers to the entire population of Broward County, Florida. So, there is no statistical inference here.

c. Descriptive statistics. The average age of 34 years old is a descriptive statistic, since it describes the sample. But no inference is made regarding a larger population.

d. Statistical inference. The survey found that 57% of the sample of 1000 citizens would pay higher prices in order to reduce greenhouse emissions. This 57% is a statistic. Then the city planners used this statistic in order to perform statistical inference about the population of all Seattle citizens. ▪

A Statistical Literacy Quiz

Regardless of major, every student in America (indeed, every citizen) needs to become *statistically literate* in order to survive in today's wired society. Why not take this quiz to find out if you are statistically literate? Answer each question true or false.

1. A fair coin is tossed five times and comes up heads each time. That means that tails is "due" and the chances of tails on the next toss is increased.

2. One politician says that the mean income is rising, while another politician says that the median income is falling. One of them has to be lying.

3. Jim is tested for HIV and the test comes back positive. Thus, Jim is HIV-positive.

The correct answer to each question is *false*. Question 1 deals with something called "the gambler's fallacy," and we will cover this, along with the explanation for Question 3, in Chapter 5. We will deal with Question 2, the quirks of means and medians, in Chapter 3. ■

Section 1.2 SUMMARY

1 The field of statistics is the art and science of collecting, analyzing, presenting, and interpreting data.

2 Descriptive statistics refers to methods for summarizing and organizing the information in a data set. Data sets include information collected on elements (hospital patients or homes in Florida). The types of information collected are called variables (diagnosis of patient, home sales price). Variables may be either quantitative (numerical) or qualitative (categorical). A discrete variable is a quantitative variable that can take either a finite or a countable number of possible values. A continuous variable is a quantitative variable that can take an infinite number of possible values. A population is a collection of all elements of interest, while a sample is a subset of the population.

3 Inferential statistics consists of methods for estimating and drawing conclusions about population characteristics

based on the information in the sample. The characteristics for a population are called parameters, while the characteristics for a sample are called statistics. The value of the parameter is usually unknown but constant, while the value of the statistic varies from sample to sample.

Section 1.2 EXERCISES

CLARIFYING THE CONCEPTS

1. Write a sentence describing in your own words the field of statistics.
2. True or false: Statistical inference refers to methods for summarizing and organizing the information in a data set.
3. What do we call the entities from which the data are collected?
4. Describe the difference between a qualitative and a quantitative variable.
5. What is another term for a qualitative variable?
6. True or false: The actual value of a population parameter is usually unknown.
7. What is the difference between a sample and a population?
8. Explain what a statistic is.
9. Describe one difference between a statistic and a parameter.
10. What is a census?

PRACTICING THE TECHNIQUES

For Exercises 11–24:
   a. State whether the variable is qualitative or quantitative.
   b. Identify the level of measurement represented by the data.

11. The temperature (in degrees Fahrenheit) in the room you are in right now
12. Whether you own an iPod or not
13. The price of tea in China
14. The SAT Math score of the person sitting next to you (scores range from 200 to 800)
15. The winning score in next year's Super Bowl
16. The winning team in next year's Super Bowl
17. The rank of the winning Super Bowl team in their division
18. The number of friends on a student's Facebook page
19. Your favorite video game
20. How many contacts you have on your cell phone
21. Your favorite topping on pizza
22. Your credit card balance
23. How old your car is
24. What model your car is

For Exercises 25–30, identify the population and the sample.
25. A researcher is interested in the median home sales price in Tarrant County, Texas. He collects sales data on 100 home sales.
26. A psychologist is concerned about the health of veterans returning from war. She examines 20 veterans and

assesses whether they show signs of post-traumatic stress disorder.
27. A sociologist wants to learn about the number of meetings per year of the 4-H clubs in Utah. He collects information from 10 different 4-H clubs in various parts of Utah.
28. A physical therapist would like to determine whether a new exercise method can delay the onset of osteoporosis in older women. She chooses 10 of her patients to use the new method.
29. An educator asks a sample of students at Portland Community College whether they would be interested in taking a course online.
30. A financial adviser would like to assess the effect of mergers on price/earnings ratio. She collects data on 50 companies that recently underwent a merger.

For Exercises 31–38, state whether descriptive statistics or statistical inference was used, and explain why.
31. The average price in a sample of 15 homes sold in Newington, Connecticut, for the week of April 21 was $253,200.
32. According to the Department of Transportation, 60% of all automobile passengers wear seat belts. This is based on a survey of 1000 automobile passengers, of whom 600 wore seat belts.
33. In a sample of 500 subjects, it was found that daily exercise lowered the average cholesterol level by 10%. A medical spokesperson then stated that daily exercise can lower everyone's cholesterol level by 10%.
34. In a sample of 140 traffic fatalities in New York, 75 involved alcohol.
35. The goals-against average for the Charlestown Chiefs hockey team in a sample of 20 games was 3.57 goals per game.
36. The Department of Health and Human Services conducted a survey, where it was found that the percentage of 15- to 18-year-olds using illicit drugs has dropped in the last two years. The department concluded that illicit drug use has fallen among all 15- to 18-year-olds.
37. The average on the first statistics exam for a sample of 10 students in Ms. Reynolds' class was 70.
38. A Department of Homeland Security survey of 750 Americans found that more than half of those surveyed feel less safe than they did 10 years ago. The department concluded that more than half of all Americans feel less safe than they did 10 years ago.

APPLYING THE CONCEPTS
Endangered Species. For Exercises 39 and 40, refer to the following table, which lists 4 of the 11 animal species in Minnesota identified by the U.S. Fish and Wildlife Service as threatened (T) or endangered (E).

Species	Status	Type	Scientific name
Karner blue butterfly	E	Insect	*Lycaeides melissa samuelis*
Canada lynx	T	Mammal	*Lynx canadensis*
Piping plover	E	Bird	*Charadrius melodus*
Gray wolf	E	Mammal	*Canis lupus*

39. Do these 4 species represent a sample or a population?
40. Refer to the table of endangered species.
   a. List the elements.
   b. List the variables.
   c. Identify the qualitative variables. Identify the quantitative variables.
   d. Provide the observation for the Canada lynx.

41. Top Five Employers in Santa Monica, CA. Refer to the following table.
   a. List the elements.
   b. List the variables.
   c. Which of the variables is quantitative?
   d. Which of the variables is qualitative?
   e. Provide the observation for St. John's Health Center.

Company	Employees	Industry
City of Santa Monica	1892	Government
St. John's Health Center	1755	Health services
The Macerich Company	1605	Real estate investment trust
Fremont General Corp.	1600	Property and casualty insurance
Entravision Communications Corp.	1206	Diversified media company

Source: Santa Monica Chamber of Commerce.

42. Genetically Engineered Crops. Genetically engineered (GE) crops are now planted on the majority of acreage in many states around the country. There are three varieties of GE corn: insect-resistant, herbicide-tolerant, and stacked genes. The following table contains the proportion of the corn grown in each of five states, that is GE, along with the GE type most prevalent in each state, for 2007.[5]

State	Proportion of GE corn	Most prevalent type
Texas	79%	Herbicide-tolerant
Missouri	62%	Insect-resistant
Minnesota	86%	Herbicide-tolerant
Ohio	41%	Stacked genes
South Dakota	93%	Stacked genes

   a. List the elements.
   b. List the variables.
   c. Which of the variables is quantitative?
   d. Which of the variables is qualitative?
   e. Provide the observation for Texas.

43. Crime Statistics for Stillwater, OK. Refer to the following table.
   a. List the elements.
   b. List the variables.
   c. Which of the variables is quantitative?
   d. Which of the variables is qualitative?
   e. Provide the observation for motor vehicle thefts.

Crime type	2005 Total	Per 100,000 people	National per 100,000 people	Compared to national average
Robberies	10	24.4	195.4	Better
Aggravated assaults	83	202.4	340.1	Better
Burglaries	317	772.9	814.5	Better
Larceny/thefts	1147	2796.7	2734.7	Worse
Motor vehicle thefts	55	134.1	526.5	Better

44. Web Page Statistics for Kirksville Rocks. Refer to the following table, which contains statistics for the Web site Kirksville Rocks! (which contains live-music listings for Kirksville, Missouri), logged from April 27, 2005, to February 16, 2007.
   a. List the elements.
   b. List the variables.
   c. Which of the variables are quantitative?
   d. Which of the variables are qualitative?
   e. Provide the observation for **www.kvrocks.com/venues/**.

Web page	Requests	Percent	Title of page
www.kvrocks.com/	35,012	44.71%	Home Page
www.kvrocks.com/musicians/	6,069	7.75%	Musicians Page
www.kvrocks.com/events/	3,867	4.94%	Events Page
www.kvrocks.com/venues/	3,332	4.26%	Venues Page

45. Hospitals near Jackson, MS. Refer to the following table (page 17).
   a. List the elements.
   b. List the variables.
   c. Which of the variables is quantitative?
   d. Which of the variables is qualitative?
   e. Provide the observation for Montfort Jones Memorial Hospital.

Hospital name	Beds	City	Zip
Hardy Wilson Memorial Hospital	49	Hazlehurst	39083
Humphreys County Memorial Hospital	34	Belzoni	39038
Jefferson County Hospital	30	Fayette	39069
Lackey Memorial Hospital	15	Forest	39074
Leake Memorial Hospital	25	Carthage	39051
Madison County Medical Center	67	Canton	39046
Montfort Jones Memorial Hospital	72	Kosciusko	39090
Rankin Medical Center	134	Brandon	39042
University of Mississippi Medical Center—Holmes County	25	Lexington	39095

Births and Maternal Age in Westchester County, NY.
The following table represents the number of births and the average maternal age for 10 hospitals in northwest Westchester County, New York, in 2002. Refer to the table for Exercises 46–48.

Hospital	Births	Average maternal age
Briarcliff Manor	71	34.1
Buchanan	25	31.6
Cortlandt	348	32.2
Croton-on-Hudson	93	33.5
Mount Pleasant	277	32.8
Ossining 1	80	32.1
Ossining 2	371	29.2
Peekskill	365	29.0
Pleasantville	79	32.9
Sleepy Hollow	134	29.2

46. What are the elements?
47. List the variables. Are they quantitative or qualitative?
48. Give the observation for Sleepy Hollow Hospital.
49. Radio Ratings. The ratings company Arbitron (**www.arbitron.com**) reports that, in the winter of 2004, 14% of all radio listeners listened to adult contemporary (AC) music. The breakdown within the AC category was Mainstream AC (7.8%), Hot AC (3.7%), Mod AC (0.8%), and Soft AC (1.7%).
  a. What are the elements? Can they be listed? Why or why not?
  b. What are the variables? Are they qualitative or quantitative?
  c. What proportion of AC listeners are Mainstream AC listeners?
50. Commodity Prices. The financial company Bloomberg (**www.bloomberg.com**) reported that, on July 30, 2004, the prices in dollars for the following commodities were oil ($43.80, up 1.05%), gold ($393.70, up 3.90%), and coffee ($66.45, down 1.20%).
  a. What are the elements?
  b. What are the variables? Are they quantitative or qualitative?
  c. Explain the meaning of the value of percent increase for oil.

Light Bulb Lifetime. Use the following information for Exercises 51 and 52. An electrical company has developed a new form of light bulb that it claims lasts longer than current models. The company has 1 million bulbs in its inventory.
51. How do you think the company found evidence for its claim?
52. Suppose you take a representative sample of 100 of the new light bulbs and find the average lifetime to be 2000 hours.
  a. Is this a statistic or a parameter?
  b. Write a sentence that estimates the average lifetime of all the new light bulbs.
53. Tornado Deaths. The Tornado Project (**www.tornadoproject.com**) reported the following list of the 10 years with the fewest tornado deaths.

Year	Deaths	Year	Deaths
1910	12	1996	26
1986	15	1972	26
2004	16	1980	27
1981	24	1963	27
1962	25	1951	29

  a. What are the elements?
  b. What is the variable? Is it quantitative or qualitative?
  c. Could this be considered a representative sample of the number of annual tornado deaths for all years? Explain why or why not.

1.3 Gathering Data

Objectives: By the end of this section, I will be able to...

1 Explain what a random sample is, and why we need one.

2 Identify other sampling methods.

3 Explain the main factors that go into designing a good questionnaire.

4 Understand the difference between an observational study and an experiment.

1 Random Sampling

We can use the information gathered from a sample to generalize about the population when it is impractical or impossible to take a census of the entire population. However, if we get a "bad" sample, the information gleaned from the sample will be misleading, with potentially catastrophic consequences. This section introduces a method of sampling that minimizes many potential biases, which could lead to incorrect generalizations about the population. This sampling method is called *random sampling.* Everyday examples of random sampling include (a) randomly selecting lottery numbers from a basket which continuously churns the number-balls, (b) randomly choosing one card from a deck of playing cards that has been well shuffled, and (c) randomly pulling a name out of a hat, after the names have been well stirred. Since random samples are not always practicable or desirable, this section also discusses some of the many alternative sampling methods available, including stratified sampling and cluster sampling.

What Is a Random Sample, and Why Do We Need It?

Survey sampling, or polling, has now become so widespread that hardly a day goes by without the results of some new poll or survey making the headlines. Political polls are a good example of statistical sampling at work. The pollsters canvass about 1000 or so respondents, analyze the sample results, and then report their statistical inference that, for example, "54% of Minnesotans support Garrison Keillor for President."

Today many polls are conducted quite scientifically, and their results are usually very accurate. However, such was not always the case. In 1936, the *Literary Digest* had correctly predicted the past three presidential elections and went to work to predict the winner of the contest between Republican Alf Landon and Democrat Franklin Roosevelt. The magazine sent ballots to 10 million citizens. The results ran strongly in favor of Landon, leading the *Literary Digest* to predict Landon to win the election. About 25% of the ballots were returned, giving the newsweekly a sample size of 2.5 million. George Gallup, on the other hand, was working with a sample size that was about 1000 times *smaller* than the *Literary Digest*'s. However, Gallup predicted a victory for Roosevelt. Clearly, with more data, the *Literary Digest* should have been able to give a more accurate prediction, right? Not necessarily. Roosevelt won in a landslide, and the embarrassed *Literary Digest* later declared bankruptcy.

The problem stemmed from the way that the *Literary Digest* identified its sample. It used lists of people who owned cars and had telephones, which in the 1930s excluded millions of poor and underprivileged people, who overwhelmingly supported Roosevelt. Its sample of 2.5 million therefore was highly biased toward the richer folks, who were less likely to have any great fondness for Roosevelt and his New Deal policies. Gallup, on the other hand, chose his sample more scientifically, and even though his sample size was smaller, it was more representative of the population as a whole.

One inexpensive way of eliminating many types of bias is to make sure your sample is a **random sample.**

> A **random sample** (also known as a **simple random sample**) is a sample for which every element has an equal chance of being selected.

The *Random Sample* applet allows you to produce a random sample of up to 100 elements, in the form of a lotto.

Note: When we take a sample, we usually discard any repeated elements because we already have their information.

How the Gallup Organization Obtains a Random Sample

The Gallup Organization (**www.gallup.com**) has been conducting polls since the 1930s. People often wonder how a random sample of 1000 adults can represent the sentiments of the more than 300 million American adults. How does Gallup obtain a random sample in the first place? Gallup's objective is to make sure that every American has an *equal probability of selection*, that is, an equal chance of being selected, for their poll. Now, where is the place that most Americans are likely to be found? The answer for most Americans is: at home. (Gallup is aware that those who reside on college campuses, on military bases, and in other institutions are thereby not represented, but they usually accept this compromise due to the practical problems of including these institutions.)

In the early days, Gallup conducted interviews in person, going house to house. However, today it is much less expensive to conduct telephone interviews. How does Gallup help to ensure that its telephone sample is truly random? What about the Americans whose phone number is unlisted? The first step is to construct a table of all the telephone exchanges in America, along with an estimate of the proportion of Americans living in that exchange area and the broad characteristics of that population in terms of income, age, ethnicity, education, and so on. Gallup then uses *random digit dialing,* a computer program that generates random four-digit numbers, which are then appended to the telephone exchanges. Thus, each household phone number in America has an equal chance of being included in the sample, regardless of whether it is listed or unlisted. Finally, as of January 1, 2008, Gallup added a data base of cell phone numbers, in order to contact those who can more readily be reached via cell phone.

| Example 1.9 | Do you prefer watching the Super Bowl or the commercials? |

In February 2007, the Gallup Organization used random digit dialing in a poll of Americans who planned to watch the Super Bowl (Indianapolis Colts versus Chicago Bears). One question they asked was whether the subjects preferred to watch the game or the commercials. Does this represent a random sample?

Solution

Since random digit dialing ensures that each household phone number in America has an equal chance of being included in the sample, the sample is random.

A perhaps surprising 33% of respondents reported that they preferred watching the commercials, compared with 66% who preferred watching the game. Gender and age seemed to affect how one responded to this question. Twice the proportion of female viewers (44%) as male viewers (22%) preferred watching the commercials. Among females only, more than twice as many younger (aged 18 to 49) women preferred watching the commercials (56%) as older (aged 50 and over) women (26%). ∎

Example 1.10 Online polls

An online newspaper reports that, in an online poll of its readership, 60% say that they get most of their news from online sources. Does this number accurately reflect the proportion of all Americans who get most of their news from online sources?

Solution

No, the sample is not random. Only those Americans who are online already (and already using an online news source) can respond to this online poll. Therefore, the sample is not random, and it is biased. It overestimates the proportion of Americans who get their news from online sources. Further, there is no mechanism to guard against a single person responding repeatedly and getting his or her vote counted multiple times. Online polls are not scientific, and their results should not be considered a true reflection of the sentiments of all Americans. ■

Random samples may be generated using technology, using the *Random Sample* applet, or using the random number table provided in Table A in the Appendix (page T-2). At the end of this section, we demonstrate how to generate random samples using the TI-83/84 graphing calculator, Excel, and Minitab.

Example 1.11 Generating a random sample using technology

In 2004, *Inc. Magazine* published a list of the top 25 cities for doing business, shown in Table 1.7. Use the TI-83/84, Excel, or Minitab to generate a random sample of 7 cities from this list.

Table 1.7 Top 25 cities for doing business, according to *Inc. Magazine*

1. Atlanta, GA	10. Suburban Maryland/DC	19. Austin, TX
2. Riverside, CA	11. Orlando, FL	20. Northern Virginia
3. Las Vegas, NV	12. Phoenix, AZ	21. Middlesex, NJ
4. San Antonio, TX	13. Washington, DC, metro area	22. Miami–Hialeah, FL
5. West Palm Beach, FL	14. Tampa–St. Petersburg, FL	23. Orange County, CA
6. Southern New Jersey	15. San Diego, CA	24. Oklahoma City, OK
7. Fort Lauderdale, FL	16. Nassau–Suffolk, NY	25. Albany, NY
8. Jacksonville, FL	17. Richmond–Petersburg, VA	
9. Newark, NJ	18. New Orleans, LA	

Solution

We used the instructions provided in the Step-by-Step Technology Guide at the end of this section to create three random samples, listed below. Note that each random sample is different, as yours will be. ■

Random sample 1 using the TI-83/84	Random sample 2 using Excel	Random sample 3 using Minitab
9. Newark, NJ	6. Southern New Jersey	3. Las Vegas, NV
25. Albany, NY	23. Orange County, CA	21. Middlesex, NJ
6. Southern New Jersey	11. Orlando, FL	18. New Orleans, LA
20. Northern Virginia	14. Tampa–St. Petersburg, FL	7. Fort Lauderdale, FL
24. Oklahoma City, OK	25. Albany, NY	2. Riverside, CA
10. Suburban Maryland/ DC	7. Fort Lauderdale, FL	25. Albany, NY
1. Atlanta, GA	17. Richmond–Petersburg, VA	10. Suburban Maryland/ DC

More Sampling Methods

The idea behind random sampling is quite intuitive: choosing names at random out of a hat, for instance. However, in certain circumstances, simple random sampling can also have shortcomings. A simple random sample may not provide sufficient information about subgroups within the population. For example, suppose you are interested in knowing the proportion of those of Hispanic or Latino descent in Walnut, California, who are registered Democrats. A random sample of size 100 of all the voters in Walnut may yield only 20 of Hispanic or Latino descent, which may be too small a sample to be useful for statistical inference (and which undercounts the actual proportion of registered voters who are of Hispanic or Latino descent, which is about 35%). Therefore, the researcher needs other methods for obtaining samples, depending on the situation and the research question.

Stratified Sampling

Often, researchers are interested in investigating characteristics of a certain subgroup of a population, such as those of Hispanic or Latino descent in Walnut, California. In cases like this, the researcher divides the population into subgroups, or *strata,* according to some characteristic, such as race or gender. Then a random sample is taken from each stratum. In this way, the researcher knows that a sample will be obtained from each stratum and that it will be large enough to provide reliable statistical inference for each stratum.

Systematic Sampling

Perhaps the easiest method of sampling is systematic sampling, which is used when a random sample is either not needed or unobtainable. In systematic sampling, each element of the population is numbered, and the sample is obtained by selecting every kth element, where k is some whole number. The ancient Romans understood well how to use systematic sampling. When a Roman legion showed cowardice in battle or mutinied, every 10th member was selected and summarily executed before his comrades. Literally, the legion was *decimated,* from the Latin *decem,* meaning "ten."

Cluster Sampling

Cluster sampling is used when the population is widely scattered geographically or poses other logistical difficulties. For example, if we were interested in estimating the mean income of Manhattan residents, it would be time-consuming and expensive to visit 1000 different locations in Manhattan to elicit sample information. In cluster sampling, the population is divided into *clusters,* such as precincts or city blocks. Then several clusters are chosen at random, and all of the elements within the chosen clusters are selected for the sample. One disadvantage of cluster sampling is that the respondents from within a certain cluster will tend to be more similar to each other than the elements of a random sample would be. For example, if one of the clusters in the Manhattan income survey was a Fifth Avenue block, the mean income of residents there would be at the higher end of the income scale.

Note: All of the sampling methods mentioned here involve randomness. However, only the simple random sample is used throughout the text. Therefore, whenever you see the phrase *random sample,* it should be understood as *simple random sample.*

Convenience Sampling

In convenience sampling, subjects are chosen based on what is convenient for the survey personnel. If you were to estimate the true proportion of females taking an introductory statistics course using only the people in your class, this would be considered a convenience sample. This method is to be contrasted with *probability sampling,* where each member of the population has a known probability of selection. As we shall see below, convenience sampling usually does not result in a representative sample.

Surveys and Questionnaires

When the information cannot be found from existing sources, a researcher can commission a survey or questionnaire. Because surveys and questionnaires are so widely used, it is important to know how to design them correctly. A badly designed survey can yield misleading results. Here we learn about some common pitfalls in the design and implementation of a survey or questionnaire, including the wording of the questions.

> The **target population** is the complete collection of all elements that we are interested in studying.
>
> The **potential population** is the collection of elements from the target population that had a chance of being sampled.
>
> **Selection bias** occurs when the population from which the actual sample is drawn is not representative of the target population, due to an inappropriate sampling method.

Example 1.12   Selection bias

Suppose Ashley would like to estimate the proportion of American voters who would favor abandoning the present system of Social Security in favor of a system where retirement funds would be invested in the stock market. Ashley goes to the mall with her clipboard, and canvasses as many people as she can on Monday between 9 A.M. and 5 P.M. To each person, she asked the question "Do you favor or oppose abandoning the present Social Security system in favor of a system that invests retirement funds in the stock market?"

a.   Identify Ashley's target population.

b.   Identify Ashley's potential population.

c.   Discuss any possible problems.

Solution

a.   Ashley's **target population** is the population of all American voters.

b.   The collection of all the American voters who visited the mall on Monday between 9 A.M. and 5 P.M. represent her **potential population.**

c.   It appears that Ashley's survey may suffer from **selection bias.** The population of people who went to the mall on Monday between 9 A.M. and 5 P.M. is not representative of the target population of all American voters. Since many American voters work on Mondays between 9 A.M. and 5 P.M., they are not elements of the sampled population. Further, the proportion of retirees at the mall during that time was larger than in the target population of all American voters. These retirees tend to oppose strongly any tampering with the Social Security system and probably overwhelmingly responded in the negative to the survey question. ■

Five Factors for Good Questionnaire Design

You may have heard of the aphorism "Be careful what you ask for; you may get it." This warning is certainly relevant to the issue of questionnaire design. The wording of questions can greatly affect the responses. Here are several factors to consider when designing a questionnaire.

1.   **Remember: simplicity and clarity.** Do not use four-syllable words when one-syllable words will do. Respondents will be shy about asking you to clarify the question. The result will be confused responses and muddled data.

2.   **When reporting results, include the actual question asked.** Be careful about drawing generalizations. The conclusions you draw may not have been what your respondents had in mind when they answered the questions.

3.   **Avoid leading questions.** The respondent is often eager to please and will try to tell you what he or she thinks you want to hear. For example, a researcher is interested in determining the proportion of Americans who favor preserving the welfare system. A leading question would be "A child growing up poor in America faces more than his fair share of crime and negligence. Do you support preserving the welfare safety net to help ensure that children are given a fair chance?"

4. **Avoid asking two questions in one.** Avoid questions like "Have you argued with your friends or family in the last month?" This is really two questions in one, and you will not know which question the respondents are answering.

5. **Words mean different things to different people.** Avoid using terminology like "often" or "sometimes." Instead, try to use specific terms such as "three times a week."

If you use ambiguous terms, the data you collect will be ambiguous, and any conclusions you draw will probably not be valid.

4 Experimental Studies and Observational Studies

Two major types of statistical studies are **experimental studies** and **observational studies.** We have seen that researchers can gather data by consulting existing sources, by distributing a questionnaire, or by taking a sample. However, you may not be able to obtain the information you require by using survey or sampling methods. In this case, you may have to conduct an experimental study.

Note: What is the difference between an element and a subject? *Subject* is a term usually reserved for statistical studies, while the term *element* can be used for any data set.

> **Experimental Studies**
>
> In an **experimental study,** researchers investigate how varying the predictor variable affects the response variable.
>
> A **predictor variable** (also called an **explanatory variable**) is a characteristic intended to explain differences in the response variable.
>
> A predictor variable that takes the form of a purposeful intervention is called a **treatment.**
>
> A **response variable** is an outcome, a characteristic of the subjects of the experiment presumably brought about by differences in the predictor variable or treatment.
>
> The **subjects** in a statistical study represent the elements from which the data are drawn.

We illustrate experimental studies using the following example.

| Example 1.13 | Newborn babies and a heartbeat: an experimental study |

© Mediacolor/Alamy

A psychologist wanted to test whether the sound of a human heartbeat would help newborn babies grow. A baby nursery at a hospital was set up so that the sound of a human heartbeat could be heard throughout the nursery. The heartbeat sound was played in the nursery for a large batch of newborn children, who were then weighed to determine their weight gain after four days in the nursery. Later, a second batch of children occupied the nursery, but no heartbeat sound was played. These children were also weighed after four days in the nursery. Babies were randomly placed into the two groups. Identify the following:
a. The subjects
b. The predictor variable
c. The treatment
d. The response variable

Solution
a. The babies were the **subjects** of this experimental study.
b. The **predictor variable** is whether or not the heartbeat sound was played in the nursery.

   c.   The **treatment** is the sound of the human heartbeat.

   d.   The **response variable** is the baby's weight gain, which is the outcome of the study.

   The results were consistent with the psychologist's conjecture; the babies who listened to the heartbeat sound had a greater average weight gain than the babies for whom no heartbeat sound was played. ◼

There are three main factors that should be considered when designing an experimental study: *control, randomization,* and *replication.*

Control.   A control group is necessary to compare against the treatment group, if we wish the results of our experiment to be useful. The control group in the above example is the group of babies for whom the heartbeat sound was not played. Had the psychologist omitted this control group, there would have been nothing to compare his results against. In some experiments, especially in medicine, members of the control group receive a placebo, a nonfunctioning simulated treatment. Sometimes, the symptoms of the members of the control group improve simply by taking the placebo, a phenomenon known as the *placebo effect.*

Randomization.   Many biases can be introduced into an experiment. For example, a well-meaning doctor may want to place underweight high-risk babies in the group with the heartbeat, in the hope that such babies will flourish. To eliminate biases like these, the placement of the subjects into the treatment and control groups should be done randomly.

Replication.   One major theme of statistical investigation is that larger samples are usually better, because they allow more precise inference. In a statistical study, the treatment and the control groups each must contain a large enough number of subjects to allow detection of meaningful differences between the treatment and control. For example, if a researcher examined only three babies with the heartbeat sound and three babies without the heartbeat sound, this would not be a sufficient number of replications. In Chapter 8, we will learn how large a sample size is sufficient for the needs of a particular study.

Observational Studies

There are circumstances where it is either impossible, impractical, or unethical for the researcher to place subjects into treatment and control groups. For example, suppose we are interested in whether women who work outside the home suffer less depression than women who remain at home with the children. The explanatory variable here is whether or not a woman works outside the home. However, it is not possible for the researcher to take women and randomly separate them into groups that either work outside the home or do not work outside the home. Sometimes an experimental study is not possible for ethical reasons.

Further, suppose you are interested in whether babies born to chemically dependent mothers display differences in cognitive skills from babies born to mothers who are not chemically dependent. It is clearly not ethical to randomly assign half of the mothers in the study to become chemically dependent during their pregnancy. Therefore, researchers need another type of statistical study: the observational study. In an observational study, the researcher observes whether the subjects' differences in the predictor variable are associated with differences in the response variable. No attempt is made to create differences in the predictor variable.

A sample survey is an example of an observational study. Data about a response variable may be obtained through the survey, along with information about possible predictor variables. No attempt is made to manipulate the variables. The researcher analyzes the information to determine whether differences in the predictor variable are associated with differences in the response variable.

Example 1.14   Is Ecstasy toxic to your neurons?

According to the British medical journal *The Lancet,* experimental studies carried out on animals (nonhuman primates, squirrel monkeys, and rodents) have revealed that large doses of the drug Ecstasy (methylene-dioxy-methamphetamine, or MDMA) produce "large and possibly permanent damage" to neural axons in the brain. Explain why the researchers did not carry out their experiment on humans.

Solution

It is not ethical to randomly assign half of the human subjects to receive large doses of the drug Ecstasy, especially in view of its effect on animals. The difficulty of performing experimental studies on humans concerning the effects of controlled substances is addressed by the authors of the *Lancet* study:

> *Only a prospective[experimental] study . . . could definitively show that recreational MDMA use was neurotoxic in human beings. For ethical, political, and legal reasons such a study is unlikely to ever be done. Instead, we have to rely upon evidence from observational studies of recreational MDMA users.*[6]   ■

Example 1.15   Secondhand smoking and illness in children

A 2006 Surgeon General's report found that "the evidence is sufficient to infer a causal relationship" between secondhand tobacco smoke exposure from parental smoking and respiratory illnesses in infants and children.[7] Was this report based on an experimental study or an observational study?

Solution

It would have been unethical to force the parents of a treatment group to smoke tobacco, given the health risks associated with tobacco use. Therefore, the study must have been an observational one.   ■

STEP-BY-STEP TECHNOLOGY GUIDE: Generating a Random Sample

We illustrate using Example 1.11 (page 20).

TI-83/84

Step 1   Enter a "seed," which can be any nonzero number.
Step 2   Press **STO ⇒** .
Step 3   Press **MATH**, highlight **PRB**, select **1: rand**, and press **ENTER** (see Figure 1.6, which uses 1776 for the seed). Your seed number is now in the calculator's memory.
Step 4   Press **MATH**, highlight **PRB** and select **5: randInt(**.
Step 5   Enter **1, N,** *two times* **n**, where **N** = population size and **n** = sample size. We enter twice the sample size in case there are repeats. For Example 1.11, since **n** = 7, we enter **randInt(1, 25, 14)** and press **ENTER** (Figure 1.7).
Step 6   Store the random sample in list **L1** as follows: press **STO ⇒** , then **2ND**, then **L1** (Figure 1.7). Then press **ENTER**.
Step 7   View the random sample by pressing **STAT**, highlighting **EDIT**, and pressing **ENTER** (Figure 1.8). Note that there is a repeat (**6**). We therefore select the next number, **10**, to round out our sample. The random sample for Example 1.11 is therefore **9, 25, 6, 20, 24, 10, 1** (Figure 1.9).

Figure 1.6

Figure 1.7

Figure 1.8

Figure 1.9

EXCEL

Step 1   Select cell **A1**. Click the **Insert Function** icon f_x.
Step 2   For "Search for a function," enter **randbetween**. Click **Go**, then **OK**.
Step 3   For **Bottom**, enter **1**. For **Top**, enter population size **N**. For Example 1.11, **N** = 25. Click **OK**.
Step 4   Cell **A1** now contains a random integer between **1** and **N**. Copy and paste cell **A1** into *twice* as many cells as needed for the sample size **n**, just in case there are repeats. For Example 1.11, copy and paste into cells **A2** to **A14**. The results are shown in Figure 1.10. Note that **8** is repeated, so that our random sample is **8, 2, 20, 16, 23, 7, 22**.

Figure 1.10

MINITAB

Step 1   Click on **Calc > Random Data > Integer . . .**
Step 2   In the **Generate __ rows of data** section, enter *twice* your desired sample size, just in case there are repeats. For example, if your desired sample size is 7, enter **14**.
Step 3   In the **Store in column __** section, enter whichever column is convenient for you, such as **C1**.
Step 4   For **Minimum value**, enter **1**. For **Maximum value**, enter your population size, **N**. Click **OK**.
Step 5   The random integers appear in column **C1**. Start from the top and go down the list, omitting any repeats, until you have your sample of size **n**. Our random sample (Figure 1.11) is therefore **3, 18, 2, 11, 21, 7, 25**.

Figure 1.11

Excel and Minitab base the seed on the current time, so that you need not set it yourself.

Section 1.3 SUMMARY

1 A random sample is a sample for which every element has an equal chance of being included. A random sample can minimize many potential biases, which could lead to incorrect generalizations about the population.

2 Other sampling methods include stratified sampling, systematic sampling, cluster sampling, and convenience sampling.

3 When constructing a survey, follow the five factors for good questionnaire design.

4 There are two types of statistical studies: experimental studies and observational studies. In an experimental study, researchers investigate how varying the predictor variable affects the response variable. It is not always possible to conduct an experimental study, however, and sometimes an observational study is used instead.

Section 1.3 EXERCISES

CLARIFYING THE CONCEPTS

1. Describe a situation where the information from a badly chosen sample could lead to negative consequences.
2. What type of bias did the *Literary Digest* poll (page 18) exhibit? How did it affect the results?

3. How could the *Literary Digest* have decreased the bias in its poll?
4. Was the *Literary Digest* poll a random sample?
5. Describe in your own words what a random sample is.
6. Describe the difference between an observational study and an experimental study.

PRACTICING THE TECHNIQUES

For Exercises 7–11, state which type of sampling is represented.

7. Students in your class are divided into males and females. A random sample of size 3 is then drawn from each of the groups.

8. You are interested in estimating the average number of hours dormitory residents spend studying. Three dormitories are chosen at random. In each dormitory, one floor is chosen at random and all the students on that floor are interviewed.

9. You are researching the proportion of college students who prefer hip-hop music to other forms of music. You obtain a listing of all the students at your college and contact every 20th student on the list.

10. An instructor in a large lecture course of 300 students would like to get a student sample, and he selects every 10th name from the class roster.

11. Your campus statistical consulting center uses random digit dialing to locate potential subjects for a political survey.

12. Provide an example of a survey question that lacks simplicity and clarity. Alter it so that it is simpler and clearer, as Factor 1 recommends.

13. You have probably heard of the warning "Be careful what you ask for; you may get it." Which of the five factors best describes this danger?

For Exercises 14–16, state which type of study is involved, experimental or observational. Also, identify the response variable and the predictor variable.

14. A sociologist would be interested in whether large families (at least four children) attend religious services more often than smaller families do.

15. A financial researcher would be interested in whether companies that give large bonuses to their chief executive officers (at least $1 million per year) have a higher stock price.

16. A manufacturer would be interested in whether a wireless adapter will improve the performance of its electronics equipment.

17. Select a random sample of size 7 from Table 1.7 (page 20). Why is your sample different from the one shown in the text?

APPLYING THE CONCEPTS

18. Contradicting Ann Landers. "If you had to do it over again, would you have children?" This is the question that advice columnist Ann Landers once asked her readers. It turns out that nearly 70% of the 10,000 responses she received were "No." A professional poll by *Newsday* found that 91% of respondents would have children again. Explain the apparent contradiction between these two surveys using what you have learned in this section.

19. High School Dropouts. For the following survey, describe the target population and the potential population, and discuss the potential for selection bias. Researchers are interested in the proportion of high school students in New England who drop out (leave school before graduating). A survey is made of 15 high schools in Greater Boston.

20. Living Below the Poverty Level. For the following survey, describe the target population and the potential

population, and discuss the potential for selection bias. A sociologist is interested in the proportion of people living below the poverty level in Chicago. He takes a random sample of phone numbers from the Chicago phone directory and asks each respondent his or her annual household income.

21. Rap or Hip-Hop. Describe what is wrong, if anything, with the following survey question. "Do you enjoy listening to rap or hip-hop music?"

22. Financial Ruin. Describe what is wrong, if anything, with the following survey question: "Do you think that we should tax and spend our way into financial ruin?"

23. Abortion. Suppose 67% of female respondents respond affirmatively to the question "Do you support the right of a woman to terminate a pregnancy when her life is in danger?" Would the researcher be justified in reporting, "Two-thirds of women support abortion?"

24. Cholesterol-Lowering Medication. Cholesterol researchers are investigating whether there is any difference between a new medication and a placebo (inactive pill) in lowering LDL cholesterol levels in the bloodstream.

    a. Describe what the response variable and the predictor variable are.

    b. Describe how the cholesterol researchers would go about collecting their data. Would they use an experimental study or an observational study?

    c. What is the treatment? What is the control?

25. Most Active Stocks. Here is a list of the five most active stocks on the NYSE on March 27, 2008.

Stock	Price
Citigroup	$20.98
Merrill Lynch	$40.93
Lehman Brothers	$38.86
Washington Mutual	$10.04
Ford	$ 5.63

    a. We are about to select a random sample and determine the lowest price in the sample. Do we know what this price will be before we select the sample? Why or why not?

    b. Select a random sample of size 2 from the table.

    c. If you take another sample of size 2, is it likely to comprise the same two companies? Why or why not?

    d. Which stock in your sample has the lowest price? What is that price?

26. Most Active Stocks. Refer to the list of stocks from the previous exercise.

    a. We are about to select another random sample and determine the lowest price in the sample. Do we know what this price will be before we select the sample? Do we know whether it will be the same as in the previous exercise? Why or why not?

    b. Select another random sample of size 2.

    c. Which stock in your new sample has the lowest price? What is that price?

    d. Compare your answers in (**c**) with those from (**d**) in Exercise 25. What can we say about a quantity like "the lowest price in a random sample of stocks"?

27. Mediterranean Diet. The American Heart Association reported the results of an experimental study.[8] Patients who ate a Mediterranean diet had a significantly lower risk of having a second heart attack than did patients who ate a Western diet. Identify the response variable and the predictor variable in this experimental study.

Evidence for an Alternative Therapy? Use the following information for Exercises 28–30. A company called QT, Inc. sells "ionized bracelets," called Q-Ray Bracelets, that it claims help to ease pain by balancing the body's flow of "electromagnetic energy." QT, Inc. claims that Q-Ray Bracelets can ease pain caused by cancer, restore well-being, and provide many other health benefits. The Mayo Clinic decided to conduct a statistical study to determine whether the extravagant claims for Q-Ray Bracelets were justified.[9] In the study, 305 subjects wore the Q-Ray "ionized" bracelet and 305 wore a placebo bracelet (identical to the ionized bracelet except for the ionization) for four weeks, at the end of which certain measures of pain were evaluated and compared between the treatments. The subjects, upon entry to the study, were randomly assigned to receive either the ionized bracelet or the placebo bracelet.

28. Identify the following aspects of this study.
   a. The control
   b. The randomization
   c. The replication

29. Identify the following aspects of this study.
   a. The predictor variable
   b. The treatment
   c. The response variable

30. Does this statistical study represent an experimental study or an observational study? Explain why.

 Use the *Random Sample* applet for Exercises 31–33.

31. Generate a random sample of 7 cities from Table 1.7 (page 20).

32. Generate another random sample of 7 cities from Table 1.7. Are all the cities in the two samples the same?

33. Before we generate a third sample of 7 cities, choose a city from Table 1.7.
   a. Will this city appear in the random sample?
   b. Is there any way of telling for certain in advance whether this city will appear in the random sample?
   c. Now go ahead and generate the third random sample of 7 cities. Is your city in the sample?

Try This in Class! In-class activities to enhance your understanding of statistics

Throughout this course we will work together on various activities in class. In this way, we will share in each other's experience with statistics. Also, simply talking about the problem can help put it into perspective.

 GENERATING RANDOM DATA

1. Here is a fun way to generate random samples of students from your class by rolling a pair of dice! (For large lecture classes, it is probably easier to use technology or the *Random Sample* applet to generate a random sample. See the Step-by-Step Technology Guide at the end of Section 1.3 for instructions.) Make a table constructed as follows:

	1	2	3	4	5	6
1						
2						
3						
4						
5						
6						

Print this table on a sheet of paper and pass the paper around the class, having each student choose one cell and write his or her name in the cell. The numbers across the top are for the results of tossing a red die, and the numbers down the left are for the results of tossing a black. When a certain combination of red and black numbers is tossed, the student whose cell that combination refers to becomes part of the sample. In Chapter 5 we will learn that all the cells are equally likely to be picked

if the dice are fair. If there are fewer than 36 students in your class, simply write *Roll Again* in the remaining empty cells after all students have chosen their cells. If there are more than 36 students in your class, just make more than one of these tables, and use a third colored die (for example, a green die) to choose which table is to be sampled. If a student is chosen more than once, simply roll again.

Do you think this arrangement will produce random samples? Can you suggest any other arrangements that will produce random samples from the population of students in your class?

2. Now let's take samples of the class.

   a. Let the instructor call the first 10 names on an alphabetical list of those present for class today. Is this a random sample? Why or why not? Find the proportion of females in this sample.

   b. Let the instructor ask those students who watched ESPN this week to raise their hands. Is this a random sample? Why or why not? Find the proportion of females in this group. Is this proportion the same as in (**a**)? Why doesn't it have to be the same?

   c. Get the dice out and generate a sample using the table(s) you constructed. Is this a random sample? Why or why not? Find the sample proportion of females. Is this the same as in (**a**) or (**b**)?

   d. Choose another sample as in (**c**). Is this a random sample? Find the sample proportion of females. Is it the same as in (**a**) or (**b**) or (**c**)?

CHAPTER 1 VOCABULARY

SECTION 1.2
- **Census** (p. 12)
- **Continuous variable** (p. 10)
- **Descriptive statistics** (p. 8)
- **Discrete variable** (p. 10)
- **Element** (p. 9)
- **Observation** (p. 9)
- **Parameter** (p. 12)
- **Population** (p. 12)
- **Qualitative variable** (p. 10)

- **Quantitative variable** (p. 10)
- **Sample** (p. 12)
- **Statistic** (p. 12)
- **Statistical inference** (p. 12)
- **Statistics** (p. 6)
- **Variable** (p. 9)

SECTION 1.3
- **Experimental study** (p. 23)
- **Observational study** (p. 23)

- **Potential population** (p. 22)
- **Predictor variable (explanatory variable)** (p. 23)
- **Random sample** (p. 18)
- **Response variable** (p. 23)
- **Selection bias** (p. 22)
- **Subjects** (p. 23)
- **Target population** (p. 22)
- **Treatment** (p. 23)

CHAPTER 1 REVIEW EXERCISES

SECTION 1.2
Refer to Table 1.8 for Exercises 1–3.

TABLE 1.8 World water usage

Country	Continent	Climate	Water use (per capita gallons per day)	Main use
Iraq	Asia	Arid	3311	Irrigation
United States	North America	Temperate	1565	Industry
Pakistan	Asia	Arid	1486	Irrigation
Canada	North America	Temperate	1268	Industry
Madagascar	Africa	Tropical	1212	Irrigation
North Korea	Asia	Temperate	1194	Not reported
Chile	South America	Arid	1176	Irrigation
Bulgaria	Europe	Temperate	1158	Not reported
Afganistan	Asia	Arid	1039	Irrigation
Iran	Asia	Arid	986	Irrigation

1. Use the table to find each of the following.
   a. How many elements are there? List a few.
   b. What are the variables? List them.
   c. What is the observation for Canada?
   d. Which country uses the most water per capita?
   e. What does "per capita" mean?
2. Identify the type of each of the following variables. What are the possible values?
   a. Continent.
   b. Climate.
   c. Water use.
   d. Main use. Is "not reported" a possible value of this variable?
3. Which two temperate countries share the same continent, climate, and main use?
   a. Is their water use the same?
   b. Is it likely that their water use would be exactly the same?
4. The following table contains population figures for the five most populous states.

State	Pop. (1960, in 1000s)	Pop. (2005, in 1000s)	Increase
California	15,717	36,132	20,415
Texas	9,580	22,860	13,280
New York	16,782	19,255	2,473
Florida	4,952	17,790	12,838
Illinois	10,081	12,763	2,682

   a. Identify the elements and the variables.
   b. Are the variables qualitative or quantitative?
   c. Provide the observation for the state of Florida.
   d. Which three states had the largest population increases? Which two states had the smallest population increases?
5. An electrical company has developed a new form of light bulb that it claims lasts longer than current models. The company has 1 million bulbs in its inventory. Consider the population average lifetime.
   a. What is the only way to find out the population average lifetime of the 1 million bulbs in the inventory?

b. Suppose someone who worked for you wrote you a memo suggesting that it was crucial to know the exact value of the population average lifetime of all 1 million new light bulbs. How would you respond? What might you suggest instead?

6. Consider the chapter Case Study, "Does Friday the 13th Change Human Behavior?" (page 7). There are two ways of making a decision error: (i) There is nothing to the superstition, but you believe in it and stay home. (ii) There is something to the superstition, but you venture out anyway.

    a. Which of (i) or (ii) do you think is the more serious error? Explain.

    b. Which of the two situations is more likely?

7. Suppose we are interested in the proportion of left-handed statistics students, and we take a sample to estimate the percent of students in our class who are left-handed.

    a. What is the population?

    b. What is the sample?

    c. What is the variable? Is it quantitative or qualitative?

    d. Is the sample proportion likely to be exactly the same as the population proportion? Is it likely to be very far away from the population proportion? Explain.

SECTION 1.3

8. Refer to the *Literary Digest* poll discussed in Section 1.3.

    a. What was the population? Be specific.

    b. What was the sample?

    c. Discuss whether the sample was similar to the population in all important characteristics.

9. Suppose you are interested in finding out how the statistics grades for your class compare with those of the college as a whole.

a. Would you use an experimental study or an observational study?

b. Discuss how this study situation would preclude effective randomization.

10. A long-running television advertisement claimed that "3 out of 4 dentists surveyed recommend sugarless gum for their patients who chew gum."

    a. If in fact only 4 dentists were surveyed, which of the study factors were violated?

    b. Use this situation to discuss why replication is important.

11. In the study situation described in Exercise 24 (page 27), there is a patient with severe hypercholesterolemia (very high LDL cholesterol levels), and so the doctor assigns this patient to the group of patients who receive the new medication rather than the placebo.

    a. Which of the experimental factors did the doctor violate?

    b. Use this situation to discuss why randomization is important.

12. Suppose we are interested in determining whether differences exist in the cognitive levels of children from single-parent families and those from two-parent families. Would we use an observational study or an experimental study? Clearly describe why.

13. Referring to the study in the previous exercise, suppose the children from single-parent families showed lower average cognitive skills than children from two-parent families. Does this mean that living in a one-parent family causes lower levels of cognitive skills? Why or why not?

CHAPTER 1 *Quiz*

TRUE OR FALSE

1. True or false: Statistical inference consists of methods for estimating and drawing conclusions about sample characteristics based on the information contained in the population.

2. True or false: A parameter is a characteristic of a sample.

FILL IN THE BLANK

3. Statistics is the art and science of _____, analyzing, presenting, and interpreting data.

4. An _____ is the set of values of all variables for a given element.

5. A statistic is a characteristic of a _____.

SHORT ANSWER

6. Is a sample survey examining the effects of secondhand smoke an example of an experimental study or an observational study?

7. State which type of statistical study is involved in the following. A large pharmaceutical company is interested in whether a new drug will reduce Alzheimer's disease symptoms in elderly patients.

8. For the study in the previous exercise, identify the predictor variable and the response variable.

CALCULATIONS AND INTERPRETATIONS

Crop Damage from Pesticides. Use the following information for Exercises 9–11. Agricultural researchers are investigating whether a new form of pesticide will lead to lower levels of insect damage to crops than the traditional pesticide.

9. Describe the response variable and the predictor variable.

10. Would the researchers use an experimental study or an observational study?

11. What is the treatment? What is the control?

12. Describe what is wrong, if anything, with the following survey question. "How often would you say that you attend the movie theater: often, occasionally, sometimes, seldom, or never?"

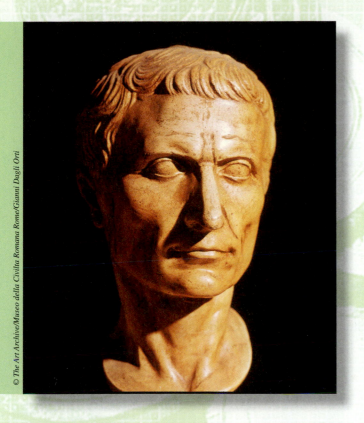

© The Art Archive/Museo della Civiltà Romana Rome/Gianni Dagli Orti

Describing Data Using Graphs and Tables

CASE STUDY | **The Caesar Cipher**

Over two thousand years ago, Julius Caesar developed the Caesar Cipher, which was a means of encoding his messages so that enemy generals would not be able to understand them if they were intercepted. He did this quite simply by shifting each letter in the message a certain number of places. For example, if each letter is shifted one place to the right, then:

The message	Would be encoded as
DEAR BRUTUS,	**EFBS CSVUVT,**
ALL GAUL IS HERE,	**BMM HBVM JT IFSF,**
WISH YOU WERE BEAUTIFUL	**XJTI ZPV XFSF CFBVUJGVM**

Where does statistics come in? Well, what if you were an enemy general and you intercepted a message from Caesar to one of his generals? You would not know which shift was being used, so *how could you use statistics to decode the message?* The answer is to make use of your knowledge of modern English letter frequencies (for simplicity, we assume that Caesar was fluent in English, a language that wouldn't develop until hundreds of years later). Now, in a typical English text 1000 letters long, the frequencies (counts) of the letters in the English alphabet are approximately the following:

A	B	C	D	E	F	G	H	I	J	K	L	M
73	9	30	44	130	28	16	35	74	2	3	35	25

N	O	P	Q	R	S	T	U	V	W	X	Y	Z
78	74	27	3	77	63	93	27	13	16	5	19	1

Armed with this knowledge, you as the enemy general could decide on the most likely shift of a given encoded message. For example, suppose you had intercepted the following message:

LI ZH ZLQ, SLCCD IRU HYHUBRQH (HAWUD SHSSHURQL).

Your statistical knowledge of the frequency of letters would help you to decode the message and perhaps change history (or perhaps not). For now, however, we will wait until the Case Study on page 37 to decode the secret message, at which time we will discuss how to solve this problem.

The Big Picture

Where we are coming from, and where we are headed…

In Chapter 1, we became acquainted with the building blocks of data analysis, such as the concepts of population, sample, types of variables, and the importance of a random sample. In Chapter 2, we apply the adage "A picture is worth a thousand words." The human mind can assess information presented pictorially, better than it can through words and numbers alone. We can recognize the entire scope of the interrelationships within a picture and somehow formulate a synthesis. Psychologists sometimes call this innate pattern recognition ability the "gestalt" experience. Statistical graphics take advantage of these abilities to quickly summarize data.

Before a researcher can draw conclusions from a data set, he or she must first *organize* the data, usually into some sort of a table or graph. The purpose of this chapter is to introduce graphical and tabular methods for summarizing data. We construct and interpret graphical descriptive tools such as bar graphs and tabular descriptive tools such as frequency distributions. Later, in Chapter 3, we learn how to describe a data set using statistics rather than tables and graphs. ■

2.1 Graphs and Tables for Categorical Data

Objectives: **By the end of this section, I will be able to...**

1 Construct and interpret a frequency distribution and a relative frequency distribution for qualitative data.

2 Construct and interpret bar graphs and Pareto charts.

3 Construct and interpret pie charts.

1 Frequency Distributions and Relative Frequency Distributions

Frequency Distributions

Recall from Chapter 1 that categorical (qualitative) data take values that are nonnumeric and are usually classified into categories. In this section we learn graphical and tabular methods for handling categorical data. Let us begin with an example.

The January 2004 issue of *Trends and Tudes,* the newsletter from Harris Interactive (**www.harrisinteractive.com**), described a survey that asked teenagers which career they considered to have "very great prestige." Suppose that, as part of her thesis project, a college student conducted a survey of the students in her sister's high school homeroom, asking, "Which of the following careers do you think has the highest prestige: Actor, Athlete, Doctor, Entertainer, Firefighter, Lawyer, Member of Congress, Military Officer, Police Officer, Scientist?" Each student's chosen career was recorded. Note that *career* is a qualitative variable, not quantitative. The results from a random sample of 20 students are presented in Table 2.1.

Table 2.1 Prestigious career survey data set

Student	Career	Student	Career	Student	Career
1	Doctor	8	Athlete	15	Lawyer
2	Scientist	9	Doctor	16	Military Officer
3	Military Officer	10	Scientist	17	Doctor
4	Military Officer	11	Doctor	18	Scientist
5	Doctor	12	Military Officer	19	Doctor
6	Scientist	13	Scientist	20	Lawyer
7	Military Officer	14	Lawyer		

It is not immediately clear from this data set which career is the most popular choice among the 20 students in the sample. That is why we need ways to summarize the values in a data set. One popular method used to summarize the values in a data set is the **frequency distribution** (or *frequency table*).

> The **frequency,** or **count,** of a category refers to the number of observations in each category. A **frequency distribution** for a qualitative variable is a listing of all the values (for example, categories) that the variable can take, together with the frequencies for each value.

Example 2.1   What careers do teenagers admire?

Create a frequency distribution for the variable *career* from Table 2.1.

Solution

For each career choice, we compute the **frequency;** that is, we **count** how many students preferred that particular career. Table 2.2 shows the frequency distribution. For example, five students chose Scientist. The frequency distribution summarizes the data set so that quick observations can be made, such as "Doctor was preferred by the greatest number of students in the survey." ■

Table 2.2 Frequency distribution of career

Variable: career	Frequency
Doctor	6
Scientist	5
Military Officer	5
Lawyer	3
Athlete	1

As the data set gets larger, the need for summarization gets more and more acute. (Imagine if our survey consisted of 20,000 students rather than 20.) Take a moment to add up the frequencies in Table 2.2. What do they add up to? This number is the sample size: $n = 20$. Now, is this just a coincidence, or does this happen every time? Actually, this happens every time: the sum of the frequencies equals the sample size, n. One way to check if you made a mistake in forming your frequency distribution table is to add up the frequencies and see if the sum equals the sample size.

Relative Frequency Distributions

Next, suppose you didn't know the size of the sample in the survey. Suppose you were told only that 6 students preferred Doctor. What would your response be? The logical question is "Is that a lot?" If our sample size was 10 students, then 6 students preferring Doctor is certainly a lot. However, if our sample size was 1000 students, then only 6 students preferring Doctor is *not* a lot. So, the number's significance depends on what you compare the 6 students to—that is, "relative to what?" or "compared to what?" In statistics, we compare the frequency of a category with the total sample size to get the **relative frequency.**

The **relative frequency** of a particular category of a qualitative variable is its frequency divided by the sample size. A **relative frequency distribution** for a qualitative variable is a listing of all values that the variable can take, together with the relative frequencies for each value.

Example 2.2   Relative frequency of career preferences

Create a **relative frequency distribution** for the variable *career* using the information in Table 2.2.

Solution

The relative frequency of the students who preferred Doctor is the frequency 6 divided by the sample size 20:

$$\text{relative frequency of Doctor} = \frac{\text{frequency}}{\text{sample size}} = \frac{6}{20} = 0.30$$

The relative frequency of the students who preferred Doctor is 0.30, or 30%. So, if someone told you that 30% of the students preferred Doctor, without telling you the sample size, you would have a better idea of the relative strength of preference for that group than if you were given the frequency alone.

To construct the relative frequency distribution, shown in Table 2.3, for the career survey data, divide each frequency in the frequency distribution by the sample size 20. Note that the relative frequencies always add up to 1.0, which represents 100%. ■

Table 2.3 Relative frequency distribution of career preference by students

Variable: career	Relative frequency
Doctor	6/20 = 0.30
Scientist	5/20 = 0.25
Military Officer	5/20 = 0.25
Lawyer	3/20 = 0.15
Athlete	1/20 = 0.05
Total	20/20 = 1.00

Bar Graphs and Pareto Charts

Frequency distributions and relative frequency distributions are tabular, and thus useful for summarizing data sets. The graphical equivalent of a frequency distribution or a relative frequency distribution is called a **bar graph** (or **bar chart**).

> A **bar graph** is used to represent the frequencies or relative frequencies for categorical data. It is constructed as follows:
>
> 1. On the horizontal axis, provide a label for each category.
> 2. Draw rectangles (bars) of equal width for each category. The height of each rectangle represents the frequency or relative frequency for that category. Ensure that the bars are not touching each other.

Example 2.3   Bar graphs of career preferences

Construct a frequency bar graph and a relative frequency bar graph for the career preference distributions in Tables 2.2 and 2.3.

Solution

The bar graphs are provided in Figures 2.1a and 2.1b. Across the horizontal axis are the five career categories. Next, draw rectangles, the heights of which represent either the frequency or the relative frequency for that category, represented on the vertical axis. For example, in Figure 2.1a, the first rectangle (Doctor) reaches a height of 6, while the second rectangle reaches only to 5. Note that the rectangles are of equal width, and none of them touch each other. Also notice that the two bar graphs are exactly alike except for the scale indicated on the vertical axis. This is because we divide each frequency by the same number, the sample size, to get the relative frequency. ■

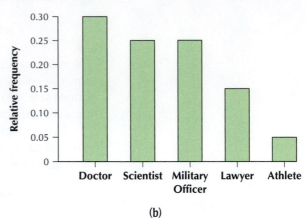

(a)   (b)

FIGURE 2.1 (a) Frequency bar graph; (b) relative frequency bar graph.

Both Figure 2.1a and Figure 2.1b are examples of **Pareto charts.** Figures 2.3a and 2.3b (page 38) are examples of bar graphs that are not Pareto charts.

> A **Pareto chart** is a bar graph in which the rectangles are presented in decreasing order from left to right.

3 Pie Charts

Pie charts are a common graphical device for displaying the relative frequencies of a categorical variable.

> A **pie chart** is a circle divided into sections (that is, slices or wedges), with each section representing a particular category. The size of the section is proportional to the relative frequency of the category.

Pie charts are typically made using technology. However, one can construct a pie chart using a protractor and a compass. Since a circle contains 360 degrees, we need to multiply the relative frequency for each category by 360°. This will tell us how large a slice to make for each category, in terms of degrees.

Example 2.4   Pie chart for the careers data

Construct a pie chart for the careers data from Example 2.2.

Solution

The relative frequencies from Example 2.2 are shown in Table 2.4. We multiply each relative frequency by 360° to get the number of degrees for that section (slice) of the pie chart.

Our pie chart will have five slices, one for each career category. Use the compass to draw a circle. Then use a protractor to construct the appropriate angles for each section. From the center of the circle, draw a line to the top of the circle. Measure your first angle using this line. For Doctor, we need an angle of 108°. This angle is shown in Figure 2.2. Then, from there, measure your second angle—in this case, the 90° right angle for Scientist. Continue until your circle is complete. ■

Table 2.4 Finding the number of degrees for each slice of the pie chart

Variable: career	Relative frequency	Multiply by 360°	Degrees for that section
Doctor	6/20 = 0.30	0.30 × 360° =	108°
Scientist	5/20 = 0.25	0.25 × 360° =	90°
Military Officer	5/20 = 0.25	0.25 × 360° =	90°
Lawyer	3/20 = 0.15	0.15 × 360° =	54°
Athlete	1/20 = 0.05	0.05 × 360° =	18°
Total	20/20 = 1.00		360°

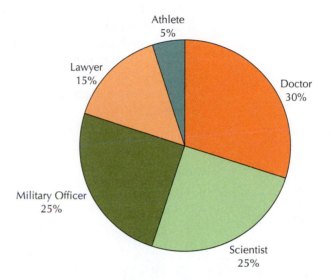

FIGURE 2.2 Pie chart of the careers data.

The Caesar Cipher

Recall the Caesar Cipher from the chapter introduction. It's time to decipher that secret message from Caesar to one of his generals that we intercepted:

LI ZH ZLQ, SLCCD IRU HYHUBRQH (HAWUD SHSSHURQL).

We will make a frequency distribution and bar graph of the letters in the message and then compare them with the frequency distribution and bar graph of the letters in the English language. The letter frequencies in a sample of modern English text that is 1000 letters long are given in Table 2.5.

Table 2.5 Frequency distribution of English letters in a sample of 1000 letters

A	B	C	D	E	F	G	H	I	J	K	L	M
73	9	30	44	130	28	16	35	74	2	3	35	25

N	O	P	Q	R	S	T	U	V	W	X	Y	Z
78	74	27	3	77	63	93	27	13	16	5	19	1

Using this frequency distribution, we can observe that the letter **E** far outstrips all other letters in the alphabet in frequency. Other high-frequency letters are **A, I, N, O, R, S,** and **T.** Compare this with the frequency distribution of the letters in the coded

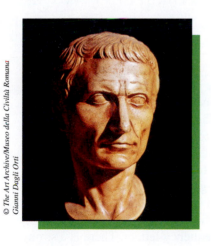

Table 2.6 Frequency distribution of letters in coded message

A	B	C	D	E	F	G	H	I	J	K	L	M
1	1	0	2	0	0	0	7	2	0	0	4	0

N	O	P	Q	R	S	T	U	V	W	X	Y	Z
0	0	0	3	3	4	0	4	0	3	0	1	2

message, shown in Table 2.6. From this frequency distribution, we can see that **H** is the most frequently occurring letter in the coded message. Other frequently occurring letters are **L, Q, R, S, U,** and **W.** Since **E** is the most frequently occurring letter in English, perhaps this means that **E** is encoded as **H,** the most common letter in our message. We turn to bar graphs to provide us with further evidence. The frequency bar graph for English letters, constructed from Table 2.5, appears in Figure 2.3a, while the frequency bar graph of letters in the coded message, from Table 2.6, is shown in Figure 2.3b.

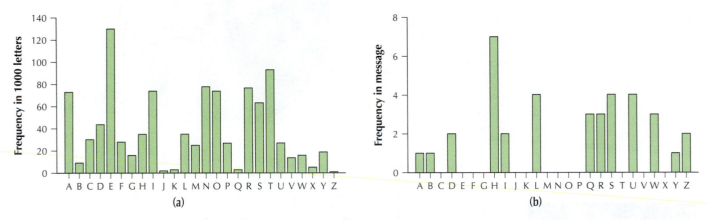

FIGURE 2.3 (a) Frequency bar graph of English letters; (b) frequency bar graph of letters in coded message.

Caesar used a simple shift of the letters for his code. If we substitute **H** for **E,** then the original letters have been shifted three places to the right (**E → F → G → H**). But this may just be an aberration. Is there further evidence for a "right shift of 3"? Let's see if this "right shift of 3" makes sense for the other high-frequency letters in the coded message. To undo a "right shift of 3," we would need to shift the letters in the coded message back three to the left to get the original letters. If the letter **L** is shifted back three places to the left, you get **I,** one of the high-frequency letters in English. Shift the letter **Q** three places, and you get **N,** another letter of high frequency in English. Shift the other letters of high frequency in our coded message, and you get **O, P, R,** and **T,** respectively, all high-frequency letters. There is a strong probability that we have found the correct decoding mechanism.

Let us now proceed to decode the message by shifting every letter in the coded message three places to the left (for example, **L → K → J → I**). It turns out that the decoded message reads

IF WE WIN, PIZZA FOR EVERYONE (EXTRA PEPPERONI).

Small wonder that Caesar went on to win an empire! We have gotten a taste of how the analysis of frequency distributions and bar graphs can be useful for solving problems.

STEP-BY-STEP TECHNOLOGY GUIDE: Frequency Distributions, Bar Graphs, and Pie Charts

We use the data set in Table 2.7 to demonstrate how to use technology to construct a frequency distribution, relative frequency distribution, bar graph, and pie chart. Table 2.7 lists the declared majors of 25 randomly selected students at a local business school. (MIS stands for management information systems.)

TABLE 2.7   Declared majors of business school students

Management	MIS	Management	MIS	Marketing
Marketing	Marketing	Management	Finance	Accounting
Accounting	Accounting	MIS	Management	MIS
Management	MIS	Management	Economics	Accounting
Finance	Management	Economics	Marketing	Finance

EXCEL

Frequency Distributions

Step 1   Enter the data in Column A, with the topmost cell indicating the variable name, *Major*.
Step 2   Select cells A1–A26, click **Insert > PivotTable**, and click **OK**.
Step 3   Under **Choose fields to add to report**, select *Major*.
Step 4   Click on *Major* and drag to the **Values** box at the lower right of the screen. The resulting frequency distribution is shown in Figure 2.4. In Excel, this takes the form of a *pivot table,* which is an interactive tabular format.

Figure 2.4

Bar Graphs and Pie Charts

Note: Excel can make bar graphs or pie charts using frequency distributions but not from the raw data.
Step 1   Enter the frequency distribution as shown in Figure 2.5.
Step 2   Select cells A1 to B7. For a bar graph, click **Insert > Column**. For a pie chart, click **Insert > Pie**.
Step 3   The resulting frequency bar graph and pie chart are shown in Figures 2.6 and 2.7.

Figure 2.5

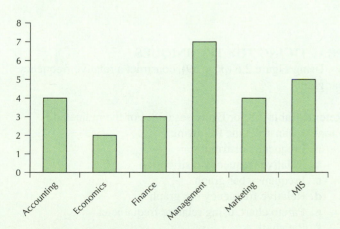

Figure 2.6 Excel frequency bar graph.

Figure 2.7 Excel pie chart.

(Continued)

MINITAB

Frequency Distributions

Step 1 Name your variable *Major* and enter the data into the C1 column.

Step 2 Click **Stat > Tables > Tally**.

Step 3 Under **Display**, select **Counts** and **Percents**.

Step 4 Click inside the **Variables** box until you see your variable *major* listed. Select the variable **C1 *Major***, and click **Select**. Then click **OK**.

Bar Graphs

Step 1 Name your variable *Major* and enter the data into the C1 column.

Step 2 Click **Graph > Bar Chart**. For raw data select **Bars Represent: Counts of Unique Values**, select **Simple**, and click

OK. (For summarized data such as a frequency distribution, select **Bars Represent: Values from a Table**, and select **Simple**. Then click **OK**.)

Step 3 In the **listing of variables** box, click on the *Major* variable to select it for analysis. Then click **OK**.

Pie Charts

Step 1 Name your variable *Major* and enter the data into the C1 column.

Step 2 Click **Graph > Pie Chart**. For raw data select **Chart Counts of unique values**. Then click in the **Variables** box to

select the variable *Major*, and click **OK**. (For summarized data such as a frequency distribution, select **Chart Data from a Table**. Then select the category variable for **Categorical variable**, and select the variable with the frequencies or relative frequencies for the **Summary variable**. Then click **OK**.)

Section 2.1 SUMMARY

In this section, we learned about tabular and graphical methods for summarizing qualitative (categorical) data.

1 Frequency distributions and relative frequency distributions list all the values that a qualitative variable can take, along with the frequencies (counts) or relative frequencies for each value.

2 A bar graph is the graphical equivalent of a frequency distribution or a relative frequency distribution. When the

rectangles are presented in decreasing order from left to right, the result is a Pareto chart.

3 Pie charts are a common graphical device for displaying the relative frequencies of a categorical variable. A pie chart is a circle divided into sections (that is, slices or wedges), with each section representing a particular category. The size of the section is proportional to the relative frequency of the category.

Section 2.1 EXERCISES

CLARIFYING THE CONCEPTS

1. Why do we use graphical and tabular methods to summarize data? What's wrong with simply reporting the raw data?

2. What's the difference between a frequency distribution and a relative frequency distribution?

3. True or false: For a given data set, a frequency bar graph and a relative frequency bar graph look alike except for the scale on the vertical axis.

4. True or false: A pie chart is used to represent quantitative data.

PRACTICING THE TECHNIQUES

5. Using Figure 2.6 (page 39), construct a relative frequency bar graph.

Refer to Table 2.8 for Exercises 6–8. For the indicated variable, construct the following:

    a. Frequency distribution
    b. Relative frequency distribution
    c. Frequency bar graph
    d. Relative frequency bar graph
    e. Pareto chart, using relative frequencies
    f. Pie chart

TABLE 2.8   World water usage

Country	Continent	Climate	Main use
Iraq	Asia	Arid	Irrigation
United States	North America	Temperate	Industry
Pakistan	Asia	Arid	Irrigation
Canada	North America	Temperate	Industry
Madagascar	Africa	Tropical	Irrigation
North Korea	Asia	Temperate	Not reported
Chile	South America	Arid	Irrigation
Bulgaria	Europe	Temperate	Not reported
Afghanistan	Asia	Arid	Irrigation
Iran	Asia	Arid	Irrigation

6. The variable *continent*.
7. The variable *climate*.
8. The variable *main use*.

9. The table below shows energy-related carbon dioxide emissions (in millions of metric tons) for 2004, by end-use sector, as reported by the U.S. Energy Information Administration.

Sector	Emissions
Residential	1213.9
Commercial	1034.1
Industrial	1736.0
Transportation	1939.2

Use the table to construct the following:
   a. Relative frequency distribution
   b. Frequency bar graph
   c. Relative frequency bar graph
   d. Pareto chart, using relative frequencies
   e. Pie chart of the relative frequencies

10. Table 2.9 shows the numbers of children enrolled in Head Start programs, as reported by the U.S. Department of Health and Human Services in 2000, by ethnic group.

TABLE 2.9   Ethnicity of children enrolled in Head Start

Census category	No. of Head Start children
White	235,945
Black	259,004
Hispanic	169,909
American Indian/Alaskan Native	27,128
Asian/Pacific Islander	21,917
Total	713,903

Use the table to construct the following:
   a. Relative frequency distribution
   b. Frequency bar graph

   c. Relative frequency bar graph
   d. Pareto chart, using frequencies
   e. Pie chart of the relative frequencies

APPLYING THE CONCEPTS
Musical Activities. Use the following information for Exercises 11–13. Every year, *USA Weekend* conducts a survey of the nation's teenagers, asking them various questions about lifestyle and music. In 2002, nearly 60,000 teenagers responded to the poll, conducted in part through *USA Weekend's* Web site. One question was, "Do you listen to music while you are . . . ?" and listed several options. Respondents were asked to select all that apply. The most common responses are shown in the table.

Doing chores	79%
On the computer	73%
Doing homework	72%
Eating meals at home	33%
In the classroom	18%

11. Do you think that the sample is representative of all U.S. teenagers? How might the sample systematically introduce bias?
12. Assuming that the sample size is 6000, construct a *frequency* bar graph of the five activities listed in the table.
13. Can you construct a pie chart of the five activities listed in the table? Explain precisely why or why not.

Music and Violence. Another poll question asked by *USA Weekend* was "Do you think shock rock and gangsta rap are partly to blame for violence such as school shootings or physical abuse?" The results are shown in the table. Use this information to answer Exercises 14–17.

Yes	31%
No	45%
I've never thought about it	24%

14. Do you think that there may have been other possible responses besides the three above? List at least two other possible responses.
15. Assuming the sample size is 6000, construct a frequency distribution of the responses.
16. Construct a frequency bar graph of the responses.
17. Construct a relative frequency pie chart of the responses.

Astrological Signs. Use the following information for Exercises 18 and 19. The General Social Survey collects data on social aspects of life in America. Here, 1464 respondents reported their astrological sign. A pie chart of the results is shown on the following page.

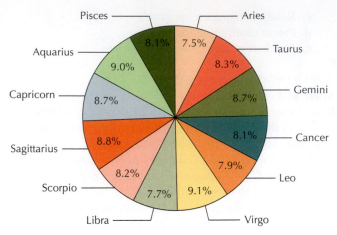

Pie chart of astrological signs.

18. Use the percentages in the pie chart to do the following:
   a. Construct a relative frequency bar graph of the astrological signs.
   b. Construct a relative frequency bar graph, but this time have the y axis begin at 7% instead of zero. Describe the difference between the two bar graphs. When would this one be used as opposed to the earlier bar graph?

19. Construct a frequency distribution of the astrological signs. Which sign occurs the least? The most?

Satisfaction with Family Life. Use the following information for Exercises 20–22. The General Social Survey also asked respondents about their amount of satisfaction with family life. Here, 1002 respondents reported their satisfaction levels with family life, as shown in the frequency distribution below.

Satisfaction	Frequency
Very great deal	415
Great deal	329
Quite a bit	99
A fair amount	90
Some	27
A little	25
None	17
Total	1002

20. What is your opinion of the categories? Do you think that everyone interprets these phrases in the same way? For example, what is the difference between "quite a bit" and a "great deal"?

21. Construct a Pareto graph of the categories. Do you think that, overall, people are happy with their family lives?

22. Construct a frequency or a relative frequency bar graph of the categories. Do you think that this bar graph reinforces the positive message in the data?

Venture Capital Deals. Use the following information for Exercises 23–25. **TheDeal.com** reports that the most active

venture capital investors, for the 12 months ending July 17, 2007, are as shown in the table.

Investor	No. of deals
Intel Capital Corp.	41
New Enterprise Associates	22
Sequoia Capital	19
Venrock Associates	19
Canaan Partners	18
Draper Fisher Jurvetson	18

23. Construct a relative frequency distribution of the number of deals.

24. Construct the following:
   a. Frequency bar graph of the number of deals
   b. Relative frequency bar graph of the number of deals

25. Construct a pie chart of the number of deals.

Health Insurance Coverage. Use the following information for Exercises 26–27. The health insurance coverage of Americans who were younger than 65 years of age in 2005 was reported by the Centers for Disease Control, as shown in the following table.[1]

Type of insurance	Count (in millions)
Private	126.8
Medicaid	13.7
Other	5.7
Uninsured	35.8

26. Construct a relative frequency distribution.

27. Construct a relative frequency bar graph.

Abortion and Rape. Use the following information for Exercises 28 and 29. Another question asked by the General Social Survey was "Please tell me whether or not you think it should be possible for a pregnant woman to obtain a legal abortion if she became pregnant as a result of rape?" Here, 934 respondents reported their responses as shown in the bar graph below.

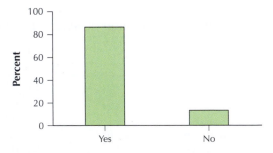

Bar graph of responses: possible for abortion if result of rape?

28. Estimating the percentages as best you can from the bar graph, construct a frequency or relative frequency distribution of the responses.

29. Estimating the percentages as best you can from the bar graph, construct a frequency or relative frequency pie chart of the responses.

9/11 Commemorative Web Sites. In the wake of the September 11 events, many people reached out through the Internet to find solace. The Pew Internet and American Life Project collected data on how people used the Internet to communicate and learn more about the tragic terrorist attacks. One question from their survey, taken from October 19 to November 18, 2001, asked: "Did you visit a memorial or commemorative web site as a result of the terrorist attacks in New York City and Washington?" Use the following table of the responses and their frequency to answer Exercises 30–34.

Response	Frequency
Yes, visited site because of attacks	475
No, have not visited this kind of site	2003
Don't know/Refused to answer	6
Total	2484

30. Construct a relative frequency distribution of the responses.

31. Construct a relative frequency bar graph of the responses.

32. If we wanted to change our relative frequency bar graph to a frequency bar graph, would we have to redraw the bars? Explain why or why not.

33. Construct a pie chart of the responses.

34. *What if* we doubled the frequencies of each response. How would that affect the following?
  a. Relative frequency distribution of the responses
  b. Relative frequency bar graph of the responses
  c. Pie chart of the responses

Brisbane Babies. Use the following information for Exercises 35–38. The *Brisbane Sunday Mail* newspaper reported that on December 18, 1997, 44 babies were born at the Brisbane Hospital in Australia. Their genders are given below.

Girl	Girl	Boy	Boy	Boy	Girl	Girl	Boy	Boy
Boy	Boy	Boy	Girl	Girl	Boy	Girl	Girl	Boy
Boy	Boy	Boy	Girl	Girl	Girl	Girl	Boy	Boy
Boy	Girl	Boy	Girl	Boy	Boy	Boy	Boy	Boy
Girl	Boy	Boy	Boy	Boy	Girl	Girl	Girl	

35. Construct a pie chart of the numbers of boys and girls born that day.

36. Construct a frequency distribution and a relative frequency distribution for the gender of the babies born that day.

37. Construct a relative frequency pie chart for the gender of the babies born that day.

38. Construct a relative frequency bar graph of gender. If we wanted to change our relative frequency bar graph to a frequency bar graph, would we have to redraw the bars? Explain why or why not.

Educational Goals in Sports. Use your knowledge of Excel or Minitab to solve Exercises 39 and 40. Open the **Goals** data set. The subjects are students in grades four, five, and six from three school districts in Michigan. The students were asked which of the following was most important to them: good grades, sports, or popularity. Information about the students' age, gender, race, and grade was also gathered, as well as whether their school was in an urban, suburban, or rural setting.[2]

39. Generate bar graphs for the following variables.
  a. *Gender*. Estimate the relative frequency of girls in the sample. Of boys.
  b. *Goals*. About what percentage of the students chose "grades" as most important? About what percentage chose "popular"? About what percentage chose "sports"?

40. Generate relative frequency distributions for the following variables.
  a. *Gender*. How close were your estimates in the previous exercise?
  b. *Goals*. How close were your estimates in the previous exercise?

Analysis of Households. For Exercises 41–43, use your knowledge of Excel or Minitab. Open the data set **Household.**

41. How many observations are in this data set? How many variables?

42. Which of the variables are qualitative? Which of the variables are quantitative?

43. What would a relative frequency distribution of the variable *state* look like? A pie chart? A bar graph?

CONSTRUCT YOUR OWN DATA SETS

Environmental Club. Use the following information for Exercises 44–46. You are the president of the College Environmental Club, which has members among all four classes, freshmen, sophomores, juniors, and seniors. The total number of members in the club is 20.

44. Set the frequency of each class so that each class has an equal number of members.
  a. Construct a frequency distribution of the variable *class*.
  b. Construct a relative frequency distribution of the variable *class*.

45. Set the frequency of each class so that there are more sophomores than freshmen, more juniors than sophomores, and more seniors than juniors.
  a. Construct a Pareto chart of the variable *class*.
  b. Construct a pie chart of the variable *class*.

46. Set the frequency of each class so that there are more seniors than any other class while the other three classes have equal numbers.
  a. Construct a frequency bar graph of the variable *class*.
  b. Construct a relative frequency bar graph of the variable *class*.

2.2 Graphs and Tables for Quantitative Data

Objectives: **By the end of this section, I will be able to...**

1 Construct and interpret a frequency distribution and a relative frequency distribution for quantitative data.

2 Use histograms and frequency polygons to summarize quantitative data.

3 Construct and interpret stem-and-leaf displays and dotplots.

4 Recognize distribution shape, symmetry, and skewness.

1 Frequency Distributions and Relative Frequency Distributions

In Section 2.1, we introduced tables and graphs for summarizing qualitative data. However, most of the data sets that we will encounter in this book are quantitative rather than qualitative. Recall from Chapter 1 that quantitative data take on numerical values. We can apply frequency and relative frequency distributions to quantitative data, just as we did for the qualitative data in Section 2.1.

Example 2.5   Ages of missing children in California

The National Center for Missing and Exploited Children (**www.missingkids.com**) keeps an online searchable data base of missing children nationwide. Table 2.10 contains a listing of the 50 children who have gone missing from California and who would have been between 1 and 9 years of age as of March 4, 2007. Suppose we are interested in analyzing the ages of these missing children. Use the data to construct a **frequency distribution** and a **relative frequency distribution** of the variable *age*.

Table 2.10 Missing children and their ages

Child	Age	Child	Age	Child	Age	Child	Age
Amir	5	Carlos	7	Octavio	8	Christian	8
Yamile	5	Ulisses	6	Keoni	6	Mario	8
Kevin	5	Alexander	7	Lance	5	Reya	5
Hilary	8	Adam	4	Mason	5	Elias	1
Zitlalit	7	Sultan	6	Joaquin	6	Maurice	4
Aleida	8	Abril	6	Adriana	6	Samantha	7
Alexia	2	Ramon	6	Christopher	3	Michael	9
Juan	9	Amari	4	Johan	6	Carlos	2
Kevin	2	Joliet	1	Kassandra	4	Lukas	4
Hazel	5	Christopher	4	Hiroki	6	Kayla	4
Melissa	1	Jonathan	8	Kimberly	5	Aiko	3
Kayleen	6	Emil	7	Diondre	4	Lorenzo	9
Mirynda	7	Benjamin	5				

Solution

We can construct the frequency distribution for the variable *age* and can construct the relative frequency distribution by dividing the frequency by the total number of observations, 50. See Table 2.11. ■

Table 2.11 Frequency distribution and relative frequency distribution of *age*

Age	Tally	Frequency	Relative frequency
1	|||	3	0.06
2	|||	3	0.06
3	||	2	0.04
4	⊞ |||	8	0.16
5	⊞ ||||	9	0.18
6	⊞ ⊞	10	0.20
7	⊞		6	0.12
8	⊞		6	0.12
9	|||	3	0.06
Total		50	1.00

We can combine several ages together into "classes," in order to produce a more concise distribution. **Classes** represent a range of data values and are used to group the elements in a data set.

Example 2.6 Frequency and relative frequency distributions using classes

Combine the age data from Example 2.5 into three classes, and construct frequency and relative frequency distributions.

Solution

Let us define the following classes for the age data: 1–3 years old, 4–6 years old, and 7–9 years old. For each class, we group together all the ages in the class. Table 2.12 provides the frequency distribution and relative frequency distribution for these three age classes. ■

Table 2.12 Distributions for the variable *age*, after combining into three classes

Class	Frequency	Relative frequency
1–3	8	0.16
4–6	27	0.54
7–9	15	0.30
Total	50	1.00

Developing Your Statistical Sense

Choosing Which Distribution to Use

So which frequency distribution is the "right" one, Table 2.11 or Table 2.12? There is no absolute answer. It depends on the goals of the analysis, as well as other factors. For example, from Table 2.12, we can see that the majority (0.54 = 54%) of missing children are aged 4–6, an observation that was not immediately apparent from Table 2.11. So, combining data values into classes can lead to interesting overall findings. However, whenever data values are combined into classes, some information is lost. For example, it is not possible, using Table 2.12 alone, to determine that age 6 has the highest proportion of missing children. ■

We use the following to construct frequency distributions and histograms (for a discussion of histograms, see pages 47–48).

> The **lower class limit** of a class equals the smallest value within that class.
> The **upper class limit** of a class equals the largest value within that class.
> The **class width** equals the difference between the lower class limits of two successive classes.

Example 2.7   Class limits and class widths for the age data

Find the **lower class limits,** the **upper class limits,** and the **class width** for the classes in Example 2.6.

Solution

The first class consists of all children aged 1, 2, and 3 years. The smallest value in this class is 1 year old. Therefore, the lower class limit for this class equals 1. The upper class limit for this class equals 3, the largest value in the class. The second class consists of all children aged 4, 5, and 6 years. So the lower class limit for this class is 4 years old, and the upper class limit is 6 years old. Similarly, for the class containing 7-, 8-, and 9-year-olds, the lower class limit is 7 and the upper class limit is 9. The class width is the difference in lower class limits between successive classes. Since our lower class limits are 1, 4, and 7, the class width of each class is 3 because the lower class limits differ by 3. For example, $4 - 1 = 3$. ■

Next, we show how to construct frequency distributions for quantitative data.

> To construct a frequency distribution for quantitative data:
> 1. Determine how many classes you will use.
> 2. Determine the class width. It is best (though not required) to use the same width for all classes.
> 3. Determine the upper and lower class limits. Make sure the classes are nonoverlapping.

Example 2.8   Constructing a frequency distribution: the management aptitude test

Twenty management students, in preparation for graduation, took a course to prepare them for a management aptitude test. A simulated test provided the following scores:

| 77 | 89 | 84 | 83 | 80 | 80 | 83 | 82 | 85 | 92 |
| 87 | 88 | 87 | 86 | 99 | 93 | 79 | 83 | 81 | 78 |

Construct a frequency distribution of these management aptitude test scores.

Solution

Step 1   **Choose the number of classes.** It is generally recommended that between 5 and 20 classes be used, with the number of classes increasing with the sample size. A small data set such as this will do just fine with 5 classes. In general, choose the number of classes to be large enough to show the variability in the data set, but not so large that many classes are nearly empty.

Step 2   **Determine the class widths.** First, find the *range* of the data, that is, the difference between the largest and smallest data points. Then, divide this range by the number of classes you chose in Step 1. This gives an estimate of the class width. Here, our largest data value is 99 and our smallest is 77, giving us a *range* of $99 - 77 = 22$.

In Step 1, we chose 5 classes, so that our estimated class width is 22/5 = 4.4. We will use a convenient class width of 5. It is recommended that each class have the same width.

Step 3   **Determine the upper and lower class limits.** Choose limits so that each data point belongs to only one class. For example, suppose we chose one class to be 75–80 and the next class to be 80–85. Then, to which class would a data value of 80 belong? The classes should not overlap. Therefore, we define the following classes:

75–79   80–84   85–89   90–94   95–100

Note: In this example, we have data values that are integers. If the data values, instead, had decimal values, then we would choose the class limits accordingly. For example, if the data values ranged from 75 to 100 but were of the form 75.6, we could choose the class limits of the first class to be 75.0–79.9, the second class to be 80.0–84.9, and so on.

Note that the lower class limit of the first class, 75, is slightly below that of the smallest value in the data set, 77. Also note that the class width equals 80 − 75 = 5, as desired.

Using these five classes, we now proceed to construct the frequency and relative frequency distributions for the management aptitude test scores (see Table 2.13). We count the number of data values that fall into each class, and we divide each frequency by the sample size (20) to obtain the relative frequency. We see that the majority of the students (0.40 + 0.30 = 0.70) received scores between 80 and 89 and that only one received a score above 94. ■

Table 2.13 Distributions for the management aptitude test scores

Class	Tally	Frequency	Relative frequency
75–79	|||	3	0.15
80–84	|||| |||	8	0.40
85–89	||||		6	0.30
90–94	||	2	0.10
95–100			1	0.05
Total		20	1.00

2 Histograms and Frequency Polygons

Histograms

There are many different methods of summarizing numeric data graphically. One example of a graphical summary for quantitative data is a **histogram.**

> A **histogram** is constructed using rectangles for each class of data. The heights of the rectangles represent the frequencies or relative frequencies of the class. The widths of the rectangles represent the class widths of the corresponding frequency distribution. The lower class limits are placed on the horizontal axis (along with the upper class limit of the rightmost class), so that the rectangles are touching each other.

Example 2.9   Histogram of management aptitude test scores

Construct a histogram of the frequency of the management aptitude test scores from Example 2.8.

Solution

Step 1   **Find the class limits and draw the horizontal axis.** Note that the class limits for these data were found in Example 2.8 and are given in Table 2.13. The lower class limits are 75, 80, 85, 90, and 95. The upper class limit of the rightmost class is 100.

Draw the horizontal axis, with the numbers 75, 80, 85, 90, 95, and 100 equally spaced along it. The numbers indicate where the rectangles will touch each other.

Step 2   **Determine the frequencies and draw the vertical axis.** Use the frequencies given in Table 2.13. These will indicate the heights of the five rectangles along the vertical axis. Find the largest frequency, which is 8. It is a good idea to provide a little bit of extra vertical space above the tallest rectangle, so make 9 your highest label along the vertical axis. Then provide equally spaced labels along the vertical axis between 0 and 9.

Step 3   **Draw the rectangles.** Draw your first rectangle. Its width is from 75 to 80, and its height is 3, the first frequency. Draw the remaining rectangles similarly.

The resulting frequency histogram is shown in Figure 2.8a. The relative frequency histogram is shown in Figure 2.8b. Note that the two histograms have identical shapes and differ only in the labeling along the vertical axis. ■

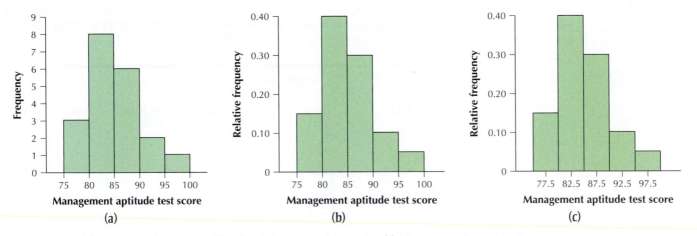

FIGURE 2.8 (a) Frequency histogram; (b) relative frequency histogram; (c) histogram using midpoints.

Note: Histograms are often presented using class midpoints rather than class limits. The class limits can be inferred by splitting the difference between the class midpoints. In Figure 2.8c, the upper class limit for the leftmost class is halfway between 77.5 and 82.5, that is, 80. Otherwise, Figure 2.8c is equivalent to Figure 2.8b.

Note that the histogram, unlike the frequency distribution, provides us with an overall impression of the data distribution. This characteristic will be crucial later on, when we evaluate the fitness of data sets to undergo certain data analysis methods. Also, notice that the rectangles are contiguous (touching), unlike the rectangles of the bar graphs in Section 2.1. Since the data are numeric, the horizontal axis in a histogram should be considered as the number line. A **class midpoint** is the average of two consecutive lower class limits. For example, the class midpoint for the leftmost class in Figure 2.8c is $(75 + 80) / 2 = 77.5$.

The *One Variable Statistics and Graphs* applet can display histograms for a selection of data sets in this textbook, including the management aptitude test scores. The applet allows you to experiment with different class widths.

Frequency Polygons

Frequency polygons provide the same information as histograms, but in a slightly different format.

> A **frequency polygon** is constructed as follows. For each class, plot a point at the class midpoint, at a height equal to the frequency for that class. Then join each consecutive pair of points with a line segment.

Example 2.10   Constructing a frequency polygon

Construct a frequency polygon for the management aptitude test data in Example 2.8.

Solution

The midpoints for the classes were calculated for Figure 2.8c. Plot a point for each frequency above each midpoint, and join consecutive points. The result is the frequency polygon in Figure 2.9. ■

FIGURE 2.9   Frequency polygon.

3 Stem-and-Leaf Displays and Dotplots

Stem-and-Leaf Displays

Stem-and-leaf displays were developed by Professor John Tukey of Princeton University in the late 1960s. This type of display generally contains more information than either a frequency distribution or a histogram. We will demonstrate how to construct a stem-and-leaf display in Example 2.11.

Example 2.11   Stem-and-leaf display of psychology final-exam scores

Construct a stem-and-leaf display for the final-exam scores of 20 psychology students, given below.

75	81	82	70	60	59	94	77	68	98
86	68	85	72	70	91	78	86	51	67

Solution

First, find the leading digits of the numbers. Each number has one of the following as its leading digit: 5, 6, 7, 8, 9. Place these five numbers, called the **stems,** in a column:

5
6
7
8
9

Each number represents the tens place of the test scores. For example, 5 represents 5 tens. Now consider the ones place of each data value. For example, the first score, 75, has 5 in the ones place. Place this number, called the **leaf,** next to its stem:

```
5
6
7 | 5
8
9
```

The second score, 81, has 1 in the ones place, and the third score, 82, has 2 in the ones place. Write the leaves 1 and 2 next to the stem 8:

```
5
6
7 | 5
8 | 12
9
```

Continue this process with the remaining data, placing each ones value next to its stem. Then, for each stem, order the leaves from left to right in increasing order. This produces the stem-and-leaf display:

```
5 | 19
6 | 0788
7 | 002578
8 | 12566
9 | 148
```

Notice that the two 68s refer to two different students who happened to get the same grade on the exam. In general, the leaf units represent the smallest decimal place represented in the data values. Then the stem unit consists of the remainder of the number. For example, suppose we have a data value of 127. Then the 7 is the leaf unit, and the 12 is the stem. Or else, suppose our data value is 0.146. Then our leaf unit is the 6 and the stem is the 14. Note that the stem-and-leaf display contains all the information that a histogram turned on its side does. But it also contains more information than a histogram, because the stem-and-leaf display shows the original values.

Split stems may sometimes be used in a stem-and-leaf display to provide a clearer idea of the data distribution when too many data points fall on just a few stems. When using split stems, each stem appears twice, with the leaves 0 to 4 on the upper stem and the leaves 5 to 9 on the lower stem. The above stem-and-leaf display of psychology scores would appear as follows when using splits stems:

```
5 | 1
5 | 9
6 | 0
6 | 788
7 | 002
7 | 578
8 | 12
8 | 566
9 | 14
9 | 8
```

The *One Variable Statistics and Graphs* applet can display stem-and-leaf displays for a selection of data sets in this textbook, including the psychology final-exam scores. The applet allows you to experiment with split stems if you like. ■

Dotplots

A simple but effective graphical display is a **dotplot.** In a dotplot, each data point is represented by a dot above the number line. When the sample size is large, each dot may represent more than one data point. Figure 2.10 is a dotplot of the 20 management aptitude test scores.

Dotplots are useful for comparing two variables. For example, suppose that an instructor taught two different sections of a management course and gave a simulated management aptitude exam in each section (MAT-1 and MAT-2). The instructor could then compare these two groups of scores directly, using a Minitab comparison dotplot, as in Figure 2.11. Although there is much overlap, Section 1 had the highest score, while Section 2 had the three lowest scores. Therefore, it looks as if Section 1 might have done better.

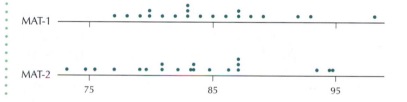

4 Distribution Shape, Symmetry, and Skewness

Frequency distributions are tabular summaries of the set of values that a variable takes. We now generalize the concept of **distribution.**

> The **distribution** of a variable is a table, graph, or formula that identifies the variable values and frequencies for all elements in the data set.

For example, a frequency distribution is a distribution since it is a table that specifies each of the values that a variable can take, along with the frequencies. However, our definition of "distribution" also includes histograms, stem-and-leaf displays, dotplots, and other graphical summaries. (In Chapter 6, we will introduce distributions defined by formulas.) These graphical distributions invite us to consider the shape of a distribution. The *shape* of a distribution is the overall form of a graphical summary, approximated by a smooth curve.

The Bell-Shaped Curve

Figure 2.12 contains the relative frequency histogram of the heights of 1000 college women. Note that there are relatively fewer women in both the left-hand tail (shorter women) and the right-hand tail (taller women). Instead, as height increases from left to right, the relative frequency gradually increases until it reaches a peak near 65 inches tall and then gradually decreases. Thus, the distribution of heights is said to be *bell-shaped*.

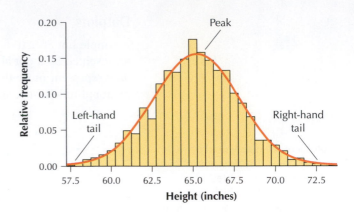

FIGURE 2.12
The bell-shaped curve superimposed on a histogram.

The rectangles represent the actual data. However, the smoothed curve represents an approximation of the overall form of the distribution, and thus the smoothed curve represents the shape of the distribution, which is bell-shaped. The formal name of the bell-shaped distribution is the *normal* distribution. In Chapter 6 we will learn much more about this important distribution. In Chapter 7, we will learn how to assess whether or not a particular distribution is normal (bell-shaped). Starting in Chapter 8, many of the methods for statistical inference we will learn depend on this assessment.

FIGURE 2.13
This butterfly is symmetric.
© *Burke/ Triolo/Jupiterimages*

Analyzing the Shape of a Distribution

We next learn some tools for analyzing the shape of a distribution. An image has *symmetry* (or is **symmetric**) if there is a line (axis of symmetry) that splits the image in half so that one side is the mirror image of the other. For example, the butterfly in Figure 2.13 has symmetry, since a line drawn down the middle of the butterfly would create two mirror images of each other. Symmetry is important in the world of statistics as well, and it is important to develop the talent for recognizing which distribution shapes are symmetric.

For example, is the distribution in Figure 2.12 symmetric? The smoothed curve in Figure 2.12 is in fact perfectly symmetric. However, the histogram rectangles reflecting the actual data are only nearly symmetric, since a vertical line drawn down the middle of the distribution would not result in two perfect mirror images. *Due to random variation, data from the real world rarely exhibit perfect symmetry*. With this in mind, the data analyst is usually content with the approximate symmetry exhibited by the data in Figure 2.12.

However, not all distributions are symmetric, or even nearly symmetric. In Chapter 8 we will discuss a distribution called the chi-square distribution, which is not symmetric but is **skewed.** It often has a longer "tail" on the right than on the left (see Figure 2.14). Since the right-hand tail is longer, we say that this distribution is *right-skewed*. Examples of right-skewed data are usually found whenever one deals with money. For example, if we graph the incomes of the families in your home state, the graph will probably be right-skewed. Most of us will lie somewhere in the middle or left with the bulk of the data, while the incomes of folks like Donald Trump and Bill Gates lie far out on the right of the graph, in the right-hand tail. Figure 2.15 shows a *left-skewed* distribution. Good examples of left-skewed data are retirement ages or death ages. Often, exam grade data can be left-skewed, as several students bump up against the 100% boundary on the right, most students are somewhere in the middle, and a few students stagger in with 40s and 50s in the left-hand tail.

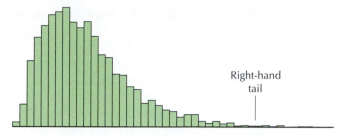

FIGURE 2.14 The chi-square distribution is right-skewed.

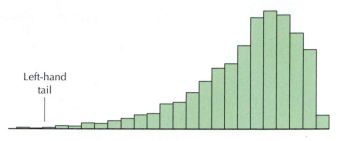

FIGURE 2.15 Some distributions are left-skewed.

Example 2.12 Choosing the appropriate graphical summary

Statistically literate citizens recognize that one may select different graphical summaries, depending on the intention of the presenter. Figures 2.16a, 2.16b, and 2.16c contain a dotplot, a histogram, and a stem-and-leaf display of the average size of households in the 50 states and the District of Columbia. Which graphical summary—the dotplot, the histogram, or the stem-and-leaf display—is most useful if our primary objective is

a. to assess symmetry and skewness?

b. to be able to construct it quickly using paper and pencil?

c. to retain complete knowledge of the original data set?

d. to give a presentation to people who have never had a stats course before?

e. to be given maximum freedom in choosing how the data are to be interpreted?

Average size

(a)

(b)

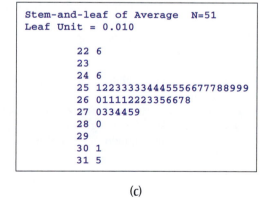

(c)

FIGURE 2.16 (a) Dotplot; (b) histogram; (c) stem-and-leaf display.

Solution

a. All three graphics are good at assessing symmetry and skewness.

b. The dotplot's great asset is its simplicity. It can be quickly drawn, with minimal preparation, in contrast to the other two summaries, which require some organization or calculation.

    c. The stem-and-leaf display was invented in order to retain complete knowledge of the data set. Histograms are the least effective in this regard.

    d. The histogram is widely used in the real world and is probably the best choice for a presentation in front of those who have not had a stats course before.

    e. We will learn in Section 2.4, "Graphical Misrepresentations of Data," that two histograms representing the same data set can look completely different, depending on the number of classes and class widths. Histograms therefore provide maximum freedom for the presenter to choose how the data are to be presented. Statistically literate citizens should be ready to examine the data more closely, in order to debunk possible misrepresentation. ■

STEP-BY-STEP TECHNOLOGY GUIDE: Quantitative Data

Suppose we would like to produce a histogram of the management aptitude test scores from Example 2.8 (page 46).

TI-83/84

Entering a Data Set
Step 1 Press **STAT**, then press **ENTER**. Highlight the **L1** list.
Step 2 Clear out any old data in **L1**. Press the **up arrow** key, then **CLEAR**, then **ENTER**.
Step 3 Enter the first data value **77** and press **ENTER**.
Step 4 Continue entering data until the entire data set is in **L1** (Figure 2.17).

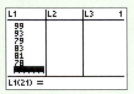

Figure 2.17 All data entered.

Constructing a Histogram
Step 1 Press **2nd**, then **Y=**. In the STAT PLOTS menu, select **1**, and press **ENTER**.
Step 2 Select **ON**, and press **ENTER**. Select the histogram icon (Figure 2.18), and press **ENTER**.
Step 3 Press **WINDOW**. Set the following values:
a. Xmin: lower class limit of leftmost class (75)
b. Xmax: upper class limit of rightmost class (100)
c. Xscl: class width (5)
d. Ymin: 0
e. Ymax: Some value slightly larger than the highest expected frequency (10).
Step 4 Press **GRAPH** (shows histogram without any details).
Step 5 Press **TRACE**. Selecting each class in turn provides class limits and class frequency. The histogram is given in Figure 2.19.

Figure 2.18 Selecting the histogram icon.

Figure 2.19 Histogram with leftmost class selected.

EXCEL

Constructing a Histogram
Make sure the Data Analysis package has been installed on your version of Excel.
Step 1 Click **Data > Data Analysis**.

Step 2 Select **Histogram** and click **OK**.
Step 3 For the *input range,* select the cells in which the data set resides. Then click **OK**.

(Continued)

MINITAB

Constructing a Histogram

Step 1 Enter the management aptitude test scores into column **C1**.

Step 2 Click **Graph > Histogram**.

Step 3 In the **Graph Variables** section, choose **Simple** and click **OK**. Select **C1** *Scores*, and click **Select**. Then click **OK**.

Step 4 The histogram is shown in Figure 2.20. Note that by default Minitab uses midpoints rather than class limits to define the classes. Double-clicking anywhere on the midpoint values (78, 81, . . .) brings up a dialog box providing a wide range of options for changing the number of classes, class limits, etc.

Constructing a Stem-and-Leaf Display

Step 1 Enter the management aptitude test scores into column **C1**.

Step 2 Click **Graph > Stem-and-Leaf**.

Step 3 Click inside the space indicated **Variables**, select **C2** *Scores*, and click **Select**. Then click **OK**.

Step 4 The output shown in Figure 2.21 tells us that the leaf unit is defined to be ones (1.0). Therefore, the stem unit is tens. (Ignore the leftmost column, which simply provides a cumulative count of the data points from the minimum and maximum.) The first row shows 7 7, indicating a single data point, 77. The second row shows 7 89, indicating two data points, 78 and 79.

Management aptitude test score

Figure 2.20

```
Stem-and-leaf of MAT
Leaf Unit = 1.0
  1    7   7
  3    7   89
  6    8   001
 10    8   2333
 10    8   45
  8    8   677
  5    8   89
  3    9
  3    9   23
  1    9
  1    9
  1    9   9
```

Figure 2.21

Section 2.2 SUMMARY

In this section, we learned about using graphs and tables for summarizing quantitative (numerical) data.

1 Quantitative variables can be summarized using frequency and relative frequency distributions.

2 Histograms are a graphical display of a frequency or a relative frequency distribution with class intervals on the horizontal axis and the frequencies or relative frequencies on the vertical axis. A frequency polygon is constructed as follows: for each class, plot a point at the class midpoint, at a height equal to the frequency for that class; then join each consecutive pair of points with a line segment.

3 Stem-and-leaf displays contain more information than either a frequency distribution or a histogram, since they retain the original data values in the display. In a dotplot, each data point is represented by a dot above the number line.

4 An image has symmetry (or is symmetric) if there is a line (axis of symmetry) that splits the image in half so that one side is the mirror image of the other. Nonsymmetric distributions with a long right-hand tail are called right-skewed, while those with a long left-hand tail are called left-skewed.

Section 2.2 EXERCISES

CLARIFYING THE CONCEPTS

1. Which of the methods for displaying data introduced in this section (frequency and relative frequency distributions, histograms, frequency polygons, stem-and-leaf displays, and dotplots) can be used with both quantitative and qualitative data? Which can be used for quantitative data only?

2. Describe at least one potential benefit of combining classes when constructing a frequency distribution. Describe at least one potential benefit from retaining a larger number of classes.

3. In general, how many classes should be used when constructing a frequency distribution?

4. Describe at least one drawback of choosing class limits that overlap.

5. Describe at least one way that a dotplot may be useful.

6. In your own words, describe what is meant by "symmetry." Provide an example of a shape that is symmetric and an example of a shape that is not symmetric.

7. What are some examples of data sets that are often right-skewed? Left-skewed?

8. For a bar graph (not specifically a Pareto chart), does it matter which order the bars are in? What does this mean for the relevance of symmetry and skewness for summaries of categorical data (such as we studied in Section 2.1)?

PRACTICING THE TECHNIQUES

9. A fair die was thrown 100 times, and the values were recorded. The accompanying histogram shows the results.

    a. Which value occurred most frequently?
    b. Which values occurred least frequently?
    c. How often was a 3 observed?
    d. What percentage of times was a 3 observed?

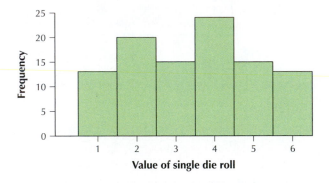

10. A random sample of 1000 police officers was taken, and the number of motor vehicle citations each handed out in a particular week was recorded. The results are shown in the accompanying histogram.

    a. What was the most number of citations issued?
    b. What was the fewest number of citations issued?

    c. What is the most frequent number of citations issued? About how many police officers issued this many citations?
    d. Describe the shape of the distribution.

11. A campus-wide statistics midterm worth 50 points resulted in the scores provided in the histogram below.

    a. Which score occurred with the greatest frequency?
    b. Which score occurred with the lowest frequency?
    c. What is the highest score? Lowest score?
    d. Describe the shape of the distribution.

12. A portfolio contains stocks of 19 technology firms. The stock prices are shown in the accompanying histogram.

    a. How many classes are there?
    b. What is the class width? Is it the same for each class?
    c. Is this a frequency histogram or a relative frequency histogram?

13. Refer to the histogram of stock prices.

    a. How could we turn this into a relative frequency histogram? Would the classes or the rectangles be affected?
    b. Suppose we were given a relative frequency histogram instead. How could we turn it into a frequency histogram?
    c. What is the sample size?

14. Refer to the histogram of stock prices.
   a. How many stocks were priced above $27.50?
   b. What is the relative frequency of stocks priced above $27.50?
   c. How many stocks had a price below $15?
   d. What is the relative frequency of stocks with a price below $15?

15. Refer to the histogram of stock prices.
   a. How many stocks are priced between $17.50 and $20?
   b. What is the relative frequency of stocks priced below $5?
   c. Which class has the largest relative frequency? Calculate this relative frequency.
   d. What is the frequency of stocks priced between $10 and $15?
   e. How many stocks had a price of $40?

16. Would you characterize the shape of the stock prices distribution as (a) tending to be symmetric, (b) tending to be right-skewed, (c) tending to be left-skewed, or (d) too close to call?

17. Refer to the accompanying stem-and-leaf display. Reconstruct the data set.

```
Stem-and-leaf of Data   N  = 20
Leaf Unit = 1.0

   2   3
   2   45
   2   67
   2   889
   3   011
   3   2223
   3   5
   3   67
   3   9
   4   0
```

18. Refer to the stem-and-leaf display. Construct a relative frequency distribution, using appropriate values for the class width and the lower class limit of the leftmost class.

19. Refer to the stem-and-leaf display. Construct a frequency histogram.

20. Refer to the stem-and-leaf display. Construct a dotplot.

21. The frequency polygon below represents the quiz scores for a course in introductory statistics.

a. What is the class width?
b. What is the lower class limit of the class that has 45 as its midpoint?
c. What is the upper class limit of the class that has 45 as its midpoint?
d. Which class has the highest frequency?
e. Which class has the lowest frequency?

22. Refer to the frequency polygon of quiz scores.
   a. About how many students scored higher than 82.5?
   b. About how many students scored lower than 52.5?
   c. Can we say how many students scored in the 90s? Why or why not?

APPLYING THE CONCEPTS

23. Small Businesses. The U.S. Census Bureau tracks the number of small businesses per city. The accompanying frequency polygon represents the numbers of small businesses per city (in thousands) for 266 cities nationwide.
   a. What is the class width?
   b. What is the lower class limit of the leftmost class? (*Hint:* Don't forget about the units.)
   c. Which class has the highest frequency?
   d. Which class has the lowest frequency?

24. Refer to the frequency polygon of small businesses per city.
   a. About how many cities have between 1000 and 3000 small businesses?
   b. About how many cities have more than 19,000 small businesses?
   c. About how many cities have between 9000 and 11,000 small businesses?

25. Countries and Continents. Suppose we are interested in analyzing the variable *continent* for the ten countries in Table 2.14. Construct each of the following tabular or graphical summaries. If not appropriate, explain clearly why we can't use that method.
   a. Frequency distribution
   b. Relative frequency distribution
   c. Frequency histogram
   d. Dotplot
   e. Stem-and-leaf display

TABLE 2.14 Countries and continents

Country	Continent
Iraq	Asia
United States	North America
Pakistan	Asia
Canada	North America
Madagascar	Africa
North Korea	Asia
Chile	South America
Bulgaria	Europe
Afghanistan	Asia
Iran	Asia

Santa Monica Employers. Answer Exercises 26–29 using the information in Table 2.15, which lists the number of employees for the ten largest employers in Santa Monica, California.

TABLE 2.15 Santa Monica employers

Company	No. of employees	Company	No. of employees
City of Santa Monica	1892	Santa Monica Hospital	1165
St. John's Health Center	1755	Santa Monica College	1050
Macerich Co.	1605	Metro-Goldwyn Mayer	1050
Fremont General Corp.	1600	Rand Corp.	1038
Entravision Communications Corp.	1206	Santa Monica School District	1008

26. Construct a relative frequency distribution of the number of employees. Use class width of 200 employees, with the lower class limit of the leftmost class equal to 900.
27. Construct a frequency histogram, using the same classes as in the previous exercise.
28. Construct a relative frequency histogram. What is the only difference between the frequency and relative frequency histograms?
29. Construct a dotplot.

Miami Arrests. Answer Exercises 30–32 using the information in the following table. The table gives the monthly number of arrests made for the year 2005 by the Miami-Dade Police Department.

Jan.	751	May	919	Sept.	802
Feb.	650	June	800	Oct.	636
Mar.	909	July	834	Nov.	579
Apr.	881	Aug.	789	Dec.	777

30. Construct a relative frequency distribution of the monthly number of arrests. Use class width of 50 arrests, with the lower class limit of the leftmost class equal to 550.
31. Construct a frequency histogram and relative frequency histogram, using the same classes as in the previous exercise. Which class or classes have the highest frequency? Lowest?

32. Construct a dotplot.

New York Townspeople. Use the following information for Exercises 33–35. For towns in New York State, the following histogram provides information on the percentage of the townspeople who are between 18 and 65 years old (data set **New York**).

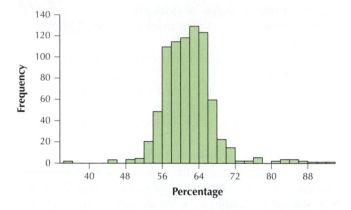

33. Would you characterize the distribution as left-skewed, right-skewed, or fairly symmetrical?
34. Provide an estimate of the "typical" percentage of townspeople who are between 18 and 65 years old. Is this typical value near the middle or near one of the "tails" of the distribution?
35. Provide a rough estimate of the sample size.

Psychology Exam Data. Use the following psychology exam data set from Example 2.11 for Exercises 36–39.

75	81	82	70	60	59	94	77	68	98
86	68	85	72	70	91	78	86	51	67

36. Without using the computer, construct the following:
   a. A frequency distribution
   b. A relative frequency distribution
   c. A relative frequency histogram
37. Without using a computer, construct a dotplot.
38. Compare and contrast the relative usefulness of each of three graphical presentation methods—dotplot, histogram, and stem-and-leaf display—if our primary objective is
   a. to assess symmetry and skewness.
   b. to be able to construct it quickly using paper and pencil.
   c. to give a presentation to people who have never had a stats course before.
   d. to be given maximum leeway in choosing how the data are to be presented.
39. *What if* we subtract the same amount (say, 10) from each psychology exam score. Explain how this would affect the following. What would change? What would stay the same?
   a. Relative frequency histogram
   b. Dotplot
   c. Stem-and-leaf display

Fats and Cholesterol. For Exercises 40–44, use your knowledge of Excel or Minitab. Open the **Nutrition** data set.

40. How many observations are there in the data set? How many variables?

41. The variable *FAT_GRAM* contains the fat content in grams for each food. Construct a histogram of *FAT_GRAM*. Comment on the symmetry or the skewness of the histogram.

42. Is there a particular type of food whose fat content is particularly large? Which type of food item is this (actually, a set of similar food items)?

43. The variable *CHOLEST* contains the cholesterol content in milligrams for each food. Construct a histogram of *CHOLEST*. Comment on the symmetry or the skewness of the histogram.

44. Which food item is highest in cholesterol?

Use the *One Variable Statistics and Graphs* applet for Exercises 45–50. Work with the **Earthquakes** data set, which shows the magnitude on the Richter scale of 57 earthquakes that occurred during the week of October 15–22, 2007.

45. Click on the **Histogram** tab.
  a. How many classes are there in the histogram?
  b. What is the class width?

46. Click on the leftmost rectangle in the histogram.
  a. What is the frequency for this class?
  b. What are the lower and upper class limits?

47. Click on the number line and drag slowly all the way to the left.
  a. What happens to the number of classes as you drag to the left?

b. What happens to the class widths as you drag to the left?

48. Click on the number line and drag slowly all the way to the right.
  a. What happens to the number of classes as you drag to the right?
  b. What happens to the class widths as you drag to the right?

49. Click on the **Stem-and-Leaf** tab.
  a. How many stems are there?
  b. Without counting, state how many leaves there are. How do we know this?

50. Select **Split Stems**.
  a. Now how many stems are there?
  b. How many leaves are there?
  c. Which stem-and-leaf display is preferable for the **Earthquakes** data, regular or split stems?

CONSTRUCT YOUR OWN DATA SETS

51. Construct your own right-skewed data set of about 20 values. Just make up the data points, but be sure you know what the data represent (income, housing costs, etc.).
  a. Construct a stem-and-leaf display of your data set.
  b. Construct a dotplot of your data set.

52. Construct your own symmetric data set of about 20 values. Just make up the data points, but be sure you know what the data represent (for example, runs in a baseball game, number of right answers on a quiz).
  a. Construct a stem-and-leaf display of your data set.
  b. Construct a dotplot of your data set.

2.3 Further Graphs and Tables for Quantitative Data

Objectives: By the end of this section, I will be able to…

1 Build cumulative frequency distributions and cumulative relative frequency distributions.

2 Create frequency ogives and relative frequency ogives.

3 Construct and interpret time series graphs.

1 Cumulative Frequency Distributions and Cumulative Relative Frequency Distributions

Since quantitative data can be put in ascending order, we can keep track of the accumulated counts at or below a certain value using a **cumulative frequency distribution** or **cumulative relative frequency distribution**. For example, if we list the prices of homes for sale in a neighborhood, a cumulative frequency distribution tells us how many homes are priced at $300,000 or less.

For a discrete variable, a **cumulative frequency distribution** shows the total number of observations *less than or equal to* the category value. For a continuous variable, a **cumulative frequency distribution** shows the total number of observations *less than or equal to* the upper class limit.

A **cumulative relative frequency distribution** shows the proportion of observations less than or equal to the category value (for a discrete variable) or the proportion of observations less than or equal to the upper class limit (for a continuous variable).

Example 2.13 Constructing cumulative frequency and relative frequency distributions

The first three columns in Table 2.16 contain the frequency distribution and relative frequency distribution for the total 2007 attendance for 25 Major League Baseball teams. Construct a cumulative frequency distribution and a cumulative relative frequency distribution for the attendance figures.

Solution
To find the cumulative frequency for a class, add the frequencies of the classes equal to or below the upper class limit of that class. For example, the cumulative frequency for the class 2.70–3.09 is the sum of the frequency for this class and the frequencies for the classes 1.90–2.29 and 2.30–2.69. The procedure for the cumulative relative frequencies is similar. The results are shown in the last two columns of Table 2.16, where we can see that more than two-thirds (0.68) of these teams had attendance of 3.09 million or less. ■

Table 2.16 Cumulative frequency distribution and cumulative relative frequency distribution

Attendance (millions)	Frequency	Relative frequency	Cumulative frequency	Cumulative relative frequency
1.90–2.29	5	0.20	5	0.20
2.30–2.69	6	0.24	5 + 6 = 11	0.20 + 0.24 = 0.44
2.70–3.09	6	0.24	5 + 6 + 6 = 17	0.44 + 0.24 = 0.68
3.10–3.49	4	0.16	5 + 6 + 6 + 4 = 21	0.68 + 0.16 = 0.84
3.50–3.89	3	0.12	5 + 6 + 6 + 4 + 3 = 24	0.84 + 0.12 = 0.96
3.90–4.29	1	0.04	5 + 6 + 6 + 4 + 3 + 1 = 25	0.96 + 0.04 = 1.00
Total	25	1.00		

Ogives

Just as histograms and frequency polygons are the graphical equivalent of frequency distributions, we have the following graphical equivalent of a cumulative frequency distribution.

An **ogive** (pronounced "oh jive") is the graphical equivalent of a cumulative frequency distribution or a cumulative relative frequency distribution. Like a frequency polygon, an ogive consists of a set of plotted points connected by line segments. The x coordinates of these points are the upper class limits; the y coordinates are the cumulative frequencies or cumulative relative frequencies.

Example 2.14 Constructing an ogive

Construct a relative frequency **ogive** for the attendance data in Table 2.16.

Solution

For the *x* coordinates, we use the upper class limits for attendance, and for the *y* coordinates we use the cumulative relative frequencies. The result is shown in Figure 2.22. ■

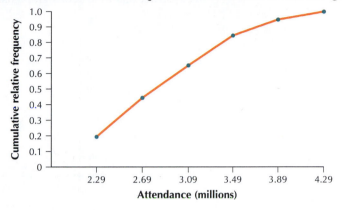

FIGURE 2.22 Ogive for baseball attendance.

③ Time Series Graphs

Data analysts are often interested in how the value of a variable changes over time. Data that are analyzed with respect to time are called *time series data*.

> A graph of time series data is called a **time series plot.** The horizontal axis of a time series plot represents time (for example, hours, days, months, years). The values of the time series data are plotted on the vertical axis, and line segments are drawn to connect the points.

Example 2.15 Constructing a time series plot

Table 2.17 contains the amount of carbon dioxide in parts per million (ppm) found in the atmosphere above Mauna Loa in Hawaii, measured monthly from October 2006 to September 2007. Construct a **time series plot** of these data.

Table 2.17 Atmospheric carbon dioxide at Mauna Loa, October 2006 to September 2007

Month	Carbon dioxide (ppm)	Month	Carbon dioxide (ppm)
Oct.	379.03	Apr.	386.37
Nov.	380.17	May	386.54
Dec.	381.85	June	385.98
Jan.	382.94	July	384.35
Feb.	383.86	Aug.	381.85
Mar.	384.49	Sept.	380.58

Source: Dr. Pieter Tans, Earth System Research Laboratory, National Oceanic and Atmospheric Administration, **www.esrl.noaa.gov/gmd/ccgg/trends**.

Solution

We indicate the twelve months October through September on the horizontal axis of the time series plot (Figure 2.23). Then, for each month, we plot the amount of carbon dioxide. Finally, we join the points using line segments. Note that the carbon dioxide level increases from the fall through the winter and peaks in the spring. It then decreases through the summer. In the Step-by-Step Technology Guide, we illustrate how to construct this time series graph using technology. ■

FIGURE 2.23 Time series plot. Carbon dioxide levels at Mauna Loa, Hawaii.

Example 2.16 Constructing a time series plot using technology

The data set **Mauna Loa** contains the carbon dioxide levels at Mauna Loa from September 1999 to September 2007. Use technology to construct a time series plot of the data.

Solution

We use the instructions provided in the Step-by-Step Technology Guide at the end of this section. The resulting time series plot is shown in Figure 2.24. (The year on the horizontal axis indicates September of each year. For example "1999" refers to September 1999.) We observe both a seasonal pattern and a long-term trend. Every autumn and winter, the carbon dioxide level increases, and every summer it decreases. In autumn and winter, leaves and other deciduous vegetation decays, releasing its store of carbon back into the atmosphere. In the spring and summer, the new year's leaves require carbon to grow and extract it from the atmosphere, thereby reducing the atmosphere's carbon dioxide level. Thus, the Earth "inhales" carbon each summer and "exhales" it each winter. However, the carbon dioxide level of each successive September does not quite reach the low level of the previous September. This leads to an overall increasing trend in the amount of carbon dioxide in the atmosphere as we move from 1999 to 2007. ■

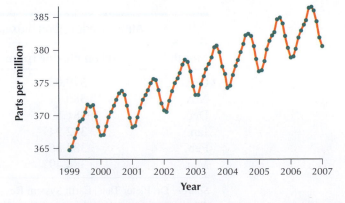

FIGURE 2.24 Watching the Earth breathe. Carbon dioxide levels at Mauna Loa, Hawaii.

STEP-BY-STEP TECHNOLOGY GUIDE: Time Series Plots

We illustrate how to construct a time series plot using Example 2.15 (page 61).

TI-83/84

Step 1 Enter your time index (integers 1, 2, . . .) into list **L1**.
Step 2 Enter the values of your time series variable into list **L2**.
Step 3 Press **2nd**, then **Y=**. In the STAT PLOTS menu, select **1**, and press **ENTER**.
Step 4 Select **ON**, and press **ENTER**. Select the time series icon (Figure 2.25), and press **ENTER**.
Step 5 Press **WINDOW**. Set the following values:
   a. **Xmin: 0**
   b. **Xmax: 13**
   c. **Xscl: 1**
   d. **Ymin: 375**
   e. **Ymax: 390**

Step 6 Press **ZOOM > 9:ZOOMSTAT** and press **ENTER**. The time series plot is shown in Figure 2.26.

Figure 2.25 Selecting the time series icon.

Figure 2.26 TI-83/84 time series plot.

EXCEL

Step 1 Enter the month data into column **A** (see Figure 2.27).
Step 2 Enter the values of your time series variable into column **B** (see Figure 2.27).
Step 3 Select cells A1–B12 and click **Insert > Line** (in the **Chart** section).
Step 4 Choose the type labeled "Line with markers."

	A	B
1	Oct	379.03
2	Nov	380.17
3	Dec	381.85
4	Jan	382.94
5	Feb	383.86
6	Mar	384.49
7	Apr	386.37
8	May	386.54
9	Jun	385.98
10	Jul	384.35
11	Aug	381.85
12	Sep	380.58

Figure 2.27

MINITAB

Step 1 Enter the values of your time series variable into column **C1**.
Step 2 Click **Graph > Time Series Plot . . .**
Step 3 Select **Simple** and click **OK**.
Step 4 For **Series**, double-click on **C1**.

Step 5 Click **Time/Scale**. Select **Calendar > Month**.
Step 6 For **Start value**, enter 10 (for October). For **Increment**, enter **1**.
Step 7 Click **OK** and **OK**.

Section 2.3 SUMMARY

1 A cumulative frequency distribution shows the total number of observations less than or equal to the category value (for a discrete variable) or the upper class limit (for a continuous variable). A cumulative relative frequency distribution shows the proportion of observations less than or equal to the category value (for a discrete variable) or the upper class limit (for a continuous variable).

2 An ogive is the graphical equivalent of a cumulative frequency distribution or a cumulative relative frequency distribution. The *x* coordinates of the points are the upper class limits; the *y* coordinates are the cumulative frequencies or cumulative relative frequencies.

3 Data that are analyzed with respect to time are called time series data. A graph of time series data is called a time series plot. The horizontal axis of a time series plot represents time (for example, hours, days, months, years). The values of the time series data are plotted on the vertical axis, and line segments are drawn to connect the points.

Section 2.3 EXERCISES

CLARIFYING THE CONCEPTS

1. Explain the difference between a frequency distribution and a cumulative frequency distribution.

2. Explain the difference between a cumulative frequency distribution and a cumulative relative frequency distribution.

3. What is the graphical equivalent of a cumulative frequency distribution?

4. Explain how to construct an ogive.

5. What do we call data that are analyzed with respect to time?

6. Explain how to construct a time series plot.

PRACTICING THE TECHNIQUES

For Exercises 7 and 8, use the histograms for the indicated exercises from Section 2.2 to

   a. construct a frequency distribution.
   b. construct a cumulative frequency distribution.
   c. construct a cumulative relative frequency distribution.

7. Exercise 9.

8. Exercise 12.

For Exercises 9 and 10, use the frequency distributions for the indicated exercises from Section 2.2 to

   a. construct a frequency ogive.
   b. construct a relative frequency ogive.

9. Exercise 25.

10. Exercise 26.

11. The frequency ogive below represents the unemployment rate (in percentages) for 367 cities nationwide.[3]

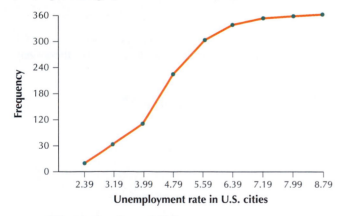

   a. What is the class width?
   b. What is the upper class limit of the leftmost class?
   c. What is the class midpoint of the leftmost class?

12. Refer to the frequency ogive of unemployment rates.
   a. About how many cities have unemployment rates 3.99 and below?
   b. About how many cities have unemployment rates 5.59 and below?

   c. About how many cities have unemployment rates 5.6 and above?

13. The following graph is a time series plot of the number of HIV/AIDS cases in children under 13 in the United States from 1992 to 2004.[4]

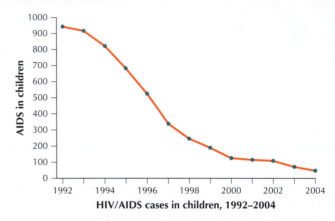

HIV/AIDS cases in children, 1992–2004

   a. About how many children had HIV/AIDS in 1992?
   b. About how many children had HIV/AIDS in 2004?
   c. Between which two years do we see the steepest decline in the number of cases?
   d. Overall, does this graph represent encouraging or discouraging news?

APPLYING THE CONCEPTS

Agricultural Exports. For Exercises 14–16, refer to Table 2.18. The table gives the value of agricultural exports (in billions of dollars) from the top 20 U.S. states in 2006.

TABLE 2.18   Agricultural exports (in billions of dollars)

State	Exports	State	Exports
California	10.5	Arkansas	1.9
Iowa	4.2	North Dakota	1.9
Texas	3.8	Ohio	1.7
Illinois	3.8	Florida	1.7
Nebraska	3.3	Wisconsin	1.5
Kansas	3.2	Missouri	1.5
Minnesota	3.0	Georgia	1.4
Washington	2.2	Pennsylvania	1.4
North Carolina	2.1	Michigan	1.2
Indiana	2.0	South Dakota	1.2

Source: U.S. Department of Agriculture.

14. Construct a cumulative frequency distribution of agricultural exports. Start at $0 and use class widths of $1.5 billion.
   a. Which class has the highest frequency? How many states belong to this class?

b. Of the classes that have nonzero frequency, which class has the lowest frequency? Which state does this represent?

c. List the states that belong to the leftmost class.

15. Construct a cumulative relative frequency distribution of agricultural exports. Start at $0 and use class widths of $1.5 billion.

a. What proportion of states belong to the class with the highest frequency?

b. What proportion of states have agricultural exports between $4.5 billion and $8.9 billion?

16. Use your cumulative relative frequency distribution to construct a relative frequency ogive of agricultural exports.

17. Interest Rates. The following data represent the prime lending rate of interest, as reported by the Federal Reserve, every six months from January 2003 to July 2007.

Jan. 2003	4.25
July 2003	4.00
Jan. 2004	4.00
July 2004	4.25
Jan. 2005	5.25
July 2005	6.25
Jan. 2006	7.25
July 2006	8.25
Jan. 2007	8.25
July 2007	8.25

a. Construct a time series plot of the prime lending rate of interest.

b. What trend do you see?

18. Rainfall in Fort Lauderdale. The following data represent the total monthly rainfall (in inches) in 2005 in Fort Lauderdale, Florida, as reported by the U.S. Historical Climatology Network.

Jan.	1.65
Feb.	0.52
Mar.	4.84
Apr.	1.64
May	3.94
June	15.64
July	7.13
Aug.	8.67
Sept.	10.40
Oct.	10.36
Nov.	5.20
Dec.	2.72

a. Construct a time series plot of the data.

b. Is it wetter in summer or winter in Fort Lauderdale?

19. In Exercise 18, *what if* we add 3 inches to each month's rainfall amount. Describe how this would affect the time series plot. What would change? What would stay the same?

20. Cigarette Use Among 12th-Graders. Table 2.19 presents the percentages of 12th-graders who smoke cigarettes, for the years 1975–2004.

a. Construct a time series plot of the data.

b. Describe any trends that you see.

21. Trade Deficit. Table 2.20 presents the annual trade deficits (imports minus exports) for the United States from 1991 to 2005, in billions of dollars.

a. Construct a time series plot of the data.

b. Describe any trends that you see.

TABLE 2.19   12th-graders who smoke

Year	Percent	Year	Percent
1975	36.7	1990	29.4
1976	38.8	1991	28.3
1977	38.4	1992	27.8
1978	36.7	1993	29.9
1979	34.4	1994	31.2
1980	30.5	1995	33.5
1981	29.4	1996	34.0
1982	30.0	1997	36.5
1983	30.3	1998	35.1
1984	29.3	1999	34.6
1985	30.1	2000	31.4
1986	29.6	2001	29.5
1987	29.4	2002	26.7
1988	28.7	2003	24.4
1989	28.6	2004	25.0

Source: Monitoring the Future Study, University of Michigan.

TABLE 2.20   U.S. trade deficits

Year	Trade deficit in $ billions
1991	7,767
1992	9,796
1993	17,641
1994	24,460
1995	23,355
1996	25,614
1997	27,380
1998	43,387
1999	69,250
2000	96,605
2001	93,988
2002	109,975
2003	125,734
2004	150,519
2005	168,650

Source: U.S. Census Bureau.

2.4 Graphical Misrepresentations of Data

Objective: **By the end of this section, I will be able to...**

1 Understand what can make a graph misleading, confusing, or deceptive.

In the Information Age, when our world is awash in data, it is important for citizens to understand how graphics may be made misleading, confusing, or deceptive. Such an understanding enhances our statistical literacy and makes us less prone to being deceived by misleading graphics.

Eight Common Methods for Making a Graph Misleading

1. Graphing/selecting an inappropriate statistic.
2. Omitting the zero on the relevant scale.
3. Manipulating the scale.
4. Using two dimensions (area) to emphasize a one-dimensional difference.
5. Careless combination of categories in a bar graph.
6. Inaccuracy in relative lengths of bars in a bar graph.
7. Biased distortion or embellishment.
8. Unclear labeling.

Example 2.17   Inappropriate choice of statistic

The United Nations Office on Drugs and Crime reports the statistics, given in Table 2.21, on the top 5 nations in the world ranked by numbers of cars stolen in 2000. The car thieves seem to be preying on cars in the United States, which has endured nearly as many cars stolen as the next four highest countries put together. (See also the bar graph in Figure 2.28.) However, the United States has a much greater population than these other countries. Is it possible that, *per capita* (per person), the car theft rate in the United States is not so bad?

Table 2.21 Top five nations for total number of cars stolen in 2000

Country	Cars stolen
1. United States	1,147,300
2. United Kingdom	338,796
3. Japan	309,638
4. France	301,539
5. Italy	243,890

FIGURE 2.28 Bar graph of the top five nations for number of cars stolen in 2000.

Solution

In this case, the total number of cars stolen is an *inappropriate statistic* since the population of the United States is greater than the populations of the other countries. To find the per capita car theft rate, divide the number of cars stolen in a country by that country's population. The resulting list in Table 2.22 of the top 5 countries for per capita car theft contains a few surprises. Note that the United States has disappeared from the list. It is found in ninth place, with 0.00409 car thefts per capita. Also, let's face it, Denmark and Norway are not exactly hotbeds of crime, are they? Do Danes and Norwegians individually have a higher risk of having their cars stolen than Americans? Most likely not. Another variable not factored in here is the number of cars in the country, which we would need for such an assessment of risk. ■

Table 2.22 Top five nations for total number of cars stolen per capita in 2000

Country	Cars stolen per capita
1. Australia	0.00712
2. Denmark	0.00600
3. United Kingdom	0.00567
4. New Zealand	0.00563
5. Norway	0.00516

Developing Your Statistical Sense

Choose the Appropriate Statistic

The bottom line is that *we need to be careful how we use statistics.* Put in an extreme form, "Figures don't lie, but liars figure." One table of statistics tells us the car theft epidemic is striking the United States with special vehemence. The other table asserts the contrary. An American insurance company looking to increase car insurance rates could point to the first table to support its rate request. A citizens group opposing the request could cite the second table. Which table of statistics is true? They both are! We need to be careful how we phrase our research questions and how we choose the type of statistical evidence to investigate the research question. ■

Example 2.18 Omitting the zero

MediaMatters.com reported that **CNN.com** used a misleading graph, reproduced here as Figure 2.29, to exaggerate the difference between the percentages of Democrats and Republicans who agreed with the Florida court's decision to remove the feeding tube from Terri Schiavo in 2005. Explain how Figure 2.29 is misleading.

Solution

Figure 2.29 is misleading because the vertical scale does not begin at zero. **MediaMatters.com** published an amended graphic, reproduced here as Figure 2.30, which includes the zero on the vertical axis and much reduces the apparent difference among the political parties. ■

FIGURE 2.29

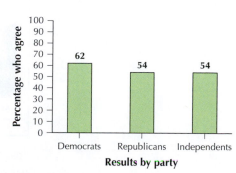

FIGURE 2.30

Example 2.19   Manipulating the scale

Figure 2.31 shows a Minitab relative frequency bar graph of the majors chosen by 25 business school students. Explain how we could manipulate the scale to de-emphasize the differences.

Solution

If we wanted to de-emphasize the differences, we could extend the vertical scale up to its maximum, 1.0 = 100%, to produce the graph in Figure 2.32. ■

FIGURE 2.31

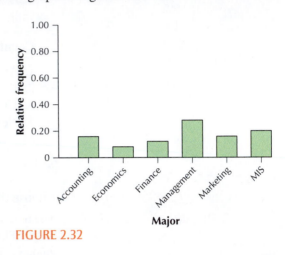

FIGURE 2.32

Example 2.20   Using two dimensions for a one-dimensional difference and unclear labeling

Figure 2.33 compares the leaders in career points scored in the NBA All-Star Game among players active in 2007. Explain how this graphic may be misleading.

Solution

The height of the players is supposed to represent the total points, but this is not clearly labeled. Points should be indicated using a vertical axis, but there is no vertical axis at all. Further, note that Shaquille O'Neal dominates the graphic, because his body *area* is larger than the body areas of the other players. This is misleading. All four players should have the same body width, just as all bars in a bar graph have the same width. This graph uses two dimensions (height and width) to emphasize a one-dimensional (points) difference. ■

FIGURE 2.33   *(O'Neal: AP Photo/Alan Diaz. Bryant: AP Photo/Mark J. Terrill. Garnett: AP Photo/David Zalubowski. Iverson: AP Photo/ David Zalubowski.)*

Example 2.21 Careless combination of categories in a bar graph and biased embellishment

Figure 2.34 shows a graphic of how often people have observed drivers running red lights. Explain how this graphic may be considered both confusing and biased.

Solution

One problem with this graphic is that the categories of *seldom* and *never* have been combined, which may not be appropriate. Also, as we learned in Chapter 1, what is "seldom" to one person may not be "seldom" to someone else. A third problem is that the red light of the "Seldom/never" category is lit up, which may be evidence of bias on the part of the designer of the graphic. Finally, note the absence of a vertical axis. ■

FIGURE 2.34

Example 2.22 Inaccuracy in relative lengths of bars in a bar graph and unclear labeling

Figure 2.35 is a horizontal bar graph of the three teams with the most World Series victories in baseball history. Explain what is unclear or misleading about this graph.

Solution

Note that 127 is more than twice as many as 52, and so the Yankees' bar should be more than twice as long as the Cardinals' bar, which it is not. ■

FIGURE 2.35

When constructing a histogram, changing the number of classes or the width of the interval can sometimes lead to a completely different-looking distribution. Thus, we need to exercise care when someone shows us a histogram, since it presents, not the data themselves, but one of many ways of classifying the data.

Example 2.23   Presenting the same data set as both symmetric and left-skewed

The National Center for Education Statistics sponsors the Trends in International Mathematics and Science Study (TIMSS). In 2003, science tests were administered to eighth-grade students in countries around the world (see Table 2.23). Construct two different histograms, one that shows the data as almost symmetric and one that shows the data as left-skewed.

Table 2.23 Science test scores

Country	Score	Country	Score	Country	Score
Singapore	578	New Zealand	520	Bulgaria	479
Taiwan	571	Lithuania	519	Jordan	475
South Korea	558	Slovak Republic	517	Moldova	472
Hong Kong	556	Belgium	516	Romania	470
Japan	552	Russian Federation	514	Iran	453
Hungary	543	Latvia	513	Macedonia	449
Netherlands	536	Scotland	512	Cyprus	441
United States	527	Malaysia	510	Indonesia	420
Australia	527	Norway	494	Chile	413
Sweden	524	Italy	491	Tunisia	404
Slovenia	520	Israel	488	Philippines	377

Solution

Figure 2.36 is nearly symmetric. But Figure 2.37 is clearly left-skewed. It is important to realize that *both figures are histograms of the very same data set*. Clever choices for the number of classes and the class limits can affect how a histogram presents the data. The reader must therefore beware! The histogram represents a summarization of the data set, and not the data set itself. Analysts may wish to supplement the histogram with other graphical methods, such as dotplots and stem-and-leaf displays, in order to gain a better understanding of the distribution of the data. ■

FIGURE 2.36 Nearly symmetric histogram.

FIGURE 2.37 Left-skewed histogram.

The *One Variable Statistics and Graphs* applet allows you to experiment with the class width and number of classes when constructing a histogram.

Section 2.4 SUMMARY

✦ Understanding how graphics are constructed will help you avoid being deceived by misleading graphics. Some common methods for making a graph misleading include manipulating the scale, omitting the zero on the relevant scale, and biased distortion or embellishment. The savvy reader should understand that the graphics represent a summary of the data set, and not the data set itself.

Section 2.4 EXERCISES

CLARIFYING THE CONCEPTS

1. Explain in your own words why it is important to be aware of the methods that can be used to make graphics misleading.
2. True or false: What we have learned in this chapter proves that all statistics are misleading.

PRACTICING THE TECHNIQUES

Refer to Example 2.17 for the following exercises.
3. Which do you think is more effective at convincing the American public that a problem exists, Table 2.21 or Figure 2.28?
4. How would factoring in the *number of cars per country* affect the rankings, in your view?
5. If you were an insurance claims adjuster arguing for higher car insurance rates, would you prefer Table 2.21 or Table 2.22? Why?

APPLYING THE CONCEPTS

6. **Eating Bread.** Consider the accompanying graphic (similar to one found in *USA Today*) of the types of bread people eat.
   a. What type of graph is it supposed to represent, among the graphs that we have learned in this chapter?
   b. Consider how the *wheat* category dominates the graph. Which of the eight common methods for misrepresenting data is present here?
   c. Construct a graphic that is not misleading in this way.

7. **Child-Rearing Costs.** Consider the accompanying graphic (similar to one found in *USA Today*) of child-rearing costs by type of cost.
   a. Identify one problem with the graphic that makes it misleading.
   b. Construct a graphic that is not misleading in this way.

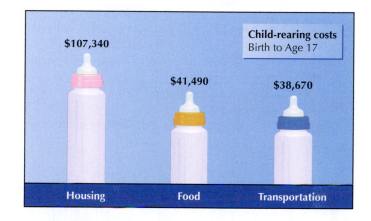

8. **Going to the Game.** Consider the accompanying graphic (similar to one found in *USA Today*) of the proportions of people who go to see professional sports events.
   a. Identify two problems with the graphic that make it misleading.
   b. Construct a graphic that is not misleading in these ways.

9. **Family Practice Doctors.** Consider the accompanying graphic (see next page).
   a. What point is the graphic trying to make?
   b. Which of the eight common problems is most obviously present here?
   c. Construct a graphic that is not misleading in this way.

LA Times

10. What's Your Sign?

The General Social Survey collects data on social aspects of life in America. Consider the accompanying bar graph of the results of asking 1464 people what their astrological sign is.

a. Which of the eight common problems is most obviously present here?

b. Construct a graphic that is not misleading in this way.

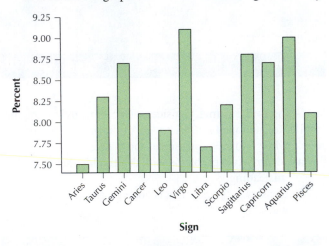

11. American League Shutouts.

The bar graph below shows the number of team shutouts in the American League in the 2002 season.

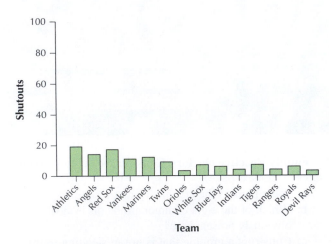

a. Which of the eight common problems is most obviously present here?

b. Construct a graphic that is not misleading in this way.

12. Cancer Pamphlet Readability.

The readability of 30 cancer pamphlets was examined and recorded in terms of school grade level. The following table displays the grade level of the 30 pamphlets. Let each grade level represent a category.

a. Construct a bar graph that overemphasizes the difference among the grade levels.

b. Which of the common methods for making graphics misleading are you using in **(a)**?

c. Construct a bar graph that underemphasizes the difference among the grade levels.

d. Which of the common methods for making graphics misleading are you using in **(c)**?

e. Construct a bar graph that fairly represents the data.

6	6	6	7	7	7	8	8	8	8
8	8	8	8	9	9	9	9	10	11
12	12	12	12	13	13	14	15	15	16

13. Music and Violence.

USA Weekend conducted a poll that asked, "Do you think shock rock and gangsta rap are partly to blame for violence such as school shootings or physical abuse?" The results are shown in the following table.

Yes	31%
No	45%
I've never thought about it	24%

a. Construct a bar graph that overemphasizes the difference among the responses.

b. Construct a bar graph that underemphasizes the difference among the responses.

c. Construct a bar graph that fairly represents the data.

Use the *One Variable Statistics and Graphs* applet for Exercises 14–16. Work with the TIMSS scores from Example 2.23.

14. Click on the **Histogram** tab. Experiment with the class widths by clicking and dragging on the number line. Produce a histogram that is nearly symmetric, like Figure 2.36.

15. Produce a histogram that is somewhat left-skewed, like Figure 2.37.

16. **Click** on the **Stem-and-Leaf** tab. The previous two exercises left us with two different ideas as to the shape of the distribution.

a. Now produce a stem-and-leaf display of the TIMSS scores.

b. Compare the regular stem-and-leaf display with the split-stem stem-and-leaf display. Which is preferable for this data set?

c. Use your preferred stem-and-leaf display from **(b)** to describe the shape of the distribution.

d. Which of the two histograms does your description in **(c)** support?

Try This in Class! In-class activities to enhance your understanding of statistics

1. Construct a frequency distribution and a relative frequency distribution of the genders of the students in your classroom.

2. Where do you think your favorite sport would fall in the distribution of favorite sports? Construct a frequency distribution and a relative frequency distribution of the favorite sports of the students in your classroom. Where do you fall in this distribution?

3. Where do you think your height would fall in a stem-and-leaf display of the heights of students in your class? Construct such a stem-and-leaf-display, using 1s as leaf units. Did you fall about where you thought you would in this display? How would you describe the shape of the distribution?

For the following activities, you may randomize using (a) technology, (b) the *Random Sample* applet, or (c) two dice and a 6 by 6 table of students' names. (See Chapter 1's Try This in Class!, page 28, for instructions.)

4. Generate a random sample of ten students, recording the gender for each student.

a. Is the distribution of genders the same as for the entire class (viewed as a population)?

b. If we generated another random sample of ten students, would the distribution of genders be the same? Why or why not? Would you expect it to be roughly the same or wildly different? How could we make the sample distribution more similar to the population distribution?

5. Generate a random sample of ten students, recording the height for each student. Construct a dotplot or histogram of the heights.

a. Is the shape similar to the distribution for the entire class?

b. If we generated another sample of ten students, would the shape of the dotplot or histogram be the same as previously? Why or why not? Would you expect it to be roughly the same or wildly different? How could we make the sample distribution more similar to the population distribution?

CHAPTER 2 VOCABULARY

SECTION 2.1
- **Bar graph (bar chart)** (p. 35)
- **Frequency (count)** (p. 34)
- **Frequency distribution** (for a qualitative variable) (p. 33)
- **Pareto chart** (p. 36)
- **Pie chart** (p. 36)
- **Relative frequency** (for a qualitative variable) (p. 34)
- **Relative frequency distribution** (for a qualitative variable) (p. 34)

SECTION 2.2
- **Class** (p. 45)
- **Class limit (lower)** (p. 46)
- **Class limit (upper)** (p. 46)
- **Class midpoint** (p. 48)
- **Class width** (p. 46)
- **Dotplot** (p. 51)

- **Frequency distribution** (for quantitative data) (p. 44)
- **Frequency polygon** (p. 48)
- **Histogram** (p. 47)
- **Relative frequency distribution** (for quantitative data) (p. 44)
- **Skewed distribution** (p. 52)
- **Stem-and-leaf display** (p. 49)
- **Symmetric** (p. 52)

SECTION 2.3
- **Cumulative frequency distribution** (p. 60)
- **Cumulative relative frequency distribution** (p. 60)
- **Ogive** (pronounced "oh jive") (p. 61)
- **Time series plot (time series graph)** (p. 61)

SECTION 2.4
- **Eight Common Methods for Making a Graph Misleading** (p. 66)

CHAPTER 2 REVIEW EXERCISES

SECTION 2.1

1.   Parts of Speech. The accompanying bar graph summarizes the frequencies for the various parts of speech. Should we be interested in determining whether this graph is symmetric or skewed? Clearly explain why or why not.

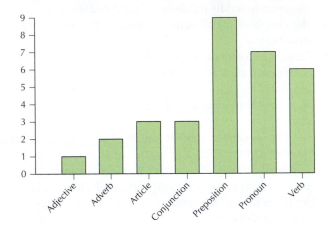

Best Sellers. Refer to Table 2.24 for Exercises 2–5.

2.   Suppose you categorize the price of books in this way: books less than $7 are inexpensive, books between $7.01 and $24.00 are moderately priced, and books more than $24 are expensive.

   a. Construct a frequency distribution for the books that are inexpensive, moderately priced, and expensive.
   b. Using this categorization, construct a relative frequency distribution for the books that are inexpensive, moderately priced, and expensive.

3.   Construct a relative frequency bar graph for the following variables:
   a. Book
   b. Gender

4.   Refer to the previous exercise. Construct pie charts for the same variables.

5.   Can you reconstruct the original data set showing the list price of the books just by using the frequency distribution? Why or why not?

SECTION 2.2

Management Aptitude Test Scores. Use Figures 2.8a and 2.8b, reproduced from Section 2.2, for Exercises 6 and 7.

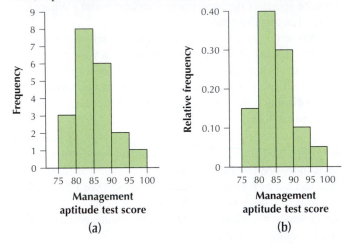

FIGURE 2.8 (a) Frequency histogram; (b) relative frequency histogram.

6.   Suppose we were given the histogram in Figure 2.8a but did not have the original data set. Would we have enough information to construct the histogram in Figure 2.8b? Explain why or why not.

7.   Suppose we were given the histogram in Figure 2.8b but did not have the original data set. Would we have enough information to construct the histogram in Figure 2.8a? Explain why or why not. If not, what piece of information is needed?

New York Townspeople. For towns in New York State, the accompanying histogram provides information on the percentage of the townspeople who are between 18 and 65 years old (data set **New York**). Refer to the histogram for Exercises 8–10.

TABLE 2.24   Top 10 best sellers from *USA Today* (March 3, 2007)

Rank	Title	Author	Type	Book	Gender	Price ($)
1	The Secret	Rhonda Byrne	NF	H	Female	23.95
2	Innocent in Death	Nora Roberts	F	H	Female	25.95
3	Step on a Crack	James Patterson	F	H	Male	27.99
4	Bridge to Terabithia	Katherine Paterson	F	P	Female	6.99
5	Sisters	Danielle Steel	F	H	Female	27.00
6	The Measure of a Man: A Spiritual Autobiography	Sidney Poitier	NF	P	Male	14.95
7	The Memory Keeper's Daughter	Kim Edwards	F	P	Female	14.00
8	The Audacity of Hope	Barack Obama	NF	H	Male	25.00
9	You: On a Diet	Michael Roizen	NF	H	Male	25.00
10	The Best Life Diet	Bob Greene	NF	H	Male	26.00

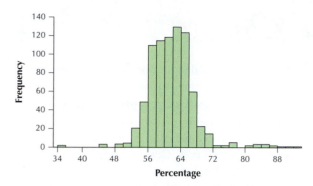

8. Using only the histogram, construct a frequency distribution and a relative frequency distribution of the proportion of townspeople between the ages of 18 and 65. Estimate the frequencies from the histogram.

9. If you had to place a fulcrum under the histogram so that it would balance perfectly like a seesaw, where would you place it?

10. Provide a rough estimate of the relative frequency of towns having 80% or more between 18 and 65 years old.

Households. Use the following information for Exercises 11–13. The data set **Household** contains eight variables' worth of information about the households in all 50 states plus the District of Columbia. The average size of the households is plotted in the accompanying dotplot, reproduced from Figure 2.16a.

Dotplot of average household size.

11. Would you characterize the dotplot as (a) tending to be symmetric, (b) tending to be right-skewed, (c) tending to be left-skewed, or (d) too difficult to say?

12. Refer to the dotplot of average household size.
  a. What is the range of the data set?
  b. How many states have an average household size of 2.55?
  c. Which average household size is the most common?
  d. How many states seem to have an unusually large average household size? How many states seem to have an unusually small average household size? What is your evidence for saying so?

13. *What if* the data were faulty, and each data point should have had 0.5 added to it. How would that affect the shape of the distribution?

SECTION 2.3

14. Countries and Continents. Use the frequency distribution you created in Exercise 25 in Section 2.2 (page 57) to
  a. construct a cumulative frequency distribution.
  b. construct a cumulative relative frequency distribution.

15. Psychology Exam Data. Use the data from Exercise 36 in Section 2.2 (page 58) to
  a. construct a frequency ogive.
  b. construct a relative frequency ogive.

16. Miami Arrests. The Miami-Dade Police Department published the monthly number of arrests made for the year 2005, given in the following table. Construct a time series graph of the data.

Jan.	751	May	919	Sept.	802
Feb.	650	June	800	Oct.	636
Mar.	909	July	834	Nov.	579
Apr.	881	Aug.	789	Dec.	777

SECTION 2.4

17. Sports Clothing. Consider the accompanying graphic of the types of sports clothing that children own.
  a. What type of graph does it represent, among the graphs that we have learned about in this chapter?
  b. Describe the difference between the representation of the NFL category versus the other categories.
  c. Which of the eight common methods for misrepresenting data is present here?
  d. Construct a graphic that is not misleading in this way.

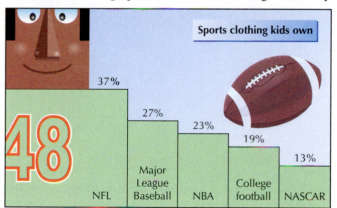

18. Carbon Dioxide Emissions. The following table contains the energy-related carbon dioxide emissions (in millions of metric tons) for 2004, by end-use sector, as reported by the U.S. Energy Information Administration.

Sector	Emissions
Residential	1213.9
Commercial	1034.1
Industrial	1736.0
Transportation	1939.2

  a. Construct a bar graph that overemphasizes the differences among the sectors.
  b. Which of the common methods for making misleading graphics are you using in **(a)**?
  c. Construct a bar graph that underemphasizes the differences among the sectors.
  d. Which of the common methods for making graphics misleading are you using in **(c)**?
  e. Construct a bar graph that fairly represents the data.

CHAPTER 2 *Quiz*

TRUE OR FALSE

1. True or false: Histograms are superior to stem-and-leaf displays because histograms retain the information contained in the data set.

2. True or false: A histogram always provides a realistic summary of the symmetry or skewness of a data set.

FILL IN THE BLANK

3. The frequencies in a frequency distribution must add up to the _____ _____ [two words].

4. A _____ _____ [two words] for a qualitative variable is a listing of all values that the variable can take, together with the frequencies for each value.

SHORT ANSWER

5. If there is a line that splits an image in half so that one side is the mirror image of the other, we say that the image is what?

6. If the right tail of a distribution is longer than the left tail, we say that the distribution is what?

CALCULATIONS AND INTERPRETATIONS

For the following exercises, refer to Table 2.5 from the Case Study (page 37). Ignore the consonants and consider the vowels (A, E, I, O, U) only. Construct the following:

7. Frequency distribution

8. Relative frequency distribution

9. Cumulative frequency distribution

10. Cumulative relative frequency distribution

11. Frequency bar graph

12. Relative frequency bar graph

13. Pie chart of the relative frequencies

14. Ogive of the frequencies

15. Relative frequency ogive of the frequencies

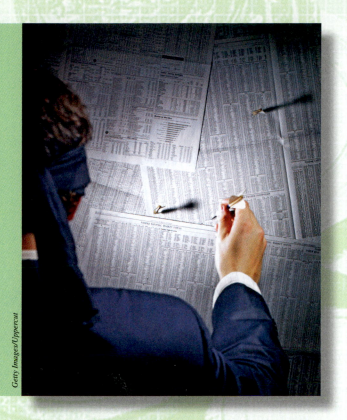

Getty Images/Uppercut

Describing Data Numerically

Can the Financial Experts Beat the Darts?

Have you ever wondered whether a bunch of monkeys throwing darts to choose stocks could select a portfolio that performed as well as the stocks carefully chosen by Wall Street experts? The *Wall Street Journal* (**www.wsj.com**) apparently believes that the comparison is worth a look. Starting in 1988, the *Journal* has run a contest between stocks chosen randomly by *Journal* staff members (rather than monkeys) throwing darts at the *Journal* stock pages (mounted on a board) and stocks chosen by a team of four professional financial experts.

At the end of six months, the *Journal* compares the percentage change in the price of the experts' stocks and the dartboard's stocks. So, who do you think did better? Did the six-figure-salary financial experts put the random dart selections to shame? We examine the results in the Case Study on pages 87 and 104.

The Big Picture

Where we are coming from, and where we are headed...

The first three chapters of the book prepare you to find out all you ever really wanted to know (and probably more) about a particular data set. Chapter 1 introduced the language of descriptive statistics. Chapter 2 showed ways to summarize data sets using tables and graphs. In Chapter 3, we will "crunch the numbers." That is, we will develop some numerical summaries to help us discover important characteristics about the data set. We expand our role as data analyst, to dig in and uncover the numerical side of the "vital information" about the data set. This vital information includes questions such as

- Where is the center of the data?

- How spread out are the data?

- How can we determine whether a particular result is unusual or not?

Here in Chapter 3, we also become acquainted with some powerful and widespread methodologies for applying the tools of descriptive statistics. These applications of descriptive statistics enable us to extract more information from a particular data set, in order to arrive at a deeper meaning of the story behind the data. These tools include percentiles, z-scores for detecting outliers, and Chebyshev's Rule and the Empirical Rule for estimating the proportion of data contained in an interval.

In Chapter 4, we will learn ways of describing the relationship between two variables. However, as we learned in Chapter 1, making inferences from a sample to the general population carries with it a measure of uncertainty. That is why we will need to study *probability* in Chapters 5 and 6. Probability is the bridge between descriptive statistics (Chapters 1–4) and inferential statistics (Chapters 7–13).

So, in this chapter, we learn how to summarize an entire data set with a few numbers. For example, one summary statistic in baseball is a player's batting average (the ratio of hits to at-bats). Suppose we want to know how good a hitter Derek Jeter of the New York Yankees is. Instead of examining thousands of individual at-bats over Jeter's long career, we simply consider one number, his lifetime batting average. This one number powerfully summarizes a very large data set, the many thousands of at-bats over Jeter's career. Since Jeter's lifetime batting average is well over .300, we can see that the Yankee shortstop is a very good hitter. Using summary statistics like lifetime batting average, we can compare his performance with those of some of the greatest players who ever played the game. For more fun statistics about Derek Jeter and all of Major League Baseball, check out **www.baseball-reference.com**. ■

3.1 Measures of Center

Objectives: **By the end of this section, I will be able to...**

1. Calculate the mean for a given data set.

2. Find the median, and describe why the median is sometimes preferable to the mean.

3. Find the mode of a data set.

4. Describe how skewness affects these measures of center.

As data analysts, we often encounter unfamiliar data sets. We can examine a data set using the tabular and graphical methods we learned in Chapter 2. In addition, we can use *numerical summaries* of the data set to help us uncover characteristics of the data. We would first like to know: Where is the middle, or *center,* of the data set? In this section, we will learn about three statistics that can tell us where the center of the data set lies: the mean, the median, and the mode.

1 The Mean

The most well known and widely used **measure of center** is the **mean.** In everyday usage, the word *average* is often used to denote the mean.

> To find the **mean** of the values in a data set, simply add up all the numbers and divide by how many numbers you have.

Example 3.1 — Bird flu cases, 2006

Health officials are carefully monitoring the number of humans who contract avian influenza (bird flu) type A (H5N1) worldwide. Table 3.1 contains the number of cases of bird flu for the five countries with the highest number of cases in 2006.[1] Find the mean number of cases of bird flu for these five countries.

Table 3.1 Bird flu cases by country, 2006

Country	Cases
Vietnam	93
Indonesia	26
Thailand	22
China	12
Turkey	12

Solution

To find the mean, we add up the number of cases for all the countries and divide by the number of countries, 5:

$$\text{mean number of cases of bird flu} = \frac{93 + 26 + 22 + 12 + 12}{5} = \frac{165}{5} = 33$$

That is, these five countries had a mean of 33 bird flu cases in 2006. ■

Notation

Statisticians like to use specialized notation. The notation is worth learning because it saves a lot of writing and because certain concepts can best be understood using this special notation.

- The **sample size,** which refers to how many observations you have in your sample data set, is always denoted by n. Here, the five countries from Table 3.1 can be considered a sample taken from the population (which in this case is all the countries in the world). Thus, here, $n = 5$.

- We denote the *i*th data value by x_i, where i is simply an index or counter indicating which data point we are specifying. For example, in Table 3.1, $x_1 = 93$, $x_2 = 26$, $x_3 = 22$, and $x_4 = 12$. The last data value is $x_n = x_5 = 12$.

- The notation for "add them together" is Σ (capital sigma), the Greek letter for "S," because it stands for "Summation." To add up the number of cases for all five countries, we could write out $93 + 26 + 22 + 12 + 12$, or we could simply represent this sum as Σx_i or, even more simply, as Σx.

- The **sample mean** is called $\bar{x}$ (pronounced "x-bar"). You should try to commit this to long-term memory, since $\bar{x}$ may be the most important symbol used in this book and will return again and again in nearly every chapter.

> The **sample mean** can be written as $\bar{x} = \Sigma x / n$. In plain English, this just means that, in order to find the mean $\bar{x}$, we
> 1. Add up all the data values, giving us Σx.
> 2. And divide by how many observations are in the data set, giving us $\Sigma x / n$.

So, for example, the sample mean number of cases of bird flu can be written as

$$\bar{x} = \frac{\Sigma x}{n} = \frac{93 + 26 + 22 + 12 + 12}{5} = \frac{165}{5} = 33$$

What Does This Number Mean?

The Mean as the Balance Point of the Data

Let's explore the bird flu data a bit further. Consider the dotplot of the bird flu cases for each country, given in Figure 3.1. To find out where the mean of the number of cases of bird flu lies on this number line, imagine that the dots are little blocks on a ruler or a seesaw and that you must decide where to place the fulcrum so that the ruler balances perfectly. *The place where the data set balances perfectly is the location of the mean.* Placing the fulcrum too far to the right or left would create an imbalance. This data set balances precisely at the sample mean, $\bar{x} = 33$. ■

FIGURE 3.1 Where is the balance point for this dotplot?

Checking Your Results Against Experience and Common Sense

When you have found the balance point, you have found the mean. Keep this idea in mind throughout your study of statistics, as a reality check for calculations of the mean. When you calculate the mean, or have a computer or calculator do it for you, don't just naively accept whatever value pops out. Since the mean always indicates the place where the data values are in balance, the mean is often near the center of the data. If the value you have calculated lies nowhere near the center of the data, then you may want to check your calculations.

For example, suppose we were finding the mean of the bird flu data, and we accidentally entered 930 for the first value instead of 93 for the number of cases for Vietnam. Then our value for the mean resulting from this *incorrect* calculation would be

$$\bar{x} = \frac{\sum x}{n} = \frac{930 + 26 + 22 + 12 + 12}{5} = \frac{1002}{5} = 200.4$$

The mean number of cases cannot equal 200.4 because all the values in the data set are less than 200.4. The mean can never be larger or smaller than all the values in the data set.

Don't automatically assume that the result you get from a computer or calculator is right. Remember GIGO: Garbage In Garbage Out. If you enter the wrong data, the calculator or computer will not bail you out. Human error is one reason for the explosion of faulty statistical analysis in the newspapers and on the Internet. Now more than ever data analysts must use good judgment. When you calculate a mean, always have an idea of what you *expect* the sample mean to be, that is, at least a ballpark figure. ■

The Population Mean μ

The mean value of a population is usually unknown. For example, we cannot know the mean systolic blood pressure of *all* the residents in your hometown at noon today. Instead, data analysts use *estimation*. We could select a random sample of, say, 30 residents, find the mean systolic blood pressure $\bar{x}$ of this sample, and use this $\bar{x}$ as an estimate of the unknown population mean systolic blood pressure. Generally, Greek letters are used to represent the (usually unknown) population parameters (such as the population mean). We denote the **population mean** with μ (mu), which is the Greek letter for "m." The **population size** is denoted by N. When all the values of the population are known, the population mean is calculated as

$$\mu = \frac{\sum x}{N}$$

The Mean Is Sensitive to Extreme Values

One drawback of using the mean to measure the center of the data is that the mean is sensitive to the presence of extreme values in the data set. We illustrate this phenomenon with the following example.

Example 3.2   Mean of home sales prices

Table 3.2 contains a sample of 6 home sales prices for Broward County, Florida, as listed in the *New York Times* Online Real Estate section (**realestate.nytimes.com**) for December 14, 2006. We would like to get an idea of the typical home sales price in Broward County. Find the mean sales price of the homes in this sample.

Table 3.2 Home sales prices in Broward County, Florida

Location	Price
Coral Springs	$415,000
Weston	$449,000
Coconut Creek	$472,000
Davie	$649,000
Hallandale	$889,000
Fort Lauderdale	$975,000

Solution

$$\bar{x} = \frac{\sum x}{n} = \frac{415,000 + 449,000 + 472,000 + 649,000 + 889,000 + 975,000}{6}$$

$$= \frac{3,849,000}{6} = \$641,500$$

So the sample mean home sales price is $641,500. Now, suppose that we append a seventh home to our sample, a home in Hillsboro Beach listed for $5,900,000, which is much more expensive than any of the other homes in the sample. Recalculating the mean, we get

$$\bar{x} = \frac{\sum x}{n}$$

$$= \frac{415,000 + 449,000 + 472,000 + 649,000 + 889,000 + 975,000 + 5,900,000}{7}$$

$$= \frac{9,749,000}{7} \approx \$1,392,714$$

Note that the mean sales price more than doubled from $641,500 to $1,392,714 when we added this extreme value. Also, this new mean is much higher than every price in the original sample. Thus, it is highly unlikely that this new mean of about $1.4 million is representative of the *typical* sales price of homes in Broward County.

This example shows why we say that the mean is sensitive to the presence of extreme values. For situations like this, we prefer a measure of center that is not so sensitive to extreme values. Fortunately, the *median* is just such a measure. ■

2 The Median

Recall that the median strip on a highway is the slice of land in the *middle* of the two lanes of the highway. In statistics, the **median** of a data set is the *middle data value* when the data are put into ascending order. There are two cases, depending on whether the sample size is odd or even.

The Median

The **median** of a data set is the *middle data value* when the data are put into ascending order. Half of the data values lie below the median, and half lie above.

- If the sample size n is odd, then the median is a unique middle value. That is, the median is the $\left(\frac{n+1}{2}\right)^{th}$ observation when the data are put in ascending order.

- If the sample size n is even, then the median is the mean of the two data values in the middle. That is, the median is the mean of the two data values that lie on either side of the $\left(\frac{n+1}{2}\right)^{th}$ position.

The case when the sample size is even is clear if you hold up four fingers on one hand. Notice that there is no unique finger in the middle. Since there is no middle value when the sample size is even, we take the two data values in the middle and split the difference.

The Median Is Not Sensitive to Extreme Values

News reports prefer the median to the mean when reporting home sales prices. For example, the *San Jose Mercury News* reported on December 14, 2006, that the median California home sales price was $469,000. Unlike the mean, the median is not sensitive to extreme values. For example, if someone purchases a very expensive house this month, the mean home sales price will jump, but the median home sales price will be less affected. Let's look at an example of how this would occur.

| Example 3.3 | Median of home sales prices |

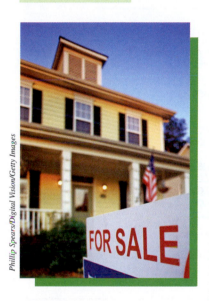

Find the median home sales price for the Broward County data from Table 3.2.

Solution

Fortunately, the data are already presented in ascending order in the table. Since $n = 6$ is even, the median is the mean of the two data values that lie on either side of the $\left(\frac{n+1}{2}\right)^{th} = \left(\frac{6+1}{2}\right)^{th} = 3.5$th position. That is, the median is the mean of the 3rd and 4th data values, $472,000 and $649,000. Splitting the difference between these two, we get

$$\text{median price} = \frac{472,000 + 649,000}{2} = \$560,500$$

We note that in Table 3.2 there are exactly as many homes with prices lower than $560,500 as there are homes with prices higher than $560,500.

Now, what happens to the median when we add in the $5,900,000 home from Hillsboro Beach? Since $n = 7$ is odd, the median is the unique $\left(\frac{n+1}{2}\right)^{th} = \left(\frac{7+1}{2}\right)^{th} = 4$th observation, given by the home in Davie for $649,000. The extreme value increased the median only from $560,500 to $649,000. Recall that the mean more than doubled to almost $1.4 million. Thus, the median home sales price is a better measure of center because it more accurately reflects the typical sales prices of homes in Broward County. Figure 3.2 shows how the mean (red triangles) changes significantly with the addition of the extreme value, while the median (green triangles) changes relatively little. ■

FIGURE 3.2 The mean (red triangles) is sensitive to extreme values, but the median (green triangles) is not.

The *Mean and Median* applet allows you to insert your own data values and see how changes in these values affect both the mean and the median.

Note that the formula $\frac{n+1}{2}$ gives the *position*, not the *value*, of the median. For example, the median home sales price for Table 3.2 is *not* $\frac{n+1}{2} = \frac{6+1}{2} = 3.5$.

Example 3.4   Using technology to find the mean and median

Find the mean and median of the home sales prices in Table 3.2, using (a) the TI-83/84, (b) Excel, and (c) Minitab.

Solution

Using the instructions in the Step-by-Step Technology Guide on page 89, we get the following output.

a.   The first TI-83/84 screen shows $\bar{x} = 641{,}500$ and $n = 6$. The second screen shows the median Med $= 560{,}500$.

b.   The mean, median, and sample size ("Count") n are highlighted in the Excel output.

Home Price	
Mean	641500
Standard Error	98255.36457
Median	560500
Mode	#N/A
Standard Deviation	240675.5077
Sample Variance	57924700000
Kurtosis	-1.878590247
Skewness	0.616113252
Range	560000
Minimum	415000
Maximum	975000
Sum	3849000
Count	6

c.   Minitab gives us N $= 6$. (This is actually n the sample size, not N the population size.) The mean and median are highlighted. ■

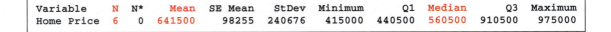

Variable	N	N*	Mean	SE Mean	StDev	Minimum	Q1	Median	Q3	Maximum
Home Price	6	0	641500	98255	240676	415000	440500	560500	910500	975000

3 The Mode

A third measure of center is called the **mode.** French speakers will recognize that the term *mode* in French refers to *fashion*. The popularity of clothing, cosmetics, music, and even basketball shoes often depends on just which style is in fashion. In a data set, the value that is most "in fashion" is the value that occurs the most.

> The **mode** of a data set is the data value that occurs with the greatest frequency.

Example 3.5   Cost of mathematical journals

The rising cost of research journals has been taking an increasing bite out of library and research budgets. Table 3.3 contains the annual subscription cost of ten research journals in mathematics and statistics for 2006. Find the following.

a. The mean journal subscription cost
b. The median journal subscription cost
c. The mode journal subscription cost

Table 3.3 Annual subscription cost for ten research journals

Journal	Annual cost (dollars)
Annals of Probability	250
Annals of Statistics	250
Bulletin of the American Mathematical Society	402
Mathematical Computation	467
Annals of Applied Probability	850
Proceedings of the American Mathematical Society	1022
Topology	1582
Journal of Mathematical Economics	1653
Transactions of the American Mathematical Society	1744
Journal of Pure and Applied Algebra	3631

Source: American Mathematical Society Journal Survey, **www.ams.org/membership/journal-survey.html**.

Solution

a. The sample mean journal cost is

$$\bar{x} = \frac{\sum x}{n} = \frac{250 + 250 + 402 + 467 + 850 + 1022 + 1582 + 1653 + 1744 + 3631}{10}$$

$$= \frac{11{,}851}{10} = \$1185.10$$

b. Since we have $n = 10$ journals, the median is the mean of the two data values that lie on either side of the $\left(\frac{n+1}{2}\right)^{th} = \left(\frac{10+1}{2}\right)^{th} = 5.5$th position. That is, the median is the mean of the 5th and 6th data values, $850 and $1022. The mean of these two journal costs is

$$\text{median journal cost} = \frac{850 + 1022}{2} = \$936$$

c. The mode is the data value that occurs with the greatest frequency. There are two journals that cost $250 each. No other cost occurs more than once. Therefore, the mode cost is $250. Note that this value of $250 for the mode is not a very good measure of center for this data set. It is the minimum value. This illustrates a weakness of using the mode as a measure of center. ■

One of the strengths of using the mode is that it can be used with categorical, or qualitative, data. Suppose you asked your friends what their favorite flower was. Six of them answered "rose," five answered "iris," three answered "lily," and one answered "daffodil." Note that these data are categorical, not numerical. What is the most popular flower among your friends? That is, what is the mode of the variable *favorite flower*? The most frequently occurring answer is "rose," with six friends preferring roses. Since this is the highest frequency, the rose represents the mode of the variable *favorite flower*. Now think about the following question for a moment: What is the mean of the variable favorite flower? Actually, because the data values are qualitative (non-numeric), the idea of a mean or median for the variable *favorite flower* is essentially meaningless. We cannot use arithmetic with categorical variables, and so means and medians cannot be found.

Finally, it may happen that no data value occurs more than once, in which case we say there is *no mode*. Or more than one data value could occur with the greatest frequency; in which case we would say there is more than one mode. Data sets with one mode are *unimodal;* data sets with more than one mode are *multimodal*.

What If Scenario ? Give the Calculator a Rest

What If Scenarios offer you a chance to reflect on how changes in the initial conditions will percolate through the various aspects of a problem. The only requirement is to put your calculator down and think through the problem. No data are given. You are asked to find the answers by using your knowledge of what the statistics *represent*.

Consider Example 3.5 once again. Now imagine: *what if* there was a data entry error (a typo), and the subscription cost for the most expensive journal, the *Journal of Pure and Applied Algebra,* was not $3631 but actually some *unspecified* dollar amount greater than $3631. Describe how and why this change would have affected the following, if at all:

   a. The mean subscription cost

   b. The median subscription cost

   c. The mode subscription cost

Solution

   a. Consider Figure 3.3, a dotplot of the subscription costs, with the triangle indicating the mean cost of $1185.10. Recall that this represents the balance point of the data. As the cost of the most expensive journal increases (blue arrow), the point at which the data balance (that is, the mean) also moves somewhat to the right. Thus, the mean cost will increase.

   b. Recall from Example 3.5 that the median is the mean of the 5th and 6th data values, $850 and $1022. In other words, the median *ignores* most of the data values, including the most expensive journal, the one whose cost has increased. Therefore, the median will remain unchanged.

   c. The mode also remains unchanged, since the only data value that occurs more than once is the original mode, $250.

FIGURE 3.3
As the cost of one journal increases, so does the mean cost.

Cost ($)

---

4 Skewness and Measures of Center

The skewness of a distribution can often tell us something about the relative values of the mean, median, and mode (see Figure 3.4).

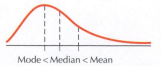

Mean < Median < Mode   Mean = Median = Mode   Mode < Median < Mean

FIGURE 3.4 How skewness affects the mean and median.

> **How Skewness Affects the Mean and Median**
> - For a right-skewed distribution, the mean is larger than the median.
> - For a left-skewed distribution, the median is larger than the mean.
> - For a symmetric unimodal distribution, the mean, median, and mode are fairly close to one another.

Example 3.6   Household size

The histogram of the average size of households in the 50 states and the District of Columbia from Example 2.12 (page 53) is reproduced here as Figure 3.5.

a. Based on the skewness of the distribution, state the relative values of the mean, median, and mode.

b. Use Minitab to verify your claim in (a).

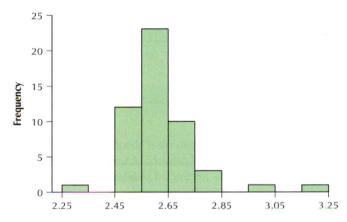

FIGURE 3.5 Household size is somewhat right-skewed.

Solution

a. The distribution of average household size is somewhat right-skewed. Thus, from Figure 3.4, we would expect the mean to be greater than the median, which is greater than the mode.

b. The Minitab descriptive statistics are shown here. Note that the mean is greater than the median, which is greater than the mode.

Descriptive Statistics: Size			
Variable	Mean	Median	Mode
Size	2.619	2.590	2.530

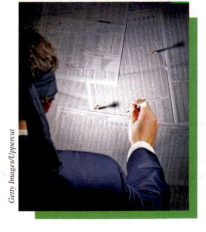

Getty Images/Uppercut

CASE STUDY | **Can the Financial Experts Beat the Darts?**

Recall the contest held by the *Wall Street Journal* to compare the performance of stock portfolios chosen by financial experts and stocks chosen at random by throwing darts at the *Journal* stock pages. By October 1998, 100 such contests had been held. We will examine the results of these 100 contests in various ways, using the methods we have learned thus far, and will return to examine them further as we acquire more analysis tools. Let's start by reporting the raw result data. The percentage increase or decrease in stock prices was calculated for the portfolios chosen by the professional financial advisers and by the randomly thrown darts and was compared with the percentage net change in the Dow Jones Industrial Average (DJIA).

Clearly, these measures of location do not give us the whole picture. We need **measures of variability** (or **measures of spread**) that will describe how spread out the data values are. Figure 3.10 illustrates that the heights of the WMU team are *more spread out* than the heights of the NCU team. ■

FIGURE 3.10
Comparative dotplots of the heights of two volleyball teams.

Just as there were several measures of the center of a data set, there are also a variety of ways to measure how spread out a data set is. The simplest measure of variability is the **range.**

> The **range** of a data set is the difference between the largest value and the smallest value in the data set:
>
> range = largest value – smallest value

A larger range is an indication of greater variability, or greater spread, in the data set.

Example 3.8 Range of the volleyball teams' heights

Calculate the range of player heights for each of the WMU and NCU teams.

Solution

From Figure 3.10, it is intuitively clear that the heights of the WMU team are more spread out than the heights of the NCU team. This should be reflected in the ranges. The range of the WMU team should be larger than the range of the NCU team, reflecting its greater variability.

$$\text{range}_{\text{WMU}} = \text{largest value} - \text{smallest value} = 75 - 60 = 15 \text{ inches}$$

$$\text{range}_{\text{NCU}} = \text{largest value} - \text{smallest value} = 72 - 66 = 6 \text{ inches}$$

The range for WMU is indeed larger than the range for NCU, reflecting WMU's greater variability in height. ■

The range is quite simple to calculate. However, it does have its drawbacks. For example, the range is quite sensitive to extreme values, since it is calculated from the difference of the two most extreme values in the data set. *It completely ignores all of the other data values in the data set*. We would prefer our measure of variability to quantify spread with respect to the center, as well as to actually use all of the available data values. Such a measure is the **standard deviation.**

What Is a Deviation?

Before we learn about the standard deviation, we need to get a firm understanding of what a **deviation** means, in the statistical sense. We shall use an example to do this.

Example 3.9   Amanda's and Bethany's delivery service

Amanda and Bethany have contracted with a local document dispatch service to deliver documents in midtown Manhattan. Their clients would like to be able to count on a consistent delivery time schedule. Not only do the documents have to arrive in a timely fashion, but the delivery times should not vary a lot. One particularly important client kept careful track of the delivery times (in minutes) for documents delivered by Amanda and Bethany. See Table 3.9 and Figure 3.11. Describe the variability in Amanda's and Bethany's delivery times.

Table 3.9 Delivery times (in minutes) for Amanda and Bethany

Amanda	15	17	12	15	10	16	14	20	16	15
Bethany	5	22	15	25	4	18	10	29	17	5

FIGURE 3.11 Comparative dotplots of Amanda's and Bethany's delivery times.

Solution

Note that Bethany's delivery times vary more than Amanda's. Bethany has some long delivery times, some short, and some of moderate length. On the other hand, Amanda's delivery times are mostly of moderate length. We can say that Bethany's delivery times

- are *more spread out,*
- show *greater variability,* and
- have *more variation.* ◾

Developing Your Statistical Sense

Develop Your Intuitive Sense

For each new statistic that you learn, try to get an intuitive feel for what it means. That is just as important as being able to calculate it—even more so nowadays because we have calculators and computers to do the calculations. For example, suppose you had miscalculated Amanda's range as being larger than Bethany's range. A glance at Figure 3.11, combined with your intuitive grasp of the range as a measure of spread, would have helped you to catch your error. ◼

In this section, we will learn three different ways of measuring the spread of a data set (further methods await later sections): the range, the variance, and the standard deviation. Now, we know that the greater the spread of a data set, the greater the range. This holds true for all of the measures of spread that we will discuss in this book. In other words, the greater the spread of a data set, the greater the value of the range, the standard deviation, and the variance.

You might think that, if the variation in Bethany's delivery times is larger, doesn't that mean that the mean length of her delivery times is longer as well? Not at all. The two sample means are equal, as shown by the red triangles in Figure 3.11.

$$\bar{x}_{\text{Amanda}} = \frac{\sum x}{n} = \frac{15 + 17 + 12 + 15 + 10 + 16 + 14 + 20 + 16 + 15}{10} = \frac{150}{10} = 15$$

$$\bar{x}_{\text{Bethany}} = \frac{\sum x}{n} = \frac{5 + 22 + 15 + 25 + 4 + 18 + 10 + 29 + 17 + 5}{10} = \frac{150}{10} = 15$$

Some measures of spread, such as the variance or standard deviation, rely on the concept of *deviation*.

Deviation

A **deviation** for a given data value x is the difference between the data value and the mean of the data set. For a sample, the deviation equals $x - \bar{x}$. For a population, the deviation equals $x - \mu$.

- If the data value is larger than the mean, the deviation will be positive.
- If the data value is smaller than the mean, the deviation will be negative.
- If the data value equals the mean, the deviation will be zero.

The deviation can roughly be thought of as the distance between a data value and the mean, except that the deviation can be negative while distance is always positive.

Example 3.10   Calculating the deviations

Calculate the deviations for Amanda's and Bethany's delivery times.

Solution

- Amanda's mean delivery time was $\bar{x} = 15$ minutes. Her first delivery (first data value) also happened to be $x = 15$ minutes long. So the deviation for this first delivery is $x - \bar{x} = 15 - 15 = 0$. A deviation of zero occurs when a data value has the same value as the mean.
- Amanda's second delivery was $x = 17$ minutes long. So the deviation for this data value is $x - \bar{x} = 17 - 15 = 2$. A deviation of 2 implies that this data value lies two units (in this case, minutes) above the mean.
- Amanda's third delivery was 12 minutes long, resulting in a deviation of $x - \bar{x} = 12 - 15 = -3$.

Continuing in this way, we find the deviations for all of Amanda's and Bethany's delivery times, as recorded in Table 3.10. ■

Table 3.10 Delivery times and their deviations

	Delivery									
	1	2	3	4	5	6	7	8	9	10
Amanda's times	15	17	12	15	10	16	14	20	16	15
Amanda's deviations	0	2	-3	0	-5	1	-1	5	1	0
Bethany's times	5	22	15	25	4	18	10	29	17	5
Bethany's deviations	-10	7	0	10	-11	3	-5	14	2	-10

Figure 3.12 illustrates these deviations in comparative dotplots. The green arrows indicate positive deviations, and the red arrows indicate negative deviations from the mean $\bar{x} = 15$. For example, the longest green arrow for Amanda is for her eighth delivery, which took 20 minutes. The deviation for this delivery is $x - \bar{x} = 20 - 15 = 5$. The shortest red arrow for Amanda is for her seventh delivery, which took 14 minutes. The deviation for this delivery is $x - \bar{x} = 14 - 15 = -1$. The lower dotplot shows Bethany's delivery times, with positive and negative deviations from the mean $\bar{x} = 15$. (Three of Amanda's deviations and one of Bethany's deviations are not indicated in the figure. Which deviations are these, and why are they not shown?) Bethany's arrows do seem longer, on average. Let us consider using the "mean deviation" as a way of measuring the spread. The mean of Amanda's deviations is

$$\text{Amanda's mean deviation} = \frac{0 + 2 + (-3) + 0 + (-5) + 1 + (-1) + 5 + 1 + 0}{10}$$

$$= \frac{0}{10} = 0$$

Note: Another way to avoid the negative deviations is to take their absolute value. The *mean absolute deviation (MAD)* is a measure of spread that looks at the average of the absolute values of the deviations:

$$\text{MAD} = \sum \left(\frac{|x - \bar{x}|}{n} \right)$$

You can confirm that Bethany's mean deviation is zero as well. It is not a coincidence that the mean deviation for both Amanda and Bethany turned out to be zero. In fact, *for any conceivable data set, the mean deviation is zero,* since the mean is the unique point on the number line where the negative deviations (the red arrows in Figure 3.12) precisely balance the positive deviations (the green arrows). Unfortunately, then, the mean deviation is not a useful measure of spread. To avoid this problem of the positive and negative deviations canceling each other out, statisticians work with the squared deviations. The next two measures of spread use the squared deviations.

FIGURE 3.12 Comparative dotplots of delivery time deviations.

3 The Standard Deviation and the Variance

By far the most common measure of variability is the *standard deviation*. To compute the standard deviation and variance, we consider the squared deviations instead of the absolute deviations. Table 3.11 shows the squared deviations for Amanda and

Table 3.11 Squared deviations of delivery times for Amanda and Bethany

	Delivery									
	1	**2**	**3**	**4**	**5**	**6**	**7**	**8**	**9**	**10**
Amanda's times	15	17	12	15	10	16	14	20	16	15
Amanda's deviations	0	2	−3	0	−5	1	−1	5	1	0
Amanda's squared deviations	$0^2 = 0$	$2^2 = 4$	$(-3)^2 = 9$	$0^2 = 0$	$(-5)^2 = 25$	$1^2 = 1$	$(-1)^2 = 1$	$5^2 = 25$	$1^2 = 1$	$0^2 = 0$
Bethany's times	5	22	15	25	4	18	10	29	17	5
Bethany's deviations	−10	7	0	10	−11	3	−5	14	2	−10
Bethany's squared deviations	$(-10)^2 = 100$	$7^2 = 49$	$0^2 = 0$	$10^2 = 100$	$(-11)^2 = 121$	$3^2 = 9$	$(-5)^2 = 25$	$14^2 = 196$	$2^2 = 4$	$(-10)^2 = 100$

Bethany. Note that Bethany's squared deviations are on average larger than Amanda's, reflecting the greater spread in Bethany's delivery times. It is therefore logical to build our measure of spread using the *mean squared deviation*.

The Population Variance σ^2

Suppose that we consider Amanda's and Bethany's delivery times to be populations rather than samples. Then the mean of each population is a population mean ($\mu = 15$) rather than a sample mean ($\bar{x} = 15$). Each deviation is ($x - \mu$), and the mean squared deviation is called the **population variance** and is symbolized by σ^2. This is the lowercase Greek letter sigma, not to be confused with the uppercase sigma (Σ) used for summation.

Population Variance σ^2

The **population variance σ^2** is the mean of the squared deviations in the population and is found by

$$\sigma^2 = \frac{\sum(x - \mu)^2}{N}$$

Example 3.11 Calculating the population variances for Amanda and Bethany

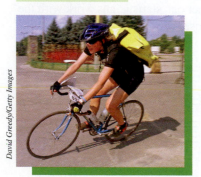

David Greedy/Getty Images

Calculate the population variances of the delivery times for Amanda and Bethany.

Solution

Using the squared deviations from Table 3.11, we have

$$\sigma^2 = \frac{\sum(x - \mu)^2}{N} = \frac{0 + 4 + 9 + 0 + 25 + 1 + 1 + 25 + 1 + 0}{10} = \frac{66}{10} = 6.6$$

for Amanda, and

$$\sigma^2 = \frac{\sum(x - \mu)^2}{N} = \frac{100 + 49 + 0 + 100 + 121 + 9 + 25 + 196 + 4 + 100}{10} = \frac{704}{10} = 70.4$$

for Bethany. Clearly, the population variance of the delivery times for Bethany is greater than that for Amanda, thus indicating that Bethany's delivery times are more variable than Amanda's.

However, what is the *meaning* of the values we got for σ^2, 6.6 and 70.4, apart from their comparative value? How would you explain these results in layman's terms to someone who has never taken a statistics course? The problem is that the units of these values are *minutes squared,* certainly not a useful measure. Unfortunately, the intuitive meaning of the variance is not self-evident. ■

The Population Standard Deviation σ

Working with the variance is quite useful in the realm of mathematical statistics, but the lack of a clear-cut interpretation of its meaning leads us to de-emphasize the variance (whether population variance or sample variance) in this text. Instead, we shall focus on the *standard deviation,* which has a much clearer interpretation in practice. The standard deviation is simply the square root of the variance. By taking the square root, we return the units of measure back to the original data unit (for example, "minutes" rather than "minutes squared"). The symbol for the **population standard deviation** is σ: conveniently, $\sqrt{\sigma^2} = \sigma$.

> **Population Standard Deviation σ**
> The **population standard deviation σ** is the square root of the population variance and is found by
>
> $$\sigma = \sqrt{\frac{\sum(x-\mu)^2}{N}}$$

Example 3.12 Calculating the population standard deviations for Amanda and Bethany

Calculate the population standard deviations of the delivery times for Amanda and Bethany.

Solution

Since Bethany's population variance of 70.4 is larger than Amanda's population variance of 6.6, Bethany's population standard deviation will also be larger, since we are simply taking the square root. We have

$$\sigma = \sqrt{\frac{\sum(x-\mu)^2}{N}} = \sqrt{\frac{66}{10}} = \sqrt{6.6} \approx 2.6$$

for Amanda, and

$$\sigma = \sqrt{\frac{\sum(x-\mu)^2}{N}} = \sqrt{\frac{704}{10}} = \sqrt{70.4} \approx 8.4$$

for Bethany.

The population standard deviation of Bethany's delivery times is 8.4 minutes, which is larger than Amanda's 2.6 minutes. As expected, the greater variability in Bethany's delivery times leads to a larger value for her population standard deviation σ. ■

The Standard Deviation

So how do we interpret these values for σ? One quick thumbnail interpretation of the standard deviation is that it represents a "typical" deviation. That is, the value of σ *represents a distance from the mean that is representative for that data set.* Remember that it is not the mean deviation, which is always zero. ■

Communicating the Results

As you study statistics, keep in mind that during your career you will likely need to explain your results to others who have never taken a statistics course. Therefore, you should always keep in mind *how to interpret your results to nonspecialists.* Communication and interpretation of your results can be as important as the results themselves. ■

The Sample Variance s^2 and the Sample Standard Deviation s

In the real world, we usually cannot determine the exact value of the population mean or the population standard deviation. We use the sample mean and **sample standard deviation** to estimate the population parameters. In this text, we will work with sample statistics unless the data set is identified as a population. The **sample variance** also depends on the concept of the mean squared deviation. If the sample mean is $\bar{x}$, and the sample size is n, it would be reasonable, working by analogy with the population variance, to expect that the formula for the sample variance would be

$$\frac{\sum(x-\bar{x})^2}{n}$$

However, data analysts would like this statistic (the sample variance) to be an unbiased estimator of the parameter (the population variance). An *unbiased estimator* is a statistic whose mean value equals the parameter it is trying to estimate. The above formula has been found to underestimate the population variance, so that we need to replace the n in the denominator with $n - 1$. We therefore have the following.

Sample Variance s^2

The **sample variance s^2** is approximately the mean of the squared deviations in the sample and is found by

$$s^2 = \frac{\sum(x - \bar{x})^2}{n - 1}$$

The sample standard deviation is perhaps the second most important statistic you will encounter in this book (after the sample mean $\bar{x}$). It is the most commonly used measure of spread. The sample standard deviation is simply the square root of the sample variance and takes as its symbol the letter s, which is the Roman letter for the Greek σ. Again, $s = \sqrt{s^2}$.

Sample Standard Deviation s

The **sample standard deviation s** is the square root of the sample variance s^2:

$$s = \sqrt{s^2} = \sqrt{\frac{\sum(x - \bar{x})^2}{n - 1}}$$

The value of s may be interpreted as the typical difference between a data value and the sample mean, for a given data set.

Example 3.13   Calculating the sample variance and the sample standard deviation

Calculate the sample variances and sample standard deviations of Amanda's and Bethany's delivery times.

Solution

Using the squared deviations from Table 3.11, we have the sample variance

$$s^2 = \frac{\sum(x - \bar{x})^2}{n - 1} = \frac{0 + 4 + 9 + 0 + 25 + 1 + 1 + 25 + 1 + 0}{9} = \frac{66}{9} \approx 7.3$$

for Amanda, and the sample variance

$$s^2 = \frac{\sum(x - \bar{x})^2}{n - 1} = \frac{100 + 49 + 0 + 100 + 121 + 9 + 25 + 196 + 4 + 100}{9} = \frac{704}{9} \approx 78.2$$

for Bethany.

We find the sample standard deviations for Amanda's and Bethany's delivery times, as follows. For Amanda, we have

$$s = \sqrt{\frac{\sum(x - \bar{x})^2}{n - 1}} = \sqrt{\frac{0 + 4 + 9 + 0 + 25 + 1 + 1 + 25 + 1 + 0}{9}} = \sqrt{\frac{66}{9}} \approx 2.7$$

and for Bethany, we have

$$s = \sqrt{\frac{\sum(x - \bar{x})^2}{n - 1}} = \sqrt{\frac{100 + 49 + 0 + 100 + 121 + 9 + 25 + 196 + 4 + 100}{9}} = \sqrt{\frac{704}{9}} \approx 8.8$$

Amanda's delivery times typically differ from the mean $\bar{x} = 15$ by only 2.7 minutes, while Bethany's delivery times typically differ from the mean by 8.8 minutes. ∎

*Developing
Your Statistical Sense*

Less Variation Is Better

Amanda's sample standard deviation of $s = 2.7$ shows a remarkable consistency in her delivery times. Amanda's clients would certainly appreciate this consistency. In most real-world applications, consistency is a great advantage. In statistical data analysis, less variation is often better. Throughout the text, you will find that smaller variability will lead to

- more precise estimates and
- higher confidence in conclusions. ■

A Note on Rounding: In this textbook, we typically round to one or two decimal places when presenting results. However, when results are to be used for later calculations, students should not round until the last calculation.

The above formulas for the variance and standard deviation can sometimes be time-consuming, since each individual deviation needs to be calculated. You may use the following *computational formulas* to simplify your calculations. The computational formulas are equivalent to the above definition formulas.

Computational Formulas for the Variance and Standard Deviation

Population variance:

$$\sigma^2 = \frac{\sum x^2 - \left(\sum x\right)^2 / N}{N}$$

Population standard deviation:

$$\sigma = \sqrt{\frac{\sum x^2 - \left(\sum x\right)^2 / N}{N}}$$

Sample variance:

$$s^2 = \frac{\sum x^2 - \left(\sum x\right)^2 / n}{n - 1}$$

Sample standard deviation:

$$s = \sqrt{\frac{\sum x^2 - \left(\sum x\right)^2 / n}{n - 1}}$$

where $\sum x^2$ means that you square each data value and then add up the squared data values, and $(\sum x)^2$ means that you add up all the data values and then square the sum.

Example 3.14   Calculating the population variance and population standard deviation using the computational formulas

Table 3.12 lists the amount of farmland (in 1000s of acres) in each county in the state of Connecticut. Since the data set contains *all* $N = 8$ counties in Connecticut, it can be considered a population. Calculate the population variance and population standard deviation using the computational formulas.

Table 3.12 Farmland in Connecticut

County	Farmland (1000s of acres)
Fairfield	12.8
Hartford	50.2
Litchfield	93.6
Middlesex	17.9
New Haven	26.0
New London	58.9
Tolland	36.8
Windham	61.1

Source: U.S. Bureau of Economic Analysis.

Solution

Construct a table similar to Table 3.13. The data values (farmland = x) go in the first column, and the squared values go in the second column. Notice that we don't have to find the deviations. Add up the columns to get $\sum x = 357.3$ and $\sum x^2 = 20{,}997.91$, respectively. Then substitute these values into the computational formula for the population variance:

$$\sigma^2 = \frac{\sum x^2 - (\sum x)^2/N}{N} = \frac{20{,}997.91 - (357.3)^2/8}{8} = 629.9998438 \approx 630.0$$

The population standard deviation is therefore

$$\sigma = \sqrt{\sigma^2} = \sqrt{629.9998438} \approx 25.1$$

The standard deviation of farmland for all counties in Connecticut is about 25,100 acres. ■

Table 3.13 Calculating $\sum x$ and $\sum x^2$

x	x^2
12.8	163.84
50.2	2520.04
93.6	8760.96
17.9	320.41
26.0	676.00
58.9	3469.21
36.8	1354.24
61.1	3733.21
$\sum x = 357.3$	$\sum x^2 = 20{,}997.91$

Example 3.15   Calculating the sample variance and sample standard deviation using the computational formulas

Note: Do you see how this sample is not a random sample?

Suppose we take a sample of the three counties from Table 3.12 that have the most farmland. Calculate the sample variance and sample standard deviation using the computational formulas.

Solution

Table 3.14 shows the calculations required to get $\sum x = 213.6$ and $\sum x^2 = 15{,}963.38$.

Table 3.14 Calculating $\sum x$ and $\sum x^2$

x	x^2
93.6	8760.96
61.1	3733.21
58.9	3469.21
$\sum x = 213.6$	$\sum x^2 = 15{,}963.38$

Then, substituting these values into the computational formula for the sample variance gives us

$$s^2 = \frac{\sum x^2 - (\sum x)^2/n}{n-1} = \frac{15{,}963.38 - (213.6)^2/3}{3-1} \approx 377.53$$

The sample standard deviation is therefore

$$s = \sqrt{s^2} = \sqrt{377.53} \approx 19.4$$

The standard deviation of farmland for this sample of three counties in Connecticut is almost 19,400 acres. ■

Example 3.16 Using technology to find the variance and standard deviation

Find the sample standard deviation and sample variance of the city gas mileage for the six 2004 cars shown in the table below. Use (a) the TI-83/84, (b) Excel, and (c) Minitab.

Vehicle	City mpg
Honda Civic	36
Toyota Camry	24
Ford Taurus	20
Pontiac Grand Prix	20
Jaguar X-Type	18
Lincoln Town Car	17

Solution

Using the instructions in the Step-by-Step Technology Guide on page 89, we get the following output. The TI-83/84 does not report the sample variance. Instead, use the following: variance = (standard deviation)^2.

a. The TI-83/84 output is shown below. The sample standard deviation s is given as $Sx = 7.03562364$.

The sample variance is

$$s^2 = (7.03562364)^2 \approx 49.5$$

```
1-Var Stats
x̄=22.5
Σx=135
Σx²=3285
Sx=7.03562364
σx=6.422616289
↓n=6
■
```

CAUTION Do not confuse Sx with σx, which the TI-83/84 uses to label the population standard deviation.

b. The Excel output is shown below. The sample standard deviation and sample variance are highlighted.

City MPG	
Mean	22.5
Standard Error	2.872281
Median	20
Mode	20
Standard Deviation	7.035624
Sample Variance	49.5
Kurtosis	3.641384
Skewness	1.873582
Range	19
Minimum	17
Maximum	36
Sum	135
Count	6

c. The Minitab output below gives us the sample standard deviation "StDev," highlighted here. (Minitab rounds s to 7.04.) We then find the sample variance using $s^2 = 7.04^2 = 49.5616$. (Because of the rounding of 7.04, this value of s is less accurate than the value of 49.5 found in **(b)**.) ■

Variable	N	N*	Mean	SE Mean	StDev	Minimum	Q1	Median	Q3	Maximum
City MPG	6	0	22.50	2.87	7.04	17.00	17.75	20.00	27.00	36.00

CASE STUDY | ### Can the Financial Experts Beat the Darts?

Recall from Section 3.1 the *Wall Street Journal* competition between stocks chosen randomly by *Journal* staff members throwing darts and stocks chosen by a team of four financial experts. Note from Figure 3.13 that the DJIA exhibits less variability than the other two portfolios. This smaller variability is due to the fact that the DJIA is made up of 29 component stocks, whereas each portfolio is made up of only 4 stocks.

FIGURE 3.13
Comparative dotplots of the net change in prices.

Smaller sample sizes can be associated with increased variability, since an unusual result in one value has a relatively strong effect on the mean when it is not offset by a large sample. We will learn more about this in Chapter 7, "Sampling Distributions."

Which of the portfolios, pros or darts, shows greater variability? It is difficult to determine, just by examining Figure 3.13, which has the greater range or standard deviation. We therefore turn to the Minitab descriptive statistics in Figure 3.14. The Max for the pros was a 75% increase, while the Min was a 37.8% *decrease*. Thus, the range for the pros is $75 - (-37.8) = 112.8$. The Max for the darts was a 72.9% increase, while the Min was a 43.0% decrease. So the range for the darts is $72.9 - (-43.0) = 115.9$. This means that the range for the darts is greater than that for the pros.

From Figure 3.14, the standard deviations are 22.25, 19.39, and 8.031 for the pros, the darts, and the DJIA, respectively. Thus, the standard deviation for the pros is greater than for the darts, while the range for the darts is greater. *Measures of spread may disagree about which data set is more variable.* However, as we stated earlier, as a measure of spread, the range cannot give us the whole picture, since it takes into account only the two most extreme data values. Our conclusion, therefore, is that the returns for the professionals exhibit the greater variability.

FIGURE 3.14
Descriptive statistics
for the portfolios.

Variable	N	Mean	Median	TrMean	StDev	SE Mean
pros	100	10.95	9.60	10.19	22.25	2.22
Darts	100	4.52	3.25	3.73	19.39	1.94
DJIA	100	6.793	7.000	7.026	8.031	0.803

Variable	Minimum	Maximum	01	03
pros	-37.80	75.00	-6.23	26.63
Darts	-43.00	72.90	-6.53	14.88
DJIA	-13.100	22.500	1.525	13.250

FIGURE 3.14
Descriptive statistics
for the portfolios.

Why all the effort to find which portfolio has the most variation? Because in finance, as in most other fields, high variability is not necessarily advantageous. Would you want to put your money into a fund that had high variability in its returns? Perhaps not, since high variability in finance is associated with greater *risk*. The professionals evidently chose higher-risk stocks with greater potential for high returns—but also greater potential for losing money.

Section 3.2 SUMMARY

1 The simplest measure of variability, or measure of spread, is the range. The range is simply the difference between the maximum and minimum values in a data set, but the range has drawbacks because it relies on the two most extreme data values.

2 A deviation is the difference between a data value and the mean of the data values.

3 The variance and standard deviation are measures of spread that utilize all available data values. The population variance can be thought of as the mean squared deviation. The standard deviation is the square root of the variance. We interpret the value of the standard deviation as the typical deviation, that is, the typical difference between a data value and the mean.

Section 3.2 EXERCISES

Unless a data set is identified as a population, you can assume that it is a sample.

CLARIFYING THE CONCEPTS

1. Explain what a deviation is.
2. What is the interpretation of the value of the standard deviation?
3. Which is better, less variability or more variability? Why?

PRACTICING THE TECHNIQUES

Use the following data set for Exercises 4–6.

    10   25   0   15   10

4. Find the range.
5. Find the variance.
6. Find the standard deviation.

Use the following data set for Exercises 7–9.

    −5   −10   −15   −20

7. Find the range.
8. Find the variance.
9. Find the standard deviation.

For the following exercises, make sure to state your answers in the proper units, such as "years" or "years squared."

Refer to Table 3.4 (page 90) for Exercises 10 and 11.

10. Find the range for *total*.
11. Find the variance and standard deviation of *total*. Which measure do you find to be more easily understood and interpreted?

Use the following information for Exercises 12 and 13.

Vehicle	City mpg
Honda Civic	36
Toyota Camry	24
Ford Taurus	20
Pontiac Grand Prix	20
Jaguar X-Type	18
Lincoln Town Car	17

12. Find the variance and standard deviation of the city mpg for the six automobiles.

13. Find the range for the city mpg for the six automobiles.

Use the following information for Exercises 14 and 15.

State	Median age
Utah	28.5
Texas	33.2
Alaska	33.9
Georgia	34.3
California	34.4
Arizona	34.5
Idaho	34.6
Colorado	34.7

14. Find the variance and standard deviation of median age. Which measure do you find to be more easily understood and interpreted?

15. Find the range for median age.

APPLYING THE CONCEPTS

NASCAR Winners. Refer to Table 3.4 (page 90) for Exercises 16–19.

16. Construct separate dotplots of the short tracks winners and the road courses winners, using the same scale. Which data set would you say has the greater variability?

17. Construct separate dotplots of the super speedways winners and the road courses winners, using the same scale. Which data set would you say has the greater variability?

18. Now find the range of each of the two data sets from Exercise 17. Which data set has the greater variability? Do these results agree with your judgment above?

19. Now find the variance and standard deviation of each of the two data sets from Exercise 17. Which data set has the greater variability? Do these results agree with your judgment above?

Zooplankton and Phytoplankton. Refer to the table below for Exercises 20 and 21. *Meta-analysis* refers to the statistical analysis of a set of similar research studies. In a meta-analysis, each data value represents an effect size calculated from the results of a particular study. The table contains effect sizes calculated in a meta-analysis for zooplankton and phytoplankton.[5]

Zooplankton		Phytoplankton	
−2.37	−3.00	10.61	3.04
−0.64	−0.68	2.97	0.65
−2.05	−1.39	1.58	2.55
−1.54	−0.64	2.55	1.05
−6.60	−3.88	5.67	2.11
0.26		1.57	

20. Calculate the ranges for the zooplankton and the phytoplankton.

  a. Which has the greater range?

  b. Which plankton group has the greater variability according to the range?

21. Calculate the standard deviations for the zooplankton and the phytoplankton.

  a. Which has the greater standard deviation?

  b. Which plankton group has the greater variability according to the standard deviation? Does this concur with your answer from the previous exercise?

  c. Without calculating the variances, say which group has the greater variance. How do you know this?

Ant Size. Use the following information for Exercises 22 and 23. A study compared the size of ants from different colonies. The masses (in milligrams) of samples of ants from two different colonies are shown in the accompanying table.[6]

Colony A		Colony B	
109	134	148	115
120	94	110	101
94	113	110	158
61	111	97	67
72	106	136	114

22. Calculate the range for each ant colony.

  a. Which has the greater range?

  b. Which colony has the greater variability according to the range?

23. Calculate the variance for each colony.

  a. Which has the greater variance?

  b. Which colony has the greater variability according to the variance? Does this concur with your answer from the previous exercise?

  c. Without calculating the standard deviations, say which colony has the greater standard deviation. How do you know this?

SAT Scores. Refer to Table 3.5 (page 90) for Exercises 24–26.

24. Construct dotplots of the SAT I Verbal and the SAT I Math scores. Would you say that the two data sets have similar variability or that one data set varies much more than the other?

25. Find the range, variance, and standard deviation for both the SAT I Verbal and the SAT I Math scores. Do your findings agree with your judgment from the previous exercise? Explain.

26. Now suppose we omit New Hampshire from the calculations.

  a. *Without recalculating them,* describe how this would affect the values of the measures of spread you found for the SAT I Verbal scores above. How about the SAT I Math scores?

  b. Now recalculate the three measures of spread for the SAT I Verbal scores with the New Hampshire data removed. Was your judgment in **(a)** supported?

c. Recalculate the three measures of spread for the SAT I Math scores with the New Hampshire data removed. Was your judgment in **(a)** supported?

Volleyball Team Heights. Refer to Table 3.8 (page 93) for Exercises 27–29.

27. Consider the deviations for each team.
   a. Would you say that a typical distance from the mean for the WMU team is about 70 inches, about 6 inches, or about 3 inches?
   b. What would you say a typical distance from the mean is for the NCU team?

28. Consider the standard deviations.
   a. Find the standard deviation for each team.
   b. Interpret the meaning of the values you obtained for the standard deviations so that someone who has never studied statistics would understand it.
   c. Which team has the greater variability according to the standard deviation?

29. Suppose that the shortest player on the NCU team was not 66 inches but only 62 inches tall.
   a. *Without recalculating them,* what effect would this change have on the three measures of spread for the NCU data?
   b. Now recalculate the three measures of spread for the NCU data, with the shortest player now 62 inches instead of 66 inches tall. Was your judgment in **(a)** supported?

30. Pulse Rates for Males and Females. The comparative dotplots in Figure 3.15 show pulse rates for a sample of males and females.

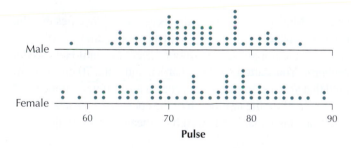

Figure 3.15 Comparative dotplots of pulse rates, by gender.

   a. What is your impression of which gender has the greater amount of variability in pulse rates?
   b. Estimate the maximum pulse rate, minimum pulse rate, and range of pulse rates for males.
   c. Estimate the maximum pulse rate, minimum pulse rate, and range of pulse rates for females. According to the range, which gender has more variability in pulse rates.
   d. Would the typical distance from the mean for females be about 78, about 28, about 18, about 8, or about 0.8?

e. What is your estimate of the typical distance from the mean for females? For males? Is your estimate larger for females or males, and why?

31. Figure 3.16 gives descriptive statistics for the pulse rates in Figure 3.15.

Variable	Gender	N	Mean	Median	TrMean	StDev
Pulse	female	65	74.15	76.00	74.27	8.11
	male	65	73.369	73.000	73.407	5.875

Variable	Gender	SE Mean	Minimum	Maximum	Q1	Q3
Pulse	female	1.01	57.00	89.00	68.00	80.00
	male	0.729	58.000	86.000	70.000	78.000

Figure 3.16 Descriptive statistics for the pulse rates, by gender.

   a. How close (in terms of beats per minute) were your earlier estimates of the maximum, minimum, and range of the pulse rates for each gender?
   b. What is the standard deviation of the pulse rates for the females?
   c. How close (in terms of beats per minute) is this value of the standard deviation to your earlier estimate of the typical distance from the mean for the females?
   d. What is the standard deviation of the pulse rates for the males?
   e. How close (in terms of beats per minute) is this value of the standard deviation to your earlier estimate of the typical distance from the mean for the males?
   f. For which gender is the standard deviation greater? Does this agree with your earlier surmise as to the relative sizes of the standard deviations?
   g. Overall, which gender has greater variability in pulse rates?

CONSTRUCT YOUR OWN DATA SETS

32. Construct two data sets, A and B, that you make up on your own, so that the range of A is greater than the range of B. Verify this.

33. Construct two data sets, A and B, that you make up on your own, so that the standard deviation of A is greater than the range of B. Verify this.

34. Construct two data sets, A and B, that you make up on your own, so that the mean of A is greater than the mean of B, but the standard deviation of B is greater than that of A. Verify this.

35. Construct two data sets, A and B, that you make up on your own, so that the mean of A is greater than the mean of B, and the standard deviation of A is greater than that of B. Verify this.

36. Construct two data sets, A and B, that you make up on your own, so that the range of A is greater than the range of B, but the standard deviation of B is greater than that of A. Verify this.

3.3 Working with Grouped Data

Objectives: **By the end of this section, I will be able to...**

1. Calculate the weighted mean.

2. Estimate the mean for grouped data.

3. Estimate the variance and standard deviation for grouped data.

1 The Weighted Mean

Note: Before tackling this section, you may wish to review Section 2.2, "Graphs and Tables for Quantitative Data" (page 44).

Sometimes not all the data values are of equal importance. Rather, certain data values are assigned greater weight than other data values when calculating the mean. For example, have you ever figured out what your final grade for a course was based on the percentages listed in the syllabus? What you actually found was the **weighted mean** of your grades.

> **Weighted Mean**
>
> To find the **weighted mean:**
>
> 1. Multiply each data point x_i by its respective weight w_i.
> 2. Sum these products.
> 3. Divide the result by the sum of the weights:
>
> $$\bar{x}_w = \frac{\sum w_i x_i}{\sum w_i} = \frac{w_1 x_1 + w_2 x_2 + \cdots + w_n x_n}{w_1 + w_2 + \cdots + w_n}$$

Example 3.17   Weighted mean of course grades

The syllabus for the Introduction to Management course at a local college specifies that the midterm exam is worth 30%, the term paper is worth 20%, and the final exam is worth 50% of your course grade. Now, say you did not get serious about the course until Halloween, so that you got a 20 on the midterm. You started working harder, and got a 70 on the term paper. Finally, you remembered that you had to pay for the course again if you flunked and had to retake it, and so you worked hard for the last month of the course and got a 90 on the final exam. Calculate your course average, that is, the weighted mean of your grades.

Solution

The data values are $x_1 = 20$, $x_2 = 70$, and $x_3 = 90$. The weights are $w_1 = 0.30$,

Note: The weights w_i do not have to be percentages that add up to 1.

$w_2 = 0.20$, and $w_3 = 0.50$. Your course weighted mean is then calculated as follows:

$$\bar{x}_w = \frac{\sum w_i x_i}{\sum w_i} = \frac{w_1 x_1 + w_2 x_2 + w_3 x_3}{w_1 + w_2 + w_3} = \frac{(0.30)(20) + (0.20)(70) + (0.50)(90)}{0.30 + 0.20 + 0.50} = \frac{65}{1.0} = 65$$

Despite your failing the midterm miserably, you got serious and barely managed to pass the course with a 65 average, which is a D. ■

2 Estimating the Mean for Grouped Data

Thus far in Chapter 3, we have computed measures of center and spread from a raw data set. However, data are often reported using frequency distributions. Without the original data, we cannot calculate the exact values of the measures of center and spread. The remainder of this section examines methods for approximating the mean, variance, and standard deviation of *grouped data*—that is, population data summarized using frequency distributions.

For each class in the frequency distribution, we estimate the class mean using the class *midpoint*. The class midpoint is defined as the mean of two adjoining lower class limits. The midpoint for the *i*th class is denoted m_i.

Example 3.18 Finding the class midpoints

There were 1150 children adopted in the state of Georgia in 2006, according to the Administration for Children and Families.[7] The frequency distribution of the ages of the children at adoption is shown in Table 3.15. Find the class midpoints.

Table 3.15 Frequency distribution of children adopted in Georgia, by age

Class: age	Midpoint m_i	Frequency f_i
0–0.99	0.5	12
1–5.99	3.5	611
6–10.99	8.5	320
11–15.99	13.5	161
16–17.99	17.0	46

Solution

The midpoint for the first class (ages 0–0.99) is the mean of the lower class limits for this class (0) and the adjoining class (1). That is, the midpoint is $(0 + 1)/2 = 0.5$.

Similarly, the midpoint for the second class (ages 1–5.99) is $(1 + 6)/2 = 3.5$. The remainder of the class midpoints are shown in Table 3.15. ■

The product of the class frequency f_i and class midpoint m_i is used as an estimate of the sum of the data values within that class. Summing these products across all classes and dividing by the total population size thus provides us with an **estimated mean for data grouped into a frequency distribution.**

Note: Notice the difference between $\hat{\mu}$ and $\bar{x}$. The estimated mean $\hat{\mu}$ is used only when the data are grouped into a frequency distribution. The sample mean $\bar{x}$ is used when all the data values in the sample are known. Both $\hat{\mu}$ and $\bar{x}$ are used to estimate the population mean μ.

Estimated Mean for Data Grouped into a Frequency Distribution

Given a frequency distribution with k classes, the **estimated mean** for the variable is given by

$$\hat{\mu} = \frac{\sum m_i f_i}{N} = \frac{m_1 f_1 + m_2 f_2 + \cdots + m_k f_k}{N}$$

where m_i and f_i represent the midpoint and frequency of the ith class, respectively, and N represents the total population size, which is the sum of the class frequencies: $N = \Sigma f_i$. The "hat" notation for $\hat{\mu}$, called "mu-hat," indicates that $\hat{\mu}$ is an estimate and not the exact value of the mean.

Example 3.19 Calculating the estimated mean for grouped data

Calculate the estimated mean age of the adopted children in Table 3.15.

Solution

The midpoints m_i and frequencies f_i are provided in Table 3.15. For example, $m_1 = 0.5$ and $f_1 = 12$. We thus calculate the sum of the products as follows:

$$\sum m_i f_i = (0.5)(12) + (3.5)(611) + (8.5)(320) + (13.5)(161) + (17)(46)$$
$$= 6 + 2138.5 + 2720 + 2173.5 + 782 = 7820$$

Next we calculate the total population size:

$$N = \sum f_i = 12 + 611 + 320 + 161 + 46 = 1150$$

The estimated mean is therefore

$$\hat{\mu} = \frac{\sum m_i f_i}{N} = \frac{7820}{1150} = 6.8$$

The estimated mean age of the children adopted in Georgia in 2006 is 6.8 years. ■

3 Estimating the Variance and Standard Deviation for Grouped Data

We also use class midpoints and class frequencies to calculate the **estimated variance for data grouped into a frequency distribution** and the **estimated standard deviation for data grouped into a frequency distribution.**

> **Estimated Variance and Standard Deviation for Data Grouped into a Frequency Distribution**
>
> Given a frequency distribution with k classes, the estimated variance for the variable is given by
>
> $$\hat{\sigma}^2 = \frac{\sum (m_i - \hat{\mu})^2 \cdot f_i}{N}$$
>
> and the estimated standard deviation is given by
>
> $$\hat{\sigma} = \sqrt{\hat{\sigma}^2} = \sqrt{\frac{\sum (m_i - \hat{\mu})^2 \cdot f_i}{N}}$$
>
> where m_i and f_i represent the midpoint and frequency of the ith class, respectively, and N represents the total population size, which is the sum of the class frequencies: $N = \sum f_i$. The "hat" notation for $\hat{\sigma}$, called "sigma-hat," indicates that $\hat{\sigma}$ is an estimate and not the exact value of the standard deviation.

Note: Notice the difference between $\hat{\sigma}$ and s. The estimated standard deviation $\hat{\sigma}$ is used only when the data are grouped into a frequency distribution. The sample standard deviation s is used when all the data values in the sample are known. Both $\hat{\sigma}$ and s are used to estimate the population standard deviation σ.

You should carry as many decimal places as you can for the value of $\hat{\mu}$ when calculating $\hat{\sigma}^2$, and for $\hat{\sigma}^2$ when calculating $\hat{\sigma}$.

Example 3.20   Calculating the estimated variance and standard deviation for grouped data

Calculate the estimated variance and standard deviation of the ages of the adopted children in Table 3.15.

Solution
Table 3.16 contains the calculations required for finding $\sum (m_i - \hat{\mu})^2 \cdot f_i = 20{,}068$. The variance is therefore estimated as

$$\hat{\sigma}^2 = \frac{\sum (m_i - \hat{\mu})^2 \cdot f_i}{N} = \frac{20{,}068}{1150}$$

and the standard deviation is estimated as

$$\hat{\sigma} = \sqrt{\hat{\sigma}^2} = \sqrt{\frac{20{,}068}{1150}} \approx 4.177371755 \approx 4.2$$

Table 3.16 Calculating $\sum(m_i - \widehat{\mu})^2 \cdot f_i$

Class: age	Midpoint m_i	Frequency f_i	$\widehat{\mu}$	$m_i - \widehat{\mu}$	$(m_i - \widehat{\mu})^2 \cdot f_i$
0–0.99	0.5	12	6.8	−6.3	476.28
1–5.99	3.5	611	6.8	−3.3	6653.79
6–10.99	8.5	320	6.8	1.7	924.8
11–15.99	13.5	161	6.8	6.7	7227.29
16–17.99	17.0	46	6.8	10.2	4785.84

$$\sum(m_i - \widehat{\mu})^2 \cdot f_i = 20{,}068$$

In other words, the age of the adopted children typically differs from the mean age of 6.8 years by about 4.2 years. ■

Example 3.21 **Using technology to find the estimated mean, variance, and standard deviation for grouped data**

Use the TI-83/84 calculator to find the estimated mean, variance, and standard deviation for the frequency distribution in Table 3.16.

Solution
Following the instructions in the Step-by-Step Technology Guide, we get the estimated mean $\widehat{\mu}$ (shown in the output as $\bar{x}$) = 6.8 and the estimated standard deviation $\widehat{\sigma}$ (shown in the output as σx) = 4.177371755. ■

```
1-Var Stats
x̄=6.8
Σx=7820
Σx²=73244
Sx=4.179189189
σx=4.177371755
↓n=1150
```

STEP-BY-STEP TECHNOLOGY GUIDE: Estimating the Mean, Variance, and Standard Deviation for Grouped Data

TI-83/84

Step 1 Press **STAT** and select **1:Edit**. Enter the class midpoints in **L1** and the frequencies or relative frequencies in **L2**.
Step 2 Press **STAT**, select the **CALC** menu, and choose **1: 1-Var Stats**.

Step 3 Press **2nd 1 Comma 2nd 2**, so that the following appears on the home screen: **1-Var Stats L1, L2**.
Step 4 Press **ENTER**.

Section 3.3 SUMMARY

1 The weighted mean is the sum of the products of the data points with their respective weights, divided by the sum of the weights.

2 Since we do not have access to the original raw data, it is not possible to find exact values for the mean, variance, and standard deviation of data that have been grouped into a

frequency distribution. The estimated mean $\widehat{\mu}$ in this case is the sum of the products of the class frequencies f_i and class midpoints m_i, divided by the total sample size N.

3 Class midpoints and class frequencies are also used to find the estimated variance $\widehat{\sigma}^2$ and estimated standard deviation $\widehat{\sigma}$ of grouped data.

Section 3.3 EXERCISES

CLARIFYING THE CONCEPTS

1. Explain why the formulas for the mean, variance, and standard deviation of grouped data will provide estimates only and not the exact values for these measures.
2. Explain what the "hat" notation indicates for $\hat{\mu}$ and $\hat{\sigma}$.
3. What quantity do both $\hat{\mu}$ and $\bar{x}$ estimate?

PRACTICING THE TECHNIQUES

For Exercises 4–8, the data values and weights are provided. Find the weighted mean.

4. $x_1 = 50, x_2 = 60; x_3 = 70; w_1 = 0.25, w_2 = 0.50, w_3 = 0.25$.
5. $x_1 = 50, x_2 = 80, x_3 = 70; w_1 = 0.25, w_2 = 0.40, w_3 = 0.35$.
6. $x_1 = 100, x_2 = 120, x_3 = 150; w_1 = 10, w_2 = 20, w_3 = 5$.
7. $x_1 = 3.0, x_2 = 2.5, x_3 = 3.5, x_4 = 4.0, x_5 = 3.0; w_1 = w_2 = w_3 = w_4 = 3, w_5 = 4$.
8. $x_1 = 70, x_2 = 80, x_3 = 85, x_4 = 95; w_1 = 0.20, w_2 = 0.30, w_3 = 0.25, w_4 = 0.25$.

For Exercises 9 and 10, the class limits are provided. Find the class midpoints.

9.

0–1.99	6–7.99
2–3.99	8–9.99
4–5.99	

10.

0–4.99	20–29.99
5–9.99	30–49.99
10–14.99	50–99.99
15–19.99	100–199.99

For Exercises 11 and 12, find the estimated mean for the frequency distribution.

11.

Midpoint m_i	Frequency f_i
5	10
10	20
15	20
20	10
25	10

12.

Midpoint m_i	Frequency f_i
−10	3
−5	2
0	5
5	12
10	8
15	10

For Exercises 13 and 14, find the estimated variance and standard deviation

13. For the frequency distribution in Exercise 11.
14. For the frequency distribution in Exercise 12.

APPLYING THE CONCEPTS

15. **Dupage County Age Groups.** The Census Bureau reports the following 2006 frequency distribution of population by age group for Dupage County, Illinois, residents less than 65 years old.

Age	Residents
0–4.99	63,422
5–17.99	240,629
18–64.99	540,949

a. Find the class midpoints.
b. Find the estimated mean age of residents of Dupage County.
c. Find the estimated variance and standard deviation of ages.

16. **Capacity of Lakes and Reservoirs.** The National Water and Climate Center tracks the percentage capacity of lakes and reservoirs (how full the lake or reservoir is). The table below contains the percentage capacity for a random sample of lakes and reservoirs in Arizona, California, and Colorado.[8] Find the weighted mean percentage capacity.

Arizona	California	Colorado
$n = 4$	$n = 6$	$n = 5$
$\bar{x} = 48.5$	$\bar{x} = 81$	$\bar{x} = 47$

17. **Broward County House Values.** Table 3.17 gives the frequency distribution of the dollar value of the owner-occupied housing units in Broward County, Florida.
a. Find the class midpoints.
b. Find the estimated mean dollar value for housing units in Broward County.
c. Find the estimated variance and standard deviation of the dollar value.

TABLE 3.17   Broward County house values

Dollar value	Housing units
0–$49,999	5,430
50,000–$99,999	90,605
100,000–$149,999	90,620
150,000–$199,999	54,295
200,000–$299,999	34,835
300,000–$499,999	15,770
500,000–$999,999	5,595

Source: Census Bureau.

18. Lightning Deaths. Table 3.18 gives the frequency distribution of the number of deaths due to lightning nationwide over a 67-year period. Find the estimated mean and standard deviation of the number of lightning deaths per year.

TABLE 3.18   Lightning deaths

Deaths	Years
20–59.99	13
60–99.99	21
100–139.99	10
140–179.99	6
180–259.99	10
260–459.99	7

Source: National Oceanic and Atmospheric Administration.

19. Calculating a Course Grade. An introductory statistics syllabus has the following grading system. The weekly quizzes are worth a total of 25% toward the final course grade. The midterm exam is worth 32%; the final exam is worth 33%; and attendance/participation is worth 10% toward the final course grade. Anthony's weekly quiz average is 70. He got an 80 on the midterm and a 90 on the final exam. He got 100 for attendance/participation. Calculate Anthony's final course grade.

20. Wages for Computer Managers. The U.S. Bureau of Labor Statistics (BLS) publishes wage information for various occupations. For the occupation "computer and information systems management," Table 3.19 gives the wages reported by the BLS for the top-paying states in May 2006. Find the weighted mean wage across all five states, using the employment figures as weights.

TABLE 3.19   Wages for computer managers

State	Employment	Hourly mean wage
New Jersey	12,380	$60.32
New York	18,580	$60.25
Virginia	9,540	$59.39
California	35,550	$57.98
Massachusetts	10,130	$55.95

21. Salaries of Scientists and Engineers. The National Science Foundation compiles statistics on the annual salaries of full-time employed doctoral scientists and engineers in universities and four-year colleges. The mean 2004 annual salary for the fields of science, engineering, and health are $67,000, $82,200, and $70,000, respectively. Suppose we have a sample of 10 professors, 5 of whom are in science, 2 in engineering, and 3 in health, and each of whom is making the mean salary for his or her field. Find the weighted mean salary of these 10 professors.

22. Challenger Exercise. Assign the weights w_i to show that the formula for the sample mean from Section 3.1 $\bar{x} = \sum x_i / n$ is a special case of the formula for the weighted mean $\bar{x}_w = \sum w_i x_i / \sum w_i$.

3.4 Measures of Position

Objectives: By the end of this section, I will be able to…

1 Find percentiles for both small and large data sets.

2 Calculate z-scores, and explain why we use them.

3 Use z-scores to detect outliers.

1 Percentiles

In this section we learn about *measures of position,* which tell us the position that a particular data value has relative to the rest of the data set. For example, a prestigious nursing school may grant admission to only the top 10% of applicants. How high a score would you need to enter? This is one type of question we will answer in this section. The first measure of position we consider is the percentile. A **percentile** gives the percentage of data values that are less than or equal to a certain value.

> **Percentile** Let *p* be any integer between 0 and 100. The *p*th **percentile** of a data set is the data value at which *p* percent of the values in the data set are less than or equal to this value.

Example 3.22   Jasmine's Math SAT percentile

Many students take the Scholastic Aptitude Test (SAT) when applying to college. The test results that you receive include your score and the percentile that your score represents. Suppose Jasmine's Math SAT score was 650, which represented the 90th percentile. What does "90th percentile" mean?

Solution

The 90th percentile simply means that 90% of all scores on the Math SAT fell at or below Jasmine's score of 650. We call the percentile a *measure of position* since it indicates the position of Jasmine's Math SAT score relative to all other Math SAT scores. Clearly, Jasmine is good at math. Figure 3.17 indicates the position of Jasmine's score relative to the rest of the test takers. ■

90% of SAT scores are less than or equal to 650.

Only 10% of SAT scores are greater than 650.

SAT scores   Jasmine's SAT score of 650 represents the 90th percentile.

FIGURE 3.17 The 90th percentile is the score with 90% of the data values at or below its value.

For large data sets, calculation of the percentiles is best left to computers. However, for small data sets, we can use the following step-by-step method to calculate the position of any percentile.

Step 1 Sort the data into ascending order (from smallest to largest).

Step 2 Calculate

Note: These steps do not give the value of the *p*th percentile itself, but rather the *position* of the *p*th percentile in the data set when the data set is in ascending order.

$$i = \left(\frac{p}{100}\right)n$$

where *p* is the particular percentile you wish to calculate, and *n* is the sample size.

Step 3 **a.** If *i* is an integer (a whole number with no decimal part), the *p*th percentile is the mean of the data values in positions *i* and *i* + 1.   **b.** If *i* is not an integer, round up and use the value in this position.

Example 3.23   Finding percentiles of a small data set

Yolanda would like to go to a prestigious graduate school of the arts. She knows that this school accepts only those students who score at the 75th percentile or higher in a grueling dance audition. The following data represent the dance audition scores of Yolanda's group. Yolanda scored 85. Find the 75th percentile of the data set. Will Yolanda be accepted at the prestigious graduate school of the arts?

78   56   89   44   65   94   81   62   75   85   30   68

Solution

Step 1 Sort the data into ascending order:

30   44   56   62   65   68   75   78   81   85   89   94

Step 2 The particular percentile we wish to calculate is the 75th percentile, so $p = 75$. There are 12 scores in our data set, so $n = 12$. Calculate

$$i = \left(\frac{p}{100}\right)n = \left(\frac{75}{100}\right)12 = 9$$

So, $i = 9$.

Step 3 Here, since i is an integer, the 75th percentile is the mean of the data values in positions 9 and 10.

Position	1	2	3	4	5	6	7	8	9	10	11	12
Score	30	44	56	62	65	68	75	78	81	85	89	94

(81 + 85)/2 = 83

Counting from left to right, the data value in the ninth position is 81, and the data value in the tenth position is 85. The mean of these two values is 83. Thus, the 75th percentile is 83. Yolanda's dance score of 85 is therefore above the 75th percentile. She will be accepted to the prestigious graduate school. ■

Example 3.24   Finding percentiles of a large data set

How old are the oldest Major League Baseball players? The data set **Baseball2007** is found on the CD. It contains data on the 516 American League baseball players in the 2007 season, including the age of each player. Find the 95th percentile of the ages of the American League baseball players in 2007.

Solution

Minitab provides a frequency distribution for the age variable, along with the cumulative counts and percentages (Figure 3.18). The sample size n (N in Minitab) is 516.

Age	Count	CumCnt	CumPct
20	1	1	0.19
21	8	9	1.74
22	16	25	4.84
23	27	52	10.08
24	34	86	16.67
25	54	140	27.13
26	51	191	37.02
27	47	238	46.12
28	33	271	52.52
29	40	311	60.27
30	44	355	68.80
31	38	393	76.16
32	24	417	80.81
33	25	442	85.66
34	9	451	87.40
35	21	472	91.47
36	16	488	94.57
37	10	498	96.51
38	5	503	97.48
39	4	507	98.26
40	5	512	99.22
41	1	513	99.42
42	2	515	99.81
44	①	516	100.00
N=	516		

The oldest player: Roger Clemens

FIGURE 3.18 Finding the 95th percentile of player ages.

The particular percentile we wish to calculate is the 95th percentile, so we use $p = 95$ in the percentile formula:

$$i = \left(\frac{p}{100}\right)n = \left(\frac{95}{100}\right)516 = 490.2$$

Since $i = 490.2$ is not an integer, we round up to the next integer. The 95th percentile is in the 491st position, when the data are in ascending order. Now, in the "CumCnt" (cumulative count) column in Figure 3.18, note that there are 488 players who are 36 years old or less, and there are 498 players who are 37 years old or less. We are looking for the 491st. Since there are 10 players that are 37 years old, the 491st must be one of these 10 guys who are 37 years old. Hence, the 95th percentile age of American League baseball players in 2007 is 37 years old. The oldest player is Roger Clemens of the New York Yankees, at 44 years old. ∎

✎ z-Scores

Recall that the standard deviation is a common measure of the variability, or spread, of a data set. The value of the standard deviation is interpreted as a typical deviation from the mean. The SAT is designed so that the distribution of scores is bell-shaped with a mean of 500 and a standard deviation of 100. Note in Figure 3.19 that we can measure the distance from a particular SAT score to the mean in terms of standard deviations. For example, an SAT score of 600 lies one standard deviation above the mean, while an SAT score of 300 lies two standard deviations below the mean.

The term **z-score** indicates how many standard deviations a particular data value is from the mean. If the z-score is positive, then the data value is above the mean. If the z-score is negative, then the data value is below the mean.

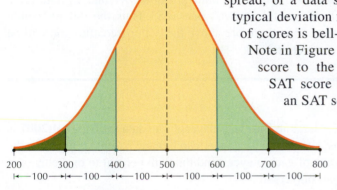

FIGURE 3.19 The distribution of SAT scores.

---

z-Score

The **z-score** for a particular data value from a *sample* is

$$z\text{-score} = \frac{\text{data value} - \text{mean}}{\text{standard deviation}} = \frac{x - \bar{x}}{s}$$

where $\bar{x}$ is the sample mean, and s is the sample standard deviation.

The z-score for a particular data value from a *population* is

$$z\text{-score} = \frac{\text{data value} - \text{mean}}{\text{standard deviation}} = \frac{x - \mu}{\sigma}$$

where μ is the population mean, and σ is the population standard deviation.

In this section, we will use the sample z-score unless otherwise indicated.

Example 3.25 Jasmine's Math SAT z-score

Suppose the mean score on the Math SAT is $\mu = 500$, with a standard deviation of $\sigma = 100$ points. Note that here we have population values. How many standard deviations is Jasmine's score from the mean?

Solution

Here $\mu = 500$, $\sigma = 100$, and Jasmine's score is $x = 650$. Her z-score is

$$z\text{-score} = \frac{\text{data value} - \text{mean}}{\text{standard deviation}} = \frac{x - \mu}{\sigma} = \frac{650 - 500}{100} = 1.5$$

Jasmine's z-score of 1.5 indicates that her Math SAT is 1.5 standard deviations from the mean of 500. Z-scores can be positive or negative. Jasmine's z-score is positive (1.5), which means that her Math SAT score falls above the mean. Bright lady! Consider Figure 3.20, which shows the distribution of SAT scores, with a mean of 500 and a standard deviation of 100. The black arrows represent "units" of one standard deviation each, that is, each arrow is 100 SAT points long. Counting the arrows as you go above or below the mean is thus the same as counting the number of standard deviations above or below the mean. Jasmine's SAT score lies between 600 and 700, an area with z-scores ranging from 1 to 2. ■

FIGURE 3.20 Jasmine's z-score of 1.5 places her 1.5 standard deviations above the mean.

In Example 3.25, since the standard deviation equals 100, the z-score represents units of 100. That is, a z-score of 1 represents 1 standard deviation above the mean, which is 100 points above the mean. Thus, the *scale* of the z-scores for the SAT scores in Figure 3.20 is in units of 100, since the standard deviation equals 100. However, if the standard deviation was, say, $\sigma = 15$, then the scale would be different.

Example 3.26 Lead levels in blood

Note: This use of "μ" for "micro" in the measure "micrograms per deciliter" is not related to our use of μ as the population mean.

A study of workers who were exposed to lead at their jobs found that their mean blood lead level was 31.4 μg/dl (micrograms per deciliter) with a standard deviation of 14.2 μg/dl.[9]

a. If we calculate z-scores, what is the scale?

b. Calculate the z-scores for the following workers:

    i. Ryan, with a blood lead level of 78.26 μg/dl.

ii. Megan, with a blood lead level of 1.58 μg/dl.

iii. Kyle, with a blood lead level of 55.54 μg/dl.

c. For each worker, interpret the value of the z-score.

Solution

a. If we calculate z-scores for the workers' lead levels, the scale of the z-scores will be 14.2 μg/dl, since that is the value of the standard deviation.

b. Here are the workers' lead levels.

    i. Ryan:

$$z\text{-score} = \frac{x - \bar{x}}{s} = \frac{78.26 - 31.4}{14.2} = \frac{46.86}{14.2} = 3.3$$

   ii. Megan:

$$z\text{-score} = \frac{x - \bar{x}}{s} = \frac{1.58 - 31.4}{14.2} = \frac{-29.82}{14.2} = -2.1$$

  iii. Kyle:

$$z\text{-score} = \frac{x - \bar{x}}{s} = \frac{55.54 - 31.4}{14.2} = \frac{24.14}{14.2} = 1.7$$

c. Ryan's lead level lies 3.3 standard deviations above the mean; Megan's lead level lies 2.1 standard deviations below the mean; and Kyle's lead level lies 1.7 standard deviations above the mean. ■

Example 3.27   **Using the z-score to compare data from different data sets**

Andrew is bragging to his friend Brittany that he did better than she did on the last statistics test. Andrew got a 90 while Brittany got an 80. Andrew's class mean was 80 with a standard deviation of 10. Brittany's class mean was 60 with a standard deviation of 10. The professors in both classes grade "on a curve" using z-scores. Who did better relative to his or her class?

Solution

Brittany can use z-scores to show that she did better *relative to her class*. Figure 3.21 shows comparative dotplots of the scores in the two classes. The blue dots represent Brittany's and Andrew's scores. Brittany found her z-score by subtracting her class mean from her score of 80 and then dividing by the standard deviation $s = 10$:

$$z\text{-score}_{\text{Brittany}} = \frac{x - \bar{x}}{s} = \frac{80 - 60}{10} = 2$$

FIGURE 3.21 Brittany actually did better relative to her class.

Brittany's z-score is 2. What does that mean? It means that *Brittany scored 2 standard deviations above the mean* of 60. Brittany then found the z-score for Andrew:

$$z\text{-score}_{\text{Andrew}} = \frac{x - \bar{x}}{s} = \frac{90 - 80}{10} = 1$$

Andrew's z-score was 1, which means that Andrew scored one standard deviation above the mean. From Figure 3.21 we can observe that Andrew's exam score of 90 lies closer to the mean exam score of 80 for his class. That is, the blue arrow is shorter for Andrew than for Brittany. Finally, note that 10 of the 100 students who took the exam in his class did better than he did, whereas only 2 did better than Brittany in her class. So Brittany is right. Relative to her class, she did better than Andrew, even though Andrew got a higher score. The z-scores allowed her to compare their grades, even though they were in different classes. ■

3 Detecting Outliers Using z-Scores

An **outlier** is an extremely large or extremely small data value relative to the rest of the data set. It may represent a data entry error, or it may be genuine data. Usually an outlier can be identified because it is much farther than three standard deviations from the mean, that is, its z-score is less than −3 or greater than 3. How can we use this information? Although we will investigate this question more later on in this chapter, we can give the following rough guidelines here. Look at Figure 3.22.

FIGURE 3.22
z-Scores help to identify outliers.

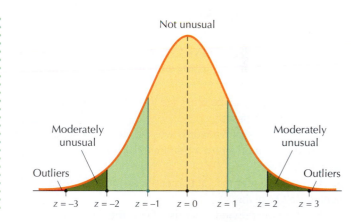

Guidelines for Identifying Outliers

Note: If an outlier is detected, it does not automatically follow that it should be discarded. Outliers often indicate the presence of something interesting going on in the data that would call for further investigation. On the other hand, it could simply be a typo. The analyst should check with the data source. Finally, the analyst may prefer to use the *robust method of detecting outliers* that we will learn in Section 3.6.

1. A data value whose z-score lies in the following range is not considered to be unusual:

$$-2 < z\text{-score} < 2$$

2. A data value whose z-score lies in either of the following ranges may be considered moderately unusual:

$$-3 < z\text{-score} \leq -2 \quad \text{or} \quad 2 \leq z\text{-score} < 3$$

3. A data value whose z-score lies in either of the following ranges may be considered an outlier:

$$z\text{-score} \leq -3 \quad \text{or} \quad z\text{-score} \geq 3$$

Example 3.28 Detecting outliers

For the three workers in Example 3.26, determine whether each of their blood lead levels represents an outlier.

Solution

Ryan's z-score is 3.3, which is greater than 3. Thus, Ryan's lead level of 78.26 μg/dl represents an outlier. Megan's z-score is -2.1, which lies between -3 and -2. Hence, Megan's lead level of 1.58 μg/dl may be considered moderately unusual but is not an outlier. Kyle's z-score is 1.7, which lies between -2 and 2. Thus, Kyle's lead level of 55.54 is not considered unusual. ■

Section 3.4 SUMMARY

1 In this section we learned about measures of position, which tell us the position that a particular data value holds relative to the rest of the data set. The *p*th percentile of a data set is the value at which *p* percent of the values in the data set are less than or equal to this value.

2 The z-score indicates how many standard deviations a particular data value is from the mean. The z-score equals the data value minus the mean, divided by the standard deviation.

3 An outlier is an extremely large or extremely small data value relative to the rest of the data set. An outlier can be identified when its z-score is less than -3 or greater than 3.

Section 3.4 EXERCISES

CLARIFYING THE CONCEPTS

1. Explain in your own words what the 95th percentile of a data set means.

2. Why doesn't it make sense for there to be a 120th percentile of a data set?

3. Explain in your own words why we need z-scores.

4. What does it mean for a z-score to be positive? Negative? Zero?

5. Is it possible for the 1st percentile to equal the 99th percentile? Explain.

6. Refer to the previous exercise. Describe the data set where this might happen. Describe the spread of the data set using any measure you like.

PRACTICING THE TECHNIQUES

Refer to the following table for Exercises 7–11.

Vehicle	City mpg
Honda Civic	36
Toyota Camry	24
Ford Taurus	20
Pontiac Grand Prix	20
Jaguar X-Type	18
Lincoln Town Car	17

7. Find the following percentiles:
  a. 50th
  b. 75th
  c. 25th

8. Find the following percentiles:
  a. 10th
  b. 95th
  c. 5th

9. Find the z-scores for the city mpg for the following automobiles:
  a. Toyota Camry
  b. Lincoln Town Car
  c. Jaguar X-Type

10. Find the z-scores for the city mpg for the following automobiles:
  a. Honda Civic
  b. Ford Taurus
  c. Pontiac Grand Prix

11. Does the mileage for the Honda Civic represent an outlier? Explain.

APPLYING THE CONCEPTS

Refer to Table 3.20 for Exercises 12–15.

TABLE 3.20 Calories in 12 breakfast cereals

Cereal	Calories
Apple Jacks	110
Basic 4	130
Bran Chex	90
Bran Flakes	90
Cap'n Crunch	120
Cheerios	110
Cinnamon Toast Crunch	120
Cocoa Puffs	110
Corn Chex	110
Corn Flakes	100
Corn Pops	110
Count Chocula	110

12. Find the following percentiles:
   a. 50th   **b.** 75th

13. Find the following percentiles:
   a. 25th   **b.** 95th

14. Find the z-scores for the calories for the following cereals:
   a. Corn Flakes   **b.** Basic 4

15. Find the z-scores for the calories for the following cereals:
   a. Bran Flakes   **b.** Cap'n Crunch

Refer to Table 3.21 for Exercises 16–20. The table gives the number of American adults who have used the indicated "nonvitamin, nonmineral, natural products."

TABLE 3.21 Use of dietary supplements

Product	Usage (in millions)	Product	Usage (in millions)
Echinacea	14.7	Ginger supplements	3.8
Ginseng	8.8	Soy supplements	3.5
Ginkgo biloba	7.7	Ragweed/chamomile	3.1
Garlic		Bee pollen or	
supplements	7.1	royal jelly	2.8
Glucosamine	5.2	Kava kava	2.4
Saint-John's-		Valerian	2.1
wort	4.4	Saw palmetto	2.0
Peppermint	4.3		
Fish oil	4.2		

Source: Centers for Disease Control and Prevention, Vital and Health Statistics, 2004.

16. Find the following percentiles:
   a. 5th   **b.** 95th

17. Find the following percentiles:
   a. 50th   **b.** 75th

18. Find the z-scores for usage for the following products:
   a. Fish oil   **b.** Valerian

19. Find the z-scores for usage for the following products:
   a. Ginseng   **b.** Garlic supplements

20. Does the usage for echinacea represent an outlier? Explain.

21. Females SAT Performance. Females tend to perform well on the Verbal SAT. The College Board reported that the mean score on the Verbal SAT for females in 2003 was 503. Assume that the standard deviation is 100 and that we do not know the distribution. Brandy scored in the 50th percentile on the Verbal SAT.
   a. Draw a distribution curve where the 50th percentile is above the mean.
   b. Draw a distribution curve where the 50th percentile is below the mean.
   c. If the data distribution is bell-shaped, what was Brandy's score?

22. Expenditure per Pupil. The 5th percentile expenditures per pupil nationwide in 2005 was $6381, the 50th percentile was $8998, and the 95th percentile was $17,188.[10]
   a. Determine whether the distribution of expenditures is symmetric, left-skewed, or right-skewed.
   b. Would we expect the mean expenditure per pupil to be less than, equal to, or greater than $8998? Explain.
   c. Draw a distribution curve that matches this information.

Use the following information for Exercises 23–26.

State	Median age
Utah	28.5
Texas	33.2
Alaska	33.9
Georgia	34.3
California	34.4
Arizona	34.5
Idaho	34.6
Colorado	34.7

23. Find the following percentiles for state median age:
   a. 25th   **b.** 50th   **c.** 75th

24. The mean for the 8 states is 33.5125 with a standard deviation of 2.0822. Use these statistics to find the z-scores for the following states:
   a. Utah   **b.** Texas   **c.** Colorado

25. Identify any outliers.

26. Identify any states that are moderately unusual.

3.5 Chebyshev's Rule and the Empirical Rule

Objectives: **By the end of this section, I will be able to...**

1. Calculate percentages using Chebyshev's Rule.
2. Find percentages and data values using the Empirical Rule.

1 Chebyshev's Rule

P. L. Chebyshev (1821–94, Russia) derived a result, called **Chebyshev's Rule,** that can be applied to any continuous data set whatsoever.

> **Chebyshev's Rule**
>
> The proportion of values from a data set that will fall within k standard deviations of the mean will be *at least*
>
> $$\left(1 - \frac{1}{k^2}\right)100\%,$$
>
> where $k > 1$. Chebyshev's Rule may be applied to either samples or populations.

Portrait of Pafnuty Tchebyshev-Yaroslav Sergeyevich (1899–1978)/State Central Artillery Museum, St. Petersburg, Russia/The Bridgeman Art Library

---

Example 3.29   Finding the scores within 3 standard deviations of the mean

An instructor giving an exam expects that the mean will be 75 and the standard deviation will be 5. What exam scores represent the scores within 3 standard deviations of the mean exam score?

Solution

We have $\bar{x} = 75$ and $s = 5$. The instructor is interested in estimating the proportion *within 3 standard deviations,* so that $k = 3$. In general, "within 3 standard deviations of the mean" means

a. the mean minus 3 standard deviations ($\bar{x} - 3s$), up to
b. the mean plus 3 standard deviations ($\bar{x} + 3s$), as shown in Figure 3.23.

$\bar{x} - 4s$	$\bar{x} - 3s$	$\bar{x} - 2s$	$\bar{x} - 1s$	$\bar{x}$	$\bar{x} + 1s$	$\bar{x} + 2s$	$\bar{x} + 3s$	$\bar{x} + 4s$
55	60	65	70	75	80	85	90	95

FIGURE 3.23 Within 3 standard deviations of the mean.

Since $\bar{x} = 75$ and $s = 5$, "within 3 standard deviations of the mean" means from $75 - 3(5) = 60$ to $75 + 3(5) = 90$.

Developing Your Statistical Sense

The "At Least" Is Not Optional

Chebyshev's Rule states that at least a certain percentage of the distribution lies between these two numbers. Data analysts say that Chebyshev's Rule gives a *lower bound* to this percentage, since it uses the phrase "at least." When giving percentages using Chebyshev, make sure you use the words "at least." There is a big difference between "about 84%" and "at least 84%." "At least 84%" could mean 100%, 95%, 84%, or anything in between. It is rather imprecise. ■

A key to solving a Chebyshev-type problem lies in finding the value of k, the number of standard deviations that a data value lies from the mean. The following example demonstrates how to find the value of k, as well as how to calculate the lower bound.

Example 3.30   Finding k and calculating the lower bound

Continuing the previous example, find a lower bound on the percentage of students who will score between 60 and 90 on the upcoming exam.

Solution

First, we need to find k. This is done as follows:

$$k = \left| \frac{\text{numerical value} - \text{mean}}{\text{standard deviation}} \right| = \left| \frac{60 - 75}{5} \right| = |-3.0| = 3$$

$$k = \left| \frac{\text{numerical value} - \text{mean}}{\text{standard deviation}} \right| = \left| \frac{90 - 75}{5} \right| = |3.0| = 3$$

Thus, $k = 3$. You may have noticed that the k is found in a manner similar to finding the z-score. What is the difference, then, between k and a z-score? The difference is that z-scores are always associated with an existing data value, whereas k need not be.

Once you have found the value of k, plug it into the *at least* $\left(1 - \frac{1}{k^2} \right) 100\%$ formula from Chebyshev's Rule to solve the problem, as follows:

$$at\ least \left(1 - \frac{1}{k^2} \right) 100\% = at\ least \left(1 - \frac{1}{3^2} \right) 100\% = at\ least \left(1 - \frac{1}{9} \right) 100\%$$

$$= at\ least \left(\frac{8}{9} \right) 100\% \approx at\ least\ 88.8889\% \approx at\ least\ 88.9\%$$

That is, at least 88.9% of the students will score between 60 and 90 on the upcoming exam. Note that you were told nothing about the shape of the distribution. Chebyshev's Rule can be applied to distributions of any shape, even unknown distributions, as long as the data are continuous.

The Empirical Rule

Metaphorically, the **Empirical Rule** is a high-performance Porsche compared to Chebyshev's Rule, which is a go-anywhere ATV (all-terrain vehicle). If your path calls for dirt roads and mud, you wouldn't use the Porsche but would hop in the ATV. Same here. If you have a skewed data distribution or an unknown distribution, use Chebyshev's Rule. However, in the right circumstances, you should use the Empirical Rule. What are those circumstances? *Use the Empirical Rule when you know that your distribution is bell-shaped (normal).* Of course, on the autobahn, the Porsche would leave the ATV in the dust. Same here. The Empirical Rule, when the data distribution is bell-shaped, outperforms Chebyshev. The more information you have, the more precise and powerful your statistical analysis will be.

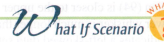

hat If Scenario ? **Robust Detection of an Outlier**

What if the minimum dance score of 30 is changed to 23. Based on Example 3.43, this new value should be detected as an outlier. Note that changing the minimum value does not affect the calculation of Q1, Q3, the IQR, or the thresholds for outlier detection.

Figure 3.43 shows that the box, hinges, and whiskers are all located at precisely the same spots as in the boxplot of the original dance score data. However, the software has calculated, using the robust detection method, that the new data value of 23 is an outlier and indicates it as such with a blue dot. Comparing this boxplot to the earlier one (see Figure 3.36), we notice that the lower whisker is shorter. In Figure 3.43, the whisker terminates at the dance score of 44 instead of 30.

FIGURE 3.43
Boxplot of dance score data showing presence of outlier, after change.

The next example shows how comparison boxplots may be used to compare two data sets side-by-side.

Example 3.44   Comparison boxplots: comparing body temperatures for women and men

Determine whether the body temperatures of women or men exhibit greater variability.

Solution
Consider the comparison boxplots in Figure 3.44. The box for females (on top) lies slightly to the right of that for the males, meaning that the first quartile, the median, and the third quartile are each higher for the women than the men. Therefore, the middle 50% of the body temperatures is higher for women than men.

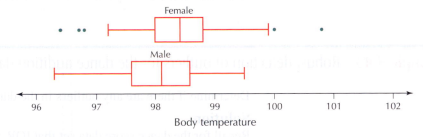

FIGURE 3.44 Comparison boxplots of female and male body temperatures.

This figure seems to offer some evidence that the mean body temperature for women may be higher than that for men. We will formally test whether there is a difference in the true mean body temperature between women and men in Chapter 10. The location of the box is an indication of the center of the data. But where would we look for a difference in the variability of body temperatures between women and men? From Figure 3.45, for the females we have

$$IQR = Q3 - Q1 = 98.8 - 98.0 = 0.8$$

For the males we have

$$IQR = Q3 - Q1 = 98.6 - 97.6 = 1.0$$

So the IQR for males is greater.

Let's determine which data set has greater variability based on the three different measures of spread that we have learned: the range, the standard deviation, and the IQR.

Gender	N	Mean	Median	StDev	Min	Max	Q1	Q3
female	65	98.394	98.4	0.743	96.4	100.8	98.0	98.8
male	65	98.105	98.1	0.699	96.3	99.5	97.6	98.6

FIGURE 3.45 Descriptive statistics for body temperature, by gender.

Range for women = 100.8 − 96.4 = 4.4   Range for men = 99.5 − 96.3 = 3.2

Standard deviation for women = 0.743   Standard deviation for men = 0.699

IQR for women = 0.8   IQR for men = 1.0

When Measures of Spread Disagree

Two measures of spread that are sensitive to the presence of extreme values—range and standard deviation—find that the female body temperatures are more variable. The measure of spread that is resistant to the effects of extreme values—IQR—finds that the male body temperatures are more variable. How do we resolve this apparent inconsistency? What appears to be happening is that, for the middle 50% of each data set, the men are more variable, but as we move toward the tails, the women are more spread out.

Note that there are outliers for the women but not for the men. In part, this may be because the IQR for the women is smaller, and thus the distance 1.5(IQR) is smaller as well. For example, the woman whose body temperature is 100 degrees is identified as an outlier because 100 is the same as the outlier threshold Q3 + 1.5(IQR) = 98.8 + 1.5(0.8) = 100. *The same temperature in a man would not be classified as an outlier, even though the male temperatures are lower overall* (and Q3, specifically, is lower). This is because the temperature of 100 is not higher than Q3 + 1.5(IQR) = 98.6 + 1.5(1.0) = 100.1, the male outlier threshold. Thus, the measures of spread that are sensitive to outliers indicate that women have greater variability, while the measure of spread that is not sensitive to outliers indicates that men have greater variability. ■

STEP-BY-STEP TECHNOLOGY GUIDE: Boxplots

We will make boxplots for the data in Example 3.37 (page 133). Use the Step-by-Step Technology Guide at the end of Section 3.1 (page 89), and choose the appropriate statistics. Note that Minitab's values for Q1 and Q3 differ from those we calculated earlier because Minitab uses a different way of calculating the quartiles. If you wish to check your hand calculations, do so with the TI-83/84, which uses the method we learned here.

TI-83/84

Step 1 Enter the data in list **L1**.
Step 2 Press **2nd Y=**, and choose 1: **Plot 1**.
Step 3 Turn plots **On**. Highlight the boxplot icon, as shown in Figure 3.46.
Step 4 Press **ZOOM**, and choose **9: ZoomStat**. A boxplot similar to Figure 3.36 in Example 3.39 is then produced.

Figure 3.46

MINITAB

Step 1   Enter the data in column **C1**, and name your data *Scores*.
Step 2   Click **Graph > Boxplot**.
Step 3   Select **Simple** and click **OK**.
Step 4   Select the variable *Scores,* and click **OK**, as shown in Figure 3.47. A boxplot similar to Figure 3.36 in Example 3.39 is then produced.

Figure 3.47

Section 3.6 SUMMARY

1 Section 3.6 presents robust measures and methods, which are not sensitive to the presence of outliers. Quartiles divide the data set into approximately equal quarters. The interquartile range is a measure of variability found by taking the difference between the third and first quartiles.

2 The five-number summary is a robust alternative to the usual mean-and-standard-deviation method of summarizing a data set. It consists of simply reporting the minimum, first quartile, median, third quartile, and maximum of the data set.

3 A boxplot is a graphical representation of the five-number summary and is useful for investigating skewness and the presence of outliers.

4 The robust method of detecting outliers is to consider a data value an outlier if it is located 1.5(IQR) or more below Q1, or it is located 1.5(IQR) or more above Q3.

Section 3.6 EXERCISES

CLARIFYING THE CONCEPTS

1. Why do we need robust measures? Why can't we just use the mean and standard deviation all the time?
2. Consider whether the following scenarios are possible. If it is possible, then clearly describe what the data set would look like. What would the boxplot for such a data set look like? If it is not possible, why not?
    a. A scenario where the first and second quartiles of a data set are equal.
    b. A scenario where all five numbers in the five-number summary are equal.
    c. A scenario where the mean of a data set is larger than Q3.
    d. A scenario where the median of a data set is smaller than Q1.
3. True or false: The five-number summary consists of the following: Minimum, Q1, Mean, Q3, Maximum.
4. True or false: The IQR is sensitive to the presence of outliers.

Here are the final-exam scores for 20 psychology students:

| 75 | 81 | 82 | 70 | 60 | 59 | 94 | 77 | 68 | 98 |
| 86 | 68 | 85 | 72 | 70 | 91 | 78 | 86 | 51 | 67 |

Use this information to answer Exercises 5–9.
5. Find the following quartiles for *final-exam score:*
    a. Q1
    b. Q2
    c. Q3
6. Find the interquartile range of *final-exam score.*
7. Find the five-number summary for *final-exam score.*
8. Use the robust method to investigate the presence of outliers.
9. Construct a boxplot for *final-exam score.*

APPLYING THE CONCEPTS

Women's Volleyball Team Heights. Refer to Table 3.8 (page 93) for Exercises 10–12.
10. Find five-number summaries for the volleyball team heights for WMU and NCU. Comment on whether one should use a five-number summary for data sets of size 5.
11. Find the interquartile range for each team. Interpret what this value actually means, so that a nonspecialist could understand it.
12. By hand, construct a boxplot for each team. Comment on why your boxplots look unusual.

Most Active Stocks. Use Table 3.22 for Exercises 13–20. These companies represent the 11 most actively traded stocks on the New York Stock Exchange for February 5, 2007. Variables include the stock price and the net change in stock price, with both variables in dollars.

TABLE 3.22   The most active stocks on the NYSE

Company	Price	Change
Pfizer Inc.	26.88	+0.08
Slm Corp.	42.37	−4.09
Ford Motor Co.	8.33	+0.10
General Electric Co.	36.37	+0.10
Equity Office Properties Trust	55.46	+0.08
AT&T Inc.	37.79	−0.33
EMC Corp.	13.69	−0.05
LSI Logic Corp.	9.14	−0.14
Motorola, Inc.	19.87	+0.02
Advanced Micro Devices	15.60	−0.09

13. Find the five-number summary for *price*.
14. Find the interquartile range for *price*. Interpret what this value actually means, so that a nonspecialist could understand it.
15. Use the robust method to investigate the presence of outliers in *price*.
16. Construct a boxplot for *price*.
17. Find the five-number summary for *change*.
18. Find the interquartile range for *change*. Interpret what this value actually means, so that a nonspecialist could understand it.

19. Use the robust method to investigate the presence of outliers in *change*.
20. Construct a boxplot for *change*.

Dietary Supplements. Refer to Table 3.21 (page 121) for Exercises 21–26.
21. Find the five-number summary for *usage*.
22. Find the interquartile range for *usage*. Interpret what this value actually means, so that a nonspecialist could understand it.
23. Use the robust method to investigate the presence of outliers in *usage*.
24. Construct a boxplot for *usage*.
25. Calculate the mean and standard deviation of *usage*.
26. Find the *z*-score for echinacea, and use it to determine whether the product is an outlier. Compare the result with that from the robust method.

Nutrition. Use the data set **Nutrition** for Exercises 27–30.
27. Open the data set **Nutrition.**
    a. How many observations are in the data set?
    b. How many variables?
28. Use a statistical computing package (like Minitab) to explore the variable *iron*.
    a. Find the mean and standard deviation for the amount of iron in the food.
    b. Find the five-number summary, the range, and the interquartile range.
29. Which food item has the maximum amount of iron? Does this surprise you?
30. Use the computer to generate a boxplot. Also, comment on the symmetry or the skewness of the boxplot.

Try This in Class! In-class activities to enhance your understanding of statistics

MEASURES OF CENTER

1. What is your guess of the typical height of all students in your class?
2. Make a dotplot of the heights of the students in your class.
3. Discuss where to place the center of this distribution of student heights. Without crunching any numbers, form a consensus on the location of the center.
4. Calculate the mean, median, and mode of the student heights.
5. Which measure (mean, median, or mode) comes closest to the consensus of where the center is located in Activity 3?
6. What is the relation between these measures and your guess of the typical height in Activity 1?
7. Which measure (mean, median, mode, class consensus, your guess) do you think is the best measure of the center of student heights?

MEASURES OF SPREAD

8. Do you think that the distribution of the heights of all students in your class is more spread out or less spread out than the distribution of the heights of only the females in your class?
9. Would the values of our measures of spread (range, standard deviation) be larger for the entire class or for only the females?
10. Make a dotplot of the heights of only the females in the class. Make sure it uses the same scale as the dotplot for the heights of all the students in the class.
11. Use the two dotplots to assess which group has greater variability.
12. Back up your intuition by calculating and comparing our measures of spread (range, standard deviation) for the two groups.

CHAPTER 3 FORMULAS AND VOCABULARY

SECTION 3.1
- **Mean** (p. 79)
- **Measure of center** (p. 79)
- **Median** (p. 82)
- **Mode** (p. 84)
- **Population mean** (p. 81). $\mu = \sum x/N$.
- **Population size** (p. 81). Denoted by N.
- **Sample mean** (p. 80). $\bar{x} = \sum x/n$.
- **Sample size** (p. 80). Denoted by n.

SECTION 3.2
- **Deviation** (p. 96). $x - \bar{x}$.
- **Measure of variability (measure of spread)** (p. 94)
- **Population standard deviation** (p. 98).

$$\text{Definition formula: } \sigma = \sqrt{\frac{\sum(x - \mu)^2}{N}}$$

$$\text{Computational formula: } \sigma = \sqrt{\frac{\sum x^2 - \left(\sum x\right)^2/N}{N}}$$

- **Population variance** (p. 98).

$$\text{Definition formula: } \sigma^2 = \frac{\sum(x - \mu)^2}{N}$$

$$\text{Computational formula: } \sigma^2 = \frac{\sum x^2 - \left(\sum x\right)^2/N}{N}$$

- **Range** (p. 94)
- **Sample standard deviation** (p. 99).

$$\text{Definition formula: } s = \sqrt{\frac{\sum(x - \bar{x})^2}{n - 1}}$$

$$\text{Computational formula: } s = \sqrt{\frac{\sum x^2 - \left(\sum x\right)^2/n}{n - 1}}$$

- **Sample variance** (p. 99).

$$\text{Definition formula: } s^2 = \frac{\sum(x - \bar{x})^2}{n - 1}$$

$$\text{Computational formula: } s^2 = \frac{\sum x^2 - \left(\sum x\right)^2/n}{n - 1}$$

- **Standard deviation** (p. 94)

SECTION 3.3
- **Estimated mean for data grouped into a frequency distribution** (p. 109).

$$\hat{\mu} = \frac{\sum m_i f_i}{N} = \frac{m_1 f_1 + m_2 f_2 + \cdots + m_k f_k}{N}$$

- **Estimated standard deviation for data grouped into a frequency distribution** (p. 110).

$$\hat{\sigma} = \sqrt{\hat{\sigma}^2} = \sqrt{\frac{\sum(m_i - \hat{\mu})^2 \cdot f_i}{N}}$$

- **Estimated variance for data grouped into a frequency distribution** (p. 110).

$$\hat{\sigma} = \frac{\sum(m_i - \hat{\mu})^2 \cdot f_i}{N}$$

- **Weighted mean** (p. 108).

$$\bar{x}_w = \frac{\sum w_i x_i}{\sum w_i} = \frac{w_1 x_1 + w_2 x_2 + \cdots + w_n x_n}{w_1 + w_2 + \cdots + w_n}$$

SECTION 3.4
- **Outlier** (p. 119)
- **Percentile** (p. 113)
- **z-Score** (p. 116)

  a. The z-score for a particular data value from a *sample* is

$$z\text{-score} = \frac{\text{data value} - \text{mean}}{\text{standard deviation}} = \frac{x - \bar{x}}{s}$$

  b. The z-score for a particular data value from a *population* is

$$z\text{-score} = \frac{\text{data value} - \text{mean}}{\text{standard deviation}} = \frac{x - \mu}{\sigma}$$

SECTION 3.5
- **Chebyshev's Rule** (p. 122). The proportion of values from a data set that will fall within k standard deviations of the mean will be *at least* $\left(1 - \frac{1}{k^2}\right)100\%$, where $k > 1$.
- **Empirical Rule** (p. 123). If the data distribution is bell-shaped:

  About 68% of the data values will fall within one standard deviation of the mean.

  About 95% of the data values will fall within two standard deviations of the mean.

  About 99.7% of the data values will fall within three standard deviations of the mean.

SECTION 3.6
- **Boxplot** (p. 133)
- **Five-number summary** (p. 132)
- **Interquartile range (IQR)** (p. 132). $\text{IQR} = Q3 - Q1$.
- **Quartiles** (p. 128)
- **Robust measures** (p. 128)
- **Robust method of detecting outliers** (p. 137)

CHAPTER 3 REVIEW EXERCISES

SECTION 3.1

Phytoplankton. Refer to the phytoplankton data from Exercise 20 in Section 3.2 (page 106) for Exercises 1–3.
1. Find the mean.
2. Find the median.
3. Find the mode, if any.

Calories in Cereal. For Exercises 4–6, refer to the histogram of calories, below and to Table 3.20 (page 121).

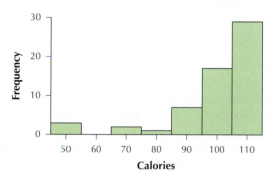

Histogram of calories contained in a sample of breakfast cereals.

4. Which is largest, the mean, median, or mode? How do you know?
5. If we eliminated the cereals with less than 60 calories from the sample, which measure would not be affected at all? Why?
6. How would adding 10 calories to each cereal affect the mean, median, and mode? Would it affect each of the measures equally?

SECTION 3.2

Women's Volleyball Team Heights. Refer to Table 3.8 (page 93) for Exercises 7–9.
7. Suppose a new player joins the NCU team. She is 7 feet tall (84 inches) and replaces the 72-inch-tall player.
  a. Would you expect the standard deviation to go up or down, and why?
  b. Now find the standard deviation for the team including the new player. Was your intuition correct?
8. Add 4 inches to the height of each player on the WMU team.
  a. Recalculate the range and standard deviation.
  b. Formulate a rule for the behavior of these measures of variability when a constant (like 4) is added to each member of the data set.
  c. By hand, draw a dotplot of the heights of the original WMU team. Now draw a dotplot of the heights of the team after you added 4 inches to each player.
    i. Does the spread of the data set appear to have increased, decreased, or stayed the same?
    ii. Does this concur with your answer to **(b)**?
9. Starting with the original data, double the height of each player on the NCU team.

a. Recalculate the range and standard deviation.
b. Formulate a rule for the behavior of these measures of variability when each member of the data set is multiplied by a constant (such as α).
c. Now compare a dotplot of the original NCU heights with a dotplot of the heights after you doubled each player's height.
  i. Does the spread of the data set appear to have increased, decreased, or stayed the same?
  ii. Does this concur with your answer to **(b)**?

SECTION 3.3

10. Calculating a Grade Point Average. At a certain university in Texas, student grade point averages are calculated as follows. For each credit hour, an A is worth 4.0 quality points, an A− is worth 3.7 quality points, a B+ is worth 3.3 quality points, a B is worth 3.0, a B− is worth 2.7, a C+ is worth 2.3, and so on. To find the grade point average, the number of credits for each course is multiplied by the quality points earned for that course; the results are added together; and the sum is divided by the number of credits. This semester, Yolanda's grades are as follows. She got an A in her four-credit honors biology course, an A− in her three-credit calculus course, a B+ in her three-credit English course, a B− in her three-credit anthropology course, and a C+ in her two-credit physical education course. Calculate Yolanda's grade point average for this semester.

11. AIDS Cases by Age. The National Center for Health Statistics reported the number of cases of acquired immunodeficiency syndrome (AIDS) by age of patient in 2004.[12] Find the estimated mean and standard deviation of the age of AIDS patients.

Class: age	Frequency f_i
0–12.99	48
13–14.99	60
15–24.99	2,114
25–34.99	9,361
35–44.99	16,778
45–54.99	10,178
55–64.99	3,075
65–74.99	901

SECTION 3.4

Research Journals. Use the research journal data in Table 3.3 (page 85) for Exercises 12–19.

Find the following percentiles of subscription costs.
12. 10th percentile
13. 50th percentile
14. 90th percentile

Find the z-scores for the following journals.
15. *Journal of Pure and Applied Algebra*

16. *Annals of Statistics*
17. *Bulletin of the American Mathematical Society*
18. Identify any outliers in the research journal data.
19. Identify any moderately unusual observations in the research journal data.

SECTION 3.5

Use the following information for Exercises 20–22. The vehicles on Interstate 95 have a mean of 70 miles per hour with a standard deviation of 5 miles per hour.

20. If the distribution of vehicle speeds is unknown, find the proportion of vehicles traveling between 60 and 80 miles per hour.

21. If the distribution of vehicle speeds is bell-shaped, find the proportion of vehicles traveling between 60 and 80 miles per hour.

22. If the distribution of vehicle speeds is bell-shaped, find an approximation of the 84th percentile.

SECTION 3.6

Ragweed Pollen. Use Table 3.23 for Exercises 23–27. Do you suffer from ragweed pollen? You are not alone. The American Academy of Allergy maintains the ragweed pollen index, which details the severity of the pollen problem for hundreds of communities across the nation. The following table contains the ragweed pollen index on a particular day for ten localities in New York State.

23. Find the first, second, and third quartiles of the ragweed pollen index.

24. Find the interquartile range. Interpret what this value actually means, so that a nonspecialist could understand it.

TABLE 3.23   Ragweed pollen index in New York localities

Locality	Ragweed pollen index
Albany	48
Binghamton	31
Buffalo	59
Elmira	43
Manhattan	25
Rochester	60
Syracuse	25
Tupper Lake	8
Utica	26
Yonkers	38

25. Let's draw a boxplot of the ragweed pollen index.
 a. What is the five-number summary?
 b. By hand, draw a boxplot.
 c. Is the data set left-skewed, right-skewed, or symmetric?
 d. What should the symmetry or skewness mean in terms of the relative values of the mean and median?
 e. Find the mean and standard deviation. Is your prediction in **(d)** supported?

26. Detect any outliers using the z-score and the robust method.
 a. First use the z-score method.
 b. Now use the robust method of outlier detection.
 c. Do the two methods concur or disagree?

27. Suppose the ragweed pollen index in Rochester were 600 instead of 60. How would this outlier affect the quartiles and the IQR? What property of these measures is this behavior an example of?

CHAPTER 3 *Quiz*

TRUE OR FALSE

1. True or false: If two data sets have the same mean, median, and mode, then the two data sets are identical.
2. True or false: The variance is the square root of the standard deviation.
3. True or false: The Empirical Rule applies for any data set.

FILL IN THE BLANK

4. An _____ is an extremely large or extremely small data value relative to the rest of the data set.
5. The mean can be viewed as the _____ point of the data.
6. The measure of center that is sensitive to the presence of extreme values is the _____.

SHORT ANSWER

7. What do we call summary descriptive measures that are not sensitive to the presence of outliers?

8. Which of the mean, median, and mode may be used for categorical data?
9. For any data set, what is the average of the deviations?
10. What do we use to estimate the mean for each class in a frequency distribution?

CALCULATIONS AND INTERPRETATIONS

Airline Passengers. Refer to Table 3.24 for Exercises 11 and 12.

TABLE 3.24   Passengers arriving at Portland International Airport, January–April 2007, by airline

Airline	Passengers
Alaska Airlines	98,008
Delta Air Lines	31,054
Horizon Air	117,964
Southwest Airlines	106,178
United Airlines	84,059

11. Calculate the following:
   a. Sample mean
   b. Sample median
12. Calculate the following:
   a. Range
   b. Sample standard deviation

13. Deaths Due to Heat. The following frequency distribution contains the numbers of deaths due to heat, by age group, as reported by the National Weather Service for 2006. Find the estimated mean and standard deviation of age.

Age	Deaths due to heat
0–39.99	22
40–49.99	31
50–59.99	51
60–69.99	47
70–79.99	44
80–89.99	44

14. The Department of Agriculture reports on the consumption of carbonated beverages by Americans. A sample of 30 Americans yielded a sample mean of 60 gallons. Assume the standard deviation is 40 gallons. Find the z-scores for the following amounts of carbonated beverage consumption.
   a. 120 gallons
   b. 20 gallons
   c. 100 gallons
   d. 0 gallons
   e. 60 gallons

15. Refer to the information in Exercise 14. Assume the distribution is bell-shaped. Use your knowledge about the empirical rule to give a range for the proportion.
   a. Find the 50th percentile.
   b. Estimate the proportion of Americans who drink between 29 and 109 gallons per year.
   c. Discuss whether we could find the estimate in **(b)** without assuming that the distribution is bell-shaped.
   d. Estimate the proportion of Americans who drink more than 109 gallons per year.

Use the information in Table 3.25 for Exercises 16–20.

TABLE 3.25 2004 SAT scores for the states with the highest participation rate

State	SAT I Math
New York	510
Connecticut	515
Massachusetts	523
New Jersey	514
New Hampshire	521
Maine	501
Pennsylvania	502
Delaware	499

16. Find the following quartiles for SAT I Math score:
   a. Q1
   b. Q2
   c. Q3
17. Find the interquartile range of SAT I Math score.
18. Find the five-number summary for SAT I Math score.
19. Use robust methods to investigate the presence of outliers.
20. Construct a boxplot for SAT I Math score.

FIGURE 4.19 Actual versus predicted high temperature.

Extrapolation

The y intercept is the estimated value for y when x equals zero. However, in many regression problems, a value of zero for the x variable would not make sense. For example, suppose we were trying to predict freshman GPA (y) based on SAT Verbal scores. However, there is no zero score for the SAT Verbal exam, so the y intercept would not be meaningful. With our temperature data set, however, a value of zero for the low temperature does make sense. Therefore, we would be tempted to predict 10.78°F as the high temperature for a city with a low of zero degrees. However, $low = 0°F$ is not within the range of the data set. Making predictions based on x-values that are beyond the range of the x-values in our data set is called *extrapolation*. It may be dangerous and should be avoided.

Interpreting the Slope

What is the meaning of the slope coefficient b_1? In data analysis, we interpret the slope of the regression line as the *estimated change in y per unit increase in x*. In our temperature example, the units are degrees Fahrenheit, and the value of b_1 is 0.9865. Therefore, $b_1 = 0.9865$ means the following: "For each increase of 1°F in low temperature, the estimated high temperature increases by 0.9865°F." Also note that the slope is positive, which concurs with our assessment of the overall relationship between high and low temperatures. Further, is it a coincidence that both the slope and the correlation coefficient are positive? Not at all.

> **Relationship Between Slope and Correlation Coefficient**
>
> The slope b_1 of the regression line and the correlation coefficient r always have the same sign.
>
> - b_1 is positive if and only if r is positive.
> - b_1 is negative if and only if r is negative.

Hence, when we found in Section 4.2 that the correlation coefficient between high and low temperatures was positive, we could have immediately concluded that the slope of the regression line was also positive. ∎

David Greedy/Getty Images

CASE STUDY | Female Literacy and Fertility Worldwide

Overpopulation is a serious concern for planet Earth in the twenty-first century. However, there are some encouraging signs of how to combat overpopulation, especially in developing countries. Figure 4.20 shows a scatterplot of the fertility rate versus the female literacy rate in 204 countries around the world.[3] The fertility rate is the mean number of children born to a typical woman in the country, and the female literacy rate is the percentage of women at least 15 years old who can read and write.

FIGURE 4.20

Scatterplot of fertility rate versus female literacy rate for 204 countries.

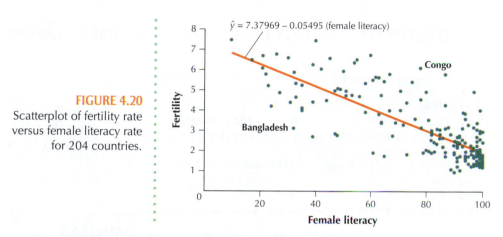

Based on the scatterplot in Figure 4.20, we can state that there is a *negative* relationship between female literacy and fertility. That is, as female literacy increases, fertility tends to decrease. Since the relationship is negative, the correlation coefficient r must therefore be negative, $-1 < r < 0$. Excel provides the correlation coefficient in Figure 4.21, $r = -0.7962$.

As expected, the correlation coefficient is negative, so that female literacy and fertility are negatively correlated. An increase in female literacy is associated with a decrease in fertility. The y intercept and slope of the regression line are shown in the Excel output in Figure 4.22, giving us the regression line

$$\hat{y} = 7.37969 - 0.05495(\text{female literacy})$$

The estimated fertility rate is 7.37969 children minus 0.05495 times the female literacy rate. We interpret the slope b_1 as follows: an increase of 1% in the female literacy rate is associated with a decrease of 0.05495 children in the fertility rate. Thus, for example, an increase of 20% in the female literacy rate is associated with a decrease in the estimated number of births per woman of $20(0.05495) = 1.099$ children. The y intercept is interpreted as follows: when the female literacy rate is zero percent, the estimated fertility rate equals $b_0 = 7.37969$ children.

	Female literacy	Fertility
Female Literacy	1	
Fertility	**−0.7962**	1

FIGURE 4.21

	Coefficients	
Intercept	7.37969	*y* intercept
Female Literacy	−0.05495	slope

FIGURE 4.22

For a country with a 50% literacy rate, the estimated number of births per woman is

$$\hat{y} = 7.37969 - 0.05495(50) \approx 4.6$$

Note the two countries singled out in Figure 4.20. The Republic of Congo has a relatively high female literacy rate, so one would expect the fertility rate to be low. But the fertility rate in the Republic of Congo is 6.45 births per woman, which is higher than expected. Bangladesh has a relatively low female literacy rate, so one would expect the fertility rate to be high. But the fertility rate in Bangladesh is 3.11 births per woman, which is lower than expected. Note that the data points for both Congo and Bangladesh lie far from the regression line. In general, unusual observations in a linear regression problem are points that lie at an unusually large vertical distance from the regression line.

STEP-BY-STEP TECHNOLOGY GUIDE: Correlation and Regression

We illustrate using Example 4.5, the temperature data (page 159).

TI-83/84
Step 1   Turn diagnostics on as follows. Press **2nd 0**. Scroll down and select **DiagnosticOn** (Figure 4.23). Press **ENTER** twice to turn diagnostics on.
Step 2   Enter the **X (Low Temp)** data in **L1**, and the **Y (High Temp)** data in **L2**.
Step 3   Press **STAT** and highlight **CALC**.
Step 4   Press **8** (not **4**) to choose **LinReg(a+bx)**, as shown in Figure 4.24.
Step 5   On the home screen, **LinReg(a+bx)** appears. Press **ENTER**.

Figure 4.23   **Figure 4.24**

MINITAB
Correlation Coefficient *r*
Step 1   Select **Stat > Basic Statistics > Correlation…**
Step 2   Select the variables you want to analyze and click **OK**.

EXCEL
Step 1   Enter the *x* variable in column **A** and the *y* variable in column **B**.
Step 2   Click on **Data > Data Analysis > Regression** and click **OK**.
Step 3   For **Input Y Range**, select cells **B1–B10**. For **Input X Range**, select cells **A1–A10**. Click **OK**.

Regression
Step 1   Enter the *x* variable variable in **C1** and the *y* variable in **C2**.
Step 2   Click on **Stat > Regression > Regression**.
Step 3   Select the *y* variable for the **Response Variable** and the *x* variable for the **Predictor Variable**. Click **OK**.

Section 4.3 SUMMARY

Section 4.3 introduces regression, where the linear relationship between two numerical variables is approximated using a straight line, called the regression line. The equation of the regression line is written as $\hat{y} = b_0 + b_1 x$ where the regression coefficients are the *y* intercept, b_0, and the slope, b_1.

The regression equation can be used to make predictions about values of *y* for particular values of *x*.

Section 4.3 EXERCISES

CLARIFYING THE CONCEPTS
1. What is the objective of regression analysis?
2. What is the regression equation?
3. Describe how we use the regression equation to make predictions.
4. Explain the difference between *y* and $\hat{y}$.

5. Describe what is meant by extrapolation.

6. What is the relationship between the slope of the regression line and the correlation coefficient?

PRACTICING THE TECHNIQUES
Refer to the following table for Exercises 7–9.

Driver	y = super speedway wins	x = short tracks wins
Darrell Waltrip	18	47
Dale Earnhardt	29	27
Jeff Gordon	15	15
Cale Yarborough	15	29
Richard Petty	19	23

7. Calculate the values for the regression coefficients b_0 and b_1, using the following steps.
 a. Compute the slope b_1.
 b. Calculate the y intercept b_0.
 c. Write down the regression equation for the regression of y = Super Speedways wins versus x = Short Tracks wins. Express this equation in words that a nonspecialist would understand.

Note: When we say, "Perform a regression of Variable 1 *versus* Variable 2," Variable 1 is <u>always</u> the y variable, and Variable 2 is <u>always</u> the x variable. Similarly, if we say, "Regress Variable 1 *on* Variable 2, Variable 1 is the y variable and Variable 2 is the x variable.

8. Clearly interpret the meaning of the following:
 a. Slope b_1
 b. y intercept b_0

9. If appropriate, estimate the number of Super Speedways wins for drivers with the following numbers of Short Tracks wins. If not appropriate, clearly explain why not.
 a. 30
 b. 50
 c. 20

Refer to the following table for Exercises 10–12.

State	x = Mean SAT I Verbal	y = Mean SAT I Math
New York	497	510
Connecticut	515	515
Massachusetts	518	523
New Jersey	501	514
New Hampshire	522	521

10. Calculate the values for the regression coefficients b_0 and b_1, using the following steps.
 a. Compute the slope b_1.
 b. Calculate the y intercept b_0.
 c. Write down the regression equation for the regression of y = SAT I Math versus x = SAT I Verbal scores. Express this equation in words that a nonspecialist would understand.

11. Clearly interpret the meaning of the following:
 a. Slope b_1
 b. y intercept b_0
 c. Comment on the usefulness of the literal interpretation of the y intercept in this case.

12. If appropriate, estimate the SAT I Math scores for states with the following SAT I Verbal scores. If not appropriate, clearly explain why not.
 a. 500
 b. 510
 c. 490

APPLYING THE CONCEPTS
Brain and Body Weight. A study compared the body weight (in kilograms) and brain weight (in grams) for a sample of mammals. We are interested in estimating brain weight (y) based on body weight (x).

Body weight (kg)	Brain weight (g)
52.16	440
60	81
27.66	115
85	325
36.33	119.5
100	157
35	56
62	1320
83	98.2
55.5	175

13. Calculate the values for the regression coefficients b_0 and b_1.
 a. Write down the regression equation for the regression of brain weight versus body weight.
 b. Express this equation in words that a nonspecialist would understand.

14. Clearly interpret the meaning of the following:
 a. Slope b_1
 b. y intercept b_0
 c. Comment on the usefulness of the literal interpretation of the y intercept in this case.

15. If appropriate, estimate the brain weight for mammals with the following body weights. If not appropriate, clearly explain why not.
 a. 50 kg
 b. 100 kg
 c. 200 kg

Scrabble. Use the information in the following table and in Figure 4.25 for Exercises 16–18. Scrabble® (**www.hasbro.com/scrabble/home.cfm**) is one of the most popular games in the world. We are interested in approximating the relationship between the frequency (x) of the letter tiles in the game and their point value (y). Figure 4.25 is a scatterplot of the point value versus the letter frequency, along with the regression line, as plotted by Minitab.

Letter	Freq.	Points	Letter	Freq.	Points
A	9	1	N	6	1
B	2	3	O	8	1
C	2	3	P	2	3
D	4	2	Q	1	10
E	12	1	R	6	1
F	2	4	S	4	1
G	3	2	T	6	1
H	2	4	U	4	1
I	9	1	V	2	4
J	1	8	W	2	4
K	1	5	X	1	8
L	4	1	Y	2	4
M	2	3	Z	1	10

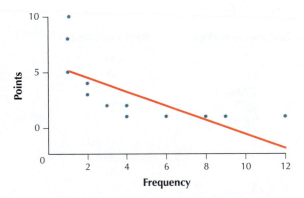

Figure 4.25 Scatterplot of Scrabble® point value versus letter frequency, with regression line.

16. Does the relationship between frequency and points seem to be positive or negative?

17. Does the relationship between frequency and points seem to be a straight-line relationship or a curved relationship?

18. Do you think that we should use a regression line to approximate the relationship between frequency and points? If not, explain clearly what you think is wrong.

DC Households. Use the following information for Exercises 19–21. The data set **Households,** located on your CD and companion Web site, contains information on the number and type of households in the fifty states and the District of Columbia. For each state, there are seven variables. Two of these variables are the percentage of households headed by women ($y = HHLD\_WOMEN$) and the total number of households in the state ($x = TOT\_HHLD$). Minitab provides the following regression equation:

Regression Analysis
```
The regression equation is
HHLD_Women = 10.5 + 2.82E-07 TOT_HHLD
```

Note: The notation 2.82E-07 refers to the scientific notation method of writing numbers. Often, software and calculators will present you with this type of notation, so you need to know how to read it. The number 2.82E-07 represents 2.82 times 10^{-7}, or 0.000000282.

19. In this exercise, we explore the regression coefficients and the regression equation.
  a. Find and interpret the meaning of the value for the y intercept. Does it make sense?
  b. Would the estimate in (a) be considered extrapolation? Why or why not?
  c. Find and interpret the meaning of the slope coefficient as the total number of households in the state increases.
  d. Write the regression equation. Now state in words what the regression equation means.
  e. Is the correlation coefficient positive or negative? How do you know?

20. Estimate the increase or decrease in the percentage of households headed by women, using a sentence, for the following situations.
  a. Suppose State A has one million more households than State B.
  b. Suppose State C has five million fewer households than State D.

21. The number of households per state ranges from about 170,000 to about ten million.
  a. Estimate the percentage of households headed by women for a state with seven million households, if appropriate.
  b. Estimate the percentage of households headed by women for a state with 100,000 households, if appropriate.

Chapter 3 Case Study (Continued). Use the following information for Exercises 22–25. Shown here is the regression equation for the linear relationship between the randomly selected Darts portfolio and the Dow Jones Industrial Average (DJIA), from the Chapter 3 Case Study.

Regression Analysis
```
The regression equation is
Darts = -2.49 + 1.032 DJIA
```

22. In this exercise, we examine the y intercept.
  a. Which variable is the x variable and which is the y variable? Work by analogy with the previous exercises.
  b. What is the value of the y intercept?
  c. Interpret the meaning of this value for the y intercept. Does it make sense?
  d. Would the estimate in (c) be considered extrapolation? Why or why not?

23. In this exercise, we look at the slope and the regression equation.
  a. Find and interpret the meaning of the value of the slope coefficient, as the DJIA increases.
  b. Write the regression equation. Now state in words what the regression equation means.
  c. Is the correlation coefficient positive or negative? How do we know?

24. Estimate the increase or decrease in the net price of the Darts portfolio, using a sentence, for the following situations.

a. Suppose that for Contest A the DJIA increased by 10% more than for Contest B.

b. Suppose that for Contest C the DJIA decreased by 5% more than for Contest D.

25. The net change in the DJIA ranged from –13.1% to 22.5%.

   a. Estimate the net price change for the Darts portfolio when the DJIA is up by 22%, if appropriate.

   b. Estimate the net price change for the Darts portfolio when the DJIA is down by 10%, if appropriate.

   c. Estimate the net price change for the Darts portfolio when the DJIA is down by 22%, if appropriate.

26. Consider again the temperature data in Example 4.8. *What if* there was a typo, and all the low temperatures in the data set needed to be adjusted downward by the same amount. Explain how this change would affect the following, and why. Increase, decrease, or no change?

   a. $\bar{x}$

   b. $\bar{y}$

   c. y intercept b_0

   d. Slope b_1

   e. Correlation coefficient r

CREATE YOUR OWN DATA SETS

27. Describe two variables from real life whose regression line would have a positive slope b_1.

a. Explain why the y variable depends on the x variable.

b. Explain why the slope is positive.

28. Create a sample of five observations from each of your variables from Exercise 27, and put them into a table similar to Table 4.8 in Section 4.2.

   a. Construct a scatterplot of the variables.

   b. Draw a single straight line through the data points in the plot in a manner that you think best approximates the relationship between the variables.

   c. Using your regression line from **(b)**, estimate the slope b_1 and the y intercept b_0.

   d. Write your results from **(c)** in the form of a regression equation.

Use the *Correlation and Regression* applet for Exercises 29 and 30.

29. Create a set of $n = 10$ points such that the slope of the regression line has the following characteristic. (Note that you can drag points up or down to adjust your regression line.)

   a. The slope is positive.

   b. The slope is negative.

   c. The slope is neither positive nor negative.

30. Describe the relationship between the variables for each of the sets of points in the previous exercise.

4.4   Further Topics in Regression Analysis

Objectives: **By the end of this section, I will be able to…**

1   Explain the prediction error, calculate SSE, and utilize the standard error s as a measure of a typical prediction error.

2   Describe how total variability, prediction errror, and improvement are measured by SST, SSE, and SSR.

3   Explain the meaning of the coefficient of determination r^2 as a measure of the usefulness of the regression.

In Section 4.3 we were introduced to regression analysis, which uses an equation to approximate the linear relationship between two quantitative variables. Here in Section 4.4, we learn some further topics that will enable us to better apply the tools of regression analysis for a deeper understanding of our data.

1 Prediction Error, SSE, and Standard Error of the Estimate *s*

To delve deeper into the tools of regression analysis, we will use the following example.

Copyright David Young-Wolff/Photo Edit

Example 4.12 Short-term memory test

Table 4.13 shows the results for ten student subjects who were given a set of short-term memory tasks to perform within a certain amount of time. These tasks included memorizing nonsense words and random patterns. Later, the students were asked to repeat the words and patterns, and the students were scored according to the number of words and patterns memorized and the quality of the memory. Partially remembered words and patterns were given partial credit, so the score was a continuous variable.

Table 4.13 Results of short-term memory test

Student	Time to memorize (in minutes) (x)	Short-term memory score (y)
1	1	9
2	1	10
3	2	11
4	3	12
5	3	13
6	4	14
7	5	19
8	6	17
9	7	21
10	8	24

a. Draw the scatterplot and find the regression line.
b. Estimate the score for a subject given 3.5 minutes to study.

Solution

a. Figure 4.26 displays the scatterplot of y = score versus x = time, together with the regression line $\hat{y} = 7 + 2x$, as calculated by Minitab.

FIGURE 4.26 Scatterplot with regression line.

```
The regression equation is
Score = 7.00 + 2.00 Time
```

Minitab regression results (excerpt).

b. Substituting $x = 3.5$ in the equation $\hat{y} = 7 + 2x$ gives $\hat{y} = 7 + 2(3.5) = 14$. The estimated score when a subject is given 3.5 minutes to study is 14 points. ■

Figure 4.27 shows the regression line for the data in Example 4.12. The (x, y) data values are represented by the blue dots, while the points whose y-values are the $\hat{y}$'s, the predicted values of y, lie exactly on the regression line. The student (7) who was

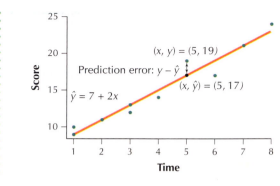

FIGURE 4.27
Prediction error for Student 7.

given $x = 5$ minutes obtained a short-term memory score of $y = 19$. However, the predicted score $\hat{y}$ for this student was somewhat lower than the actual score of $y = 19$. Using the regression equation, we can calculate this predicted score:

$$\hat{y} = 7 + 2x = 7 + 2(5) = 17$$

> The **prediction error** or **residual** $(y - \hat{y})$ measures how far the predicted value $\hat{y}$ is from the actual value of y observed in the data set.

This predicted score $\hat{y} = 17$ for $x = 5$ is given by the point $(x, \hat{y}) = (5, 17)$, which falls exactly on the regression line (black dot). Thus, the prediction generated by the regression equation is "off" by two points, giving us a prediction error of $y - \hat{y} = 19 - 17 = 2$ points. The vertical distance between each dot and the regression line represents the difference between the actual value of y (19) and the predicted value $\hat{y}$ (17). The term for this difference $(y - \hat{y})$ is the **prediction error** or the **residual.**

The residual can be calculated for every observation in the data set. In the example for Student 7, the residual is $(y - \hat{y}) = (19 - 17) = 2$. If we wish to use the regression to make useful predictions, we would like to keep all our prediction errors small. To measure the prediction errors, we calculate the sum of squared prediction errors, or more simply, the **sum of squares error (SSE):**

> **Sum of Squares Error (SSE)**
>
> $$SSE = \sum(y - \hat{y})^2 = \sum(\text{residual})^2 = \sum(\text{prediction error})^2$$

Since we want our prediction errors to be small, it follows that we want SSE to be as small as possible. The least-squares criterion states that the regression line will be the *line for which the SSE is minimized.* That is, out of all possible straight lines, the least-squares criterion chooses the line with the smallest SSE to be the regression line.

Example 4.13   Calculating SSE

Calculate the sum of squares error (SSE) for Example 4.12.

Solution
Table 4.14 shows the $\hat{y}$-values and residuals for the data in Table 4.13. The sum of squares error is then found by squaring each residual and taking the sum. Thus, $SSE = \sum(y - \hat{y})^2 = 12$. Since we know that $\hat{y} = 7 + 2x$ is the regression line, according to the least-squares criterion, no other possible straight line would result in a smaller SSE. ∎

7. Data in Exercise 3
8. Data in Exercise 4
9. Data in Exercise 5
10. Data in Exercise 6

For Exercises 11–14, follow steps **(a)–(e)**. Note that SSE was calculated in Exercises 7–10.
   a. Construct a table like Table 4.15, calculating $(y - \bar{y})$ and $(y - \bar{y})^2$ for each observation.
   b. Calculate SST, the total sum of squares. What does this quantity measure?
   c. Calculate SSR = SST − SSE. What does this quantity measure?
   d. Calculate and interpret the coefficient of determination r^2.
   e. Calculate and interpret the correlation coefficient r.
11. Data in Exercise 3
12. Data in Exercise 4
13. Data in Exercise 5
14. Data in Exercise 6

APPLYING THE CONCEPTS
For Exercises 15–18, follow these steps.
   a. Construct the scatterplot.
   b. Compute the regression equation.
   c. Interpret the meaning of b_0 and b_1 in the context of the particular exercise.
   d. Calculate $s = \sqrt{SSE/(n - 2)}$, the standard error of the estimate, and interpret this statistic.

15. Volume and Weight. The table contains the volume (x, in cubic meters) and weight (y, in kilograms) of five randomly chosen packages shipped to a local village.

Volume (x)	Weight (y)
4	10
8	16
12	25
16	30
20	35

16. World Temperatures. Listed in the table are the low (x) and high (y) temperatures for a particular day in 2006, measured in degrees Fahrenheit, for a random sample of cities worldwide.

City	Low (x)	High (y)
Kolkata, India	57	77
London, England	36	45
Montreal, Quebec	7	21
Rome, Italy	39	55
San Juan, Puerto Rico	70	83
Shanghai, China	34	45

17. NCAA Power Ratings. For the NCAA basketball power ratings, is there a relationship between a school's winning percentage and the power rating? The table shows each team's winning percentage (x) and power rating (y) for the 2004 NCAA Basketball Tournament, according to **www.teamrankings.com**.

School	Win % (x)	Rating (y)
Duke	0.838	96.020
St. Joseph's	0.938	95.493
Connecticut	0.846	95.478
Oklahoma State	0.882	95.320
Pittsburgh	0.857	94.541
Georgia Tech	0.737	93.091
Stanford	0.938	92.862
Kentucky	0.844	92.692
Gonzaga	0.903	92.609
Mississippi State	0.867	91.912

18. Teenage Birth Rate. The National Center for Health Statistics publishes data on state birth rates. The table contains the overall birth rate and the teenage birth rate for ten randomly chosen states. The overall birth rate is defined by the NCHS as "live births per 1000 women," and the teenage birth rate is defined as "live births per 1000 women aged 15–19." We are interested in estimating teenage birth rate based on overall birth rate.

State	Overall birth rate (x)	Teenage birth rate (y)
Alabama	13.1	52.4
Arizona	16.3	60.1
California	15.2	39.5
Florida	12.5	42.4
Georgia	15.7	53.4
New York	13.0	26.9
Ohio	13.0	38.5
Pennsylvania	11.7	30.5
Texas	17.0	62.6
Virginia	13.9	35.2

For Exercises 19–22, follow these steps.
   a. Calculate and interpret the coefffficient of determination r^2.
   b. Calculate and interpret the correlation coefficient r.
19. Data in Exercise 15
20. Data in Exercise 16
21. Data in Exercise 17
22. Data in Exercise 18

For Exercises 23 and 24, use Minitab or Excel.
   a. Construct the scatterplot.
   b. Compute the regression equation.
   c. Calculate and interpret $s = \sqrt{\dfrac{SSE}{n - 2}}$, the standard error of the estimate.
   d. Calculate and interpret the coefficient of determination r^2.
   e. Calculate and interpret the correlation coefficient r.

23. Open the **Nutrition** data set. Estimate the number of calories per gram (y) using the amount of fat per gram (x).

24. Open the **Pulse and Temp** data set. Estimate body temperature (y) using heart rate (x).

CREATE YOUR OWN DATA SETS

Suppose we have a tiny data set with the following (x, y) pairs.

x	y
1	?
2	?
3	?

For Exercises 25–27, create a set of y-values that would fulfill each specification.

25. The slope of the line is positive.

26. The slope of the line is negative.

27. The slope of the line is 0.

APPLET Use the *Correlation and Regression* applet for Exercises 28–30.

28. In these applet exercises, use the "thermometer" above the graph (where it says "Sum of squares =") to help find the least-squares regression line interactively.

a. Select 5 points so that the correlation coefficient is about 0.8. Then select "Draw line."

b. Make your best guess about where the least-squares regression line should be, and draw the line there.

29. The blue section of the thermometer is a measure of the sum of squares error, the total squared vertical distance from the data points to the actual regression line. Recall that the least-squares regression line minimizes this distance. The green section of the thermometer tells you how much "extra" squared error you get from using the line you constructed in Exercise 28(a).

a. Adjust the line you drew in Exercise 28(a) by clicking and dragging on the points until the green section of the thermometer has disappeared.

b. What does the disappearance of the green part tell you about the adjusted line you constructed?

c. Will the line now coincide with the least-squares regression line?

30. Verify that your adjusted line from Exercise 29 coincides with the least-squares regression line by selecting "Show least-squares line."

Try This in Class! In-class activities to enhance your understanding of statistics

CORRELATION AND REGRESSION

For each student in the class, obtain the following information: the distance (in miles) his or her (off-campus) home is from the statistics classroom and the time (in minutes) it takes to commute that distance. If the class is large, take a random sample of the students in the class using *APPLET* either (a) technology, (b) the *Random Sample* applet, or (c) the method shown in the Chapter 1 Try This in Class! (page 28).

1. Discuss which variable is the predictor (x) variable, and which variable is the response (y) variable.

2. Construct a scatterplot of time versus distance.

3. Is there a positive relationship? Negative relationship? No apparent relationship?

4. Estimate the value of the correlation coefficient. Will it be positive or negative?

5. Calculate the correlation coefficient.

6. Interpret the correlation coefficient.

7. Next, consider the regression of time on distance. Will the slope be positive or negative?

8. Using the scatterplot, visually estimate the slope and y intercept of the regression line.

9. Calculate the actual values of the slope and y intercept.

10. Carefully interpret the meaning of the value for the slope, using words that someone who has not taken statistics would understand.

11. Interpret the meaning of the y intercept.

12. For the person who lives farthest from campus, use the regression equation to estimate the time it takes him or her to get to class.

13. Calculate the prediction error for this student (the difference between the actual commuting time and the estimated commuting time).

14. Can you estimate commuting time for a new student who lives farther away than any other student in the class? If not, clearly explain why not.

CHAPTER 4 FORMULAS AND VOCABULARY

SECTION 4.1
- **Crosstabulation (two-way table, contingency table)** (pages 148–149)
- **Clustered bar graph** (page 150)
- **Scatterplot** (page 151)

SECTION 4.2
- **Correlation coefficient *r*** (page 159).
Definition formula:

$$r = \frac{\sum(x - \bar{x})(y - \bar{y})}{(n - 1)s_x s_y}$$

Computational formula:

$$r = \frac{\sum xy - \left(\sum x \sum y\right)/n}{\sqrt{\left[\sum x^2 - \left(\sum x\right)^2/n\right]\left[\sum y^2 - \left(\sum y\right)^2/n\right]}}$$

- **Positive and negative correlation** (page 161).

SECTION 4.3
- **Regression equation (regression line)** (page 168).
$\hat{y} = b_0 + b_1 x$
- **Slope of the regression line** (page 168).
Definition formula:

$$b_1 = \frac{\sum(x - \bar{x})(y - \bar{y})}{\sum(x - \bar{x})^2}$$

Computational formula:

$$b_1 = \frac{\sum xy - \left(\sum x \sum y\right)/n}{\sum x^2 - \left(\sum x\right)^2/n}$$

- **y Intercept of the regression line** (page 168).
$b_0 = \bar{y} - (b_1 \cdot \bar{x})$

SECTION 4.4
- **Coefficient of determination r^2** (p. 183).

$$r^2 = SSR/SST$$

- **Prediction error, or residual** (p. 179).

$$(y - \hat{y})$$

- **SSE, sum of squares error** (p. 179).

$$SSE = \sum(y - \hat{y})^2$$

- **Standard error of the estimate *s*** (p. 180).

$$s = \sqrt{\frac{SSE}{n - 2}}$$

- **SSR, sum of squares regression** (p. 182).
Definition formula:
$SSR = \sum(\hat{y} - \bar{y})^2$

Computational formula: $SSR = \dfrac{\left[\sum xy - \left(\sum x\right)\left(\sum y\right)/n\right]^2}{\sum x^2 - \left(\sum x\right)^2/n}$

- **SST, total sum of squares** (p. 181).
Definition formula: $SST = \sum(y - \bar{y})^2$

Computational formula: $SST = \sum y^2 - \left(\sum y\right)^2/n$

CHAPTER 4 REVIEW EXERCISES

SECTION 4.1
Happiness in Marriage. The General Social Survey tracks trends in American society through annual surveys. Use the following contingency table for Exercises 1–5.

Happiness of Marriage

Respondents' gender	Very happy	Pretty happy	Not too happy	Total
Male	242	115	9	366
Female	257	149	17	423
Total	499	264	26	789

1. What proportion of the males responded that they were very happy in their marriage?
2. What proportion of the females responded that they were very happy in their marriage?
3. What proportion of the males responded that they were not too happy in their marriage?
4. What proportion of the females responded that they were not too happy in their marriage?
5. Construct a clustered bar graph of the data.

Midterm Exams and Overall Grade. Use the data in Table 4.16 to answer Exercises 6 and 7. Can you predict how you will do in a course based on the result of the midterm exam only? The midterm exam score and the overall grade were recorded for a random sample of 12 students in an elementary statistics course. The results are shown in the following table.

TABLE 4.16 Midterm exams and overall grades

Student	Midterm exam score (x)	Overall grade (y)
1	50	65
2	90	80
3	70	75
4	80	75
5	60	45
6	90	95
7	90	85
8	80	80
9	70	65
10	70	70
11	60	65
12	50	55

6. Construct a scatterplot of overall grade versus midterm exam score.
7. Refer to your scatterplot from Exercise 6.
   a. Characterize the relationship as positive, negative, or not apparent.
   b. Write a sentence that describes the behavior of the overall grade as the midterm exam score increases.

SECTION 4.2

Refer to Table 4.16, the midterm exam and overall grade data, to answer Exercises 8–9.
8. Calculate the value of the correlation coefficient between midterm exam scores and overall grades.
9. Interpret in two ways the meaning of this value of the correlation coefficient.

Income and Hours Worked. Use the following information for Exercises 10–12. Table 4.17 contains the weekly gross income (y) and the number of hours worked (x) for a random selection of 5 fast-food workers.

TABLE 4.17 Gross income (in dollars) and hours worked for five fast-food workers

	Fast-Food Worker				
	1	2	3	4	5
Gross income: y	80	120	160	200	240
Hours worked: x	10	15	20	25	30

10. Construct a scatterplot of the variables.
11. What type of relationship do these variables have: positive, negative, or no apparent linear relationship?
12. Calculating and interpreting the correlation coefficient.
    a. Before calculating the value of the correlation coefficient, try to guess what it might be, based on your scatterplot.
    b. Compute the value of the correlation coefficient, using a table similar to Table 4.9. Check your answer using a computer or calculator.
    c. Does this value for r agree with your earlier estimate?
    d. Interpret in two different ways the meaning of this value of the correlation coefficient.

SECTION 4.3

Income and Hours Worked. Use the fast-food worker data from Table 4.17 for Exercises 13–16.
13. In your scatterplot from Exercise 10, draw a single straight line that you think best approximates the relationship between the variables.
    a. Estimate the gross income for a fast-food worker with 27 hours worked.
    b. Estimate the gross income for a fast-food worker with 24 hours worked.
    c. Comment on the usefulness of using this x variable to predict this y variable.
14. Using your scatterplot only, estimate the value of the slope coefficient b_1. If you have trouble, think about its interpretation as the estimated change in y for every unit increase in x. Put this interpretation into the words of this exercise to see if you can figure out what b_1 is.
15. Using your scatterplot only, estimate the value of the y intercept b_0. If you have trouble, think about its interpretation as the estimated value for y when x equals zero. Put this interpretation into the words of this exercise, to see if you can figure out what b_0 is.
16. Now calculate the equation of the regression line. Are the actual values of b_0 and b_1 close to your estimates above?

SECTION 4.4

Refer to Table 4.16, the midterm exam and overall grade data, to answer Exercises 17–21.
17. Calculate SSE.
18. Calculate s, the standard error of the estimate. What does this number mean?
19. Calculate SST. Then use SSE and SST to find SSR.
20. Calculate r^2, the coefficient of determination. Comment on how useful midterm exam scores are for predicting overall grades.
21. Use r^2 to calculate the correlation coefficient. Comment on the relationship between midterm exam scores and overall grades.

CHAPTER 4 *Quiz*

TRUE OR FALSE

1. True or false: Scatterplots are constructed with the y variable on the horizontal axis and the x variable on the vertical axis.

2. True or false: The y intercept measures the strength of the linear relationship between two numerical variables.

FILL IN THE BLANK

3. The "hat" over the y in $\hat{y}$ indicates that it is an _____ of y.

4. We interpret the slope of the regression line as the estimated change in y per _____ increase in x.

SHORT ANSWER

5. Making predictions based on x-values that are beyond the range of the x-values in our data set is called what?

6. Values of r close to –1 indicate what type of relationship between the two variables?

CALCULATIONS AND INTERPRETATIONS

Religious Preference and Excitement in Life. Use the information in Table 4.18 for Exercises 7–9. The General Social Survey asked 967 respondents, "Do you find life exciting, routine, or dull?" along with other questions, including religious preference. Table 4.18 shows the interrelationship of these variables. Unfortunately, however, not all cell counts are provided.

TABLE 4.18   Crosstabulation of religious preference with excitement in life

Religious preference	Is Life Exciting or Dull?			
	Exciting	**Routine**	**Dull**	**Total**
Protestant	264		33	623
Catholic	107	128	6	
Jewish	12	8		20
None	38	29	2	69
Other	9	5		14
Total			41	967

7. Fill in the missing entries in the crosstabulation.

8. Convert the table to a relative frequency crosstabulation so that the "Exciting," "Routine," and "Dull" proportions in each row add up to 1.0.

9. Which religious preference (including None and Other) has the largest proportion of those who find life exciting?

Violent Crime. Use the following information for Exercises 10 and 11. The Federal Bureau of Investigation publishes crime statistics, including those in the following table, which shows the percentage of violent crime committed per month nationwide for the years 2002 and 2004.[4]

Month	2002	2004
January	7.9	7.8
February	6.8	7.0
March	7.9	8.3
April	8.1	8.2
May	8.7	9.0
June	8.8	8.6
July	9.3	9.2
August	9.3	9.0
September	9.2	8.5
October	8.6	8.6
November	7.7	7.8
December	7.7	7.9

10. Construct a scatterplot of 2004 monthly crime versus 2002 monthly crime.

11. Based on your scatterplot, would you characterize the linear relationship, if any, as positive or negative?

Equities and Investment Price Changes. Use the information in Table 4.19 for Exercises 12 and 13. The table contains the net price change for equities (y) and the net price change for guaranteed income investments (x) for a random selection of 6 investors.

TABLE 4.19   Net price change (in percent) for equities and guaranteed income investments for six investors

	Investor					
	1	**2**	**3**	**4**	**5**	**6**
Equities: y	10	−10	−5	5	8	−8
Guaranteed income: x	3	3	4	4	5	5

12. Investigating the correlation.
   a. Construct a scatterplot of the variables.
   b. What type of relationship do these variables have: positive, negative, or no apparent linear relationship?
   c. Will the correlation coefficient be positive, negative, or near zero? Will the slope of the regression line be positive, negative, or near zero?

13. Calculating and interpreting the correlation coefficient.
   a. Compute the value of the correlation coefficient, using a table similar to Table 4.9. Check your answer using a computer or calculator.
   b. Does this value for r concur with your earlier estimate?
   c. Interpret in two different ways the meaning of this value of the correlation coefficient for this data set.

Calories and Fat Content. Use the following information for Exercises 14 and 15. Minitab calculated the regression equation, shown here, for the linear relationship between the number of calories (y) and the fat content in grams (x) in a sample of breakfast cereals.

> **Regression Analysis**
> ```
> The regression equation is
> Calories = 97.1 + 9.65 Fat
> ```

14. The regression coefficients and the regression equation.
   a. What is the value of the y intercept?
   b. Interpret the meaning of this value for the y intercept. Does it make sense?
   c. In view of the fact that there are many cereals in the data set with zero fat content, would the value for the y intercept in **(b)** be considered extrapolation? Why or why not?
   d. Interpret the meaning of the slope coefficient $b_1 = 9.65$ as fat content increases.
   e. Write the regression equation. Now state in words what the regression equation means.
   f. Is the correlation coefficient positive or negative? How do you know?

15. Estimate the increase or decrease in calories, using a sentence, for the following situations.
   a. Suppose Cereal A has two more grams of fat than Cereal B.
   b. Suppose Cereal C has three fewer grams of fat than Cereal D.

Violent Crime. For Exercises 16–20, refer to the violent-crime data.

16. Calculate the three sums of squares: SSR, SST, and SSE.

17. Calculate s, the standard error of the estimate. What does this number mean?

18. Calculate r^2. Comment on how useful the 2002 percentages are in predicting the 2004 percentages.

19. Use r^2 to calculate and interpret the correlation coefficient.

20. Find the prediction error for the following percentages:
   a. 7.9   **b.** 9.3.   **c.** 8.1

Hank Morgan/Photo Researchers

chapter 15

Probability

CASE STUDY | **The ELISA Test for the Presence of HIV**

If someone suspects that he or she is at increased risk of HIV infection, then he or she might be interested in going for an HIV ELISA test. The ELISA test is used to screen blood for the presence of HIV. Sometimes called an HIV enzyme immunoassay (EIA), an HIV ELISA is the most basic test for finding out if an individual is carrying a particular pathogen, such as HIV.

Like most diagnostic procedures, the ELISA test is not foolproof. In this chapter's Case Study we study the types of errors the ELISA test can make and what this means for those who carry the HIV virus and for those who do not. For example, did you know that if your ELISA test comes back positive, then the chances are eight out of ten that you do *not* carry the virus?

The Big Picture

Where we are coming from, and where we are headed...

The first part of this textbook dealt with descriptive statistics, which help us summarize the characteristics of data sets. In later chapters, we will learn to use the information contained in a sample to make generalizations about the characteristics of a population. However, whenever you proceed from a part of the whole (the sample) to the whole (the population), there is *uncertainty* involved in drawing conclusions.

In this chapter, we learn how to discuss the likelihood of events using the language of uncertainty: **probability**. The chapter explains the tools of probability, which enable data analysts to quantify the level of uncertainty in their statistical inference.

In Section 5.1, we learn how to analyze and solve probability problems like the following: You are striding down the midway of your local town fair, when a particular game of chance catches your eye. The object of this game is to roll a 6 on a single roll of a single fair die. If you do so, you win $5. It costs $1 to play the game. What is the likelihood of winning?

To solve this and other problems, we introduce the building blocks of probability, learn how to assign probabilities, and learn how to simulate probabilities using technology. In Section 5.2, we learn about combining events using the concepts of union, intersection, and complement. In Section 5.3, we examine conditional probability, independent events, and sampling with and without replacement. Finally, in Section 5.4 we learn about the counting methods used to find the number of combinations and permutations of a set of objects.

Later, in Chapter 6, we will acquire further tools for putting probability to work for us, such as how to work with probability distributions. The two most important probability distributions, the binomial and the normal, will be our companions for the remainder of the text. ■

5.1 Introducing Probability

Objectives: By the end of this section, I will be able to...

1 Understand the meaning of an experiment, an outcome, an event, and a sample space.

2 Describe the classical method of assigning probability.

3 Explain the Law of Large Numbers and the relative frequency method of assigning probability.

1 Building Blocks of Probability: Experiments, Outcomes, Events, and Sample Spaces

Our daily lives are filled with *uncertainty,* seemingly governed by *chance.* We try to cope with uncertainty by estimating the *chances* that a particular event will occur. We are daily called on to make intelligent decisions about probabilities.

Consider the following scenarios, and think about how the italicized words all refer to uncertainty.

- What is the *chance* that there will be a speed trap on this stretch of I-95 on a particular day?

- What is the *likelihood* that this lottery ticket will make me rich?

- What is the *probability* that this throw of the dice will come up a seven?

Sometimes, the amount of uncertainty in our daily lives is so great that there appears to be no order to the world whatsoever. However, if you look closely, there are *patterns in randomness,* predictable structures in the seeming chaos. In this chapter, we learn to become better decision makers by organizing and structuring our thought process. We learn the language and rules of probability so that we can better quantify the myriad uncertainties of everyday life.

Developing Your Statistical Sense

A Different Perspective

As you read this chapter, notice that the perspective differs from that in previous chapters. Earlier, we were looking at a data set and trying to describe it graphically and numerically. Now, instead of trying to describe a data set, we are faced with an experimental situation, and our task is to calculate probabilities associated with various outcomes in the experiment. ■

> The **probability** of an outcome is defined as the long-term proportion of times the outcome occurs.

We discuss what we mean by "long-term proportion" later in this section. But first we acquaint ourselves with the building blocks of probability, starting with the concept of an experiment. In probability, an **experiment** is any activity for which the outcome is uncertain. Consider the stock market, for example. Suppose you own 100 shares of Consolidated Widgets and are interested in what the share price will be at the end of trading tomorrow. Will the share price increase or decrease? The actual result is uncertain, so this is an example of an experiment. Each of the possible results of the experiment is called an **outcome**. Another example of an experiment is when you toss a coin. In the coin toss experiment, the result may be heads or it may be tails. The collection of all possible outcomes is called the **sample space**. The sample space for the coin toss experiment is {heads, tails} or {H, T} (where the braces are used to enclose a set of outcomes). Following are some common experiments, together with their sample spaces.

Experiment	Sample space
Roll a single six-sided die	{1, 2, 3, 4, 5, 6}
Toss two coins	{HH, HT, TH, TT}
Play a video game	{win, lose}

We use the building blocks of probability to investigate the likelihood of an outcome or **event**.

> **Building Blocks of Probability**
>
> An **experiment** is any activity for which the outcome is uncertain.
>
> An **outcome** is the result of a single performance of an experiment.
>
> The collection of all possible outcomes is called the **sample space**. We denote the sample space S.
>
> An **event** is a collection of outcomes from the sample space. To find the probability of an event, add up the probabilities of all the outcomes in the event.

When we talk about the probability of some outcome, we are referring to a number that indicates how likely the particular outcome is. The notation $P(A)$ stands for "the probability that outcome A occurred." Say we define outcome W to be $W =$ "you win the video game." Then "the probability that you win the video game" can be denoted as $P(W)$. Probabilities abide by the following rules.

Rules of Probability

1. The probability $P(E)$ for any event E is always between 0 and 1, inclusive. That is, $0 \leq P(E) \leq 1$.

2. **Law of Total Probability:** For any experiment, the sum of all the outcome probabilities in the sample space must equal 1.

From the definition, the probability of an event is a proportion, so the probability cannot be negative because proportions cannot be negative and it cannot be greater than 1 (100%) because an event cannot occur more than 100% of the time. A **probability model** is a table or listing of all the possible outcomes of an experiment, together with the probability of each outcome. A probability model must follow the Rules of Probability.

If the probability that you calculated is negative or greater than 1, then you should try again.

Throughout the remainder of this book, you will often be asked to calculate the probability of various events. Following are the meanings of some probabilities.

Probability value	Meaning
Near 0	Outcome or event is very unlikely.
Equal to 0	Outcome or event cannot occur.
Near 1	Outcome or event is nearly certain to occur.
Equal to 1	Outcome or event is certain to occur. It's "a sure thing."
Low	Outcome or event is unusual.
High	Outcome or event is not unusual.

Higher probability values are associated with higher likelihood of occurrence. An outcome with probability 0.5 will happen about half of the time. An outcome with probability 0.95 is very likely. We say that an outcome or event is *unusual* if its probability is below a certain threshold, say, 0.05. (What determines this threshold depends on the specific experiment; the 0.05 is not set in stone.) When we perform an experiment, it is a "sure thing" that one of the outcomes in the sample space will occur. For example, when you toss a coin, you know that it will be either heads or tails. Put into probability terms, the sum of the probabilities of all the individual outcomes must equal 1, the **Law of Total Probability**.

The following table shows some typical events for the experiments in the probability value table.

Experiment	Sample space	Typical events
Roll a single die	{1, 2, 3, 4, 5, 6}	E: roll an even number = {2, 4, 6} L: roll a 4 or larger = {4, 5, 6}
Toss two coins	{HH, HT, TH, TT}	H: exactly one head = {HT, TH} T: at most one tail = {HH, HT, TH}
Play a video game	{win, lose}	W: win = {win} L: lose = {lose}

Classical Method of Assigning Probability

Many people have a certain degree of intuition when it comes to assigning probabilities. For example, when asked what the chances are of rolling a 6 on a single toss of a fair die, many people would quite correctly answer 1/6. However, intuition can often let us down. For example, when asked what the chances are of observing two heads when you toss a fair coin twice, many people would incorrectly respond 1/3 ("Well, it's either both heads or both tails or one of each." The correct answer is in fact 1/4.) In this section, we learn how to quantify our methods of assigning probabilities so that we don't have to depend on intuition alone.

There are three methods for assigning probabilities:

- Classical method
- Relative frequency method
- Subjective method

We first take a close look at the classical method. Later in this section we will examine the relative frequency method and the subjective method.

Many experiments are structured so that each experimental outcome is equally likely. **Equally likely outcomes** are outcomes that have the same probability of occurring. For example, if you toss a fair coin, the probability of observing either of the outcomes heads or tails is the same. The **classical method of assigning probabilities** is used when an experiment has equally likely outcomes.

Classical Method of Assigning Probabilities

Let $N(E)$ and $N(S)$ denote the number of outcomes in event E and the sample space S, respectively. If the experiment has equally likely outcomes, then the probability of event E is then

$$P(E) = \frac{\text{number of outcomes in } E}{\text{number of outcomes in sample space}} = \frac{N(E)}{N(S)}$$

Example 5.1　Probability of drawing an ace

Find the probability of drawing an ace when drawing a single card at random from a deck of cards.

Solution

The sample space for the experiment where a subject chooses a single card at random from a deck of cards is given in Figure 5.1. If the card is chosen truly at random, then it is

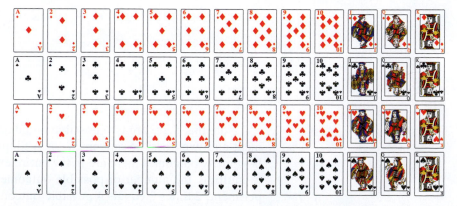

FIGURE 5.1 Sample space for drawing a card at random from a deck of cards.

reasonable to assume that each card has the same chance of being drawn. Since each card is equally likely to be drawn, we can use the classical method to assign probabilities.

There are 52 outcomes in this sample space, so $N(S) = 52$. Let E be the event that an ace is drawn. Event E consists of the four aces {A♥, A♦, A♣, A♠}, so $N(E) = 4$. Therefore, the probability of drawing an ace is

$$P(E) = \frac{N(E)}{N(S)} = \frac{4}{52} = \frac{1}{13}$$

Tree Diagrams

Often we need to count the number of outcomes of an experiment to solve a probability problem. A **tree diagram** is a device we can use to count the outcomes of an experiment like tossing a coin twice.

Think of this experiment as a two-stage process:

- Stage 1: Toss the coin the first time.

- Stage 2: Toss the coin the second time.

A tree diagram is a graphical display that helps us to visualize a multistage experiment. Figure 5.2 shows the tree diagram for the experiment of tossing a fair coin twice. Note the branches for Stage 1: the first time the coin is tossed, it can come up heads or tails. At Stage 2, the tree diagram again has branches for either heads or tails. The tree diagram helps us to construct the sample space for a multistage experiment. In this case, the sample space for the experiment of tossing a coin twice is {HH, HT, TH, TT}. There are $N(S) = 4$ outcomes in the sample space.

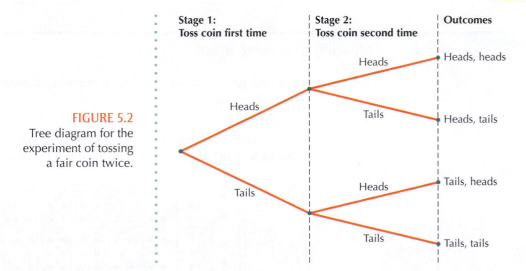

FIGURE 5.2
Tree diagram for the experiment of tossing a fair coin twice.

Note that there are two possible outcomes at Stage 1 of this two-stage experiment and two possible outcomes when flipping the coin at Stage 2. To determine how many outcomes there are in the entire experiment, the *counting rule* is simply to multiply the number of possible outcomes at each stage. In this two-stage experiment, 2 times 2 equals 4 possible outcomes, which is the number of outcomes we see in the sample space.

Example 5.2	Tossing a coin twice

Find the probability of obtaining one heads and one tails when a fair coin is tossed twice.

Solution

It is reasonable to assume that the $N(S) = 4$ outcomes in the sample space {HH, HT, TH, TT} are equally likely. The coin doesn't remember what occurred at Stage 1, so the probabilities at Stage 2 are precisely the same as at Stage 1. Also, recall from the Law of Total Probability that the sum of the probabilities of all the outcomes in the sample space must equal 1. Thus, each of the four outcomes must have probability 1/4. Let E be the event that one heads and one tails is obtained. Then $E = $ {HT, TH}, so $N(E) = 2$. Thus,

$$P(E) = \frac{\text{number of outcomes in } E}{\text{number of outcomes in sample space}} = \frac{N(E)}{N(S)} = \frac{2}{4} = \frac{1}{2}$$

Example 5.3	Probabilities when tossing two dice

Imagine that you are playing Monopoly with your dormitory roommate, and the loser has to do the laundry for both of you for the rest of the semester. You have a hotel on Boardwalk, and if your roommate lands on it, you will surely win. Right now your roommate's piece is on Short Line: if he or she rolls a 4, you get your laundry done for free for the remainder of the semester! Put into statistical terms, the experiment is to toss two fair dice and observe the sum of the two dice. Find the probability of rolling a sum of 4 when tossing two fair dice.

Solution

It is reasonable to assume that each of these $N(S) = 36$ outcomes in the sample space (Figure 5.3) is equally likely. If you wish, the experiment of tossing two dice can be viewed as a two-stage experiment, where we add the result from the first die to the result from the second die. If a 5 appears on the first (say, dark green) die, and a 3 appears on the second (light green) die, the overall outcome is (5,3), with the resulting sum equal to 8. Note that the outcome (5,3) is not the same as the outcome (3,5), where the dark green die comes up 3 and the light green die comes up 5.

Note: Did you know? People have been tossing dice for a long time. Archaeologists have dug up dice from Roman ruins looking just the same as ours. These three dice were uncovered from the ruins of Pompeii buried by the eruption of Mount Vesuvius in the first century A.D.

FIGURE 5.3 Sample space for tossing two fair dice.

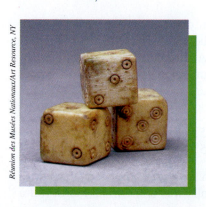

Let E denote the event that your roommate rolls a sum equal to 4. Then the outcomes that belong in this event are E: {(3,1) (2,2) (1,3)}, so $N(E) = 3$. Since the outcomes are equally likely, we can use the classical method for finding probabilities of events.

$$P(E) = \frac{\text{number of outcomes in } E}{\text{number of outcomes in sample space}} = \frac{N(E)}{N(S)} = \frac{3}{36} = \frac{1}{12}$$

The probability that your roommate will land on Boardwalk on this throw of the dice is 1/12.

Example 5.4   Inappropriate use of classical method: teenagers who own a personal computer

A recent study[1] showed that 59% of teenagers owned a computer (either a desktop or a laptop). Suppose we choose one teenager at random. Define the following events:

   C: The randomly chosen teenager owns a computer.
   D: The randomly chosen teenager does not own a computer.

Determine whether the classical method can be used to assign probability to events *C* and *D*.

Solution

Because a majority of teenagers own a computer, if we choose a teenager at random, we are more likely to select a teenager who owns a computer than to select one who does not. Therefore, the events *C* and *D* are not equally likely. It would be inappropriate to use the classical method of assigning probabilities for this experiment because the classical method can be used only when all the outcomes of an experiment are equally likely. ■

Example 5.5   Fair die toss outcomes are equally likely

Recall the town fair example from The Big Picture at the beginning of this chapter (page 194). In the game, you win if you roll a 6 on a single roll of a single fair die. Find the probability of winning the game.

Solution

The sample space for a single die toss consists of *six* outcomes, {1, 2, 3, 4, 5, 6}. When the six outcomes are equally likely, we say that the die is *fair*. If the outcomes are not equally likely, then the die is *loaded* or defective. If we assume the die is fair, then, since the sum of the probabilities of the $n = 6$ outcomes must equal 1, the probability of any particular outcome must equal 1/6, using the classical method. We write

$$\text{probability of winning} = P(W) = 1/6$$ ■

3 Law of Large Numbers and Relative Frequency Method of Assigning Probability

In Example 5.5, we need the classical method to find that the probability of rolling a 6 with a fair die is 1/6. What does this probability mean? Remember that the definition of *probability* included the phrase "long-term proportion." The next example demonstrates what we mean by "long-term."

Example 5.6   Simulating the long-term proportion of 6's in a fair die roll

Suppose we would like to investigate the proportion of 6's we observe if we roll a fair die 100 times. We can use technology, such as the TI-83/84 used here, to help us simulate rolling a fair die a large number of times. A **simulation** uses methods such as rolling dice or computer generation of random numbers to generate results from an experiment. The actual die rolls from our simulation are shown here, in order, with the 6's in boldface.

   144**6**24321343343**6**355153552131115556**6**3**6**21**6**55446541146
   42226325561131**6**5466**6**555255342464551**6**311135423336253

Thus, the first die roll was a 1, so the proportion of 6's was 0/1. The second and third die rolls were 4's, so the proportion of 6's after 3 rolls was 0/3. On the fourth roll a

6 appeared, so the proportion of 6's after the fourth roll was 1/4. Figure 5.4 provides a graph of the proportion of 6's in this simulation as the number of die rolls increased. Note that as the number of die rolls increases, the proportion of 6's tends to get closer to the horizontal line, $0.1667 \approx 1/6$.

The simulation was rerun, this time with 1000 die rolls. The resulting graph of the proportion of 6's is provided in Figure 5.5. Note that as the number of die rolls increases, the proportion of 6's approaches the line $0.1667 \approx 1/6$, and the fit is tighter with 1000 die rolls than with 100. This is what we mean by "long-term proportion." ▪

FIGURE 5.4 Proportion of 6's, 100 die rolls.

FIGURE 5.5 Proportion of 6's, 1000 die rolls.

This example leads directly to the following law.

Law of Large Numbers

As the number of times that an experiment is repeated increases, the relative frequency (proportion) of a particular outcome tends to approach the *probability* of the outcome. For quantitative data, as the number of times that an experiment is repeated increases, the mean of the outcomes tends to approach the population mean. For categorical (qualitative) data, as the number of times that an experiment is repeated increases, the proportion of times a particular outcome occurs tends to approach the population proportion.

The *Law of Large Numbers for Proportions* applet allows you to simulate coin tossing and observe the proportion of heads as the number of tosses increases.

Relative Frequency Method

Note: Tree diagrams can be used for the relative frequency method as well as the classical method of assigning probability.

If we can't use the classical method for assigning probabilities, then the **Law of Large Numbers** gives us a hint about how we can estimate the probability of an event. It often happens that previous information is available about the relative frequency of an event. Relative frequency information can be used to estimate the probability of the event.

Relative Frequency Method of Assigning Probabilities

The probability of event *E* is approximately equal to the relative frequency of event *E*. That is,

$$P(E) \approx \text{relative frequency of } E = \frac{\text{frequency of } E}{\text{number of trials of experiment}}$$

The relative frequency method is also known as the **empirical method**.

Example 5.7   Teen bloggers

A recent study found that 35% of all online teen girls are bloggers, compared to 20% of online teen boys. Suppose that the 35% came from a random sample of 100 teen girls who use the Internet, 35 of whom are bloggers. If we choose one teen girl at random, find the probability that she is a blogger.

Solution
Define the event.

B: The online girl is a blogger.

We use the relative frequency method to find the probability of event B:

$$P(B) \approx \text{relative frequency of } B = \frac{\text{frequency of } B}{\text{number of trials in experiment}} = \frac{35}{100} = 0.35 \quad \blacksquare$$

We can also use the relative frequency method to build a probability model with data that have been summarized in a table.

Example 5.8   Probability models based on tabular data

Table 5.1 contains the employment type for a sample of 1000 employed citizens of Fairfax County, Virginia.[2] Use the data to generate the relative frequencies and use the frequencies to estimate the probabilities for each employment type.

Table 5.1 Employment types

Employment type	Count
Private company	597
Federal government	141
Self-employed	97
Private nonprofit	92
Local government	59
State government	12
Other	2

Solution
We calculate the relative frequencies of each employment group by dividing the count (frequency) for each group by the sample size 1000. For example, the relative frequency for "Private Company" is $\frac{597}{1000} = 0.597$. The relative frequency is then used to estimate the probability of selecting citizens who work at private companies in Fairfax County, Virginia. Filling in the remaining calculations produces the *probability model* in Table 5.2. Note that the table follows the Rules of Probability in that (a) each outcome has probability between 0 and 1 and (b) the sum of the probabilities of all the outcomes equals 1.0. $\blacksquare$

Table 5.2 Probability model

Employment type	Probability
Private company	0.597
Federal government	0.141
Self-employed	0.097
Private nonprofit	0.092
Local government	0.059
State government	0.012
Other	0.002

Example 5.9   Random draws using a probability model

Suppose we consider the probabilities in Table 5.2 as population values. Use technology to simulate random draws using the probability model in Table 5.2.

Solution

Using the Step-by-Step Technology Guide below, we draw samples of sizes 10, 100, 1000, and 10,000 from the probability model in Table 5.2. The results are shown in Table 5.3.

Table 5.3 Relative frequencies from random draws of different sizes

Employment type	Rel freq $n = 10$	Rel freq $n = 100$	Rel freq $n = 1000$	Rel freq $n = 10,000$
Private company	0.60	0.62	0.566	0.596
Federal government	0.20	0.15	0.15	0.143
Self-employed	0.10	0.11	0.109	0.991
Private nonprofit	0.10	0.07	0.106	0.914
Local government	0.00	0.04	0.055	0.056
State government	0.00	0.01	0.012	0.012
Other	0.00	0.00	0.002	0.002

Note that each relative frequency tends to approach its respective probability as the sample sizes grow larger. ■

Subjective Method

There are cases where the outcomes are not equally likely (so classical method does not apply) and there has been no previous research (so the relative frequency approach does not apply). For example, what is the probability that the Dow Jones Industrial Average will decrease today? In cases like this, there is no absolutely correct probability. Reasonable people can disagree reasonably over these probabilities. The idea is to consider all available information, tempered by our experience and intuition, and then assign a probability value that expresses our estimate of the likelihood that the outcome will occur. For example, we might say, "The Chairman of the Federal Reserve warned against inflation in a major speech yesterday, so we expect that the probability that the Dow Jones Industrial Average will go down today is about 90%." Finally, it should be noted that the subjective method should be used when the event is not (even theoretically) repeatable.

> **Subjective probability** refers to the assignment of a probability value to an outcome based on personal judgment.

STEP-BY-STEP TECHNOLOGY GUIDE: Probability Simulations Using Technology

TI-83/84
Simulating 100 Die Rolls
Step 1 Set the random number seed as follows. (The random number seed is a number that the calculator uses to generate random numbers.) Enter any number on the home screen. Press

STO→, then **MATH**, highlight **PRB**, select **1: rand**, and press **ENTER**. On the home screen press **ENTER**.
Step 2 Press **MATH**, highlight **PRB**, select **5: randInt(**, and press **ENTER**.

(Continued)

Step 3 Enter **1**, comma, **6**, comma, **100**, close parenthesis (Figure 5.6).
Step 4 Store the data in list **L1** as follows. Press **STO→**, then **2nd**, then **1**, then press **ENTER**.
Step 5 To examine the die rolls, press **STAT**, select **1: EDIT**, and press **ENTER** (Figure 5.7).

Simulating Coin Flips

You can simulate coin flips instead of die rolls by coding "heads" as 1 and "tails" as 0. Use the instructions for simulating 100 die rolls with the following changes: Enter **0**, comma, **1**,

Figure 5.6

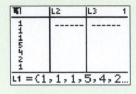

Figure 5.7

comma, **100**, close parenthesis, so that the home screen shows **randInt(0, 1, 100)**.

EXCEL
Simulating 100 Die Rolls
Step 1 Select cell **A1**. Click the **Insert Function** icon f_x.
Step 2 For **Search for a Function**, type **randbetween** and click **OK**.
Step 3 For **Bottom**, enter **1**. For **Top**, enter **6** (Figure 5.8). Click **OK**. Cell **A1** now contains a simulated random die roll.
Step 4 Select cell **A1**, copy it, and paste the contents into cells **A2** through **A100**.

Figure 5.8

Simulating the Sum of Two Dice
Step 1 Generate 100 die rolls in column A and another 100 die rolls in column B.
Step 2 Select cell **C1**. Enter **=(A1+B1)**, and press **ENTER**.

MINITAB
Simulating 100 Die Rolls
Step 1 Click on **Calc > Random Data > Integer**.
Step 2 For **Generate ___ rows of data**, enter **100**.
Step 3 For **Store in column(s)**, select **C1**.
Step 4 For **Minimum value**, enter **1**. For **Maximum value**, enter **6**.
Step 5 Click **OK**.

Simulating the Sum of Two Dice
Step 1 Generate 100 die rolls in **C1** and another 100 die rolls in **C2**.
Step 2 Click **Calc > Calculator**. For **Store result in variable**, enter **C3**. For **Expression**, enter **C1 + C2**. Click **OK**. Column **C3** then represents 100 randomly generated sums of two dice.

Step 3 Select cell **C1**, copy it, and paste the contents into cells **C2** through **C100**. Column C then represents 100 randomly generated sums of two dice.

Simulating Random Draws from a Probability Table
We illustrate using Example 5.9 (page 203). Excel and Minitab both require that the categories in the probability model be coded as numeric. We therefore code "Private company" as 1, "Federal government" as 2, and so on.
Step 1 Type the model categories (for example, "Employment type") in column A, their numeric codes in column B, and the respective probabilities in column C.
Step 2 Click **Data > Data Analysis > Random Number Generation**, then **OK**.
Step 3 For **Number of Variables**, enter **1**.
Step 4 For **Number of Random Numbers**, enter the desired sample size.
Step 5 For **Distribution**, select **Discrete**.
Step 6 For **Value & Prob. Input Range**, click and drag to select the coded categories and their probabilities, for example, **B1:C7**.

Repeat Steps 1–6 for increasing sample sizes.

Simulating Coin Flips Using Technology
You can simulate coin flips instead of die rolls by coding "heads" as 1 and "tails" as 0. Use the die roll instructions with the following changes: For **Bottom**, enter **0**. For **Top**, enter **1**.

Simulating Random Draws from a Probability Table
Step 1 Type the model categories in **C1**, their numeric codes in **C2**, and the respective probabilities in **C3** (Figure 5.9).

↓	C1-T Type	C2 Type-N	C3 Prob	C4 10	C5 100	C6 1000	C7 10000
1	Private Company	1	0.597	1	3	1	1
2	Federal Government	2	0.141	1	5	1	4
3	Self-Employed	3	0.097	1	1	1	2
4	Private Nonprofit	4	0.092	1	1	4	4
5	Local Government	5	0.059	1	1	1	1
6	State Government	6	0.012	3	1	2	3
7	Other	7	0.002	4	1	1	2
8				2	1	1	1
9				2	1	3	1
10				1	4	1	6
11					1	1	5
12					1	3	1
13					4	1	3

Figure 5.9

Step 2   Click on **Calc > Random Data > Discrete**.
Step 3   For **Generate ___ rows of data**, enter the desired sample size.
Step 4   For **Store in column(s)**, select the next available column, such as **C4**.
Step 5   For **Values in**, enter the column with the numerically coded categories, such as **C2**.
Step 6   For **Probabilities in**, enter the column with the probabilities, such as **C3**.

Step 7   Click **OK**.

Repeat Steps 1–7 for increasing sample sizes, as shown in Figure 5.9.

Simulating Coin Flips

You can simulate coin flips instead of die rolls by coding "heads" as 1 and "tails" as 0. Use the die roll instructions with the following changes: For **Minimum value**, enter **0**. For **Maximum value**, enter **1**.

Section 5.1 SUMMARY

1 Section 5.1 introduces the building blocks of probability, including the concepts of probability, outcome, experiment, and sample space. Probabilities always take values between 0 and 1, where 0 means that the outcome cannot occur and 1 means that the outcome is certain.

2 The classical method of assigning probability is used if all outcomes are equally likely. The classical method states that the probability of an event *A* equals the number of outcomes in *A* divided by the number of outcomes in the sample space.

3 The Law of Large Numbers states that, as an experiment is repeated many times, the relative frequency (proportion) of a particular outcome tends to approach the probability of the outcome. The relative frequency method of assigning probability uses prior knowledge about the relative frequency of an outcome. The subjective method of assigning probability is used when the other methods are not applicable.

Section 5.1 EXERCISES

CLARIFYING THE CONCEPTS

1. Give three examples from your own life of experiments, as the term is used in this chapter.
   a. For each experiment, what are some of the outcomes?
   b. Write out the sample space of one of these experiments.
   c. Describe how the Law of Total Probability applies to the sample.
2. What assumption do we need to make to use the classical method?
3. When can we use the relative frequency method?
4. If we can't use either the classical method or the relative frequency method, explain how we go about using the subjective method.
5. How would you find the probability that a randomly chosen student at your college likes hip-hop music? What method would you use?
6. Describe the meaning of the following probabilities.
   a. Near 0
   b. 0
   c. Near 1
   d. 1

PRACTICING THE TECHNIQUES

Determine whether each table in Exercises 7–11 is a probability model. If not, clearly explain why it is not a probability model.

7. Customers at a clothing store at the mall

Gender	Probability
Females	1.5
Males	0.2

8. Singers in the church choir

Voice	Probability
Soprano	0.25
Alto	0.25
Tenor	−0.25
Bass	0.50

9. Voters at a town meeting

Party	Probability
Democrat	0.3
Republican	0.25
Independent	0.25
Green	0.1
Libertarian	0.1
Other	0.1

10. Students taking undergraduate introductory statistics

Class	Probability
Freshmen	0.15
Sophomores	0.25
Juniors	0.40
Seniors	0.20

11. Reasons why Hurricane Katrina survivors did not evacuate

Reason	Probability
I did not have a car or a way to leave	0.36
I thought the storm and its aftermath would not be as bad as it was	0.29
I just didn't want to leave	0.10
I had to care for someone who was physically unable to leave	0.07
I waited too long	0.07
I was physically unable to leave	0.05
I worried that my possessions would be stolen or damaged if I left	0.04
I didn't want to leave my pet	0.01

For Exercises 12–17, the experiment is to roll a fair die once. Find the following probabilities.
12. Observing a 3
13. Observing an even number
14. Observing a number greater than 3
15. Observing a number less than 3
16. Observing a 3 or a 5
17. Observing a 3 and a 5

For Exercises 18–21, the experiment is to draw a card at random from a shuffled deck of 52 cards. Find the following probabilities.
18. Drawing a king
19. Drawing a heart
20. Drawing the king of hearts
21. Drawing a black card

For Exercises 22–26, consider the experiment of tossing a fair coin three times.
22. Construct the tree diagram for the experiment.
23. What is the sample space for the experiment?
24. How does the tree diagram help to construct the sample space?
25. How do we find each outcome using the tree diagram?
26. What is the probability of observing three heads? What method of assigning probability are you using?

For Exercises 27 and 28, refer to Exercises 12–17.
27. For each of Exercises 12–17, was the probability you found for an event or an outcome?
28. Explain in your own words why the probability of observing a 3 cannot be more than the probability of observing a 3 or a 5.

29. The experiment is to roll two fair dice and observe the sum.
  a. What is the most likely outcome?
  b. What is the probability of this outcome?

APPLYING THE CONCEPTS
30. Picnic Lunch. Picnickers at the Fourth of July Fair have the following preferences for grilled lunch: cheeseburger 50%, hot dog 25%, veggieburger 25%. Consider the experiment of two picnickers chosen at random choosing their preferred lunch.
  a. Construct the tree diagram for the experiment.
  b. What is the sample space?
31. Rainy Days. Students at the local middle school have been keeping track of the number of days it has rained. In the past 100 days, it rained on 33 days.
  a. What is the probability that it rains on a randomly chosen day?
  b. What is the probability that it doesn't rain on a randomly chosen day?
  c. Which method of assigning probability did you use?
32. Illegal Parking. Suppose you find that 10 out of 50 cars parked in handicapped zones did so illegally.
  a. You would like to find the probability that a car parked in the handicapped zone is parked illegally. Explain whether or not you can use the classical method of assigning probability and why.
  b. Find the probability that a car parked in a handicapped zone is illegally parked. Which method did you use?
33. Cutting Classes. Over her college career, a student has cut 25 of the total 1000 class periods in all her courses.
  a. Find the probability that the student cuts a particular class.
  b. Which method did you use?
  c. Does it make sense to use the classical method? Why or why not?
34. Basketball. Your college's basketball team is playing a game next week.
  a. What is the probability that the team will win the game?
  b. What reasoning went into your probability assessment?
  c. Which method did you use?
35. Die Rolls. A die is rolled twice. Define the following events for each roll: low = $\{1, 2\}$, medium = $\{3, 4\}$, high = $\{5, 6\}$.
  a. Construct a tree diagram for this experiment.
  b. Construct the sample space for this experiment.
  c. What is the probability of observing two low die results? What method of assigning probability are you using?
36. Refer to Exercise 35. Find the following probabilities.
  a. 2 high die results
  b. Exactly 1 medium die result
  c. Exactly 0 low die results
  d. At least 1 high die result
  e. At most 1 medium die result

37. Brisbane Babies. The table shows the births of babies at a Brisbane, Australia, hospital on a particular day.

Girl	Girl	Boy	Boy	Boy	Girl	Girl	Boy	Boy
Boy	Boy	Boy	Girl	Girl	Boy	Girl	Girl	Boy
Boy	Boy	Boy	Girl	Girl	Girl	Girl	Boy	Boy
Boy	Girl	Boy	Girl	Boy	Boy	Boy	Boy	Boy
Girl	Boy	Boy	Boy	Boy	Girl	Girl	Girl	

a. Construct a relative frequency distribution of the numbers of girls and boys born.
b. Use the relative frequencies to construct a probability model.
c. Confirm that your probability model follows the Rules of Probability.

Mozart Effect. Use this information for Exercises 38 and 39. The Mozart effect refers to the hypothesis that listening to a piano sonata by the great Austrian master improves cognitive performance. Assume that we have an experiment where two children have each listened to Mozart's Sonata for Two Pianos in D Major (K 448). The children's cognitive skills are then tested. Assume that for each child, there are three possible outcomes: cognitive performance increased, cognitive performance decreased, and cognitive performance remained unchanged.

38. a. Construct a tree diagram for the experiment.
   b. Construct the sample space for the experiment.
   c. For how many outcomes do both children improve their cognitive performance?
   d. Explain why it is not appropriate to use the classical method to assign probabilities for the Mozart experiment.

39. Challenger Exercise. Using your own subjective probabilities, find the probability that both children in the Mozart experiment will improve their cognitive performance.

Letters in English. Use Table 5.4 for Exercises 40 and 41 (it is the frequency distribution of letters in the English alphabet that we used in the Chapter 2 Case Study, The Caesar Cipher).

Imagine yourself on the television game show *Wheel of Fortune,* where contestants guess the letters contained in a hidden phase. You would like to ask for the letters that have the greatest chance of occurring, so you want to know the various probabilities of the letters in the English alphabet.

The experiment is to choose one letter at random from a sample of 1000 letters. The sample space is the English alphabet consisting of the usual 26 letters. Since the total sample size is 1000, you can find the relative frequencies of the letters in English simply by dividing each frequency by the total sample size (remember this from Chapter 2?).

40. Answer the following questions using Table 5.4.
   a. Choosing one letter at random, which is more likely, a consonant or a vowel?
   b. Of the sample of 1000 letters, what proportion are vowels?
   c. Of the 8 letters with the highest frequencies, what proportion are vowels?

TABLE 5.4   Frequency distribution of English letters

A	B	C	D	E	F	G
73	9	30	44	130	28	16
H	**I**	**J**	**K**	**L**	**M**	**N**
35	74	2	3	35	25	78
O	**P**	**Q**	**R**	**S**	**T**	**U**
74	27	3	77	63	93	27
V	**W**	**X**	**Y**	**Z**		
13	16	5	19	1		

41. Answer the following questions using Table 5.4.
   a. Compare your answers to exercise **40(b)** and **(c)**, and explain why *Wheel of Fortune* contestants are not allowed to guess vowels (they have to buy them).
   b. If you were a contestant on *Wheel of Fortune* and had no money to buy a vowel, what would be your first five letter choices?

42. Draw an Ace. If you draw the ace of spades from a deck of cards, you win $100.
   a. What is the probability of winning this game?
   b. What would be a fair price for playing this game? *Hint:* A fair price might be determined by balancing out the winnings and the price in the long run.

43. A Bazaar Game. Lenny has gone to the church bazaar with his family. In one of the games at the bazaar, if Lenny rolls two dice and gets a sum of at least 9 he wins $5; otherwise he wins nothing.
   a. Find the probability of winning $5.
   b. Find the probability of winning nothing.
   c. What would you suggest would be a fair (break-even) price for playing this game?

Teenagers' Music and Lifestyle. Every year, *USA Weekend* conducts a survey of the nation's teenagers, asking them various questions about lifestyle and music. In 2002, nearly 60,000 teenagers responded to the survey, conducted in part through a Web site. Use this information for Exercises 44 and 45.

44. One *USA Weekend* survey question asked teenagers, "Do you listen to music while you are . . . ?" and listed several options. The most common responses are shown in the table. Respondents could choose more than one response. Explain why you cannot construct a probability model with these percentages.

Doing chores	79%
On the computer	73%
Doing homework	72%
Eating meals at home	33%
In the classroom	18%

45. Another survey question asked by *USA Weekend* was "If you had to choose just one type of music to listen to exclusively, which would it be?" The results are shown in the table.

Hip-hop/rap	27%
Pop	23%
Rock/punk	17%
Alternative	7%
Christian/gospel	6%
R&B	6%
Country	5%
Techno/house	4%
Jazz	1%
Other	4%

a. Construct a probability model.
b. Is it unusual for a respondent to prefer jazz?
c. Use technology to draw random samples of sizes 10, 100, 1000, and 10,000 from your probability model.
d. What can you conclude about the relative frequencies as the sample size increases?

46. Fairfax County Income. The following table contains a probability model for the distribution of income in Fairfax County, Virginia.

Annual income	Probability
Under $25,000	0.083
$25,000 to $49,999	0.166
$50,000 to $74,999	0.169
$75,000 to $99,999	0.160
$100,000 to $149,999	0.200
$150,000 or more	0.222

a. Use technology to draw random samples of sizes 10, 100, 1000, and 10,000 from this probability model.
b. What can you conclude about the relative frequencies as the sample size increases?

Use the *Law of Large Numbers for Proportions* applet for Exercises 47 and 48.

47. Set the probability of heads to 0.5 and the number of tosses to 40. Click **Toss**.
a. Record the proportion of heads observed.
b. Without pressing **Reset**, continue to click **Toss** until the total number of tosses is 120. Again record the proportion of heads.
c. Without pressing **Reset**, continue to click **Toss** until the total number of tosses is 240. Again record the proportion of heads.
d. Without pressing **Reset**, continue to click **Toss** until the total number of tosses is 480. Again record the proportion of heads.

48. The proportions you recorded in Exercise 47 are relative frequencies of heads. What can you conclude about the relative frequencies as the sample size increases?

5.2 Combining Events

Objectives: **By the end of this section, I will be able to...**

1 Understand how to combine events using complement, union, and intersection.

2 Apply the Addition Rule to events in general and to mutually exclusive events in particular.

1 Combining Events Using Complement, Union, and Intersection

In Example 5.3, if your roommate rolled a 4, then your roommate was to do your laundry for the rest of the semester. Your roommate is keenly interested in *not* rolling a 4. If A is an event, then the collection of outcomes not in event A is called the **complement of A**, denoted A^c. The term *complement* comes from the word "to complete," meaning that any event and its complement together make up the complete sample space.

Example 5.10   Finding the probability of the complement of an event

If A is the event "observing a sum of 4 when the two fair dice are rolled," then your roommate is interested in the probability of A^C, the event that a 4 is not rolled. Find the probability that your roommate does not roll a 4.

Solution
Which outcomes belong to A^C? By the definition, A^C is all the outcomes in the sample space that do not belong in A. There are the following outcomes in A: $\{(3,1), (2,2), (1,3)\}$.

Figure 5.10 shows all the outcomes except the outcomes from A in the two-dice sample space. There are 33 outcomes in A^C and 36 outcomes in the sample space. The classical probability method then gives the probability of not rolling a 4 to be

$$P(A^C) = \frac{N(A^C)}{N(S)} = \frac{33}{36} = \frac{11}{12}$$

FIGURE 5.10 Outcomes in A^C.

So your roommate may be breathing a little easier. The probability is high that, on this roll at least, your roommate will not land on Boardwalk. (But you never know until you roll dem bones. . . .) ■

For event A in Example 5.10, note that

$$P(A) + P(A^C) = \frac{1}{12} + \frac{11}{12} = 1$$

Is this a coincidence, or does the sum of the probabilities of an event and its complement always add to 1? Recall the Law of Total Probability (Section 5.1), which states that the sum of all the outcome probabilities in the sample space must be equal to 1. Since any event A and its complement A^C together make up the entire sample space, then it always happens that $P(A) + P(A^C) = 1$.

Probabilities for Complements
For any event A and its complement A^C, $P(A) + P(A^C) = 1$. Applying a touch of algebra gives the following:

- $P(A) = 1 - P(A^C)$
- $P(A^C) = 1 - P(A)$

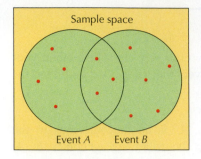

FIGURE 5.11 Union of events *A, B*.

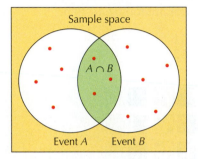

FIGURE 5.12 Intersection of Events *A, B*.

Sometimes we need to find the probability of a combination of events. For example, consider the casino game of craps where you roll two dice. One way of winning is by rolling the sum 7 or 11. We can find the probability of the following two events: the sum is 7 or the sum is 11. First, we need some tools for finding the probability of a combination of events.

The **union** of two events *A* and *B* is defined as the event containing all the outcomes that belong to *A* or *B* or both. The union of events *A* and *B* is denoted $A \cup B$. The English word associated with union is "or." So if you are asked to find the probability of "*A* or *B*," you should find the probability of $A \cup B$. Figure 5.11 shows the union of two events, with the red dots indicating the outcomes. Note from Figure 5.11 that the union of the events *A* and *B* refers to all outcomes in *A* or *B* or both.

The **intersection** of two events *A* and *B* is the event containing the outcomes that belong to both *A* and *B*. Figure 5.12 shows that the intersection of the two events is the part where *A* and *B* overlap. The intersection of events *A* and *B* is denoted as $A \cap B$ and is associated with the English word "and." Both union and intersection are commutative. That is, $A \cup B = B \cup A$ and $A \cap B = B \cap A$.

> **Union and Intersection of Events**
>
> The **union** of two events *A* and *B* is the event representing all the outcomes that belong to *A* or *B* or both. The union of *A* and *B* is denoted as $A \cup B$ and is associated with "or."
>
> The **intersection** of two events *A* and *B* is the event representing all the outcomes that belong to both *A* and *B*. The intersection of *A* and *B* is denoted as $A \cap B$ and is associated with "and."

Example 5.11 Union and intersection

Let our experiment be to draw a single card at random from a deck of cards. Define the following events:

   A: The card drawn is an ace.

   H: The card drawn is a heart.

a. Find $A \cup H$.

b. Find $A \cap H$.

Solution

a. The union of *A* and *H* is the event containing all the outcomes that are either aces or hearts or both (the ace of hearts). That is, the event $A \cup H$ consists of the set of outcomes (the cards) shown in Figure 5.13.

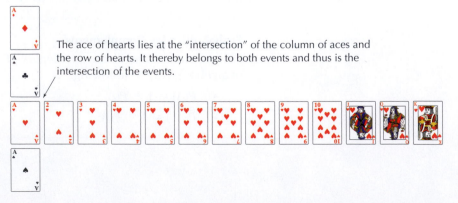

The ace of hearts lies at the "intersection" of the column of aces and the row of hearts. It thereby belongs to both events and thus is the intersection of the events.

FIGURE 5.13 The ace of hearts is the intersection of the events "ace" and "hearts."

b. The intersection of A and H is the event containing the outcomes that are common to both A and H. There is only one such outcome: the ace of hearts (see Figure 5.13). ■

Addition Rule

We are often interested in finding the probability that either one event *or* another event may occur. The formula for finding these kinds of probabilities is called the **Addition Rule**.

Addition Rule

$$P(A \text{ or } B) = P(A \cup B) = P(A) + P(B) - P(A \cap B)$$

What Does the Addition Rule Mean?

We can use Figure 5.11 to understand the Addition Rule. We are trying to find the probability of all the outcomes in A or B or both. The formula says to first count the probabilities of the outcomes in A and then add the probabilities of the outcomes in B. But what about the overlap between A and B, outcomes that belong to both events? To avoid counting the outcomes in the overlap (intersection) twice, we have to subtract the probability of the intersection, $P(A \cap B)$. ■

Example 5.12 Addition Rule applied to a deck of cards

Suppose you pay \$1 to play the following game. You choose one card at random from a deck of 52 cards, and you will win \$3 if the card is either an ace *or* a heart. Find the probability of winning this game.

Solution

Using the same events defined in Example 5.11, we find $P(A \text{ or } H) = P(A \cup H)$. By the Addition Rule, we know that

$$P(A \cup H) = P(A) + P(H) - P(A \cap H)$$

There are 4 aces in a deck of 52 cards, so by the classical method (equally likely outcomes), $P(A) = 4/52$. There are 13 hearts in a deck of 52 cards, so $P(H) = 13/52$. From Example 5.11, we know that $A \cap H$ represents the ace of hearts. Since each card is equally likely to be drawn, then $P(\text{ace of hearts}) = P(A \cap H) = 1/52$. Thus,

$$P(A \cup H) = P(A) + P(H) - P(A \cap H)$$
$$= \frac{4}{52} + \frac{13}{52} - \frac{1}{52} = \frac{16}{52} = \frac{4}{13}$$

■

The intersection of two events may be represented by the intersection of a row and a column in a two-way table. Recall from Section 4.1 (pages 148–149) that a *two-way table* (also known as a *crosstabulation* or a *contingency table*) is a tabular summary of the relationship between two categorical variables.

Example 5.13   Addition Rule applied to a two-way table

A study of online dating behavior found that users of a particular online dating service self-reported their physical appearance according to the counts given in Table 5.5.[3]

Table 5.5 Gender and self-reported physical appearance

| Gender | Physical Appearance | | | | |
	Very attractive	Attractive	Average	Prefer not to answer	Total
Female	3113	16,181	6,093	3478	28,865
Male	1415	12,454	7,274	2809	23,952
Total	4528	28,635	13,367	6287	52,817

Using this information, find the probability that a randomly chosen online dater has the following characteristics.

a.   Is female
b.   Self-reported as attractive
c.   Is a female who self-reported as attractive
d.   Is a female *or* self-reported as attractive

Solution

a.   There are a total of $N(S) = 52{,}817$ online daters in the entire data set. Of these, 28,865 are female, denoted as event F. Therefore,

$$P(F) = P(\text{Female}) = \frac{N(\text{Female})}{N(S)} = \frac{N(F)}{N(S)} = \frac{28{,}865}{52{,}817} \approx 0.5465$$

b.   There are 28,635 people who self-reported their physical appearance as attractive, denoted as event A. Therefore,

$$P(A) = P(\text{Self-reported attractive}) = \frac{N(\text{Self-reported attractive})}{N(S)} = \frac{N(A)}{N(S)}$$

$$= \frac{28{,}635}{52{,}817} \approx 0.5422$$

c.   The online daters who are both female and self-reported as attractive are shown in the highlighted cell in Table 5.5. This cell is located at the intersection of the row of females and the column of people who self-reported as attractive. Therefore, this cell reports the frequency of people who belong to both events. Thus,

$$P(F \text{ and } A) = P(F \cap A) = P(\text{Female and self-reported attractive}) = \frac{N(F \cap A)}{N(S)}$$

$$= \frac{16{,}181}{52{,}817} \approx 0.3064$$

d.   Here we seek $P(F \text{ or } A) = P(F \cup A)$. By the Addition Rule,

$$P(F \cup A) = P(F) + P(A) - P(F \cap A) = 0.5465 + 0.5422 - 0.3064 = 0.7823 \quad \blacksquare$$

Mutually Exclusive Events

When drawing a card at random from a deck of 52 cards, the events "a heart is drawn" and "a diamond is drawn" have no outcomes in common. That is, no card is both a heart and a diamond. We say that these two events are **mutually exclusive**.

> Two events are said to be **mutually exclusive** or *disjoint* if they have no outcomes in common.

Note that any event and its complement are always mutually exclusive. Other examples of mutually exclusive events are given in Table 5.6.

Table 5.6 Examples of mutually exclusive events

Experiment	Mutually exclusive events
Toss fair coin	Observe heads; observe tails
Draw a single card from a deck of 52 cards	Card is red; card is a spade
Select a student at random	Student is female; student is male
Choose a digit at random	Digit is even; digit is odd

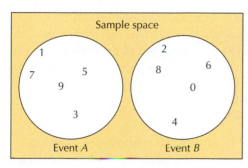

FIGURE 5.14 Mutually exclusive events.

Figure 5.14 shows how mutually exclusive events are represented graphically. It shows the events

$$A = \{1, 3, 5, 7, 9\} \quad \text{and} \quad B = \{0, 2, 4, 6, 8\}$$

Note that there is no overlap between the two events. When two events are mutually exclusive, they share no outcomes, and therefore the intersection of mutually exclusive events is empty. Since the intersection $(A \cap B)$ is empty, then for mutually exclusive events, $P(A \cap B) = 0$. Therefore, we can formulate a special case of the **Addition Rule for Mutually Exclusive Events** A and B:

$$P(A \cup B) = P(A) + P(B) - P(A \cap B) = P(A) + P(B) - 0 = P(A) + P(B)$$

> **Addition Rule for Mutually Exclusive Events**
> If A and B are mutually exclusive events, $P(A \cup B) = P(A) + P(B)$.

Example 5.14   Addition Rule for Mutually Exclusive Events

Using Table 5.5 from Example 5.13, find the probability that a randomly chosen online dater self-reported as either attractive or very attractive.

Solution

From Table 5.5, there are 28,635 online daters who self-reported as attractive and 4528 who self-reported as very attractive, yielding the following probabilities:

$$P(A) = P(\text{Self-reported attractive}) = \frac{N(A)}{N(S)} = \frac{28,635}{52,817} \approx 0.5422$$

$$P(V) = P(\text{Self-reported very attractive}) = \frac{N(V)}{N(S)} = \frac{4528}{52,817} \approx 0.08573$$

Since no online daters self-reported as both attractive and very attractive, the two groups are mutually exclusive. Thus, by the Addition Rule for Mutually Exclusive Events,

$$P(A \cup V) = P(A) + P(V) = 0.5422 + 0.08573 = 0.62793$$

Section 5.2 SUMMARY

1 Combinations of events may be formed using the concepts of complement, union, and intersection.

2 The Addition Rule provides the probability of Event *A* or Event *B* to be the sum of their two probabilities minus the probability of their intersection. Mutually exclusive events have no outcomes in common.

Section 5.2 EXERCISES

CLARIFYING THE CONCEPTS

1. Describe in your own words what it means for two events to be mutually exclusive.
2. Describe the intersection of two mutually exclusive events.
3. Is it true that the union of two events always contains at least as many outcomes as the intersection of two events? Use Figures 5.11 and 5.12 to help you visualize this problem.
4. If we choose a student at random from your college or university, is it more likely that we choose a male or a male football player? Why?
5. What is your personal estimate of the probability that it will rain on any given day? How about the probability that it won't rain? Why do these numbers have to add up to 1 (or 100%)?

PRACTICING THE TECHNIQUES

For Exercises 6–11, consider the experiment of rolling a fair die twice. Find the indicated probabilities.

6. Exactly one of the dice is a 4
7. Neither die is a 4
8. Sum of the two dice equals 3
9. Sum of the two dice equals 3 and one of the dice is a 4
10. Sum of the two dice equals 3 or one of the dice is a 4
11. Sum of the two dice equals 3 or neither of the dice is a 4

For Exercises 12–17, consider the experiment of drawing a card at random from a shuffled deck of 52 cards. Find the indicated probabilities.

12. Drawing a king and a heart
13. Drawing a king or a heart
14. Drawing a card that is neither a king nor a heart
15. Drawing a heart or a spade
16. Drawing a heart and a spade
17. Drawing a card that is not the king of hearts

For Exercises 18–23, consider the experiment of drawing a card at random from a shuffled deck of 52 cards. Find the indicated probabilities.

18. Drawing a face card (king, queen, or jack)
19. Drawing a card that is not red
20. Drawing a card that is not a face card
21. Drawing a face card that is not a diamond
22. Drawing a face card or a diamond
23. Drawing a face card and a diamond

For Exercises 24–27, consider the experiment of tossing a fair coin three times. Find the indicated probabilities. (*Hint*: Use a tree diagram similar to the one in Figure 5.2 in Section 5.1 [page 198] but going one step farther.)

24. Observing 3 heads
25. Not observing 3 heads
26. Observing 2 tails
27. Not observing 2 tails

For Exercises 28–33, imagine that your sister is going to have triplets. Assume that the probability of a baby boy or a baby girl is equally likely. (In fact, it is not quite.)

28. Construct the sample space.
29. Find the probability of 1 girl and 2 boys.
30. Find the probability of 1 boy and 2 girls.
31. Find the probability of 2 of one gender and 1 of the other gender.
32. Find the probability of 1 girl or 1 boy.
33. Find the probability of getting 3 girls.

APPLYING THE CONCEPTS

34. Game of Craps. You win the casino game of craps if you roll a 7 or 11. Find the probability of rolling a sum of 7 or 11 when two dice are rolled.

35. Acing College Courses. Jessica has gotten A's in 10 of her 30 college courses.
   a. Find the probability that Jessica gets an A in a randomly chosen college course.
   b. Find the probability that Jessica gets a grade other than an A in a randomly chosen college course.

Trout Fishing. Use the following information for Exercises 36 and 37. Of the 20 fish Brent has caught at his favorite fishing spot this season, 5 have been trout and 7 have been bass.

36. Find the following probabilities.
   a. Catching a trout
   b. Catching a bass
37. Find the following probabilities.
   a. Catching a trout or a bass
   b. Catching a fish that is not a trout
   c. Catching a fish that is neither a trout nor a bass

38. Traffic Lights. Let *A* be the event that you encounter a green light at your next traffic light.
   a. What outcomes make up A^C?
   b. What is the probability of *A*? Which method did you use?
   c. What is the probability of A^C?

39. Monopoly. You are playing Monopoly with your roommate (the loser does laundry for both for the rest of the semester). Your piece is sitting on Water Works, and you are about to roll your two fair dice. Your opponent has hotels on Park Place and Boardwalk, and if you land on either one, you will lose. Park Place lies 9 spaces away and Boardwalk lies 11 spaces away.
   a. What is the probability that you will land on Park Place and get stuck with laundry duty?
   b. What is the probability that you will land on Boardwalk and still get stuck with laundry duty?
   c. What is the probability that you will lose the game on this roll?

40. Monopoly. As well as the conditions in Exercise 39, you have hotels on the green properties (Pacific, North Carolina, and Pennsylvania Avenues), and if your roommate lands on one of them, your roommate will be sorting your laundry starting tonight. Your roommate's piece is on B & O Railroad.
   a. What is the probability that your roommate will land on Pacific Avenue, 6 spaces away?
   b. What is the probability that your roommate will land on North Carolina Avenue, 7 spaces away?
   c. What is the probability that your roommate will land on Pennsylvania Avenue, 9 spaces away?
   d. What is the probability that your roommate will lose the game on this roll?
   e. For this roll only, who has the greater chance of sorting your permanent press from your towels for the rest of the semester?

41. High School Students. In a local high school of 500 students, there are 200 females, 100 sophomores, and 50 female sophomores.
   a. If we choose a student at random, what is the probability that we choose a female or a sophomore?
   b. Find the probability that a randomly chosen student is a male or a sophomore.
   c. Find the probability that a randomly chosen student is a female or is not a sophomore.

42. Halloween Candy. At Halloween, the accent is on taste—taste of candy, that is. In a sample of 100 children, 70 like chocolate bars, 60 like peanut butter cups, and 50 like both.
   a. If we choose one child at random, find the probability that the child likes either chocolate bars or peanut butter cups.
   b. In **(a)**, suppose you forgot to subtract the probability of the intersection. How would you know that your answer is wrong?

43. Pick a Card. If we draw a single card at random from a deck of 52 playing cards, find the probability that the card is
   a. a heart or a diamond.
   b. a red card or a jack.
   c. a club or a face card (king, queen, jack).
   d. a heart and a diamond.
   e. not a spade.

Don't Mess with Texas. Don't Mess with Texas (**dontmesswithtexas.org**) is a Texas statewide antilittering organization. Its 2005 report, *Visible Litter Study 2005*, identified paper, plastic, metal, and glass as the top four categories of litter by composition. The report also identified tobacco, household/personal, food, and beverages, as the top four categories of litter by use. Assume that a sample of 12 items of litter had the following characteristics. Use Table 5.7 for Exercises 44–46.

TABLE 5.7   Litter composition and use

Litter item	Composition	Use
1	Paper	Tobacco
2	Plastic	Household/personal
3	Glass	Beverages
4	Paper	Tobacco
5	Metal	Household/personal
6	Plastic	Food
7	Glass	Beverages
8	Paper	Household/personal
9	Metal	Household/personal
10	Plastic	Beverages
11	Paper	Tobacco
12	Plastic	Food

44. A litter item is chosen at random.
   a. Find the probability that the composition of the item is paper.
   b. Find the probability that the composition of the item is not paper. Calculate this probability in two different ways.

45. A litter item is chosen at random.
   a. Find the probability that the use of the item is tobacco.
   b. Find the probability that the use of the item is not tobacco. Calculate this probability in two different ways.

46. A litter item is chosen at random.
   a. Find the probability that the composition of the item is paper *and* its use is tobacco.
   b. Find the probability that the composition of the item is paper *or* its use is tobacco.

47. A New Sonata. Music researchers have discovered a new sonata from the classical period of the late eighteenth century. The probability that the sonata was written by the following composers is as follows: Mozart 35%, Haydn

50%, Beethoven 15%. It is known that the sonata composers are mutually exclusive, meaning that no sonata was written by more than one of these composers. Find the following probabilities.

    a. The sonata was written by either Mozart or Haydn.
    b. The sonata was written by either Mozart or Beethoven.
    c. The sonata was written by either Haydn or Beethoven.
    d. The sonata was written by both Mozart and Haydn.

48. Cheeseburger or Chicken? The college football team (60 players) stopped off at a hamburger joint on the way home from a big road win. Thirty players ordered a bacon cheeseburger. Twenty players ordered a chicken deluxe. No player ordered both.

    a. Find the probability that a randomly chosen player ordered either a bacon cheeseburger or a chicken deluxe.
    b. Suppose we didn't know that no player ordered both. Would we have been able to answer **(a)**? Why or why not?

49. Baseball Strategy. Earnshaw Cook analyzed 20,000 baseball games and used the relative frequency method to assign the following probabilities.[4] The probability of scoring a run with a runner on first and no outs is 0.43. The probability of scoring a run with a runner on second and one out is 0.45. Actually, considering all the bunting and sacrifice fly strategies used by baseball managers to get from the first situation to the second, it seems hardly worth it, increasing the probability of scoring by only 0.02. Using these probabilities, answer the following questions.

    a. If there is a runner on first with no outs, what is the probability of not scoring?
    b. If there is a runner of second with one out, what is the probability of not scoring?

50. Online Dating Data. Refer to Table 5.5 (page 212). Find the probability that a randomly selected online dater has the following characteristics.

    a. Is male
    b. Prefers not to describe physical appearance
    c. Is male and prefers not to describe physical appearance
    d. Is male or prefers not to describe physical appearance

Causes of Death. Refer to Table 5.8 for Exercises 51–53.

51. Find the following probabilities.
    a. The cause of death was heart disease .
    b. The cause of death was not heart disease .

52. Find the following probabilities.
    a. The cause of death was heart disease and cancer.
    b. The cause of death was heart disease or cancer.

53. Are the causes of death mutually exclusive?

TABLE 5.8   Causes of death

Cause of death	Deaths
Heart disease	654,092
Cancer	550,270
All other causes	1,194,003
Total	2,398,365

Source: Centers for Disease Control, 2004.

5.3 Conditional Probability

Objectives: **By the end of this section, I will be able to…**

    1 Calculate conditional probabilities.

    2 Explain independent and dependent events.

    3 Solve problems using the Multiplication Rule.

    4 Recognize the difference between sampling with replacement and sampling without replacement.

1 Introduction to Conditional Probability

As we progress through this book, you will notice a recurring theme: *the more information available, the better*. Very often, when we are investigating the probability of a certain event *A,* we learn that another event *B* has occurred. If events *A* and *B* are related, then the occurrence of event *B* often influences the probability that event *A* will occur.

Example 5.15   Having more information often affects the probability of an event

In Section 5.1, we found that the probability of rolling a sum of 4 when tossing two dice is 3/36 ≈ 0.0833. But what if we were told that at least one of the dice shows a 1. How does this extra information affect the probability of rolling a 4?

Solution
Figure 5.15 shows the 11 outcomes from the two-dice sample space in which at least one die shows a 1. The extra information reduces the number of possible outcomes in the sample space from 36 to 11. We see that two of these outcomes have a sum equal to 4. Thus, the probability of observing a sum of 4, *given that* at least one of the dice shows a 1, is 2/11 ≈ 0.1818. ◼

FIGURE 5.15 Using the extra knowledge changes the probability.

The extra information about a related event changed the probability of the event of interest. This type of probability is an example of what is called **conditional probability**.

> For two related events A and B, the probability of B given A is called a **conditional probability** and denoted $P(B|A)$.

Thus, if we let A represent the event that one of the dice shows a 1, and let B represent the event that the sum of the two dice equals 4, then

$$P(B) = \frac{3}{36} \approx 0.0833 \qquad \text{but} \qquad P(B|A) = \frac{2}{11} \approx 0.1818$$

Figure 5.16 can help us visualize how conditional probability works. The idea is that, once event A has occurred, the only chance for event B to occur is in the overlap, the intersection $A \cap B$. Therefore, the conditional probability that B will occur, given that event A has already taken place, is found by taking the ratio $P(A \cap B)/P(A)$.

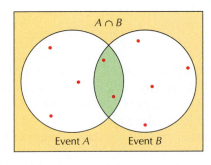

(a)   (b)

FIGURE 5.16 How conditional probability works.

> **Calculating Conditional Probability**
> The conditional probability that B will occur, given that event A has already taken place, equals
> $$P(B|A) = \frac{P(A \cap B)}{P(A)} = \frac{N(A \cap B)}{N(A)}$$

Example 5.16   Probability of responding to a direct mail marketing campaign

Punchstock/Photodisc

Market research represents a fast-growing field in need of analysts with strong statistical expertise. Table 5.9 is adapted from a study on direct mail marketing. It contains the numbers of customers who either responded or did not respond to a direct mail marketing campaign, along with whether they had a credit card on file with the company. The two events are

R: Responded to direct mail marketing campaign.
C: Has a credit card on file.

Table 5.9 Credit card status and marketing response

Response	Credit Card on File?		Totals
	No	Yes	
Did not respond	161	79	240
Did respond	17	31	48
Totals	178	110	288

Source: Daniel Larose, *Data Mining Methods and Models* (Wiley Interscience, 2006).

a. Find the probability that a randomly chosen customer responded to the marketing campaign.
b. Find the conditional probability that a randomly selected customer responded, given that the customer has a credit card on file.

Solution

a. $P(R) = \frac{N(R)}{N(S)}$. There are $N(R) = 48$ customers who did respond, and there are $N(S) = 288$ customers in this experiment. Thus,

$$P(R) = \frac{N(R)}{N(S)} = \frac{48}{288} \approx 0.1667$$

b. We will use $P(R|C) = N(R \cap C)/N(C)$ because in this example it is easier to work directly with the numbers of outcomes rather than the probabilities. Now, $R \cap C$ represent customers who did respond and had a credit card on file. From Table 5.9, there are $N(R \cap C) = 31$ such customers. Also, there are $N(C) = 110$ customers total who had a credit card on file. Therefore,

$$P(R|C) = \frac{N(R \cap C)}{N(C)} = \frac{31}{110} \approx 0.2818$$

That is, the probability that a randomly chosen customer responded to the direct mail marketing campaign, given that the customer had a credit card on file, is 0.2818. ■

hat Do These Numbers Mean?

Conditional Probability

Conditional probabilities can often be interpreted as percentages of some *subset* of a population. For example, the conditional probability that a customer responded, given that the customer has a credit card on file, may be interpreted as the percentage of customers with credit cards who responded. ■

Students sometimes confuse the meanings of $P(B \cap A)$ and $P(B \mid A)$. For $P(B \mid A)$, we *assume that the event A has occurred* and now need to find the probability of event B, given event A. On the other hand, for $P(B \cap A)$, we do not assume that event A has occurred and instead need to determine the probability that both events occurred.

2 Independent Events

Since having a credit card on file increased the probability of a customer responding from 0.1667 to 0.2818, we can therefore say that the probability of responding *depends* in part on whether the customer has a credit card on file. In other words, the events R and C are **dependent events**.

On the other hand, if the probability of responding had been unaffected by whether the customer had a credit card on file, then we would have said that R and C were **independent events**. That is, R and C would have been independent events had $P(R \mid C)$ equaled $P(R)$. In general, if the occurrence of an event does not affect the probability of a second event, then the two events are independent.

Events A and B are **independent** if

$$P(A \mid B) = P(A) \qquad \text{or if} \qquad P(B \mid A) = P(B)$$

Otherwise the events are said to be **dependent**.

Strategy for Determining Whether Two Events Are Independent

1. Find $P(B)$.
2. Find $P(B \mid A)$.
3. Compare the two probabilities. If they are equal, then A and B are independent events. Otherwise, A and B are dependent events.

Example 5.17   Are successive coin tosses independent?

Let our experiment be tossing a fair coin twice. Determine whether the following events are independent.

A: Heads is observed on the first toss.
B: Heads is observed on the second toss.

hat Results Might We Expect?

Intuitively, we may consider that the coin has no memory, and that therefore, the result of previous tosses would not affect the probability of future outcomes. Thus, we would expect that successive coin tosses are independent. ■

Solution

Figure 5.17 shows the tree diagram for this experiment.

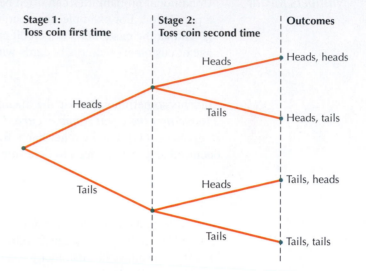

FIGURE 5.17 Tree diagram for the experiment of tossing a fair coin twice.

Step 1   To find $P(B)$, the probability of observing heads on the second toss, we concentrate on the second flip of the experiment, where the coin is tossed for the second time. At Stage 2, there are four branches, two of which are labeled "Heads." Therefore, since each outcome is equally likely, the probability of observing heads on the second toss is $P(B) = 2/4 = 1/2$.

Step 2   Next we find $P(B \mid A)$. Again consider Stage 2, but this time restrict your attention to the two "upper branches," which correspond to having observed heads on the first toss. One upper branch is labeled "Heads." Therefore, since each outcome is equally likely, $P(B \mid A) = 1/2$.

Step 3   Finally, we have $P(B) = P(B \mid A)$, and we conclude that events A and B are *independent events*. That is, *successive coin tosses are independent*. (Exercises 33–37 [page 229] show that successive die rolls are independent.)   ■

Example 5.18   Gambler's Fallacy

Suppose we have tossed a fair coin ten times and have observed "heads" come up every time. Find the probability of tails on the next toss.

Solution

Since we have observed an unusual number of heads, we might think that the probability of tails on the next toss is increased. However, the short answer is "Not so." We learned in Example 5.17 that successive tosses of a coin are independent. Thus, what happened on the first ten tosses has no effect on the next toss. Probability theory tells us that in the long run, the proportion of heads and tails will eventually even out, if the coin is fair. Therefore, the probability of tails on the next toss is 0.5. This is an example of the Gambler's Fallacy.   ■

𝟛 Multiplication Rule

Just as the Addition Rule is used to find probabilities of unions of events, the **Multiplication Rule** is used to find probabilities of intersections of events. Recall the formula for the conditional probability of event B given event A:

$$P(B \mid A) = \frac{P(A \cap B)}{P(A)}$$

We solve for $P(A \cap B)$ by multiplying each side by $P(A)$:

$$P(A \cap B) = P(A)\,P(B \mid A)$$

Similarly, consider the conditional probability of event A given event B:

$$P(A \mid B) = \frac{P(A \cap B)}{P(B)}$$

Solving for $P(A \cap B)$ gives us a second equation for $P(A \cap B)$:

$$P(A \cap B) = P(B)\,P(A \mid B)$$

The two equations for $P(A \cap B)$ lead directly to the Multiplication Rule.

Multiplication Rule

$$P(A \cap B) = P(A)\,P(B \mid A) \qquad \text{or equivalently} \qquad P(A \cap B) = P(B)\,P(A \mid B)$$

Example 5.19   Probability of heads on both tosses of a fair coin

Again suppose that our experiment consists of tossing a fair coin twice. What is the probability of observing heads on both tosses?

Solution

Define the following events:

    A: Heads is observed on the first toss.
    B: Heads is observed on the second toss.

Since the coin is fair, $P(A) = 1/2$, and from Example 5.17, $P(B \mid A) = 1/2$. Thus, using the Multiplication Rule,

$$P(A \cap B) = P(A)\,P(B \mid A) = (1/2)(1/2) = 1/4$$

This value concurs, of course, with the result in Example 5.17 for the probability of observing two heads. ◼

When events A and B are independent, $P(A \mid B) = P(A)$ or $P(B \mid A) = P(B)$. Using these identities, we can formulate a special case of the Multiplication Rule. Using $P(A \mid B) = P(A)$, we can write the Multiplication Rule as

$$P(A \cap B) = P(B)\,P(A \mid B) = P(B)\,P(A) = P(A)\,P(B)$$

Equivalently, the Multiplication Rule also states that $P(A \cap B) = P(A)\,P(B \mid A)$, but if A and B are independent, $P(B \mid A) = P(B)$, so again $P(A \cap B) = P(A)\,P(B)$.

Multiplication Rule for Two Independent Events

If A and B are any two independent events, $P(A \cap B) = P(A)\,P(B)$.

---

d. What is the probability that you will accept the batch?

e. What is the probability that you will reject the batch?

f. Usually you accept each batch of computers from this wholesaler. Do you think that is a wise move, considering that 10% of their product is defective? How could you make your test stricter so that there is a smaller chance of accepting a batch with 10% defectives?

47. 9/11 Information. Within two weeks of the terrorist attacks of September 11, 2001, the Pew Internet and American Life Project conducted a survey, asking, among other things, from which medium the respondent received most of his/her information regarding these events. The results indicated that 1.9% of Americans received most of their information about September 11 from the Internet. Assuming independence, find the following probabilities.

a. 2 randomly selected people both received most of their information from the Internet

b. 2 randomly selected people both received most of their information from sources other than the Internet

c. 5 randomly selected people received most of their information from sources other than the Internet

d. At least 1 of 3 randomly selected persons received most information from the Internet

48. Treasury Bonds. One of the most important tasks for economists is to make forecasts for the performance (up or down) of investments such as 30-year Treasury bonds. *The Journal of Investing* (Vol. 6, No. 2, page 8, 1997) reports that, in a sample of 30 six-month surveys, the consensus estimate of performance for the 30-year Treasury bond has been wrong 20 out of the 30 times!

a. Find the probability that two randomly selected consensus estimates were correct.

b. Find the probability that three randomly selected consensus estimates were wrong.

c. If we choose two consensus estimates and if we sample with replacement, find the probability that the second consensus estimate was right, given that the first consensus estimate was right. Are the successive draws independent? Why or why not?

d. Repeat (c), this time sampling without replacement. Are the successive draws independent? Why or why not?

49. Female Students. At Northern Connecticut University, 55% of the students are female. Of these, 10% are business majors. Find the following probabilities, assuming sampling with replacement.

a. A randomly chosen student is a female business major.

b. Two randomly selected students are females.

c. The second of two randomly selected students is a female, given that the first is a female.

d. The second of two randomly selected students is a female, given that the first is a male.

e. Two randomly selected students are female business majors.

f. If possible, find the probability that two randomly selected students are business majors. If not possible, explicitly explain why not.

50. Adjustable Rate Mortgages. Half of the 20 mortgages provided by a certain mortgage lending company last week are adjustable rate mortgages (ARMs). Suppose we sample three mortgages without replacement. Find the following probabilities.

a. The first mortgage is an ARM.

b. The second mortgage is an ARM, given that the first mortgage is an ARM.

c. The third mortgage is an ARM, given that the first two mortgages are ARMs.

5.4 Counting Methods

Objectives: By the end of this section, I will be able to...

1 Apply the Multiplication Rule for Counting to solve certain counting problems.

2 Use permutations and combinations to solve certain counting problems.

3 Compute probabilities using combinations.

Counting methods allow us to solve a range of problems, including how to compute certain probabilities.

1 Multiplication Rule for Counting

Let us begin with an example illustrating a general rule of counting.

Bruce Laurance/The Image Bank/Getty Images

Example 5.27   Design your own T-shirt

A store at the local mall allows customers to design their own T-shirts. The store offers the following options to its customers:

- **Sleeve type:** Long-sleeve or short-sleeve
- **Color:** White, black, or red
- **Image:** Stock picture or uploaded photo

Figure 5.18 is a tree diagram that shows all the different T-shirts that can be designed.

Solution

There are two choices for type of sleeve. For each sleeve type, there are three choices for color. For each color, there are two choices of image: stock picture or uploaded photo. All together, customers have a choice from among

$$2 \cdot 3 \cdot 2 = 12$$

different T-shirt options.

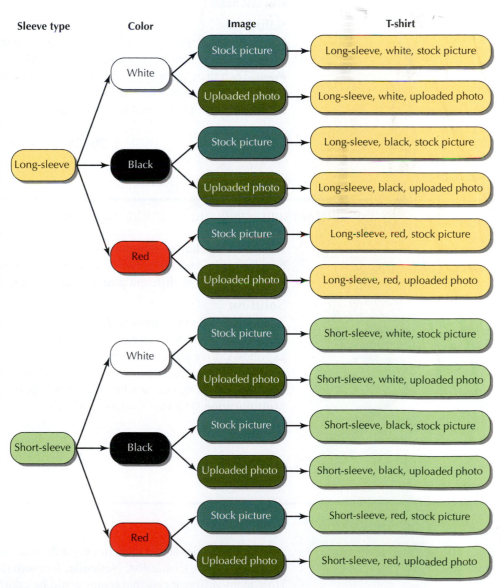

FIGURE 5.18 Tree diagram for the different T-shirt options.

We can generalize the result as the **Multiplication Rule for Counting**.

> **Multiplication Rule for Counting**
> Suppose an activity consists of a series of events in which there are *a* possible outcomes for the first event, *b* possible outcomes for the second event, *c* possible outcomes for the third event, and so on. Then the total number of different possible outcomes for the series of events is
>
> $$a \cdot b \cdot c \cdot \ldots$$

Example 5.28   Counting with repetition: Famous initials

Some Americans in history are uniquely identified by their initials. For example, "JFK" stands for John Fitzgerald Kennedy, and "FDR" stands for Franklin Delano Roosevelt. How many different possible sets of initials are there for people with a first, middle, and last name?

Solution

Let us consider the three initials as an activity consisting of three events. Note that a particular letter may be repeated, as in "AAM" for A. A. Milne, author of *Winnie the Pooh*. Then there are $a = 26$ ways to choose the first initial, $b = 26$ ways to choose the second initial, and $c = 26$ ways to choose the third initial. Thus, by the Multiplication Rule for Counting, the total number of different sets of initials is

$$26 \cdot 26 \cdot 26 = 17{,}576$$

Example 5.29   Counting without repetition: Intramural singles tennis

Note: To summarize the key difference between Examples 5.28 and 5.29, if repetitions are allowed, then $a = b = c$. If repetitions are not allowed, then the numbers being multiplied decrease by one from left to right.

A local college has an intramural singles tennis league with five players, Ryan, Megan, Nicole, Justin, and Kyle. The college presents a trophy to the top three players in the league. How many different possible sets of three trophy winners are there?

Solution

The major difference between Examples 5.28 and 5.29 is that in Example 5.29 there can be no repetition. Ryan cannot finish in first place *and* second place. So we proceed as follows. Five possible players could finish in first place, so $a = 5$. Now there are only four players left, one of whom will finish in second place, so $b = 4$. That leaves only three players, one of whom will finish in third place, giving $c = 3$. Thus, by the Multiplication Rule for Counting, the number of different possible sets of trophy winners is

$$5 \cdot 4 \cdot 3 = 60$$

Example 5.30   Traveling salesman problem

A Southeast regional salesman has eight destinations that he must travel to this month: Atlanta, Raleigh, Charleston, Nashville, Jacksonville, Richmond, Mobile, and Jackson. How many different possible routes could he take?

Solution

The salesman has $a = 8$ different choices for where to go first. Once the first destination has been chosen, there are only $b = 7$ choices for where to go second. And once the first two destinations have been chosen, there are only $c = 6$ choices for where to go third. And so on. Thus, by the Multiplication Law for Counting, the number of different possible routes for the salesman is

$$a \cdot b \cdot c \cdot d \cdot e \cdot f \cdot g \cdot h = 8 \cdot 7 \cdot 6 \cdot 5 \cdot 4 \cdot 3 \cdot 2 \cdot 1 = 40{,}320$$

The calculation in Example 5.30 leads us to introduce the **factorial symbol,** which is used for the counting rules we will learn in the remainder of this section.

> For any integer $n \geq 0$, the **factorial symbol $n!$** is defined as follows:
> - $0! = 1$
> - $1! = 1$
> - $n! = n(n - 1)(n - 2) \cdots 3 \cdot 2 \cdot 1$

For example:

$$2! = 2 \cdot 1 = 2$$
$$3! = 3 \cdot 2 \cdot 1 = 6$$
$$4! = 4 \cdot 3 \cdot 2 \cdot 1 = 24$$
$$5! = 5 \cdot 4 \cdot 3 \cdot 2 \cdot 1 = 120$$
$$6! = 6 \cdot 5 \cdot 4 \cdot 3 \cdot 2 \cdot 1 = 720$$
$$7! = 7 \cdot 6 \cdot 5 \cdot 4 \cdot 3 \cdot 2 \cdot 1 = 5040$$
$$8! = 8 \cdot 7 \cdot 6 \cdot 5 \cdot 4 \cdot 3 \cdot 2 \cdot 1 = 40{,}320, \text{ as in Example 5.30.}$$

Permutations and Combinations

Example 5.31   Traveling to some but not all of the cities

Example 5.30 calculated the number of possible routes for traveling to $n = 8$ cities. However, suppose, we are interested in traveling to *some but not all* of the cities? For example, suppose that the salesman is traveling to three of the eight cities. Find the number of possible routes.

Solution

There are eight choices for the first city, seven choices for the second city, and six choices for the third city. Since the salesman is traveling to three cities only, the number of possible routes is thus

$$8 \cdot 7 \cdot 6 = 336$$

This result may be rewritten using factorial notation, as follows:

$$8 \cdot 7 \cdot 6 = \frac{8 \cdot 7 \cdot 6 \cdot (5 \cdot 4 \cdot 3 \cdot 2 \cdot 1)}{(5 \cdot 4 \cdot 3 \cdot 2 \cdot 1)} = \frac{8!}{5!} = \frac{8!}{(8 - 3)!}$$

Example 5.31 leads us to the following definition.

Permutations

A **permutation** is an arrangement of items, such that

- r items are chosen at a time from n distinct items.
- repetition of items is not allowed.
- the order of the items is important.

The number of permutations of n items chosen r at a time is denoted as ${}_nP_r$ and given by the formula:

$$_nP_r = \frac{n!}{(n-r)!}$$

In Example 5.31, we are looking for the number of permutations of 8 cities taken 3 at a time. We have $n = 8$, $r = 3$,

$$_nP_r = {}_8P_3 = \frac{n!}{(n-r)!} = \frac{8!}{(8-3)!} = \frac{8!}{5!} = 8 \cdot 7 \cdot 6 = 336$$

Example 5.32   Counting permutations: Secret Santas

"Secret Santa" refers to a method whereby each member of a group anonymously buys a holiday gift for another member of the group. Each person is secretly assigned to buy a gift for another randomly chosen person in the group. Suppose Jessica, Laverne, Samantha, and Luisa share a dorm suite and would like to do Secret Santa this holiday season.

a. Verify that in this instance one woman purchasing a gift for another woman represents a permutation.

b. Calculate how many possible different permutations of gift buying there are for the four women.

Solution

a.
- There are $n = 4$ women, and $r = 2$ people are associated with each gift, the giver and the receiver.
- Each person can buy only one gift, so repetition is not allowed.
- Finally, there is a difference between Jessica buying for Laverne and Laverne buying for Jessica. Thus, order is important, and thus, buying a gift represents a permutation.

b. The number of permutations is calculated as follows:

$$_nP_r = {}_4P_2 = \frac{4!}{(4-2)!} = \frac{4 \cdot 3 \cdot 2!}{2!} = 12$$

Example 5.33   Calculating numbers of permutations

Find the following numbers of permutations.

a. ${}_5P_2$ **b.** ${}_6P_2$ **c.** ${}_6P_6$

Solution

a. ${}_5P_2 = \dfrac{5!}{(5-2)!} = \dfrac{5 \cdot 4 \cdot 3!}{3!} = 20$

b. $_6P_2 = \dfrac{6!}{(6-2)!} = \dfrac{6 \cdot 5 \cdot 4!}{4!} = 30$

c. $_6P_6 = \dfrac{6!}{(6-6)!} = \dfrac{6 \cdot 5 \cdot 4 \cdot 3 \cdot 2 \cdot 1}{0!} = 720$

In a permutation, order is important. For example, in Example 5.32, there was a difference between Jessica buying a gift for Laverne and Laverne buying one for Jessica. However, what if we consider shaking hands instead? Then Jessica shaking hands with Laverne is considered the same as Laverne shaking hands with Jessica. Hence, sometimes order is not important. What is important here is the **combination** of Jessica and Laverne.

Combinations

A **combination** is an arrangement of items in which

- r items are chosen from n distinct items.
- repetition of items is not allowed.
- the order of the items is not important.

The number of combinations of r items chosen from n different items is denoted as

$$_nC_r$$

Example 5.34   How many combinations in the intramural tennis league?

We return to the intramural singles tennis league at the local college. There are five players, Ryan, Megan, Nicole, Justin, and Kyle. Each player must play each other once.

a. Confirm that a match between two players represents a combination.

b. How many matches will be held?

Solution

a. Let {Ryan, Megan} denote a tennis match between Ryan and Megan. *Note:*
- There are $r = 2$ players chosen from $n = 5$ players.
- Each player plays each other player once, so repetition is not allowed.
- There is no difference between {Ryan, Megan} and {Megan, Ryan}, so order is not important.

Thus, a tennis match between two players represents a combination.

b. The list of all matches is as follows.

{Ryan, Megan}	{Megan, Nicole}	{Nicole, Justin}
{Ryan, Nicole}	{Megan, Justin}	{Nicole, Kyle}
{Ryan, Justin}	{Megan, Kyle}	{Justin, Kyle}
{Ryan, Kyle}		

Thus there are $_5C_2 = 10$ possible matches of $r = 2$ players chosen from $n = 5$ players.

We saw in Example 5.33 that $_5P_2 = 20$ and in Example 5.34 that $_5C_2 = 10$. Permutations and combinations differ only in that ordering is ignored for combinations. To calculate the number of combinations $_nC_r$, we simply do not count however many rearrangements there are of the same items. For example, in Example 5.34, there are $r! = 2! = 2$ rearrangements of the same players, such as {Ryan, Megan} and {Megan, Ryan}. Thus,

$$_5C_2 = \frac{_5P_2}{2!} = \frac{20}{2} = 10$$

In general, the **number of combinations** can be computed as the number of permutations divided by factorial of the number of items chosen.

Note: Following are some special combinations you may find useful. For any integer n:

$$_nC_n = 1$$
$$_nC_0 = 1$$
$$_nC_1 = n$$
$$_nC_{n-1} = n$$

> **Formula for the Number of Combinations**
>
> The number of combinations of r items chosen from n different items is given by
>
> $$_nC_r = \frac{n!}{r!(n - r)!}$$

For instance, in Example 5.34, the formula for the number of combinations is

$$_5C_2 = \frac{5!}{2!(5 - 2)!} = \frac{5!}{2!\,3!} = \frac{5 \cdot 4 \cdot 3!}{2 \cdot 1 \cdot 3!} = \frac{20}{2} = 10$$

Thus the relation: $_5C_2 = {_5P_2}/2!$ is verified.

Example 5.35   Calculating numbers of combinations

Find the following numbers of combinations.
a. $_6C_2$ **b.** $_6C_3$ **c.** $_6C_4$

Solution

a. $_6C_2 = \dfrac{6!}{2!(6 - 2)!} = \dfrac{6 \cdot 5 \cdot 4!}{2 \cdot 1 \cdot 4!} = \dfrac{30}{2} = 15$

b. $_6C_3 = \dfrac{6!}{3!(6 - 3)!} = \dfrac{6 \cdot 5 \cdot 4 \cdot 3!}{3 \cdot 2 \cdot 1 \cdot 3!} = \dfrac{120}{6} = 20$

c. $_6C_4 = \dfrac{6!}{4!(6 - 4)!} = \dfrac{6!}{(6 - 4)!\,4!} = \dfrac{6 \cdot 5 \cdot 4!}{2 \cdot 1 \cdot 4!} = \dfrac{30}{2} = 15$

Note that in (c) we used the commutative property of multiplication ($a \cdot b = b \cdot a$) and found that $_6C_4 = {_6C_2} = 15$. In general, $_nC_r = {_nC_{n-r}}$ for this reason. ◾

Example 5.36   Calculating the number of permutations and combinations using technology

Use the TI-83/84 and Excel to calculate the following.
a. $_9P_6$ **b.** $_{10}C_7$

Solution
We use the instructions provided in the Step-by-Step Technology Guide at the end of this section (page 240).
a. From Figures 5.19 and 5.20 we find that $_9P_6 = 60{,}480$.

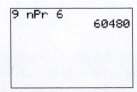

FIGURE 5.19 TI-83/84 permutation results.   **FIGURE 5.20** Excel permutation results.

b.   From Figures 5.21 and 5.22 we find that $_{10}C_7 = 120.$ ■

```
10 nCr 7
           120
```

FIGURE 5.21 TI-83/84 combination results.   **FIGURE 5.22** Excel combination results.

Sometimes we wish to find the number of permutations of items where some of the items are not distinct.

Example 5.37   Permutations with nondistinct items

How many distinct strings of letters can we make by using all the letters in the word STATISTICS?

Solution

Each string will be ten letters long and include 3 S's, 3 T's, 2 I's, 1 A, and 1 C. The ten positions shown here need to be filled.

$$\underline{\quad} \ \underline{\quad} \ \underline{\quad} \ \underline{\quad} \ \underline{\quad} \ \underline{\quad} \ \underline{\quad} \ \underline{\quad} \ \underline{\quad} \ \underline{\quad}$$
$$\mathbf{1} \quad \mathbf{2} \quad \mathbf{3} \quad \mathbf{4} \quad \mathbf{5} \quad \mathbf{6} \quad \mathbf{7} \quad \mathbf{8} \quad \mathbf{9} \quad \mathbf{10}$$

The string-forming process is as follows:

Step 1   Choose the positions for the three S's.

Step 2   Choose the positions for the three T's.

Step 3   Choose the positions for the two I's.

Step 4   Choose the position for the one A.

Step 5   Choose the position for the one C.

There are $_{10}C_3$ ways to place the three S's in Step 1. Once Step 1 is done, there are seven slots left, leaving $_7C_3$ positions for the three T's. Once Step 2 is done, there are four slots left, so there are $_4C_2$ ways to place the two I's. Once Step 3 is done, there are only 2 slots left, so there are only $_2C_1$ ways to position the A. Finally, there is only $_1C_1$ way to place the C.

Putting Steps 1–5 together, we calculate the number of distinct letter strings as

$$_{10}C_3 \cdot {}_7C_3 \cdot {}_4C_2 \cdot {}_2C_1 \cdot {}_1C_1 = \frac{10!}{3! \, 7!} \cdot \frac{7!}{3! \, 4!} \cdot \frac{4!}{2! \, 2!} \cdot \frac{2!}{1! \, 1!} \cdot \frac{1!}{1! \, 0!}$$

$$= \frac{10!}{3! \, 3! \, 2! \, 1! \, 1!} = \frac{3{,}628{,}800}{72}$$

$$= 50{,}400$$

There are 50,400 distinct strings of letters that can be made using the letters in the word STATISTICS.   ■

This example can be generalized in the following result.

Permutations of Nondistinct Items

The number of permutations of n items of which n_1 are of the first kind, n_2 are of the second kind, . . ., and n_k are of the kth kind is calculated as

$$\frac{n!}{n_1! \cdot n_2! \cdot \ldots \cdot n_k!}$$

where $n = n_1 + n_2 + \ldots + n_k$.

Example 5.38   Number of permutations of nondistinct items

Brandon brings a healthy snack to school each day, consisting of 5 carrot sticks, 4 celery sticks, and 2 cherry tomatoes. If Brandon eats one item at a time, in how many different ways can he eat his snack?

Solution

We are seeking the number of permutations of $n = 11$ items, of which $n_1 = 5$ are carrot sticks, $n_2 = 4$ are celery sticks, and $n_3 = 2$ are cherry tomatoes. Using the formula for the number of permutations of nondistinct items,

$$\frac{n!}{n_1! \cdot n_2! \cdot n_3!} = \frac{11!}{5! \cdot 4! \cdot 2!} = \frac{39,916,800}{120 \cdot 24 \cdot 2} = 6930$$

There are 6930 distinct ways in which Brandon can eat his snack. ■

Acceptance sampling refers to the process of (1) selecting a random sample from a batch of items, (2) evaluating the sample for defectives, and (3) either accepting or rejecting the entire batch based on the evaluation of the sample.

Example 5.39   Acceptance sampling uses combinations

Suppose we have a batch of 20 cell phones, of which, unknown to us, 3 are defective and 17 are non-defective. We will take a random sample of size 2 and evaluate both items once.
a. Are the arrangements in acceptance sampling permutations or combinations?
b. Find the number of ways that both sampled cell phones are defective.

Solution

a. Both permutations and combinations require the following:
   • r items are chosen from n distinct items. Here we are selecting $r = 2$ phones from a batch of $n = 20$.
   • Repetition of the items is not allowed. Each item is evaluated only once.
   The difference between permutations and combinations is that, for permutations order is important while for combinations order is not important. In acceptance sampling, the order of the items is not important. Thus, acceptance sampling uses combinations.
b. The number of ways of choosing 2 of the 3 defectives is

$$_3C_2 = \frac{3!}{2!(3-2)!} = \frac{3 \cdot 2!}{2! \cdot 1!} = 3$$

Selecting 2 defectives means that we are choosing 0 of the 17 non-defectives. The number of ways this can happen is

$$_{17}C_0 = \frac{17!}{0!(17-0)!} = \frac{17!}{1 \cdot 17!} = 1$$

By the Multiplication Rule for Counting, the number of ways that both sampled cell phones are defective is

$$_3C_2 \cdot {_{17}C_0} = 3 \cdot 1 = 3$$

3 Computing Probabilities Using Combinations

The counting methods we have learned in this section may be used to compute probabilities. We assume that each possible outcome in a random sample is *equally likely,* and thus we use the classical method for assigning the probability of an event E:

$$P(E) = \frac{\text{number of outcomes in } E}{\text{number of outcomes in sample space}} = \frac{N(E)}{N(S)}$$

Example 5.40   Probability using combinations: acceptance sampling

Continuing with Example 5.39, if both cell phones in the sample of size 2 are defective, we will reject the batch and cancel our contract with the supplier.
a. What is the number of ways that both cell phones will be defective?
b. What is the number of outcomes in this sample space?
c. What is the probability that both cell phones will be defective?

Solution
a. From Example 5.39, the number of ways that both cell phones will be defective is

$$_3C_2 \cdot {_{17}C_0} = 3 \cdot 1 = 3$$

b. The number of outcomes in the sample space is given by the number of ways of selecting 2 cell phones out of a batch of 20, that is,

$$N(S) = {_{20}C_2} = \frac{20!}{2!(20-2)!} = \frac{20 \cdot 19 \cdot 18!}{2! \cdot 18!} = \frac{380}{2} = 190$$

c. Therefore, the probability that both cell phones will be defective is given by

$$P(\text{Both defective}) = \frac{\text{number of ways both defective}}{\text{number of outcomes in sample space}} = \frac{3}{190} \approx 0.01579 \quad \blacksquare$$

Example 5.41   Florida Lotto

You can win the jackpot in the Florida Lotto by correctly choosing all 6 winning numbers out of the numbers 1–53.
a. What is the number of ways of winning the jackpot by choosing all 6 winning numbers?
b. What is the number of outcomes in this sample space?
c. If you buy a single ticket for $1, what is your probability of winning the jackpot?
d. If you mortgage your house and buy 500,000 tickets, what is your probability of winning the jackpot (assuming that all the tickets are different)?

Solution

a. The number of ways of winning the jackpot by correctly choosing all 6 of the winning numbers and none of the losing numbers is

$$N(\text{Jackpot}) = {}_6C_6 \cdot {}_{47}C_0 = 1 \cdot 1 = 1$$

b. The size of the sample space is

$$N(S) = {}_{53}C_6 = \frac{53!}{6!(53-6)!} = \frac{53 \cdot 52 \cdot 51 \cdot 50 \cdot 49 \cdot 48 \cdot 47!}{6! \cdot 47!}$$

$$= \frac{16,529,385,600}{720} = 22,957,480$$

c. Therefore, if you buy a single ticket for $1, your probability of winning the jackpot is given by

$$P(\text{Jackpot}) = \frac{1}{22,957,480} \approx 0.00000004356$$

d. If you buy 500,000 tickets and they are all unique, then your probability of winning becomes

$$P(\text{Jackpot}) = \frac{500,000}{22,957,480} \approx 0.02178$$

This is because the unique tickets are mutually exclusive, and the Addition Rule for Mutually Exclusive Events allows us to add the probabilities of the 500,000 tickets. After mortgaging your $500,000 house and buying lottery tickets with the proceeds, there is a better than 97% probability that you will *not* win the lottery. ■

STEP-BY-STEP TECHNOLOGY GUIDE: Factorials, Permutations, and Combinations

TI-83/84

Factorials *n*!

Step 1 On the home screen, enter the value of *n*.

Step 2 Press **MATH**, highlight **PRB**, and select **4: !** (Figure 5.23).

Step 3 Press **ENTER**.

```
MATH NUM CPX PRB
1:rand
2:nPr
3:nCr
4:!
5:randInt(
6:randNorm(
7:randBin(
```

Figure 5.23

Permutations $_nP_r$ and Combinations $_nC_r$

Step 1 On the home screen, enter the value of *n*.

Step 2 **a.** For permutations, press **MATH**, highlight **PRB**, and select **2:nPr**.

   b. For combinations, press **MATH**, highlight **PRB**, and select **3:nCr**.

Step 3 On the home screen, enter the value of *r*.

Step 4 Press **ENTER** (see Figure 5.19 and Figure 5.21 in Example 5.36 [pages 236–237]).

EXCEL

Factorials *n*!

Calculate **9!**

Step 1 Select an empty cell, and type = **FACT**(**9**).

Step 2 Press **ENTER**.

Permutations $_nP_r$

We illustrate Example 5.36a (page 236): $_9P_6$.

Step 1 Select an empty cell and type = **PERMUT**(**9,6**).

Step 2 Press **ENTER**. See Figure 5.20 in Example 5.36 for the result.

Combinations $_nC_r$

We illustrate Example 5.36b (page 236): $_{10}C_7$.

Step 1 Select an empty cell and type =**COMBIN**(**10,7**).

Step 2 Press **ENTER**. See Figure 5.22 in Example 5.36 for the result.

Section 5.4 SUMMARY

1 The Multiplication Rule for Counting provides the total number of different possible outcomes for a series of events.

2 A permutation $_nP_r$ is an arrangement in which
- r items are chosen from n distinct items.
- repetition of items is not allowed.
- the order of the items is important.

In a permutation, order is important. In a combination, order does not matter. A combination $_nC_r$ is an arrangement in which
- r items are chosen from n distinct items.
- repetition of items is not allowed.
- the order of the items is not important.

3 Combinations may be used to calculate certain probabilities. For such problems, use the following steps.

Step 1 Confirm that the desired probability involves a combination.
Step 2 Find $N(E)$, the number of outcomes in event E.
Step 3 Find $N(S)$, the number of outcomes in the sample space.
Step 4 Assuming that each possible combination is equally likely, find the probability of event E as follows:

$$P(E) = \frac{N(E)}{N(S)}$$

Section 5.4 EXERCISES

CLARIFYING THE CONCEPTS

1. What type of diagram is helpful in itemizing the possible outcomes of a series of events?
2. Explain in words how 5! is calculated.
3. What is the difference between a permutation and a combination?
4. Does $_8P_9$ make sense? Explain why or why not.
5. Describe in your own words what is meant by acceptance sampling.
6. The counting methods that we have learned in this section may be used to compute probabilities.
   a. For assigning probability, which method is used, classical, relative frequency, or subjective?
   b. Referring to part a, what assumption must be made to apply the method?

PRACTICING THE TECHNIQUES

For Exercises 7–12, find the value of each factorial.
7. 6!
8. 9!
9. 0!
10. 11!
11. 1!
12. 15!

For Exercises 13–20, find the value of each permutation $_nP_r$.
13. $_7P_3$
14. $_7P_4$
15. $_8P_5$
16. $_8P_3$
17. $_{100}P_1$
18. $_{100}P_0$
19. $_{100}P_{100}$
20. $_{100}P_{99}$

For Exercises 21–28, find the value of each combination $_nC_r$. Then answer Exercises 29 and 30.
21. $_7C_3$
22. $_7C_4$
23. $_{11}C_8$
24. $_{11}C_9$
25. $_{11}C_{10}$
26. $_{11}C_{11}$
27. $_{100}C_0$
28. $_{100}C_1$
29. Explain why the answers to Exercises 21 and 22 are equal. Use the commutative property of multiplication (for example, $2 \cdot 7 = 7 \cdot 2$) in your answer.
30. Use the idea behind your answer to Exercise 29 to find a combination that is equal to $_{11}C_8$. Verify your answer.
31. List all the permutations of the following people taken three at a time: Amy, Bob, Chris, Danielle. What is $_4P_3$?
32. List all the combinations of the following people taken three at a time: Amy, Bob, Chris, Danielle. What is $_4C_3$?
33. Explain in your own words why $_4P_3$ is larger than $_4C_3$.
34. What quantity do we divide $_4P_3$ by to get $_4C_3$? Express this quantity as a factorial. (*Hint:* For example, if the quantity were 120, we would express it as 5!)
35. In general, what do we divide $_nP_r$ by to get $_nC_r$?

APPLYING THE CONCEPTS

36. Fast Food. A fast-food restaurant has three types of sandwiches: chicken sandwich, fish sandwich, and beef burger. The restaurant has two types of side dishes: French fries and salad.
   a. Draw a tree diagram to find all the different meals a customer can order at this restaurant.
   b. How many different meals can a customer order at this restaurant?

37. What to Eat? A sit-down restaurant has two types of appetizers: garden salad and buffalo wings. It has three entrees: spaghetti, steak, and chicken. And it offers three kinds of desserts: ice cream, cake, and pie.
   a. Draw a tree diagram to find all the different meals a customer can order at this restaurant.
   b. How many different meals can a customer order at this restaurant?

38. Greek Alphabet. The ancient Greek alphabet had 24 letters. How many different possible initials are there for people with a first and last name?

39. Baseball Lineup. A baseball manager has to submit to the umpire the ordering of his lineup of nine players. How many ways can the manager arrange the nine players?

40. Facebook Friends. A student has 10 friends on her Facebook page. How many ways can she arrange her 10 friends top to bottom?

41. Document Delivery. A document delivery person must deliver documents to five different destinations within a particular city. How many different routes are possible?

42. Country Music Mix. Madison has just burnt 15 songs onto a country music mix CD. She chooses "random" on the CD player when she listens to the CD. How many different orderings of the 15 songs are there?

43. Traveler Fellow. A corporate sales executive must travel to the following countries this quarter: China, Russia, Germany, Brazil, India, and Nigeria. How many different routes are possible?

44. Sales Traveler. A corporate sales executive has the choice of traveling to four of the following six countries this quarter: China, Russia, Germany, Brazil, India, and Nigeria. How many different routes are possible?

45. Playing Catch. Five children are playing catch with a ball. How many different ways can one child throw a ball to another child once?

46. Chimp Grooming. Six chimpanzees are grooming each other at the city zoo. In how many different ways can one chimp groom another?

47. Shake Hands. In an ice-breaker exercise, each of 25 students is asked to shake hands with each of the other students. How many handshakes will there be in all?

48. Statistics Competition. Three students from the Honors Statistics class of 15 students will be chosen to represent the school at the state statistics competition. How many different possible groupings of three students are there?

49. How many random samples of size 1 can be chosen from a population of size 20?

50. How many random samples of size 20 can be chosen from a population of size 20?

51. How many random samples of size 10 can be chosen from a population of size 20?

52. How many distinct strings of letters can be made using all the letters in the word MATHEMATICS?

53. How many distinct strings of letters can be made using all the letters in the word BUSINESS?

54. CD Burning. You would like to use your new CD burner to make two new music-mix CDs. Your box of disks contains 14 CDs and 7 DVDs. If you choose 2 at random, find the number of ways that both disks will be CDs.

55. Grocery Produce Manager. A grocery produce manager inspects the corn supplied by a nearby farm. In a bushel of 30 ears of corn, the manager inspects 3 ears of corn. If 2 or more of the ears of corn are defective, the entire bushel is rejected. In this bushel, 5 ears of corn are defective.

   a. Explain whether the grocery produce manager is using a permutation or a combination.

   b. Find the number of ways that 2 or more ears of corn will be found defective by the produce manager.

   c. Find the probability that the bushel will be rejected.

56. Acceptance Sampling. A shipment of 25 personal digital assistants (PDAs) contains 3 that are defective. A quality control specialist inspects 2 of the 25 PDAs. If both are defective, then the shipment is rejected.

   a. Explain whether a permutation or a combination is being used.

   b. Find the number of ways that both PDAs will be defective.

   c. Find the probability of rejecting the shipment.

57. Downloading Music. You have just downloaded a new album of 12 songs from an online music retailer and made a playlist of the 12 songs. It turns out that you like 7 of the songs. Setting the playlist to random mode, find the probability that, of the first 5 songs you hear,

   a. you like all of them.

   b. you like none of them.

   c. you like exactly 2 of them.

   d. you like exactly 3 of them.

Try This in Class! In-class activities to enhance your understanding of statistics

1. Have you ever taken a multiple-choice test where you couldn't decide between the alternatives and wished you could flip a coin or roll a die to decide? Suppose you are facing a four-question multiple-choice exam, with three alternatives (A, B, or C) for each question, and you have no clue as to what the right answers are. In this activity you will calculate the probability of getting one or more answers right.

   a. Why does it make sense to use the classical method?

   b. Draw a tree diagram for this four-question test—the experiment.

   c. What is the probability that you will answer a particular question correctly?

   d. Suppose that you need to answer at least three out of the four questions correctly to pass. What is the probability that you will pass?

2. We use simulation to estimate the probability we calculated in Step **1c.** We use a single fair die to determine our responses.

   a. How should we "code" the die results so that there is an even chance of choosing A, B, or C?

b. Your answer to each exam question will be either right or wrong. Why not use a coin toss rather than a die roll for this simulation?

c. We need to code the probability of getting a correct answer. For each question, A, B, or C is correct. We want to simulate the probability of guessing correctly, which is 1/3. How do you code the die roll so that the probability of guessing correctly is 1/3?

d. Perform 20 trials of the experiment. Rolling the die one time is considered to be answering one test

question. So rolling the die four times is considered to be one "trial" of the experiment (taking the test). For each trial, keep track of the number of "right" and "wrong" guesses. Answer question (**e**) using your simulation results.

e. Estimate the probability that you will answer exactly one particular question correctly.

f. Estimate the probability that you will pass.

CHAPTER 5 FORMULAS AND VOCABULARY

SECTION 5.1
• **Classical method of assigning probabilities** (p. 197).

$$P(E) = \frac{\text{number of outcomes in } E}{\text{number of outcomes in sample space}} = \frac{N(E)}{N(S)}$$

• **Event** (p. 195)
• **Experiment** (p. 195)
• **Law of Large Numbers** (p. 201)
• **Law of Total Probability** (p. 196)
• **Outcome** (p. 195)
• **Probability** (p. 195)
• **Probability model** (p. 202)
• **Probability tree diagram** (p. 198)
• **Relative frequency method of assigning probabilities** (p. 201).

$$P(E) \approx \frac{\text{frequency of } E}{\text{number of trials of experiment}}$$

Also known as the **empirical method**.
• **Sample space** (p. 195)
• **Simulation** (p. 200)
• **Subjective probability** (p. 203)

SECTION 5.2
• **Addition Rule** (p. 211). $P(A \text{ or } B) = P(A \cup B) = P(A) + P(B) - P(A \cap B)$.
• **Addition Rule for Mutually Exclusive Events** (p. 213). If A and B are mutually exclusive, then $P(A \cup B) = P(A) + P(B)$.
• **Complement of an event** A (p. 208). Denoted as A^C.
• **Intersection of two events** A and B (p. 210). Denoted as $A \cap B$ or as "A and B."
• **Mutually exclusive (disjoint) events** (p. 213)
• **Probabilities for complements** (p. 209). $P(A) + P(A^C) = 1$, $P(A) = 1 - P(A^C)$, and $P(A^C) = 1 - P(A)$.
• **Union of two events** A and B (p. 210). Denoted as $A \cup B$ or as "A or B."

SECTION 5.3
• **Conditional probability** (p. 217).

$$P(B|A) = \frac{P(A \cap B)}{P(A)} = \frac{N(A \cap B)}{N(A)}$$

• **Independent events** (p. 219). Events A and B are *independent* if $P(A|B) = P(A)$ or if $P(B|A) = P(B)$.
• **Multiplication Rule** (p. 221). $P(A \cap B) = P(B) P(A|B)$ or, equivalently, $P(A \cap B) = P(A) P(B|A)$.
• **Multiplication Rule for Independent Events** (p. 221). If events A and B are independent, then $P(A \cap B) = P(A) P(B)$.
• **Multiplication Rule for** n **Independent Events** (p. 223). If $A, B, C, \ldots$ are independent events, then $P(A \cap B \cap C \cap \ldots) = P(A) P(B) P(C) \ldots$.
• **Sampling with replacement** (p. 225)
• **Sampling without replacement** (p. 225)

SECTION 5.4
• **Acceptance sampling** (p. 238)
• **Combination** (p. 235).

$$_nC_r = \frac{n!}{r!(n-r)!}$$

• **Factorial symbol** $n!$ (p. 233). $0! = 1$; $1! = 1$; $n! = n(n-1)(n-2)\ldots 3 \cdot 2 \cdot 1$
• **Multiplication Rule for Counting** (p. 232)
• **Permutation** (p. 234).

$$_nP_r = \frac{n!}{(n-r)!}$$

• **Permutations of nondistinct items** (p. 238).

$$\frac{n!}{n_1! \cdot n_2! \cdot \ldots \cdot n_k!}$$

SECTION 5.1
For Exercises 1–5, consider the experiment of tossing a fair coin three times and find the probabilities.
1. 2 heads
2. At least 2 heads
3. 4 heads
4. 2 tails
5. At most 1 tail
6. **A New Sonnet.** Literature researchers have unearthed a sonnet that they know to be by either William Shakespeare or Christopher Marlowe. The probability that the sonnet is by Marlowe is 25%.
   a. What is the probability that the sonnet is by Shakespeare?
   b. What method of assigning probability do you think was used here? Why was this method used, and not the others?

SECTION 5.2
7. **Farmworkers' Educational Level.** The U.S. Department of Agriculture reports on the demographics of hired farmworkers.[7] An excerpt of the results is provided in the table, showing the percentage of noncitizen and citizen farmworkers who attained various educational levels. The educational levels are mutually exclusive. Find the following probabilities.
   a. The probability that a noncitizen farmworker is a high school graduate or has some college
   b. The probability that a citizen farmworker is a high school graduate or has some college
   c. The probability that a noncitizen has less than a ninth-grade education and has some college

	Noncitizens	Citizens
Less than 9th grade	238,008	61,776
9th–12th grade (no diploma)	57,904	152,880
High school graduate	59,784	222,144
Some college	20,304	187,200

Ecstasy and Methamphetamine Use. The following table contains the numbers of 12th-grade students who said that they had used Ecstasy (MDMA) or methamphetamine in the past 30 days for data collected in 2006 and 2007, as reported by the National Institute on Drug Abuse (**www.drugabuse.gov**). Use the table to answer Exercises 8–10.

Drug	2006	2007
Ecstasy	185	232
Methamphetamine	128	87
Sample size	14,200	14,500

8. Find the indicated probabilities.
   a. A randomly selected 12th-grader in 2006 used Ecstasy in the past 30 days.
   b. A randomly selected 12th-grader in 2006 used methamphetamine in the past 30 days.
   c. A randomly selected 12th-grader in 2007 used Ecstasy in the past 30 days.
   d. A randomly selected 12th-grader in 2007 used methamphetamine in the past 30 days.
9. Assume that no student in the samples repeated twelfth grade. Answer the following questions.
   a. Are these two events mutually exclusive: $A =$ 12th-grader in 2006 used Ecstasy and $B =$ 12th-grader in 2007 used Ecstasy? Explain why or why not.
   b. Find the probability that a randomly selected 12th-grader in 2006 used Ecstasy or a randomly selected 12th-grader in 2007 used Ecstasy.
   c. Find the probability that a randomly selected 12th-grader in 2006 used methamphetamine or a randomly selected 12th-grader in 2007 used methamphetamine.
10. Answer the following questions.
   a. Are these two events mutually exclusive: $C =$ used Ecstasy in the past 30 days and $D =$ used methamphetamine in the past 30 days? Explain why or why not.
   b. Discuss why we cannot find the probability that a randomly chosen 2006 12th-grader used both Ecstasy and methamphetamine in the past 30 days. What extra information do we need?

SECTION 5.3
11. **Drug Research Studies.** The *Annals of Internal Medicine* reported that 39 of the 40 research studies sponsored by a drug company had outcomes favoring a certain drug. Find the following probabilities, assuming independence.
   a. Three randomly selected research studies all favor this drug.
   b. None of the three randomly selected research studies favors this drug.
   c. At least one of three randomly selected research studies favors this drug.
12. **Gun Control.** A Harris poll conducted in June 2004 of 2408 American adults found 52% who supported tighter gun control. Find the following probabilities.
   a. Two randomly selected respondents both supported tighter gun control.
   b. At least two of three randomly selected respondents supported tighter gun control.
   c. At most one of three randomly selected respondents did not support tighter gun control.

13. Politics of Gay Marriage. The Pew Research Center for the People and the Press (**http://people-press.org**) reported in 2004 that 322 out of 1149 people surveyed would not vote for a political candidate who disagreed with their views on gay marriage. Find the probabilities. Assume independence.
   a. Three out of three randomly selected respondents would vote for a political candidate who disagreed with his or her views on gay marriage.
   b. At least one of three randomly selected respondents would not vote for a political candidate who disagreed with his or her views on gay marriage.

14. Drug Research Studies. Use the information in Exercise 11. Suppose we sample two research studies without replacement. Find the probability that the second study does not favor this drug given that the first study does not favor this drug.

Gender and Pet Preference. Do you think your gender affects what type of pet you own? For Exercises 15–18, use the following table, showing preferences for various pets by owner gender.

Gender of owner	Cats	Dogs	Other pets	Total
Female	100	50	30	180
Male	50	50	20	120
Total	150	100	50	300

15. Find the probability that a randomly chosen person has the following characteristics.
   a. Owns a cat, $P(C)$ **b.** Owns a dog, $P(D)$

16. Find the probability that a randomly chosen person has the following characteristics.
   a. Is female and owns a dog, $P(F \cap D)$
   b. Is male and owns a dog, $P(M \cap D)$
17. Find the following conditional probabilities for a randomly chosen person.
   a. Owns a dog, given that the person is female, $P(D \mid F)$
   b. Owns a dog, given that the person is male, $P(D \mid M)$
18. If you were a dog-food manufacturer, would you advertise more on a men's TV channel or a women's TV channel? Why? Cite your evidence.

SECTION 5.4
19. How many distinguishable strings of letters can be made using all the letters in the word MISSISSIPPI?
20. Statistics Quiz. On a statistics quiz, there are 5 true/false questions, 4 fill-in-the-blank questions, and 3 short-answer questions. How many different ways are there of taking this quiz?
21. Inspection Time. A U.S. Army drill instructor will perform inspection on 2 soldiers in a squad of 18 soldiers. If both soldiers fail the inspection because their rifles are not clean, the entire squad will have to run a five-mile course in full gear. Three of the 18 soldiers have rifles that are not clean.
   a. Explain whether the drill instructor is using a permutation or a combination.
   b. Find the number of ways that both soldiers will fail the inspection.
   c. Find the probability that the entire squad will have to run a five-mile course in full gear.

CHAPTER 5 *Quiz*

TRUE OR FALSE
1. True or false: An outcome is a collection of a series of events from the sample space of an experiment.
2. True or false: For any event A (even events like A: the moon is made of green cheese) the probability of A plus the probability of A^C always add up to 1.

FILL IN THE BLANK
3. The minimum value that a probability can take is _____ and the maximum value is _____.
4. The union of two events is associated with the English word _____, and the intersection of two events is associated with the English word _____.
5. Someone has told you that there is a 50-50 chance of rain tomorrow. This means that the probability of rain tomorrow equals _____.

SHORT ANSWER
6. For any experiment, what is the sum of all the outcome probabilities in the sample space?

7. For which type of sampling are consecutive draws independent?
8. For two events A and B, what do we call the event containing only those outcomes that belong to both A and B?

CALCULATIONS AND INTERPRETATIONS
9. The experiment is to roll a fair die twice. Find the following probabilities.
   a. Pair of 1's
   b. Pair of 2's
   c. One of the dice shows 5
   d. Sum of the two dice equals 7

10. Consider the experiment of rolling a fair die twice. Find the following probabilities.
   a. Sum of the two dice equals 5
   b. Sum of the two dice does not equal 5
   c. One of the dice shows 2
   d. Sum of the two dice equals 5 and one of the dice shows 2
   e. Sum of the two dice equals 5 or one of the dice shows 2

11. Suppose that A and B are any two events, with $P(B) = 0.75$ and $P(A \cap B) = 0.15$. Find $P(A|B)$.

12. Suppose that A and B are any two events, with $P(B) = 0.85$ and $P(A|B) = 0.25$. Find $P(A \cap B)$.

13. Pick a Card. Consider the experiment of drawing a single card from a deck of 52 cards. Find the probability of observing the following events.

   a. Heart
   b. Face card (king, queen, or jack)
   c. Seven
   d. Red card
   e. Seven of hearts
   f. Red queen

14. Direct Mail Marketing. The following table was taken from a study on modeling response to direct mail marketing.

Credit Card on File?

Response	No	Yes	Total
Did not respond	16,102	7,935	24,037
Did respond	1,666	3,096	4,762
Total	17,768	11,031	28,799

Source: Daniel Larose, *Data Mining Methods and Models* (Wiley Interscience, 2006).

The events are A = "responded to direct mail marketing campaign" and B = "has a credit card on file." Find the probability that a randomly selected customer has the following characteristics.

   a. Has a credit card on file
   b. Responded to the direct mail marketing campaign
   c. Has a credit card on file and responded to the direct mail marketing campaign
   d. Has a credit card on file or responded to the direct mail marketing campaign

Happiness in Marriage. The General Social Survey tracks trends in American society through annual surveys. The married respondents were asked to characterize their feelings about being married. The results, crosstabulated with gender, are shown in the following figure. Use this information for Exercises 15–18.

RESPONDENTS SEX * HAPPINESS OF MARRIAGE Crosstabulation

Count

| | | HAPPINESS OF MARRIAGE | | | |
		VERY HAPPY	PRETTY HAPPY	NOT TOO HAPPY	Total
RESPONDENTS SEX	MALE	242	115	9	366
	FEMALE	257	149	17	423
Total		499	264	26	789

15. Find the probabilities that a randomly chosen person has the following characteristics.

   a. Is female, $P(F)$
   b. Is male, $P(M)$
   c. Is not too happily married, $P(Not)$

16. Find the probabilities that a randomly chosen person has the following characteristics.

   a. Is female and not too happily married, $P(F \cap Not)$
   b. Is male and not too happily married, $P(M \cap Not)$

17. Find the following conditional probabilities for a randomly chosen person.

   a. Is not too happily married, given that the person is female, $P(Not | F)$
   b. Is not too happily married, given that the person is male, $P(Not | M)$

18. Are gender and being not too happily married independent? Why or why not?

19. Football Teams. The four teams in the AFC South division of the National Football League are Indianapolis Colts, Jacksonville Jaguars, Tennessee Titans, and Houston Texans. Suppose the top three teams in the division this year will make the playoffs. How many different sets of teams making the playoffs are there?

20. State Lottery. In a state lottery, balls numbered 1 to 20 are placed in an urn. To win, you must choose numbers that match the three balls chosen in the order that they're chosen.

   a. Explain whether a permutation or a combination is being used.
   b. How many possible outcomes are there?
   c. Find the probability of winning this lottery if your ticket contains a single ordering of three numbers.

David Madison/Digital Vision/Getty Images

Random Variables and the Normal Distribution

CASE STUDY | **Be Careful What You Assume: Major League Baseball Salaries**

Many fans of Major League Baseball (MLB) are concerned about the continuing increase in player salaries, making a trip to the ballpark an expensive family outing. In this Case Study, we consider a scenario where the MLB owners wish to "cap" a team's top salary, with the owners suggesting the 75th percentile of all salaries. In our scenario, the players' union states that they will agree to the salary cap only if management adds a bonus to the salaries of the lowest-paid players, as measured by the 10th percentile. We will find out what transpires in Section 6.5.

The Big Picture

Where we are coming from, and where we are headed…

In Chapter 5, we were introduced to probability, which allows us to quantify the uncertainty we will face when performing statistical inference in the later chapters. Probability provides us with a framework when we are called upon to make decisions, such as:

- Should we invest in this mutual fund?
- What is a typical profit for a small capital investor?
- What would be an unusual number of runs scored in a baseball game?

However, to answer these questions we need a new set of tools in our probability toolbox: *random variables* and *probability distributions*. With these new tools, we can increase the efficiency of our decision making. Sections 6.1 and 6.2 introduce us to discrete random variables, including the binomial random variable. In Section 6.3 we learn about continuous random variables, including the normal random variable. Section 6.4 covers the special case of the standard normal distribution, and Section 6.5 provides some applications of the normal distribution to everyday problems.

Chapter 6 lies within the scope of probability, because we are still faced with an experimental situation, and our task is to calculate probabilities associated with various events in the experiment. In Chapter 7, we will shift our perspective toward sampling and learn about sampling distributions. Chapter 7 is pivotal, because all the statistical inference we will learn in the remainder of the course depends on the behavior of statistics, which is quantified using sampling distributions. ■

6.1 Discrete Random Variables

Objectives: By the end of this section, I will be able to…

1 Identify random variables.

2 Explain what a discrete probability distribution is and construct probability distribution tables and graphs.

3 Calculate the mean, variance, and standard deviation of a discrete random variable.

❶ Random Variables

Life is constantly calling upon us to make decisions in the face of uncertainty. As a way of helping us to make these decisions, we can sometimes list all the possible outcomes of the situation and the probabilities associated with each outcome. In this section, we will learn about random variables and how they can help us quantify the uncertainty.

Example 6.1   **Decision making: should Kristin invest?**

In her position as junior manager Kristin, fresh out of college, has managed to save up $2000. She wants to invest in the stock market, in particular Cape Ann Biotech, a nano-biotechnology company. Based on the company's recent performance, she is optimistic about its prospects, despite the *uncertainty* involved in the investment. Based on her research and professional estimates about the direction of the overall economy, she estimates the probabilities of the five outcomes shown in Table 6.1 for making or losing money over the next 12 months. Note that the two loss outcomes are expressed as negative gains, and that the probabilities add up to 1.0.

a.   What method is Kristin using to assign probabilities?
b.   Use Table 6.1 to find the probability that Kristin will gain $1000.
c.   Will Kristin make money or lose money?

Table 6.1 Kristin's probability estimates for five investment outcomes for a period of 12 months (initial investment $2000)

Scenario	Financial gain	Kristin's estimated probability
Cape Ann Biotech does very well.	Gain $1000	0.15
Cape Ann Biotech does fairly well.	Gain $500	0.30
Cape Ann Biotech treads water.	Gain $0	0.25
Cape Ann Biotech does not do very well.	Lose $200 (Gain −$200)	0.20
Cape Ann Biotech folds.	Lose $2000 (Gain −$2000)	0.10

Solution

a.   Kristin is using the subjective method to assign probabilities. The outcomes are not equally likely, and she is not using relative frequencies to assign the probabilities. Instead, she is assigning probabilities based on her own subjective estimates of the company's prospects and the direction of the economy.

b.   Kristin's estimated probability that her financial gain will be $1000 is 0.15.

c.   We do not know yet whether Kristin will make money or lose money, because the outcome is uncertain, which fits the definition of an *experiment*.   ◼

We can represent Kristin's financial gain with a variable. Recall from Chapter 1 that a *variable* is a characteristic of an element and can assume different values. Examples of variables include *age* and *gender*. Sometimes we use letters of the alphabet, such as X or Z, to represent variables. Kristin's financial gain is a variable that takes on different values. However, for this variable, there is an element of *chance* that influences which value it takes. The outcome of Kristin's investment is uncertain and contains elements of randomness. To help her decide how to quantify this uncertainty, we need to define a new kind of variable, called a **random variable**.

> **Random Variable**
> A **random variable** is a variable whose values are determined by chance.

Developing Your Statistical Sense

Random Variables Must Be Random!

The role of chance in the definition of a random variable is crucial. For example, a variable that is based on the results of an experiment is a random variable because an experiment by definition has outcomes that are unknown. For example, is your age a random variable? If we are just talking about you and no one else, and we know your age, then there is no chance involved. In that case, your age is not a random variable. On the other hand, what if we select students *at random* by picking names from a hat? Then the age of the person drawn is a random variable because its value depends at least partly on chance (on which name is drawn at random). ■

Example 6.2   Notation for random variables

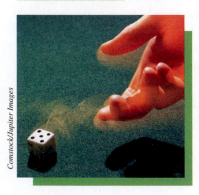
Comstock/Jupiter Images

Suppose our experiment is to toss a single fair die, and we are interested in the number rolled. We define our random variable X to be the outcome of a single die roll.
  a. Why is the variable X a random variable?
  b. What are the possible values that the random variable X can take?
  c. What is the notation used for rolling a 5?
  d. Use random variable notation to express the probability of rolling a 5.

Solution
  a. We don't know the value of X before we toss the die, which introduces an element of chance into the experiment, thereby making X a random variable.
  b. The possible values for X are 1, 2, 3, 4, 5, and 6.
  c. When a 5 is rolled, then X equals the outcome 5, and we write $X = 5$.
  d. Recall from Section 5.1 that the probability of rolling a 5 for a fair die is 1/6. In random variable notation, we denote this as $P(X = 5) = 1/6$. ■

There are two main types of random variables: **discrete random variables** and **continuous random variables**. The difference between the two types relates to the possible values that each type of random variable can assume.

> **Discrete and Continuous Random Variables**
> - A **discrete random variable** can take either a finite or a countable number of values. Since these values may be written as a list of numbers, each value can be graphed as a separate point on a number line, with space between each point.
>
> - A **continuous random variable** can take infinitely many values. Because there are infinitely many values, the values of a continuous random variable form an interval on the number line.

Examples of discrete random variables include the number of children a randomly selected person has and the number of times a randomly chosen student has been pulled over for speeding on the interstate. Continuous random variables often need to be measured, not counted. For example, the temperature in Atlanta, Georgia, at noon today may be reported as 77 degrees, but this value represents actual temperatures that may lie anywhere between 76.5 degrees and 77.5 degrees.

Example 6.3   Discrete and continuous random variables

For the following random variables, (i) determine whether they are discrete or continuous, and (ii) indicate the possible values they can take.
a.   The number of automobiles owned by a family
b.   The width of your desk in this classroom
c.   The number of games played in the next World Series
d.   The weight of model year 2007 SUVs

Solution
a.   Since the possible number of automobiles owned by a family is finite and may be written as a list of numbers, it represents a discrete random variable. The possible values are $\{0, 1, 2, 3, 4, \ldots\}$.
b.   Width is something that must be measured, not counted. Width can take infinitely many different possible values, with these values forming an interval on the number line. Thus, the width of your desk is a continuous random variable. The possible values might be $1 \text{ ft} \leq W \leq 10 \text{ ft}$.
c.   The number of games played in the next World Series can be counted and thus represents a discrete random variable. The possible values are finite and may be written as a list of numbers: $\{4, 5, 6, 7\}$.
d.   The weight of model year 2007 SUVs must be measured, not counted, and so represents a continuous random variable. Weight can take infinitely many different possible values, with these values forming an interval on the number line: $2500 \text{ lb} \leq Y \leq 7000 \text{ lb}$. ∎

We will return to continuous random variables in Section 6.3. Sections 6.1 and 6.2 concentrate on discrete random variables. Let's start with an example aimed at helping you move from the language of probability (experiments and outcomes) to the language of random variables.

Example 6.4   The number of heads in a two-coin-toss experiment is a discrete random variable

Recall the experiment of tossing two fair coins from Section 5.1. The sample space is

$$\{HH, HT, TH, TT\}$$

Now, suppose we are interested in the number of heads observed on the two tosses.
a.   Define our random variable X.
b.   What are the possible values of X?
c.   Is X a discrete or continuous random variable, and why?
d.   For each possible value of X, use random variable notation to express the probability of that value.

Solution
a.   We define our random variable X as follows:

$$X = \text{the number of heads observed on the two tosses}$$

b.   When tossing two fair coins, we may observe either 0, 1, or 2 heads. Thus, the possible values of X are 0, 1, and 2.
c. X is a discrete random variable because it takes a finite number of possible values that may be written as a list.
d.   In Section 5.1 we learned that the probability of observing two heads is 1/4. Therefore, the probability that $X = 2$ is 1/4, denoted $P(X = 2) = 1/4$. The probability of observing one head is 1/2. Therefore, the probability that $X = 1$ is 1/2, written as $P(X = 1) = 1/2$. The probability of observing zero heads is 1/4. Therefore, the probability that $X = 0$ is 1/4, written as $P(X = 0) = 1/4$. ∎

✎ Discrete Probability Distributions

In this section, we consider discrete probability distributions, which inform us of the probabilities associated with the various values that the discrete random variable can take.

> A **probability distribution of a discrete random variable** provides all the possible values that the random variable can assume, together with the probability associated with each value. The probability distribution can take the form of a table, graph, or formula. Probability distributions describe populations, not samples.

When constructing the tabular form of a **probability distribution of a discrete random variable**, create a table with two rows:

- The top row will contain all the possible values of *X*.

- The bottom row will contain the probability associated with each value of *X*.

Example 6.5   Probability distribution table

Construct the probability distribution table of the number of heads observed when tossing a fair coin twice.

Solution

The probability distribution table given in Table 6.2 uses probabilities we found in Example 6.4. ▪

Table 6.2 Probability distribution of number of heads on two fair coin tosses

X = number of heads observed	0	1	2
$P(X)$ = probability of observing that many heads	1/4	1/2	1/4

Note that the probabilities in the bottom row of Table 6.2 add up to 1. This is no coincidence. According to the Law of Total Probability from Chapter 5, the sum of the probabilities of all the outcomes in the sample space must equal 1. Put into the language of random variables, this means that the sum of the probabilities of all the possible values of a discrete random variable *X* must equal 1. This requirement is written as $\sum P(X) = 1$. Also, since each value in the bottom row is a probability, each value must be between 0 and 1, inclusive, that is, $0 \leq P(X) \leq 1$. There are no such restrictions on the values that *X* can take except that they be countable.

> **Requirements for the Probability Distribution of a Discrete Random Variable**
> - The sum of the probabilities of all the possible values of a discrete random variable must equal 1. That is, $\sum P(X) = 1$.
> - The probability of each value of *X* must be between 0 and 1, inclusive. That is, $0 \leq P(X) \leq 1$.

The probabilities assigned to the various values of *X* are found using either the classical method, the relative frequency method, or the subjective method. The probabilities in Table 6.2 were assigned using the classical method, since we assumed that tossing a fair coin would result in equally likely outcomes. In the next example, we turn to the subjective method to assign probabilities.

Example 6.6   Probability distribution for Kristin's financial gain

Recall Example 6.1, where Kristin assigned probabilities to various levels of financial gain associated with an investment of $2000 in Cape Ann Biotech. Table 6.1 is repeated here.

Table 6.1 Kristin's probability estimates for five investment outcomes for a period of 12 months (initial investment $2000)

Scenario	Financial gain	Kristin's estimated probability
Cape Ann Biotech does very well.	Gain $1000	0.15
Cape Ann Biotech does fairly well.	Gain $500	0.30
Cape Ann Biotech treads water.	Gain $0	0.25
Cape Ann Biotech does not do very well.	Lose $200 (Gain −$200)	0.20
Cape Ann Biotech folds.	Lose $2000 (Gain −$2000)	0.10

Once we define our random variable, Table 6.1 can be converted into a probability distribution table. Since Kristin is interested in what happens to her money, we define our random variable X to be

X = financial gain (the financial gain associated with the five possible scenarios)

a. Why is the variable X = financial gain a random variable?
b. What are the possible values that X can take?
c. Use random variable notation to express the probabilities associated with each possible outcome of X.
d. Construct the probability distribution of X = financial gain.
e. Find the probability that Kristin will make a profit on her investment.
f. Find the probability that Kristin will take a loss on her investment.

Solution

a. X = financial gain is a random variable because we do not know, before the investment is made, the value the variable will take.
b. The possible values of X are

$$X = \{-\$2000, -\$200, \$0, \$500, \$1000\}$$

c. Table 6.1 gives the probabilities of each of these outcomes:

$$P(X = -2000) = P(\text{gain} -\$2000) = 0.10$$
$$P(X = -200) = P(\text{gain} -\$200) = 0.20$$
$$P(X = 0) = P(\text{gain} \$0) = 0.25$$
$$P(X = 500) = P(\text{gain} \$500) = 0.30$$
$$P(X = 1000) = P(\text{gain} \$1000) = 0.15$$

d. For consistency, we retain the term "gain" when a loss is involved, inserting a negative value for the gain, which means the same as a loss of that value. Using this information, we may now proceed to construct the probability distribution as shown in Table 6.3. By convention, the values of X are usually shown in increasing order.

Table 6.3 Probability distribution of Kristin's financial gain

X = financial gain in dollars	−2000	−200	0	500	1000
$P(X)$	0.10	0.20	0.25	0.30	0.15

Note: In discrete probability distributions, the outcomes of the random variable are always mutually exclusive. For example, it is not possible for the result of Kristin's investment to be both gain $1000 and gain $500. This is why we may always use the Addition Rule for Mutually Exclusive Events to find the probability of two or more outcomes for a discrete random variable.

e. The probability distribution table makes it straightforward to combine probabilities. For example, the probability that Kristin will make a profit on the investment is

$$P(\text{gain } \$1000) + P(\text{gain } \$500) = P(X = 1000) + P(X = 500) = 0.15 + 0.30 = 0.45$$

So there is a 45% chance that Kristin will profit from this investment. Since the outcomes $X = 1000$ and $X = 500$ are mutually exclusive, we may add the probabilities of these outcomes by the Addition Rule for Mutually Exclusive Events: $P(A \text{ or } B) = P(A) + P(B)$.

f. The probability that she will take a loss is

$$P(X = -200) + P(X = -2000) = 0.20 + 0.10 = 0.30$$

Discrete Probability Distribution as a Graph

We can construct graphs of probability distributions that show all the information contained in probability distribution tables. The graphical representation allows us to identify patterns more quickly.

The horizontal axis in the probability distribution graph in Figure 6.1 is the usual *x* axis. The range is all the possible values that the random variable *X* can take. Here the random variable *X* takes on values -2000, -200, 0, 500, and 1000, from Table 6.3. The horizontal axis gives the same information as the top row of the probability distribution table. The vertical axis represents probability. A vertical bar is drawn at each value of *X*; the height of the bar represents the probability of that value of *X*. The bar of probability at $X = 0$ goes up to 0.25 and represents the probability that Kristin will neither gain nor lose money from her investment. The heights of these bars give the same information as the bottom row in the probability distribution table.

FIGURE 6.1
Graph of probability distribution for Kristin's financial gain.

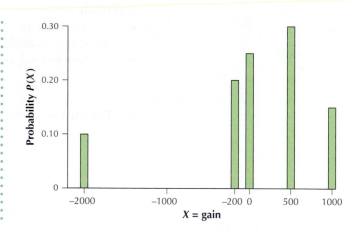

Given a graph of a probability distribution, you should know how to construct the probability distribution table, and vice versa.

3 Mean and Variability of a Discrete Random Variable

Suppose you are about to perform a new experiment, such as finding the number of siblings that a randomly chosen classmate has. You have assigned probabilities to the outcomes, but the experiment has not been performed yet. Wouldn't it be nice to know the mean number of siblings? If you had a particular outcome in mind, wouldn't it be nice to know whether this outcome was unusual? Questions like these are addressed in this section.

The **mean μ of a discrete random variable** represents the mean result when the experiment is repeated an indefinitely large number of times.

Note that this meaning of *mean* differs from that of the sample mean we learned in Chapter 3. The sample mean $\bar{x}$ is simply the arithmetic mean of a finite set of sample values.

Example 6.7   The meaning of the mean of a discrete probability distribution

The U.S. Department of Health and Human Services reports that there were 250,000 babies born to teenagers aged 15–18 in 2004. Of these 250,000 births, 7% were to 15-year-olds, 17% were to 16-year-olds, 29% were to 17-year-olds, and 47% were to 18-year-olds. The following table contains the probability distribution of the random variable X = age.

X = age	$P(X)$
15	0.07
16	0.17
17	0.29
18	0.47

a. Graph the probability distribution of the random variable X = age.
b. Estimate the mean of X using the balance point idea from page 80.
c. If one of the teenagers represented in the table is chosen at random, what is the most likely age of that teenager when her baby was born?

Solution
a. The probability distribution graph of X = age is given in Figure 6.2.

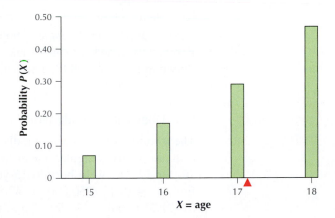

FIGURE 6.2

b. Recall that one interpretation of the mean is the *balance point* of the distribution. The balance point for the probability distribution graph appears to be somewhere near 17 years old. Hence, we would estimate the mean age to be about 17 years old, as indicated by the red fulcrum in Figure 6.2. We will calculate the actual value for the mean in Example 6.8.
c. Since $P(X = 18)$ has a larger value than any other probability in the table, and the bar for $X = 18$ is the tallest bar in the probability distribution graph, 18 is the most likely age.

> **Finding the Mean of a Discrete Random Variable**
>
> The mean μ of a discrete random variable is found as follows:
>
> 1. Multiply each possible value of X by its probability.
> 2. Add the resulting products.
>
> This procedure is denoted as
>
> $$\mu = \sum [X \cdot P(X)]$$

Example 6.8   Calculating the mean of a discrete random variable

Find the mean age of the mother for the babies born to teenagers aged 15−18 in 2004, from Example 6.7.

Note: The population mean μ need not equal any of the values of X, nor need it be an integer.

Solution

First we need to multiply each possible outcome (value of X) by its probability $P(X)$. We multiply the value $X = 15$ by its probability $P(X) = 0.07$, the value $X = 16$ by its probability $P(X) = 0.17$, and so on. Then we add these four products to find the mean:

$$\mu = 15(0.07) + 16(0.17) + 17(0.29) + 18(0.47) = 17.16$$

The mean age of the mother for the babies born to teenagers aged 15–18 is 17.16 years old, which is near 17 years old, as we estimated in Example 6.7 using the balance point method.   ■

What Does This Number Mean?

What does it mean to say that $\mu = 17.16$ is the mean of the random variable X? First of all, the mean of the random variable X is definitely not the same as the mean of a sample of teenage mothers. The latter is a sample mean. For example, suppose that, for a certain hospital, the teenage mothers' ages for the last 5 such births were 16, 18, 18, 17, 18. The mean of this sample of 5 births is $\bar{x} = 17.4$. However, if we were to consider an *infinite number* of births to mothers aged 15–18, then the mean of this large (!) sample would indeed converge to 17.16. So the mean μ of a discrete random variable is interpreted as the mean of the results from the *population* of all possible repetitions of the experiment. That is why we denote the mean of a random variable as μ, since it denotes a population mean (population of all possible results). Fortunately, since we cannot take a sample of infinite size, we have the method shown in Example 6.8 for calculating the mean.   ■

Developing Your Statistical Sense

Why Does This Formula Work?

The formula for the mean of a discrete random variable works because it is a special case of the weighted mean (page 108). Of the population of 250,000 babies, 7%, or 17,500, were born to 15-year-olds. Thus, $w_1 = 17{,}500$. Similarly, we can find, $w_2 = (0.17)(250{,}000) = 42{,}500$, $w_3 = (0.29)(250{,}000) = 72{,}500$, and $w_4 = (0.47)(250{,}000) = 117{,}500$. Thus, the population weighted mean is

$$\mu = \frac{\sum w_i x_i}{\sum w_i} = \frac{(17{,}500)(15) + (42{,}500)(16) + (72{,}500)(17) + (117{,}500)(18)}{250{,}000}$$

Dividing through and rearranging terms give us

$$\mu = (15)(0.07) + (16)(0.17) + (17)(0.29) + (18)(0.47) = \sum [X \cdot P(X)]$$ ■

The mean μ of a random variable is also called the **expected value** or the **expectation of the random variable X.** It does not necessarily follow that the expected value of X is the most likely value of X (although it does follow if the distribution is

FIGURE 6.3

Example 6.11 An unusual result?

Using the information from Example 6.10, determine whether $X = -\$2000$ is an unusual result.

Solution
Recall from Section 3.4 (page 119) that a data value farther than two standard deviations from the mean may be considered moderately unusual. How many standard deviations below the mean does $X = -\$2000$ lie? We have the mean $\mu = \$60$ and the standard deviation $\sigma = \$793.35$. The deviation for $X = -\$2000$ is

$$|X - \mu| = |-2000 - 60| = 2060$$

Expressed in terms of standard deviations, we see that the value $X = -\$2000$ lies $2060/793.35 \approx 2.6$ standard deviations below the mean. Thus, it would be considered moderately unusual for Kristin to lose $2000 on her investment.

Section 6.1 SUMMARY

1 Section 6.1 introduces the idea of random variables, a crucial concept that we will use to assess the behavior of variable processes for the remainder of the text. Random variables are variables whose value is determined at least partly by chance. Discrete random variables take values that are either finite or countable and may be put in a list. Continuous random variables take an infinite number of possible values, represented by an interval on the number line.

2 Discrete random variables can be described using a probability distribution, which specifies the probability of observing each value of the random variable. Such a distribution can take the form of a table, graph, or formula. Probability distributions describe populations, not samples.

3 We can find the mean μ, standard deviation σ, and variance σ^2 of a discrete random variable using formulas.

Section 6.1 EXERCISES

CLARIFYING THE CONCEPTS
1. Explain in your own words what a random variable is. Give an example of a random variable from your own life experience.
2. Is your height a random variable? Under what circumstances would your height be considered a random variable? Under what circumstance would your height not be considered a random variable?
3. What is the difference between a discrete random variable and a continuous random variable?
4. What is the difference between a discrete random variable and a discrete probability distribution?

symmetric with a single mode). However, the expected value of X (that is, the mean μ) is often a good indication of the center of the distribution of the random variable.

> The **expected value, or expectation, of a random variable** X is the mean μ of X. It is denoted as $E(X)$. This definition holds for both discrete and continuous random variables.

Example 6.9 Expected value of Kristin's financial gain from her investment

Calculate the expected value for Kristin's financial gain from her investment described in Example 6.1.

Solution
The following table contains the probability distribution we found earlier for the discrete random variable X = financial gain for Kristin's investment.

X = financial gain in dollars	−2000	−200	0	500	1000
$P(X)$	0.10	0.20	0.25	0.30	0.15

We calculate the expected value of Kristin's financial gain as follows. We multiply each value of X by its associated probability (entry in the top row by the entry below it), and we add up all these products.

$$\mu = (-2000)(0.10) + (-200)(0.20) + 0(0.25) + 500(0.30) + 1000(0.15) = 60$$

Since the unit for X is dollars, the mean is $60, which is the expected financial gain for Kristin's investment. Note that, according to Table 6.1, in no particular 12-month period can Kristin's earnings exactly equal the expected gain of $60. The $60 is the long-run mean.

Variability of a Discrete Random Variable

In Chapter 3 we discussed methods for detecting the presence of outliers in data sets. Similarly, in this section, we use the **variance** or **standard deviation of a random variable** to help us determine whether a particular value of that random variable is unusual. Say you are playing a game that requires you to toss two dice and you toss a pair of ones. Is this an unusual result? We can answer this question using the concepts of variation and standard deviation. Recall in Chapter 3 we examined measures of center and measures of spread. Just as a random variable has a mean (μ), which is a measure of center, so a random variable also has a standard deviation (σ) and variance (σ^2), which are measures of spread. The variance of a discrete random variable is given by

$$\sigma^2 = \sum[(X - \mu)^2 \cdot P(X)]$$

Notice that this formula includes μ as one of its terms, so that you must first find the mean of a discrete random variable before you find the variance (or standard deviation). This definition formula can be tedious in that it requires you to find each of the deviations $(X - \mu)$. There is a computational formula for the variance of a discrete random variable that is equivalent to the definition formula but sometimes is computationally simpler.

Formulas for the Variance and Standard Deviation of a Discrete Random Variable

Definition Formulas	Computational Formulas
$\sigma^2 = \sum[(X - \mu)^2 \cdot P(X)]$	$\sigma^2 = \sum[X^2 \cdot P(X)] - \mu^2$
$\sigma = \sqrt{\sum[(X - \mu)^2 \cdot P(X)]}$	$\sigma = \sqrt{\sum[X^2 \cdot P(X)] - \mu^2}$

Recall from Chapter 3 that the standard deviation is simply the square root of the variance. The definition and computational formulas for the standard deviation of a discrete random variable are the square roots of the formulas for the variance.

Example 6.10   Finding the variance and standard deviation of a discrete random variable

Find the variance and standard deviation of Kristin's financial gain, using (a) the definition formula, and (b) the computational formula. Assume we have already calculated the mean $\mu = \$60$.

Solution

a. Refer to Table 6.4. The first two columns correspond to the probability distribution of Kristin's financial gain. The next two columns are intermediate steps for the calculation in the rightmost column, $(X - \mu)^2 \cdot P(X)$. Summing the values in the rightmost column provides the variance. Taking the square root of the variance gives us the standard deviation.

Table 6.4 Using the definition formula

X	$P(X)$	$X - \mu$	$(X - \mu)^2$	$(X - \mu)^2 \cdot P(X)$
−2000	0.10	−2060	4,243,600	424,360
−200	0.20	−260	67,600	13,520
0	0.25	−60	3,600	900
500	0.30	440	193,600	58,080
1000	0.15	940	883,600	132,540

$$\sigma^2 = \sum[(X - \mu)^2 \cdot P(X)] = 629{,}400$$
$$\sigma = \sqrt{\sigma^2} = 793.3473388 \approx \$793.35$$

b. Refer to Table 6.5. The rightmost column contains the values $X^2 \cdot P(X)$. Summing the values in the rightmost column provides $\sum[X^2 \cdot P(X)]$. To find the variance, we must subtract μ^2. Taking the square root of the variance gives us the standard deviation.

Table 6.5 Using the computational formula

X	$P(X)$	X^2	$X^2 \cdot P(X)$
−2000	0.10	4,000,000	400,000
−200	0.20	40,000	8,000
0	0.25	0	0
500	0.30	250,000	75,000
1000	0.15	1,000,000	150,000

$$\sum[X^2 \cdot P(X)] = 633{,}000$$
$$\sigma^2 = \sum[X^2 \cdot P(X)] - \mu^2 = 629{,}400$$
$$\sigma = \sqrt{\sigma^2} = 793.3473388 \approx \$793.35$$

The standard deviation of $X =$ Kristin's financial gain is approximately $793. ∎

*W*hat Does This Value for the Standard Deviation Mean?

The standard deviation of a random variable is a measure of spread of the distribution of X. We have $\mu + 1\sigma \approx \$60 + \$793 = \$853$, and $\mu - 1\sigma \approx \$60 - \$793 = -\$733$, as shown in Figure 6.3. Note that most of the values of X lie between $\mu - 1\sigma$ and $\mu + 1\sigma$. Like the mean, the standard deviation σ of a random variable X is interpreted as the standard deviation of the results from the population of all possible repetitions of the experiment. ∎

Solution

a. Figure 6.4 shows the tree diagram for this experiment.

b. As we can see from Figure 6.4, there are $(_nC_X) = (_3C_2) = 3$ different way exactly two of the three online daters could be LTRers (highlighted in bl

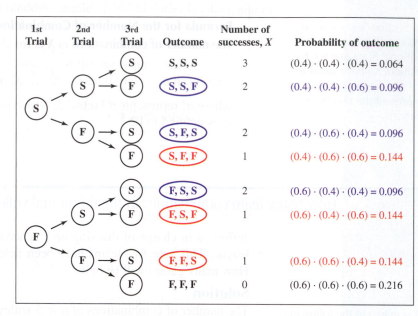

1st Trial	2nd Trial	3rd Trial	Outcome	Number of successes, X	Probability of outcome
		S	S, S, S	3	$(0.4) \cdot (0.4) \cdot (0.4) = 0.064$
	S	F	S, S, F	2	$(0.4) \cdot (0.4) \cdot (0.6) = 0.096$
S	F	S	S, F, S	2	$(0.4) \cdot (0.6) \cdot (0.4) = 0.096$
		F	S, F, F	1	$(0.4) \cdot (0.6) \cdot (0.6) = 0.144$
	S	S	F, S, S	2	$(0.6) \cdot (0.4) \cdot (0.4) = 0.096$
F		F	F, S, F	1	$(0.6) \cdot (0.4) \cdot (0.6) = 0.144$
	F	S	F, F, S	1	$(0.6) \cdot (0.6) \cdot (0.4) = 0.144$
		F	F, F, F	0	$(0.6) \cdot (0.6) \cdot (0.6) = 0.216$

FIGURE 6.4 Tree diagram and binomial probabilities for Example 6.14.

For each of these three outcomes, the probability that $X = 2$ is $(0.4)^2(0.$ 0.096. In particular, the outcome S, S, F (second row in Figure 6.4) has proba $P(S) \cdot P(S) \cdot P(F) = (0.4)(0.4)(0.6) = 0.096$. Similarly, the outcome $S, F,$ probability $P(S) \cdot P(F) \cdot P(S) = (0.4)(0.6)(0.4) = 0.096$, and the outcome S has probability $P(F) \cdot P(S) \cdot P(S) = (0.6)(0.4)(0.4) = 0.096$. Note that, $P(S) = p$, this probability is $(p)^2 (1 - p)$, with p having exponent $X = 2$ $(1 - p)$ having exponent $n - X = 3 - 2 = 1$. Thus,

$$P(X = 2) = (_3C_2)(0.4)^2(0.6)$$
$$= 3(0.096)$$
$$= 0.288.$$

c. Similarly, suppose that we are interested in whether exactly one $(X = 1)$ of the three online daters is an LTRer. Then, Figure 6.4 shows us, highlighted in red that there are $(_nC_X) = (_3C_1) = 3$ different ways this could happen. Each of these outcomes has probability $(p)(1 - p)^2 = (0.4)(0.6)^2 = 0.144$, where has exponent $X = 1$, and $(1 - p)$ has exponent $n - X = 3 - 1 = 2$. Thus

$$P(X = 1) = (_3C_1)(0.4)(0.6)^2$$
$$= 3(0.144)$$
$$= 0.432.$$

We can generalize these procedures and use the **binomial probability distribu formula** to find probabilities for the number of successes for any binomial experim

The Binomial Probability Distribution Formula

The probability of observing exactly X successes in n trials of a binomial experiment is

$$P(X) = (_nC_X)\, p^X (1 - p)^{n - X}$$

Next, a word about *cumulative probability*. Recall that Joshua will pass if he answers at least three of the four questions correctly. What is the probability that Joshua will not pass? In other words, what is the probability that Joshua answers at most two of the four questions correctly? This is what is known as a cumulative probability, since it is asking for the accumulated probability for $X = 0$, $X = 1$, and $X = 2$. Statistical software and the TI-83/84 graphing calculator each have a function that will find cumulative binomial probabilities for you.

| **Example 6.17** | Using technology to find binomial probabilities |

Recall that we have $n = 4$ questions, each with 5 possible alternatives, so that $p = $ probability of success on any given question is 0.2 (1/5). Use the TI-83/84 to find the following probabilities:

a. $P(X = 4)$, the probability that Joshua will get all four correct
b. $P(X \leq 2)$, the (cumulative) probability that Joshua will flunk the quiz

Solution

We use the instructions in the Step-by-Step Technology Guide at the end of this section (page 270).

a. Figure 6.6 shows that we use the function **binompdf** with $n = 4$, $p = 0.2$, and $X = 4$. Figure 6.7 shows the result, $P(X = 4) = 0.0016$.

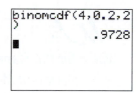

FIGURE 6.6   **FIGURE 6.7**   **FIGURE 6.8**

b. We use the function **binomcdf** with $n = 4$, $p = 0.2$, and $X = 2$. Figure 6.8 shows the result, $P(X \leq 2) = 0.9728$. ∎

4 Binomial Mean, Variance, and Standard Deviation

In Section 6.1, we examined the mean, variance, and standard deviation of a discrete random variable. Since the binomial random variable X is discrete, it also has a **mean, variance,** and **standard deviation,** shown here.

> **Mean, Variance, and Standard Deviation of a Binomial Random Variable X**
> - Mean (or expected value): $\mu = n \cdot p$
> - Variance: $\sigma^2 = n \cdot p \cdot (1 - p)$
> - Standard deviation: $\sigma = \sqrt{n \cdot p \cdot (1 - p)}$

| **Example 6.18** | Mean, variance, and standard deviation of left-handed students |

Suppose we know that the population proportion p of left-handed students is 0.10.

a. In a sample of 200 students, how many would we expect to be left-handed?
b. Would 40 left-handed students out of 200 be considered unusual?

Solution

The binomial random variable here is X = the number of left-handed students.

a. Here, $n = 200$, and $p = 0.10$. So the expected number of left-handed students in a sample of 200 is

$$E(X) = \mu = n \cdot p = (200)(0.10) = 20$$

b. To determine whether 40 lefties is unusual:
   1. Find the standard deviation of X = the number of lefties in a sample of 200 students.
   2. Calculate how many standard deviations 40 lies from the mean $\mu = 20$.

The standard deviation of X is

$$\sigma = \sqrt{n \cdot p \cdot (1 - p)} = \sqrt{(200)(0.1)(1 - 0.1)} = \sqrt{18} \approx 4.2426$$

How many standard deviations does 40 lie above the mean of 20? Simply take the difference between 40 and 20, and divide the result by the standard deviation 4.2426:

$$\frac{X - \mu}{\sigma} = \frac{40 - 20}{4.2426} \approx 4.714$$

Finding 40 lefties in a sample of 200 is unusual because this value lies 4.7 standard deviations above the mean. ■

What Does This Value for the Standard Deviation Mean?

The standard deviation of a binomial random variable X is a measure of spread of the binomial random variable X. Suppose the binomial experiment in Example 6.18 is performed an infinitely large number of times. Then the standard deviation of this infinitely large collection of values of X will be $\sigma \approx 4.2426$. ■

STEP-BY-STEP TECHNOLOGY GUIDE: Finding Binomial Probabilities

For Example 6.17 (page 269).

TI-83/84

Step 1   Press **2ⁿᵈ > DISTR** (the **VARS** key).
Step 2   Do one of **(a)** or **(b)**:
a. For individual binomial probabilities, highlight **binompdf(** and press **ENTER**.

b. For cumulative binomial probabilities, highlight **binomcdf(** and press **ENTER**.
Step 3   Enter the values for n, p, and K, separated by commas.
Step 4   Press **ENTER**.

EXCEL

Step 1   Select cell **A1**. Click the **Insert Function** icon f_x.
Step 2   In the **Search for a function** area, type **BINOMDIST**, and click **OK**.
Step 3   For **Number_s**, enter the number of successes, K. For **Trials**, enter the sample size, **n**. For **Probability_s**, enter the probability of success, **p**.
Step 4   Do one of **(a)** or **(b)**:

a. For individual binomial probabilities, next to **Cumulative**, enter **false**.
b. For cumulative binomial probabilities, next to **Cumulative**, enter **true**.
Step 5   Click **OK**. See Figures 6.9 and 6.10 for illustrations using Example 6.17.

Figure 6.9 Example 6.17a using Excel.

Figure 6.10 Example 6.17b using Excel.

MINITAB
Step 1   Click **Calc > Probability Distributions > Binomial**.
Step 2   Do one of (a) or (b):
a. For individual binomial probabilities, select **Probability** and enter the number of trials **n** and probability of success **p**.

b. For cumulative binomial probabilities, select **Cumulative Probability** and enter the number of trials **n** and probability of success **p**.
Step 3   Select **Input Constant**, enter **K** and click **OK**.

Section 6.2 SUMMARY

1 The most important discrete distribution is the binomial distribution, where there are two possible outcomes, each with probability of success *p*, and *n* independent trials.

2 The probability of observing a particular number of successes can be calculated using the binomial probability distribution formula.

3 Binomial probabilities can also be found using the binomial tables or using technology.

4 There are formulas for finding the mean, variance, and standard deviation of a binomial random variable.

Section 6.2 EXERCISES

CLARIFYING THE CONCEPTS
For Exercises 1–4, indicate whether the experiment is binomial. If it is not binomial, explain why not.
1. Asking 10 of your friends to come to your party.
(*Hint:* Remember the independence assumption.)
2. Selecting a student in the class at random until you come across a left-handed student.
3. Recording the ethnicity of the next 20 babies born at City Hospital.
4. Answering the next 10 multiple choice questions correctly or incorrectly on a test.

PRACTICING THE TECHNIQUES
For Exercises 5–10, determine whether or not the experiment is binomial. If the experiment is binomial, identify the random variable *X*, the number of trials *n*, the probability of success *p*, and the probability of failure $1 - p$. If the experiment is not binomial, say why not.
5. Toss a fair die 3 times, and note the total number of spots.

6. Toss a fair die 3 times, and note the number of sixes.
7. Toss a fair die 3 times, and note how often an "even" result is observed.
8. Toss a fair die 3 times, and note whether a six was observed.
9. Bob has paid to play 2 games at a carnival. The probability that he wins a particular game is 0.25.
10. Bob is playing a game at a carnival where he gets to play until he loses. The probability that he wins a particular game is 0.25.

For Exercises 11–14, the experiment is to toss a fair coin three times. Use the binomial formula to find the indicated probabilities.
11. No heads were observed.
12. One head was observed.
13. Two heads were observed.
14. At most two heads were observed.

Solution

Step 1   We draw a bell-shaped curve, label the horizontal line as X, and insert the mean $\mu = 3285$ in the center of the number line, as shown in Figure 6.18a.

Step 2   We are asked for the probability that X is less than 3285 grams. Note that $X = 3285$ also happens to be the mean μ. Since we need the probability of a baby weighing *less than* 3285 grams, we are looking for an area *to the left of* 3285 grams. The desired area is shaded in Figure 6.18b.

FIGURE 6.18a   FIGURE 6.18b

Step 3   Since the curve is a normal curve, the mean $\mu = 3285$ divides the area under the curve in half, because the normal distribution is symmetric about its mean. Thus, the shaded region has area 0.5, and the probability that a randomly chosen baby from the treatment group will weigh less than 3285 grams equals 0.5. ■

In Chapter 3, we learned that according to the Empirical Rule the area under the normal curve has the following properties (see Figure 6.19).

1. About 68% of the area under the curve lies within one standard deviation of the mean.

2. About 95% of the area under the curve lies within two standard deviations of the mean.

3. About 99.7% of the area under the curve lies within three standard deviations of the mean.

FIGURE 6.19
The Empirical Rule.

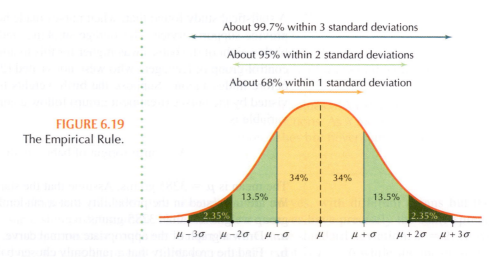

We will verify the Empirical Rule in Section 6.4.

Example 6.21   Grade inflation

Many educators are concerned about grade inflation. One study shows that one low-SAT-score high school (with mean combined SAT score = 750) had higher mean grade point average (mean GPA = 3.6) than a high-SAT-score school (with mean combined SAT score = 1050 and mean GPA = 2.6).[6] Define the following random variable:

$$X = \text{GPA at the high-SAT-score school}$$

Assume that X is normally distributed with mean $\mu = 2.6$ and standard deviation $\sigma = 0.46$.

a. What is the probability that a randomly chosen GPA at the high-SAT-score school will be between 3.06 and 3.52?

b. Find the probability that a randomly chosen GPA at the high-SAT-score school will be greater than 3.52.

Solution

a. Figure 6.20 shows the distribution of GPA at the high-SAT-score school. The area under the curve between 3.06 and 3.52 represents the area between $\mu + \sigma$ and $\mu + 2\sigma$. Courtesy of the Empirical Rule, Figure 6.19 tells us that the area between $\mu + \sigma$ and $\mu + 2\sigma$ is about 13.5% of the area under the curve.

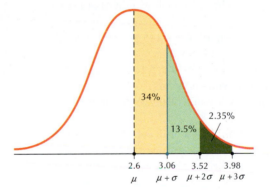

FIGURE 6.20

Therefore, the probability that a randomly chosen GPA at the high-SAT-score school will be between 3.06 and 3.52 is about 0.135.

b. The area under the normal density curve to the right of the mean equals 0.5. So the area to the right of $\mu = 2.6$ equals 0.5, or 50% of the area under the curve. To find the area to the right of $X = 3.52$, we need to subtract the yellow area (34%) and the light green area (13.5%) from 50%: 50% – 34% – 13.5% = 2.5%. Therefore, the probability that a randomly chosen GPA at the high-SAT-score school will be greater than 3.52 is about 0.025. ∎

Section 6.3 SUMMARY

1 Continuous random variables assume infinitely many possible values, with no gap between the values. Probability for continuous random variables consists of area above an interval on the number line and under the distribution curve.

2 The normal distribution is the most important continuous probability distribution. It is symmetric about its mean μ and has standard deviation σ. One should always sketch a picture of a normal probability problem to help solve it.

Section 6.3 EXERCISES

CLARIFYING THE CONCEPTS

1. For a continuous random variable X, why are we not interested in whether X equals some particular value?

2. In the graph of a probability distribution, what is represented on the number line?

3. How is probability represented in the graph of a continuous probability distribution?

4. What are the possible values for the mean of a normal distribution? For the standard deviation?

PRACTICING THE TECHNIQUES

Use the normal distribution from Example 6.20 for Exercises 5–12. Birth weights are normally distributed with a mean weight of $\mu = 3285$ grams and a standard deviation of $\sigma = 500$ grams.

5. What is the probability of a birth weight equal to 3285 grams?

6. What is the probability of a birth weight more than 3285 grams?

7. What is the probability of a birth weight of at least 3285 grams?

8. Is the area to the right of $X = 4285$ grams greater than or less than 0.5? How do you know this?

9. Is the area to the left of $X = 4285$ grams greater than or less than 0.5? How do you know this?

10. What is the probability of a birth weight between 2785 and 3785 grams?

11. What is the probability of a birth weight between 1785 and 4785 grams?

12. What is the probability of a birth weight between 785 and 5785 grams?

13. The two normal distributions in the accompanying figure have the same standard deviation of 5 but different means. Which normal distribution has mean 10 and which has mean 25? Explain how you know this.

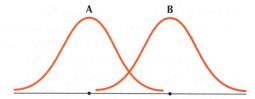

14. The two normal distributions in the figure below have the same mean of 100 but different standard deviations. Which normal distribution has standard deviation 3 and which has standard deviation 6? Explain how you know this.

For Exercises 15–22, use the graph of the normal distribution to determine the mean and standard deviation. *Hint:* The distance between dotted lines in the figures represents 1 standard deviation.

15.

16.

17.

18.

19.

20.

21.

22.

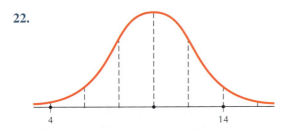

APPLYING THE CONCEPTS

For Exercises 23–32, sketch the distribution, showing μ, $\mu + \sigma$, $\mu + 2\sigma$, $\mu + 3\sigma$, $\mu - \sigma$, $\mu - 2\sigma$, and $\mu - 3\sigma$. Then answer the questions.

23. Windy Frisco. The average wind speed in San Francisco in July is 13.6 miles per hour (mph) in July, according to the U.S. National Oceanic and Atmospheric Administration. Suppose that the distribution of the wind speed in July in San Francisco is normal with mean $\mu = 13.6$ mph and standard deviation $\sigma = 4$ mph.

    a. Shade the region that represents wind speeds between 9.6 and 17.6 mph.

    b. What is the proportion of wind speeds between 9.6 and 17.6? (*Hint:* See Figure 6.19 [page 280].)

24. Magic McGrady. In the 2003–2004 National Basketball Association regular season, Tracy McGrady of the Orlando Magic led the league in points per game with 28. Suppose that McGrady's points per game follow a normal distribution with mean $\mu = 28$ and standard deviation $\sigma = 8$.

    a. Shade the region that represents more than 36 points scored in a game.

    b. What is the probability that McGrady scored more than 36 points in a game?

25. Viewers of *60 Minutes*. Nielsen Media Research reported that, for the week ending July 11, 2004, 12 million viewers watched the television show *60 Minutes*. Suppose that the distribution of viewers of *60 Minutes* is normal with mean $\mu = 12$ million and standard deviation $\sigma = 4$ million.

    a. Shade the region that represents fewer than 4 million viewers.

    b. What is the probability that fewer than 4 million viewers will watch *60 Minutes*?

26. Math Scores. The National Center for Education Statistics reports that in 2005 the mean score on the standardized mathematics test for eighth-graders increased by 7 points from 2000. The mean score in 2005 was $\mu = 273$; assume a standard deviation of $\sigma = 7$.

    a. Shade the region that represents scores between 266 and 273.

    b. What is the probability that a student scored between 266 and 273 on the test?

27. Los Angeles Temperature. The U.S. Department of Commerce reports that the mean temperature in downtown Los Angeles is 66.2 degrees Fahrenheit (°F). Assume that the distribution of temperature in Los Angeles is normal with mean $\mu = 66.2$°F and standard deviation $\sigma = 8$°F.

    a. Shade the region that represents temperatures between 50.2°F and 82.2°F.

    b. What is the probability that the temperature in Los Angeles is between 50.2°F and 82.2°F?

28. Hospital Patient Length of Stays. A study of Pennsylvania hospitals showed that the mean patient length of stay in 2001 was 4.87 days with a standard deviation of 0.97 days.[7] Assume that the distribution of patient length of stays is normal.

    a. Shade the region that represents patient length of stay of less than 3.9 days.

    b. Find the probability that a randomly selected patient has a length of stay of less than 3.9 days.

29. Engineers' Salaries. According to the U.S. Census Bureau, students in their first year after graduating with bachelor's degrees in engineering earned an average salary of $3189 per month. Assume that this distribution is normal with mean $\mu = 3189$ and standard deviation $\sigma = 600$. What proportion of salaries lie between $3789 and $4389?

30. Tobacco-Related Deaths. The World Health Organization states that tobacco is the second leading cause of death in the world. Every year, an average of 5 million people die of tobacco-related causes. Assume that the distribution is normal with mean $\mu = 5$ (in millions) and standard deviation $\sigma = 1$ (in millions). What is the probability of between 4 million and 7 million deaths?

31. Median Household Income. The Census Bureau reports that the median household income was $48,201 in 2006. Assume that the distribution of income is normal with mean $\mu = \$48,201$ and standard deviation $\sigma = \$16,000$. Find the probability that a randomly selected household has an income of greater than $80,201.

32. Stock Shares Traded. The average number of shares traded on the New York Stock Exchange in June 2004 was 2.4 billion per day. Assume that the distribution of shares traded is normal with mean $\mu = 2.4$ (in billions of shares) and standard deviation $\sigma = 0.6$ (in billions of shares). What is the probability that between 1.8 billion and 2.4 billion shares were traded?

33. Birth Weights. Can you think of a problem with the assumption that the birth weights from Example 6.20 follow a normal distribution? Is there a minimum value of X for a normal distribution? Is there a minimum birth weight?

6.4 Standard Normal Distribution

Objectives: By the end of this section, I will be able to...

1 Find areas under the standard normal curve, given a Z-value.

2 Find the standard normal Z-value, given an area.

1 Finding Areas Under the Standard Normal Curve for a Given Z-Value

Note: Understanding the techniques explained in this section will allow you to analyze a whole world of data sets, even those that are not themselves normally distributed (see the Central Limit Theorem in the next chapter). Beyond this chapter, these techniques help you to find and understand *p*-values in Chapters 9–11.

There are billions of data sets in the world that are normally distributed, from test scores to student heights. But there is one very special normal distribution called the **standard normal distribution.** To solve questions involving normally distributed test scores, for example, we *transform* the original normal distribution to the standard normal distribution, using a process called *standardizing*. More on this in Section 6.5.

The mean and standard deviation of the standard normal distribution make it unique.

> **The Standard Normal (Z) Distribution**
> The **standard normal distribution** is a normal distribution with
>
> - mean $\mu = 0$ and
> - standard deviation $\sigma = 1$.

Because of its importance, the standard normal random variable is always denoted as a capital Z. The graph of the standard normal random variable Z is given in Figure 6.21. The standard normal curve is symmetric about its mean $\mu = 0$. Since the entire area under the curve equals 1, the area to the left of 0 and the area to the right of 0 both equal 0.5.

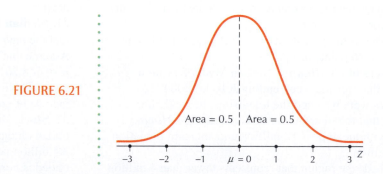

FIGURE 6.21

We will discuss two methods for finding probabilities associated with Z, using (a) the table for finding standard normal probabilities, called the **Z table,** and (b) technology. For the Z table, see Table C in the Appendix (pages T-9–T-10). *The Z table provides areas under the standard normal curve to the left of a specified value of Z, denoted as Z_1 (see Figure 6.22).*

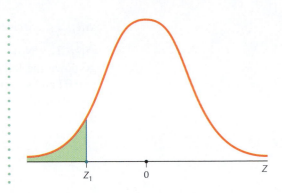

FIGURE 6.22
The Z table provides areas under the curve to the left of a specified value Z_1.

Note: Although your Z table contains only values between $Z = -3.49$ and $Z = 3.49$, there is no upper or lower limit to the values that Z may take. The curve essentially goes on forever in both the positive and the negative directions, always getting closer and closer to the horizontal axis but never quite touching it (there's a great plot for a love story in there somewhere).

Let's get acquainted with the Z table (see excerpt in Figure 6.24). Along the left side and across the top of the Z table are possible values of Z. These numbers, which in the table run from -3.49 to 3.49, are the values of Z found on the number line when you draw a graph. Down the left are the ones and tenths digits of the Z-value, and across the top is the hundredths digit. The body of the Z table contains areas (probabilities). These numbers, which run from 0.0002 to 0.9998, are areas under the standard normal curve that represent probabilities *to the left* of the specified value of Z. Table 6.7 shows the steps for finding areas under the standard normal curve, that is, for finding probabilities for specified values of Z.

Table 6.7 Steps for finding areas under the standard normal curve

Case 1	**Case 2**	**Case 3**
Find the area to the left of Z_1.	**Find the area to the right of Z_1.**	**Find the area between Z_1 and Z_2.**
Step 1 Draw the standard normal curve. Label the Z-value Z_1.	***Step 1*** Draw the standard normal curve. Label the Z-value Z_1.	***Step 1*** Draw the standard normal curve. Label the Z-values Z_1 and Z_2.
Step 2 Shade in the area to the left of Z_1.	***Step 2*** Shade in the area to the right of Z_1.	***Step 2*** Shade in the area between Z_1 and Z_2.
Step 3 Use the Z table to find the area to the left of Z_1.	***Step 3*** Use the Z table to find the area to the left of Z_1. The area to the right of Z_1 is then equal to $1 -$ (area to the left of Z_1).	***Step 3*** Use the Z table to find the area to the left of Z_1 and the area to the left of Z_2. The area between Z_1 and Z_2 is then equal to (area to the left of Z_2) $-$ (area to the left of Z_1).

Example 6.22   Case 1: Find the area to the left of a value of Z

Find the area to the left of $Z = 0.57$.

Solution

Step 1   First draw the standard normal curve and label $Z = 0.57$.

Step 2   Shade the area to the left of 0.57, as shown in Figure 6.23.

Step 3   Now, $Z = 0.57 = 0.5 + 0.07$. In the Z table, excerpted here as Figure 6.24, go down the left-hand column to 0.5 and select that row. Then go across the top row to 0.07 and select that column. The quantity at the intersection of this row and column represents the area to the left of $Z = 0.57$. That is, the area to the left of $Z = 0.57$ is 0.7157.   ■

FIGURE 6.23

Standard Normal Distribution

Z	0.00	0.01	0.02	0.03	0.04	0.05	0.06	0.07	0.08	0.09
0.0	0.5000	0.5040	0.5080	0.5120	0.5160	0.5199	0.5239	0.5279	0.5319	0.5359
0.1	0.5398	0.5438	0.5478	0.5517	0.5557	0.5596	0.5636	0.5675	0.5714	0.5753
0.2	0.5793	0.5832	0.5871	0.5910	0.5948	0.5987	0.6026	0.6064	0.6103	0.6141
0.3	0.6179	0.6217	0.6255	0.6293	0.6331	0.6368	0.6406	0.6443	0.6480	0.6517
0.4	0.6554	0.6591	0.6628	0.6664	0.6700	0.6736	0.6772	0.6808	0.6844	0.6879
0.5	0.6915	0.6950	0.6985	0.7019	0.7054	0.7088	0.7123	0.7157	0.7190	0.7224
0.6	0.7257	0.7291	0.7324	0.7357	0.7389	0.7422	0.7454	0.7486	0.7517	0.7549

FIGURE 6.24

Example 6.23   Case 2: Find the area to the right of a value of Z

Find the area to the right of $Z = -1.25$.

Solution

Step 1   First draw the standard normal curve and label $Z = -1.25$.

Step 2   Shade the area to the right of -1.25, as shown in Figure 6.25.

FIGURE 6.25

Step 3   In the Z table, excerpted here as Figure 6.26, go down the left-hand column to -1.2 and select that row. Then go across the top row to 0.05 and select that column.

The area to the left of $Z = -1.25$ is therefore 0.1056. From Case 2 in Table 6.7, the area to the right of -1.25 is then

$$1 - (\text{area to the left of } -1.25) = 1 - 0.1056 = 0.8944$$

Standard Normal Distribution

Z	0.00	0.01	0.02	0.03	0.04	0.05	0.06	0.07	0.08	0.09
−3.4	0.0003	0.0003	0.0003	0.0003	0.0003	0.0003	0.0003	0.0003	0.0003	0.0002
−3.3	0.0005	0.0005	0.0005	0.0004	0.0004	0.0004	0.0004	0.0004	0.0004	0.0003
−3.2	0.0007	0.0007	0.0006	0.0006	0.0006	0.0006	0.0006	0.0005	0.0005	0.0005
−3.1	0.0010	0.0009	0.0009	0.0009	0.0008	0.0008	0.0008	0.0008	0.0007	0.0007
−3.0	0.0013	0.0013	0.0013	0.0012	0.0012	0.0011	0.0011	0.0011	0.0010	0.0010
−1.4	0.0808	0.0793	0.0778	0.0764	0.0749	0.0735	0.0721	0.0708	0.0694	0.0681
−1.3	0.0968	0.0951	0.0934	0.0918	0.0901	0.0885	0.0869	0.0853	0.0838	0.0823
−1.2	0.1151	0.1131	0.1112	0.1093	0.1075	0.1056	0.1038	0.1020	0.1003	0.0985
−1.1	0.1357	0.1335	0.1314	0.1292	0.1271	0.1251	0.1230	0.1210	0.1190	0.1170
−1.0	0.1587	0.1562	0.1539	0.1515	0.1492	0.1469	0.1446	0.1423	0.1401	0.1379

FIGURE 6.26

CAUTION Remember that, although values of Z can be negative, *probabilities (or areas) can never be negative.*

Example 6.24   Case 3: Find the area between two Z-values (checking the accuracy of the Empirical Rule)

Recall that the Empirical Rule (page 124) states that about 68% of the area under the curve lies within one standard deviation of the mean, that is, between $\mu - \sigma$ and $\mu + \sigma$. Check this result for the standard normal distribution by using the Z table.

Solution

For the standard normal random variable Z, $\mu = 0$ and $\sigma = 1$, so that $\mu - \sigma = 0 - 1$ and $\mu + \sigma = 0 + 1 = 1$. Thus, using Case 3, we have $Z_1 = -1$ and $Z_2 = 1$.

Step 1   Draw the standard normal curve. Label the Z-values $Z_1 = -1$ and $Z_2 = 1$.

Step 2   Shade the area between -1 and 1, as shown in Figure 6.27a.

Step 3   Find the area to the left of $Z_1 = -1$ and the area to the left of $Z_2 = 1$. The Z table gives these areas as follows: area to the left of $Z_1 = -1$ is 0.1587, and area to the left of $Z_2 = 1$ is 0.8413. We subtract the smaller area from the larger to give us the area between -1 and 1, as shown in Figures 6.27a–c.

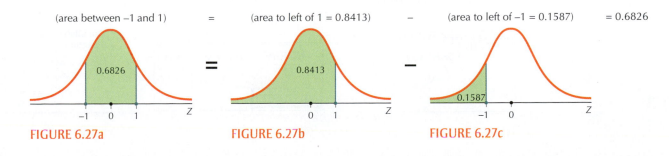

FIGURE 6.27a   FIGURE 6.27b   FIGURE 6.27c

Thus, the area under the *Z* curve within one standard deviation of the mean equals 0.6826. The Empirical Rule does very well for an approximation, missing the actual area by only 0.0026. Checking the accuracy of the Empirical Rule for other values of *Z* is left as an exercise. ■

Example 6.25   Checking the symmetry of the standard normal curve

Check the symmetry of the standard normal curve by finding the following two areas and confirming that they are equal.
a.   Area to the left of $Z = -1$
b.   Area to the right of $Z = 1$

Solution
a.   In Example 6.24, we found the area to the left of $Z = -1$ to be 0.1587.
b.   This represents a Case 2 problem from Table 6.7.

Step 1   Figure 6.28 shows the standard normal curve with $Z = 1$ indicated.

Step 2   The area to the right of $Z = 1$ is shaded.

Step 3   In Example 6.24, we found the area to the left of $Z = 1$ to be 0.8413. The area to the right of 1 is then $1 - 0.8413 = 0.1587$.

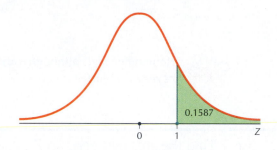

FIGURE 6.28

Thus,

$$(\text{area to the left of } Z = -1) = (\text{area to the right of } Z = 1) = 0.1587$$

Thus, the two tail areas are equal. Since this result holds for any given value of *Z*, the standard normal curve is symmetric. ■

Example 6.26   Using technology to find the area under a standard normal curve

In Example 6.22, we found the area under the standard normal curve to the left of $Z = 0.57$ to be 0.7157. Confirm this result using technology.

Solution
We follow the instructions in the Step-by-Step Technology Guide at the end of Section 6.5 (page 309). Figures 6.29a–c show the results from TI-83/84, Excel, and Minitab, respectively.

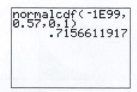

```
normalcdf(-1E99,
0.57,0,1)
          .7156611917
```

FIGURE 6.29a

=NORMDIST(0.57,0,1,TRUE)		
D	**E**	**F**
0.7156612		

FIGURE 6.29b

Cumulative Distribution Function

Normal with mean = 0 and standard deviation = 1

x	P(X <= x)
0.57	0.715661

FIGURE 6.29c

The word "cumulative" in the Minitab output means "less than or equal to." Each of these results provides the area under the standard normal curve for values of Z that are less than or equal to 0.57. Each technology rounds to a different number of decimal places. ▪

Example 6.27 **Estimating areas**

For each of the graphs in Figure 6.30,
i. estimate the shaded area, and
ii. calculate the area.

a. **b.**

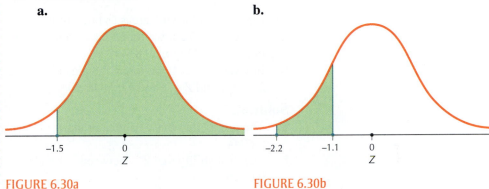

FIGURE 6.30a   FIGURE 6.30b

Solution

a. **i.** If the area to the right of $Z = 0$ is 0.5, then the shaded area must be greater than 0.5.
  ii. This is a Case 2 problem from Table 6.7. Looking up $Z = -1.5$ in the Z table, we find the area to the left of $Z = -1.5$ to be 0.0668. Therefore, the area to the right of $Z = -1.5$ must be $1 - 0.0668 = 0.9332$, which is indeed larger than 0.5.

b. **i.** This area must be less than 0.5, since the area to the left of $Z = 0$ is 0.5, and the shaded area is smaller than this.
  ii. This is a Case 3 problem from Table 6.7. We look up $Z = -1.1$ in the Z table and find 0.1357. We look up $Z = -2.2$ in the Z table and find 0.0139. Subtracting the smaller area from the larger gives us the shaded area $0.1357 - 0.0139 = 0.1218$. ▪

Developing Your Statistical Sense

Checking That Your Answer Makes Sense!

As you are finding probabilities for values of Z, you should always be checking to see that your answer makes sense. In Example 6.27a, what if we had forgotten to subtract the 0.0668 from 1? That would have given us an incorrect desired area of 0.0668. We would know that this answer is incorrect because the shaded area in the graph is large (greater than 0.5) while 0.0668 is small (less than 0.5). See how the artistic side of your brain can help out the mathematical side? ▪

The *Normal Density Curve* applet allows you to find areas associated with various values of Z.

Note that the areas we have been finding in this section may also be expressed as probabilities. For continuous distributions probabilities are represented by areas under the curve above an interval. Specifically, for the standard normal distribution, probability is represented under the standard normal curve. For instance, in Example 6.22, we found that the area under the standard normal curve to the left of $Z = 0.57$ is 0.7157. This may be reexpressed as follows:

"The probability that Z is less than 0.57 is 0.7157"

or

$$P(Z < 0.57) = 0.7157$$

Example 6.28   Expressing areas under the standard normal curve as probabilities

Reexpress the following areas as probabilities.

a. In Example 6.23, we found the area under the standard normal curve to the right of $Z = -1.25$ to be 0.8944.

b. In Example 6.24, we found the area under the standard normal curve between $Z = -1$ and $Z = 1$ to be 0.6826.

Solution

a. The probability that Z is greater than -1.25 is 0.8944. That is, $P(Z > -1.25) = 0.8944$.

b. The probability that Z is between -1 and 1 is 0.6826. That is, $P(-1 < Z < 1) = 0.6826$. ■

２ Finding Standard Normal Values for a Given Area

In the previous examples, you were asked to locate a Z-value on the outside of the Z table (that is, using the first column and the top row) and look up the area on the inside of the table. What if we turn this around, so that we know what the area is but need to find the Z-value? We will call these problems "backward" problems because we need to use the Z table in reverse. Let's check out an example.

Example 6.29   Find the value of Z with area 0.90 to the left

Note: The value of Z with area 0.90 to the left of it is $Z = 1.28$. Since the area to the left is 0.90, 90% of the Z-values are less than $Z = 1.28$. Recall that the *rth percentile* is the value in the data set such that r percent of the observations fall at or below that value. Thus, $Z = 1.28$ represents the 90th percentile of the Z distribution, since it is greater than 90% of Z-values.

Find the standard normal Z-value that has area 0.90 to the left of it.

Solution

Step 1   Draw the standard normal curve. Label the Z-value Z_1.

Step 2   Shade the area to the left of Z_1. Remember that *we are given an area and are looking for a value of Z.* Label the area to the left of Z_1 with the given area (0.90), as shown in Figure 6.31.

Area = 0.90

FIGURE 6.31

Step 3   Look for 0.90 on the inside of the Z table (that is, in the body of the table), since the values *inside the table* represent areas. Because there is no 0.90 inside the table, by convention we take the area that is closest to 0.90, which is 0.8997. Next comes the trick of the backward problems, and the reason for that name. Move from 0.8997 to the left until you reach 1.2 in the first column, and then move up from 0.8997 until you get to 0.08 (see Figure 6.32). Putting these values together, we get $Z = 1.2 + 0.08 = 1.28$. ■

Standard Normal Distribution

Z	0.00	0.01	0.02	0.03	0.04	0.05	0.06	0.07	0.08	0.09
0.0	0.5000	0.5040	0.5080	0.5120	0.5160	0.5199	0.5239	0.5279	0.5319	0.5359
0.1	0.5398	0.5438	0.5478	0.5517	0.5557	0.5596	0.5636	0.5675	0.5714	0.5753
0.2	0.5793	0.5832	0.5871	0.5910	0.5948	0.5987	0.6026	0.6064	0.6103	0.6141
0.3	0.6179	0.6217	0.6255	0.6293	0.6331	0.6368	0.6406	0.6443	0.6480	0.6517
0.4	0.6554	0.6591	0.6628	0.6664	0.6700	0.6736	0.6772	0.6808	0.6844	0.6879
0.5	0.6915	0.6950	0.6985	0.7019	0.7054	0.7088	0.7123	0.7157	0.7190	0.7224
0.6	0.7257	0.7291	0.7324	0.7357	0.7389	0.7422	0.7454	0.7486	0.7517	0.7549
0.7	0.7580	0.7611	0.7642	0.7673	0.7704	0.7734	0.7764	0.7794	0.7823	0.7852
0.8	0.7881	0.7910	0.7939	0.7967	0.7995	0.8023	0.8051	0.8078	0.8106	0.8133
0.9	0.8159	0.8186	0.8212	0.8238	0.8264	0.8289	0.8315	0.8340	0.8365	0.8389
1.0	0.8413	0.8438	0.8461	0.8485	0.8508	0.8531	0.8554	0.8577	0.8599	0.8621
1.1	0.8643	0.8665	0.8686	0.8708	0.8729	0.8749	0.8770	0.8790	0.8810	0.8830
1.2	0.8849	0.8869	0.8888	0.8907	0.8925	0.8944	0.8962	0.8980	0.8997	0.9015
1.3	0.9032	0.9049	0.9066	0.9082	0.9099	0.9115	0.9131	0.9147	0.9162	0.9177

FIGURE 6.32

Example 6.30 Find the value of *Z* with area 0.03 to the right

Find the standard normal Z-value that has area 0.03 to the right of it.

Solution

Step 1 Draw the standard normal curve. Label the Z-value Z_1. Shade the area to the right of it with the given area, as shown in Figure 6.33.

Step 2 Since the Z table contains areas to the left of values of Z, we must find the area to the left of the specific value Z_1, as follows:

$$\text{area to left of } Z_1 = 1 - \text{area to right of } Z_1$$

So, the area to the left of Z_1 is $1 - 0.03 = 0.97$.

Step 3 Look up 0.97 on the inside of the Z table. The closest area is 0.9699. Move from 0.9699 to the left until you reach 1.8, and then move up from 0.9699 until you get to 0.08 (see Figure 6.34). Putting these values together, we get $Z = 1.8 + 0.08 = 1.88$. In other words, the Z-value with area 0.03 to its right is $Z = 1.88$. ■

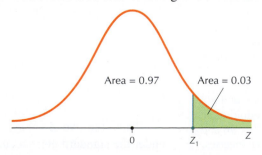

FIGURE 6.33

Standard Normal Distribution

Z	0.00	0.01	0.02	0.03	0.04	0.05	0.06	0.07	0.08	0.09
0.0	0.5000	0.5040	0.5080	0.5120	0.5160	0.5199	0.5239	0.5279	0.5319	0.5359
0.1	0.5398	0.5438	0.5478	0.5517	0.5557	0.5596	0.5636	0.5675	0.5714	0.5753
0.2	0.5793	0.5832	0.5871	0.5910	0.5948	0.5987	0.6026	0.6064	0.6103	0.6141
1.6	0.9452	0.9463	0.9474	0.9484	0.9495	0.9505	0.9515	0.9525	0.9535	0.9545
1.7	0.9554	0.9564	0.9573	0.9582	0.9591	0.9599	0.9608	0.9616	0.9625	0.9633
1.8	0.9641	0.9649	0.9656	0.9664	0.9671	0.9678	0.9686	0.9693	0.9699	0.9706
1.9	0.9713	0.9719	0.9726	0.9732	0.9738	0.9744	0.9750	0.9756	0.9761	0.9767

FIGURE 6.34

When we learn statistical inference in later chapters, we will need to identify which Z-values divide the middle 90%, 95%, or 99% of the area under the standard normal curve from the tail area.

Example 6.31   Find the values of Z that mark the boundaries of the middle 95% of the area

Find the two values of Z that mark the boundaries of the middle 95% of the area under the standard normal curve.

Solution

Step 1   Draw the standard normal curve, showing the desired middle area (95%) with boundaries labeled as Z_1 and Z_2, as shown in Figure 6.35. By symmetry, there is area $= (1 - 0.95)/2 = 0.025$ in each tail.

Step 2   Look up 0.025 on the inside of the Z table. Find Z_1 by moving to the left and up from 0.025 in the Z table, giving us $Z_1 = -1.96$.

Step 3   Since the area in the right tail is 0.025 as well, the area to the left of Z_2 is $1 - 0.025 = 0.975$. Looking up 0.975 in the Z table gives us $Z_2 = 1.96$.

Note: Is it a coincidence that the two values of Z that determine the middle 95% of the area under the standard normal curve are 1.96 and −1.96? Not at all. Since the standard normal curve is symmetric about the mean 0, the values −1.96 and 1.96 that form the boundaries of the middle 95% must be equidistant from zero. Of course, −1.96 and 1.96 are called additive inverses.

FIGURE 6.35

Thus, the two Z-values that mark the boundaries of the middle 95% of the area under the standard normal curve are −1.96 and 1.96. ■

Example 6.32   Using technology to find values of Z, given an area

In Example 6.29, we found that the value of Z with area 0.90 to its left is $Z = 1.28$. Confirm this result using technology.

Solution

We follow the instructions in the Step-by-Step Technology Guide at the end of Section 6.5 (page 309). Figures 6.36a–c show the results from TI-83/84, Excel, and Minitab, respectively. ■

FIGURE 6.36a

FIGURE 6.36b

FIGURE 6.36c

Section 6.4 SUMMARY

1 The standard normal distribution has mean $\mu = 0$ and standard deviation $\sigma = 1$. This distribution is often called the Z distribution. The Z table and technology can be used to find areas under the standard normal curve. In the Z table,

the numbers on the outside are values of Z, and the numbers inside are areas to the left of values of Z.

2 The Z table and technology can also be used to find a value of Z, given a probability or an area under the curve.

Section 6.4 EXERCISES

CLARIFYING THE CONCEPTS

1. What is the value for the mean of the standard normal distribution?

2. What is the value for the standard deviation of the standard normal distribution?

3. True or false: The area under the Z curve to the right of $Z = 0$ is 0.5.

4. True or false: $P(Z = 0) = 0$.

PRACTICING THE TECHNIQUES

For Exercises 5–14, use the graph of the standard normal distribution to

    a. estimate the area of the shaded region,

    b. find the shaded area using the Z table or technology, and

    c. check your estimate in **(a)** against your answer in **(b)**.

5.

6.

7.

8.

9.

10.

11.

12.

13.

14.

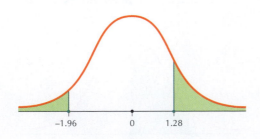

For Exercises 15–28,
  a. draw the graph,
  b. estimate the area,
  c. find the area using the Z table or technology,
  d. check your estimate against your final answer.

Find the area under the standard normal curve that lies to the left of the following.
15. $Z = 1$
16. $Z = 2$
17. $Z = -2.7$
18. $Z = -0.2$

Find the area under the standard normal curve that lies to the right of the following.
19. $Z = 1.27$
20. $Z = 2.12$
21. $Z = -3.01$
22. $Z = -0.69$

Find the area under the standard normal curve that lies between the following.
23. $Z = 0$ and $Z = 1$
24. $Z = 1$ and $Z = 2$
25. $Z = -1$ and $Z = 0$
26. $Z = -2$ and $Z = -1$
27. $Z = -1.28$ and $Z = 1.28$
28. $Z = -2.01$ and $Z = 2.37$

For Exercises 29–40, find the indicated probability for the standard normal Z.
  a. Draw the graph.
  b. Estimate the area.
  c. Find the area using the Z table or technology.
  d. Check your estimate against your final answer.
29. $P(Z = 0)$
30. $P(Z < 0)$
31. $P(Z < 10)$
32. $P(Z > 1.29)$
33. $P(Z < -2.17)$
34. $P(Z < 0.57)$
35. $P(-1.96 < Z < 1.96)$
36. $P(-2.07 < Z < 0.46)$
37. $P(-3.05 < Z < -0.94)$
38. $P(1.54 < Z < 2.20)$
39. $P(-100 < Z < 0)$
40. $P(-1.72 < Z < -1.57)$

For Exercises 41–44, find the Z-value with the following areas under the standard normal curve to its left. Draw the graph first and estimate the Z-value. Then find the value and check your estimate against your final answer.
41. 0.3336
42. 0.4602
43. 0.3264
44. 0.4247

For Exercises 45–48, find the Z-value with the following areas under the standard normal curve to its right. Draw the graph first and estimate the Z-value. Then find the value and check your estimate against your final answer.

45. 0.8078

46. 0.3085

47. 0.9788

48. 0.5120

49. Find the 50th percentile of the Z distribution.

50. Find the 75th percentile of the Z distribution.

51. Find the value of Z that is larger than 99.5% of all values of Z.

52. Find the value of Z that is smaller than 99.5% of all values of Z.

APPLYING THE CONCEPTS

53. Checking the Empirical Rule. Check the accuracy of the Empirical Rule for $Z = 2$. That is, find the area between $Z = -2$ and $Z = 2$ using the techniques of this section. Then compare your finding with the results for $Z = 2$ using the Empirical Rule.

54. Checking the Empirical Rule. Check the accuracy of the Empirical Rule for $Z = 3$. That is, find the area between $Z = -3$ and $Z = 3$ using the techniques of this section. Then compare your finding with the results for $Z = 3$ using the Empirical Rule.

55. Without Tables or Technology. Find the following areas *without* using the Z table or technology. The area to the left of $Z = -1.5$ is 0.0668.

  a. Find the area to the right of $Z = 1.5$.
  b. Find the area to the right of $Z = -1.5$.
  c. Find the area between $Z = -1.5$ and $Z = 1.5$.

56. Without Tables or Technology. Find the following areas *without* using the Z table or technology. The area to the right of $Z = 2.7$ is 0.0035.

  a. Find the area to the left of $Z = 2.7$.
  b. Find the area to the left of $Z = -2.7$.
  c. Find the area between $Z = -2.7$ and $Z = 2.7$.

57. Values of Z That Mark the Middle 99%. Find the two values of Z that contain the middle 99% of the area under the standard normal curve.

58. Values of Z That Mark the Middle 90%. Find the two values of Z that contain the middle 90% of the area under the standard normal curve.

Use the *Normal Density Curve* applet for Exercises 59 and 60.

59. Find the quartiles of the standard normal distribution. That is, find the 25th, 50th, and 75th percentiles of the standard normal distribution.

60. Use the applet to find the answers to the following exercises from this section.

  a. Exercise 6
  b. Exercise 7
  c. Exercise 15
  d. Exercise 16
  e. Exercise 22
  f. Exercise 23

6.5 Applications of the Normal Distribution

Objectives: **By the end of this section, I will be able to...**

1 Compute probabilities for a given value of any normal random variable.

2 Find the appropriate value of any normal random variable, given an area or probability.

3 Calculate binomial probabilities using the normal approximation to the binomial distribution.

1 Finding Probabilities for Any Normal Distribution

The data in problems that we face in the real world do not usually follow the standard normal distribution. Instead, a problem may be stated in terms of some normal random variable X that has a mean other than 0 or a standard deviation other than 1. In cases like these, X needs to be standardized to Z so that we can use the Section 6.4 techniques.

Standardizing X to Z

To *standardize* things means to make them all the same, uniform, or equivalent. For example, college applicants take standardized tests so that the admissions officers can compare students according to a uniform assessment tool. To standardize a normal random variable X, we *transform* that normal random variable X into the standard normal random variable Z that we worked so extensively with in Section 6.4. In this section, we will focus on real-world applications of the normal distribution.

Suppose that X is a normal random variable with population mean μ and population standard deviation σ. We standardize X by subtracting the mean μ and dividing by the standard deviation σ. The result of this transformation is the familiar **standard normal random variable Z.**

> ### Standardizing a Normal Random Variable
> Any normal random variable X can be transformed into the standard normal random variable Z by *standardizing* using the formula
>
> $$Z = \frac{X - \mu}{\sigma}$$

The key here is the following: *for a given area of interest for a normal random variable X, the corresponding area after the transformation to Z is exactly the same.* For any normal random variable X, the area between a and b is exactly the same as the area between $Z_a = (a - \mu)/\sigma$ and $Z_b = (b - \mu)/\sigma$ (see Figure 6.37).

FIGURE 6.37
Corresponding areas
are equal.

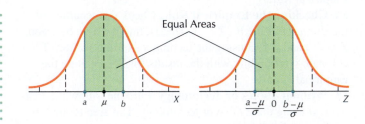

Example 6.33 April in Georgia

The state of Georgia reports that the average temperature statewide for the month of April from 1949 to 2006 was $\mu = 61.5°F$. Assume that the standard deviation is $\sigma = 8°F$ and that temperature in Georgia in April is normally distributed. Draw the normal curve for temperatures between 45.5°F and 77.5°F, and the corresponding Z curve. Find the probability that the temperature is between 45.5°F and 77.5°F in April in Georgia.

Solution
Here we have $a = 45.5$ and $b = 77.5$, giving us

$$Z_a = \frac{a - \mu}{\sigma} = \frac{45.5 - 61.5}{8} = -2 \quad \text{and} \quad Z_b = \frac{b - \mu}{\sigma} = \frac{77.5 - 61.5}{8} = 2$$

In Figure 6.38, the area between 45.5°F and 77.5°F is the same as between $Z = -2$ and $Z = 2$. In other words,

$$P(45.5 < X < 77.5) = P(-2 < Z < 2)$$

FIGURE 6.38

This is a Case 3 problem from Table 6.7. The Z table tells us that the area to the left of $Z_1 = -2$ is 0.0228, and the area to the left of $Z_2 = 2$ is 0.9772. The area between -2 and 2 is then equal to $0.9772 - 0.0228 = 0.9544$. The probability that temperature is between 45.5°F and 77.5°F in April in Georgia is 0.9544. ■

> ### Finding Probabilities for Any Normal Distribution
>
> **Step 1** Determine the random variable X, the mean μ, and the standard deviation σ. Draw the normal curve for X, and shade the desired area.
>
> **Step 2** Standardize by using the formula $Z = (X - \mu)/\sigma$ to find the values of Z corresponding to the X-values.
>
> **Step 3** Draw the standard normal curve and shade the area corresponding to the shaded area in the graph of X.
>
> **Step 4** Find the area under the standard normal curve using either (a) the Z table or (b) technology. This area is equal to the area under the normal curve for X drawn in Step 1.

Example 6.34 Ordering caps and gowns for graduation

Suppose that you are in charge of ordering caps and gowns for senior graduation in May. You know that the heights of the population of students at your college are normally distributed with a mean of 68 inches (5 feet 8 inches) and a standard deviation of 3 inches. The gown manufacturer wants to know how many students will need to special order their gowns because they are very tall. Find the proportion of students who are above 74 inches tall.

Solution

Step 1 **Determine X, μ, and σ, and draw the normal curve for X.** You know that the heights of students at your college are normally distributed with a mean of 68 inches and a standard deviation of 3 inches. Thus, the normal random variable $X =$ heights of students with $\mu = 68$ inches and $\sigma = 3$ inches. In the center of the number line, mark the mean μ. Also mark on the number line the value of X that the problem is asking about. Figure 6.39 shows the graph of X (the heights of students) with the mean of 68 inches and the height of 74 inches marked.

Since you need to know the proportion of students taller than (above) 74 inches, shade the area under the curve and above 74 inches. We can express this proportion as a probability, the probability that a randomly chosen student will be taller than 74 inches, or $P(X > 74)$. *Just by looking at Figure 6.39 you should be able to get a rough idea of what the proportion of students taller than 74 inches will be.* Certainly this proportion will be less than 50%, and probably pretty small. If you get an answer like "60%" for your proportion, you would surely know that it is wrong.

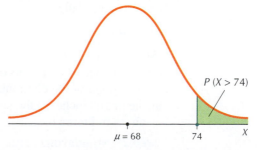

FIGURE 6.39 Graph of proportion of college students taller than 74 inches.

Step 2 **Standardize.** Now standardize the random variable X to the standard normal Z:

$$Z = \frac{X - \mu}{\sigma} = \frac{X - 68}{3}$$

Find the Z-value corresponding to the height of 74 inches:

$$\frac{74 - \mu}{\sigma} = \frac{74 - 68}{3} = 2$$

So the Z-value associated with 74 inches is 2, which indicates that the height of 74 inches is 2 standard deviations above the mean of 68 inches.

Step 3 **Draw the standard normal curve.** Heights above 74 inches are more than 2 standard deviations above the mean, so shade the area to the right of 2 in Figure 6.40. Now find the area to the right of $Z = 2$ using the methods of Section 6.4.

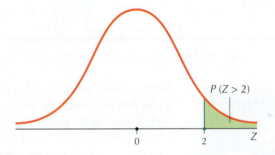

FIGURE 6.40 Graph of $P(Z > 2)$.

Step 4 **Find the area under the standard normal curve.** Figure 6.40 represents a Case 2 problem from Table 6.7. The Z table tells us that the area to the left of $Z = 2.00$ is 0.9772. Thus, the area to the right is

$$P(Z > 2) = 1 - 0.9772 = 0.0228$$

The proportion of students taller than 74 inches is 0.0228, or 2.28%. Note that this value for $P(X > 74)$ agrees with our earlier intuition that the proportion was surely less than 50% and most likely very small. ■

Example 6.35   Finding probability that X lies between two given values

Continuing the cap-and-gown problem, what percentage of students are between 60 and 70 inches tall?

Solution

Step 1 **Determine X, μ, and σ, and draw the normal curve for X.** We have already seen that X = heights of students, $\mu = 68$ inches, and $\sigma = 3$ inches. Once again, draw a graph of the distribution of heights X, with the mean 68 inches in the middle, the height 60 inches to the left of the mean, and the height 70 inches to the right of the mean, as in Figure 6.41.

Step 2 **Standardize.** This is a "between" example, where two values of X are given, and we are asked to find the area between them. In this case, just standardize both of these values of X to get a Z-value for each:

$$Z = \frac{60 - \mu}{\sigma} = \frac{60 - 68}{3} \approx -2.67 \quad \text{and} \quad Z = \frac{70 - \mu}{\sigma} = \frac{70 - 68}{3} \approx 0.67$$

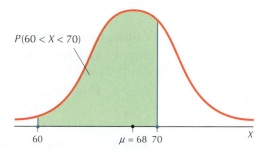

$P(60 < X < 70)$

60     $\mu = 68$ 70 X

FIGURE 6.41 Graph of percentage of students between 60 and 70 inches tall.

Step 3 **Draw the standard normal curve.** Draw a graph of Z, shading the area between $Z = -2.67$ and $Z = 0.67$, as shown in Figure 6.42. Again, the key is that the area between $Z = -2.67$ and $Z = 0.67$ is exactly the same as the area between $X = 60$ inches and $X = 70$ inches.

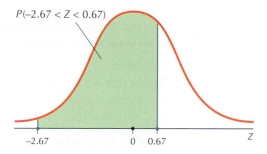

$P(-2.67 < Z < 0.67)$

-2.67 0   0.67 Z

FIGURE 6.42 Graph of percentage of Z-values between -2.67 and 0.67.

Step 4 **Find area under the standard normal curve.** Figure 6.42 looks like a Case 3 problem from Table 6.7. Find the area to the left of 0.67, which is 0.7486, and the area to the left of -2.67, which is 0.0038. Subtracting the smaller from the larger gives us

$$P(-2.67 < Z < 0.67) = 0.7486 - 0.0038 = 0.7448$$

Thus, the percentage of students who are between 60 and 70 inches tall is 74.48%. ▪

Check Your Answer! According to the Empirical Rule, almost all Z-values lie between -3 and 3, so it is unlikely that a randomly selected value of Z lies outside this range. You should remember this when you are doing your calculations. If you are standardizing a normal random variable X and get a very large Z-value (such as, say, 50), you should recheck your calculations because the probability that Z takes such a large value is very small.

Finding Normal Data Values for Specified Probabilities

Sometimes we are given a probability (or proportion or area), and we are asked to find the associated value of X. Questions like these are similar to the "backward" problems of Section 6.4, so called because we must use the Z table backward. Since the formula for standardizing X gives the value for Z, we need to use our algebra skills to find the equation for X: Start with the standard normal formula $Z = (X - \mu)/\sigma$. Multiply both sides by σ to get $Z\sigma = X - \mu$. Then add μ to both sides, giving us $X = Z\sigma + \mu$.

Finding Normal Data Values for Specified Probabilities

Step 1   Determine X, μ, and σ, and draw the normal curve for X. Shade the desired area. Mark the position of X_1, the unknown value of X.

Step 2   Find the Z-value corresponding to the desired area. Look up the area you identified in Step 1 on the *inside* of the Z table. If you do not find the exact value of your area, by convention choose the area that is closest.

Step 3   Transform this value of Z into a value of X, which is the solution. Use the formula $X_1 = Z\sigma + \mu$.

Example 6.36   How tall do you have to be before you have to special-order your cap and gown?

Suppose that we wanted only the tallest 1% of our students to have to special-order gowns. What is the height at which tall students will have to special-order their gowns?

Solution

Notice that we are not asked to find a probability (or proportion or area). Instead, we are given a percentage (1%) and asked to find the value of X (the height) that is associated with this 1%.

Step 1   Determine X, μ, and σ, and draw the normal curve for X. We already know that X = heights of students, $\mu = 68$ inches, and $\sigma = 3$ inches. The value of X we are interested in refers to very tall students, so that X_1 will be at the far right of the distribution of X. Only 1% of students will be taller than this height, so the area to the right of X_1 is 1%, as shown in Figure 6.43.

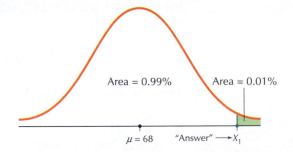

Area = 0.99%   Area = 0.01%

$\mu = 68$ "Answer" $\longrightarrow X_1$

FIGURE 6.43

Step 2   Find the Z-value corresponding to the desired area. The area to the right of X_1 equals 0.01, so that the area to the left of X_1 equals $1 - 0.01 = 0.99$. Looking up 0.99 on the inside of the Z table gives us $Z = 2.33$.

Step 3   Transform using the formula $X_1 = Z\sigma + \mu$. We calculate

$$X_1 = Z\sigma + \mu = (2.33)(3) + 68 = 74.99$$

If we want only the tallest 1% of our students to have to special-order their gowns, the height at which tall students will have to special-order their gowns is 74.99 inches. ∎

Example 6.37   Finding the *X*-values that mark the boundaries of the middle 95% of *X*-values

Edmunds.com reported that the average amount that people were paying for a 2007 Toyota Camry XLE was $23,400. Let X = price, and assume that price follows a normal distribution with μ = $23,400 and σ = $1000. Find the prices that separate the middle 95% of 2007 Toyota Camry XLE prices from the bottom 2.5% and the top 2.5%.

Solution

Step 1 **Determine X, μ, and σ, and draw the normal curve for X.** Let X = price, μ = $23,400, and σ = $1000. The middle 95% of prices are delimited by X_1 and X_2, as shown in Figure 6.44.

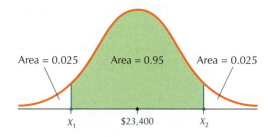

Area = 0.025   Area = 0.95   Area = 0.025

X_1 $23,400 X_2

FIGURE 6.44

Step 2 **Find the Z-values corresponding to the desired area.** The area to the left of X_1 equals 0.025, and the area to the left of X_2 equals 0.975. Looking up area 0.025 on the inside of the Z table gives us Z_1 = −1.96. Looking up area 0.975 on the inside of the Z table gives us Z_2 = 1.96.

Step 3 **Transform using the formula $X_1 = Z\sigma + \mu$.** We calculate

$$X_1 = Z_1\sigma + \mu = (-1.96)(1000) + 23,400 = 21,440$$

$$X_2 = Z_2\sigma + \mu = (1.96)(1000) + 23,400 = 25,360$$

The prices that separate the middle 95% of 2007 Toyota Camry XLE prices from the bottom 2.5% of prices and the top 2.5% of prices are $21,440 and $25,360. ■

*W*hat If Scenario **?** **How Change in Spread Affects Camry Prices**

In Example 6.37, *what if* we ask the same question again, but this time the standard deviation σ of 2007 Toyota Camry XLE prices is not $1000 but some value less than $1000. How and why would this affect the following?

    a. The values Z_1 and Z_2 found in Step 2

    b. The value X_1 separating the middle 95% of prices from the bottom 2.5%

    c. The value X_2 separating the middle 95% of prices from the top 2.5%

Solution

Figure 6.45 illustrates the distribution of 2007 Toyota Camry XLE prices, where everything is the same as in Figure 6.44 except that the standard deviation of the prices is smaller by an unknown amount. Thus, the spread of the distribution is smaller.

    a. Since we are still asking for the middle 95% of prices, the Z-values remain the same, −1.96 and 1.96.

    b. Reexpress the formula $X_1 = Z_1\sigma + \mu$ as X_1 = $23,400 − 1.96 · σ. If σ is smaller than $1000, then the quantity 1.96 · σ, which represents the difference between the mean price and X_1, will also be smaller.

FIGURE 6.45

Since X_1 is less than the mean $\mu = \$23{,}400$, the smaller difference between the mean price and X_1 leads us to conclude that X_1 will be *larger* than in Example 6.37. For example, if the new standard deviation is $\sigma = \$500$, then $X_1 = \$23{,}400 - 1.96 \cdot 500 = \$22{,}420$, which is larger than the $\$21{,}440$ in Example 6.37.

c. Similarly, a smaller σ means a smaller quantity $1.96 \cdot \sigma$, which means that $X_2 = \$23{,}400 + 1.96 \cdot \sigma$ will be closer to the mean $\mu = \$23{,}400$. Since X_2 is larger than the mean, the new value for X_2 will be *smaller* than in Example 6.37.

The *Normal Density Curve* applet allows you to find areas associated with various values of any normal random variable.

Example 6.38 Normal probabilities and percentiles using technology

Applying the information on Toyota Camry prices from Example 6.37, use the TI-83/84, Excel, or Minitab to find the following.

a. The proportion of 2007 Camry XLEs costing between $\$22{,}000$ and $\$24{,}000$, $P(22{,}000 \le X \le 24{,}000)$

b. The 99th percentile of Camry XLE prices, that is, find the value of X, namely, X_1, such that $P(X \le X_1) = 0.99$

Solution
The instructions for finding these quantities are given in the Step-by-Step Technology Guide at the end of this section (page 309).

TI-83/84
a. Figure 6.46 shows that $P(22{,}000 \le X \le 24{,}000) = 0.6449902243 \approx 0.6450$.

b. Figure 6.47 shows that the value for X_1 such that $P(X \le X_1) = 0.99$ is given by $X_1 = \$25{,}726.34788 \approx \$25{,}726.35$.

FIGURE 6.46

FIGURE 6.47

Excel
a. Excel provides the cumulative probabilities $P(X \le 22{,}000)$ in Figure 6.48 and $P(X \le 24{,}000)$ in Figure 6.49. To find $P(22{,}000 \le X \le 24{,}000)$, we subtract $P(X \le 22{,}000)$ from $P(X \le 24{,}000)$:

$$P(22{,}000 \le X \le 24{,}000) = 0.725746882 - 0.080756659 = 0.644990223$$

FIGURE 6.48   **FIGURE 6.49**

b. Excel provides the result shown in Figure 6.50, $X_1 = \$25{,}726.34787 \approx \$25{,}726.35$.

FIGURE 6.50

Minitab

a. Like Excel, Minitab asks you to take the difference of two cumulative probabilities, $P(X \le 22{,}000)$ in Figure 6.51 and $P(X \le 24{,}000)$ in Figure 6.52:

$$P(22{,}000 \le X \le 24{,}000) = 0.725747 - 0.0807567 = 0.6449903 \approx 0.6450$$

```
Cumulative Distribution Function
Normal with mean = 23400 and standard deviation = 1000

     x   P( X <= x )
 22000      0.0807567
```

```
Cumulative Distribution Function
Normal with mean = 23400 and standard deviation = 1000

     x   P( X <= x )
 24000      0.725747
```

FIGURE 6.51   **FIGURE 6.52**

b. The results are given in Figure 6.53; $X_1 = \$25{,}726.30$. ■

```
Inverse Cumulative Distribution Function
Normal with mean = 23400 and standard deviation = 1000

P( X <= x )        x
       0.99  25726.3
```

FIGURE 6.53

The Normal Approximation to the Binomial Probability Distribution

Recall from Section 6.2 that a binomial experiment satisfies the following four requirements: (1) Each trial must have two possible outcomes. (2) There is a fixed number of trials, *n*. (3) The experimental outcomes are independent of each other. (4) The probability of observing a success is the same from trial to trial.

The binomial random variable *X* represents the number of successes in *n* trials and thus depends on the sample size *n* and the probability of success *p*. For a given probability of success *p*, if the sample size *n* gets large enough, the binomial distribution begins to resemble the normal distribution. Figure 6.54 shows the binomial probability distribution for Example 6.15 (page 267), where Joshua tried to guess ($p = 0.2$) his way to success on a quiz of $n = 4$ questions. The distribution of $X =$ the number of correct responses in Figure 6.54 is clearly not normal.

FIGURE 6.54
A binomial distribution:
$n = 4$, $p = 0.2$.

If we increase the sample size to $n = 34$ while retaining the same $p = 0.2$ (Figure 6.55), the distribution of *X* becomes somewhat more bell-shaped but is still a bit right-skewed. Finally, if we further increase the sample size to $n = 64$ (Figure 6.56), the binomial distribution of *X*, which is discrete, looks like it can be nicely approximated by the normal distribution, which is continuous. We generalize this behavior as follows.

FIGURE 6.55
A binomial distribution:
$n = 34$, $p = 0.2$.

FIGURE 6.56
A binomial distribution:
$n = 64$, $p = 0.2$.

> **The Normal Approximation to the Binomial Probability Distribution**
> For the binomial random variable X with probability of success p and number of trials n: if $n \cdot p \geq 5$ and $n \cdot (1 - p) \geq 5$, the binomial distribution may be approximated by a normal distribution with mean $\mu_x = np$ and standard deviation $\sigma_x = \sqrt{np(1 - p)}$.

The *Normal Approximation to the Binomial* applet allows you to choose your own values of n and p and see how changes in these values affect the normal approximation to the binomial distribution.

Example 6.39   Childhood immunizations

Punchstock/Blend

The Centers for Disease Control and Prevention reported that 20% of preschool children lack required immunizations, thereby putting themselves and their classmates at risk. For a group of $n = 64$ children with $p = 0.2$, the binomial probability distribution is shown in Figure 6.56.

a. Verify that this distribution can be approximated by a normal distribution.
b. Find the mean and standard deviation of this normal distribution.

Solution

a. The normal approximation is valid if $n \cdot p \geq 5$ and $n \cdot (1 - p) \geq 5$. Substituting $n = 64$ and $p = 0.2$, we get

$$n \cdot p = (64)(0.2) = 12.8 \geq 5 \quad \text{and} \quad n \cdot (1 - p) = (64)(0.8) = 51.2 \geq 5$$

Thus, the normal approximation is valid.

b. The mean and standard deviation of the normal distribution are

$$\mu_x = np = (64)(0.2) = 12.8$$

$$\sigma_x = \sqrt{np(1 - p)} = \sqrt{(64)(0.2)(0.8)} = 3.2$$

Figure 6.57 reproduces Figure 6.56, with the rectangle for $X = 12$ highlighted. The height of the rectangle represents the binomial probability that exactly 12 of the 64 children lack the required immunizations, that is, $P(X = 12)$. Since the width of the rectangle equals $12.5 - 11.5 = 1$, it follows that the area of the rectangle also represents the binomial probability that $X = 12$. Now, the area under the normal curve

FIGURE 6.57

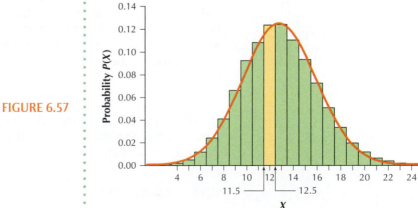

between 11.5 and 12.5 is approximately equal to this rectangle, which is $P(X = 12)$ for the binomial random variable X, with $n = 64$ and $p = 0.2$. That is

$$P(X_{\text{binomial}} = 12) \approx P(11.5 \leq Y_{\text{normal}} \leq 12.5)$$

where Y_{normal} is the normal random variable from Example 6.39b, with mean $\mu_X = 12.8$ and standard deviation $\sigma_X = 3.2$.

The 0.5 that we add and subtract from 12 when approximating the binomial distribution with the normal distribution is called the **continuity correction,** since it is an adjustment for approximating a discrete probability with a continuous one. When using the normal approximation to the binomial, the analyst must determine which binomial rectangles are included and apply the continuity correction accordingly. This is shown in Table 6.8, which provides a listing of several types of binomial probabilities and their normal probability approximations.

Table 6.8 Binomial probabilities and approximate normal probabilities

Exact binomial probability	Approximate normal probability
$P(X_{\text{binomial}} = a)$	$P(a - 0.5 \leq Y_{\text{normal}} \leq a + 0.5)$
$P(X_{\text{binomial}} \leq a)$	$P(Y_{\text{normal}} \leq a + 0.5)$
$P(X_{\text{binomial}} \geq a)$	$P(Y_{\text{normal}} \geq a - 0.5)$
$P(X_{\text{binomial}} < a)$	$P(Y_{\text{normal}} < a - 0.5)$
$P(X_{\text{binomial}} > a)$	$P(Y_{\text{normal}} > a + 0.5)$
$P(a \leq X_{\text{binomial}} \leq b)$	$P(a - 0.5 < Y_{\text{normal}} \leq b + 0.5)$
$P(a < X_{\text{binomial}} < b)$	$P(a + 0.5 < Y_{\text{normal}} \leq b - 0.5)$

Example 6.40   The normal approximation to the binomial probability distribution

For a group of $n = 64$ preschool children with probability of lack of immunization $p = 0.2$, perform the following approximations.

a.   Approximate the probability that there are at most 12 children without immunization.

b.   Approximate the probability that there are more than 12 children without immunization.

Solution

Once again we have a binomial experiment with $n = 64$ and $p = 0.2$.

a.   "At most" 12 children means 12 or fewer children. That is, $X = 12$ and $X = 11$ and $X = 10$, and so on; that is, $P(X_{\text{binomial}} \leq 12)$. In this case, we see that $X = 12$ is included in the probability we seek, as shown in Figure 6.58. From Table 6.8, we see that $P(X_{\text{binomial}} \leq 12)$ is of the form $P(X_{\text{binomial}} \leq a)$. Thus, our continuity correction takes the form $P(Y_{\text{normal}} \leq a + 0.5)$, where we add 0.5 to 12, so that

$$P(X_{\text{binomial}} \leq 12) \approx P(Y_{\text{normal}} \leq 12.5)$$

Recall that $\mu_X = 12.8$ and $\sigma_X = 3.2$. We use the TI-83/84, as shown in Figures 6.59 and 6.60, and find that the probability that at most 12 children lack immunization is $0.4626221269 \approx 0.4626$.

FIGURE 6.58

FIGURE 6.59   **FIGURE 6.60**

b. "More than" 12 children means $X = 13$ and $X = 14$, and so on. In other words, $X = 12$ is not included. That is, we want $P(X_{\text{binomial}} > 12)$. From Table 6.8, we see that $P(X_{\text{binomial}} > 12)$ is of the form $P(X_{\text{binomial}} > a)$. Thus, our continuity correction takes the form $P(Y_{\text{normal}} > a + 0.5)$, where we add 0.5 to 12, so that

$$P(X_{\text{binomial}} > 12) \approx P(Y_{\text{normal}} > 12.5)$$

Since the desired area is the complement of the green area in Figure 6.58, we can find the answer like this:

$$P(X_{\text{binomial}} > 12) \approx P(Y_{\text{normal}} > 12.5) = 1 - P(Y_{\text{normal}} \leq 12.5) = 1 - 0.4626 = 0.5374$$

The probability that more than 12 preschool children will not have the required immunizations is 0.5374. ▪

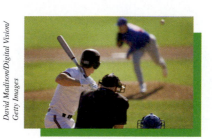

David Madison/Digital Vision/
Getty Images

CASE STUDY | **Be Careful What You Assume: Major League Baseball Salaries**

The following are the summary characteristics for the population of all baseball salaries, as of opening day, 2004:

population mean salary $\mu = \$2,482,534$

population standard deviation of salary $\sigma = \$3,520,375$

Assume that the baseball salaries are normally distributed. Suppose MLB players and owners are huddled around their statistics texts, negotiating the new MLB contract.

Problem 1. In order to decrease operating costs, MLB owners have proposed to "cap" a team's top salary at the 75th percentile of all current salaries. Find the 75th percentile of MLB player salaries on the assumption that the salaries are normally distributed.

Solution

Because of the equivalence of corresponding areas under the X curve and the Z curve, we look up 0.75 in the inside of the Z table and work backward. It turns out that the closest Z is $Z = 0.67$. Reexpressing in terms of salaries, we get

$$X_1 = Z\sigma + \mu = (0.67)(3,520,375) + 2,482,534 \approx \$4,841,185$$

The owners feel that approximately $4.841 million per year is plenty of compensation to play baseball.

Problem 2. The players' union states that they will agree to the cap in Problem 1 only if management adds a bonus to the salaries of the lowest-paid players, as measured by the 10th percentile. Find the 10th percentile of MLB player salaries on the assumption that the salaries are normally distributed.

Solution

To find the 10th percentile, we look up 0.10 in the inside of the Z table and work backward. Here, we get $Z = -1.28$. Reexpressing in terms of salaries, we get

$$X_1 = Z\sigma + \mu = (-1.28)(3,520,375) + 2,482,534 \approx -\$2,023,546$$

What went wrong? How could a salary be negative? Figure 6.61 provides a histogram (in green) of the distribution of actual player salaries, overlaid with the assumed normal distribution (in orange) of player salaries. Both are centered at $2,482,534 (red triangle), and both have standard deviation $3,520,375, but that is about all the two distributions have in common. Clearly, we cannot assume that the salaries are normally distributed.

The salaries are in fact quite remarkably right-skewed and decidedly non-normal. Our finding that the 10th percentile salary is negative is a direct result of the inappropriate assumption of normality. *Be careful what you assume.* We provide methods for assessing the normality of a data set in Chapter 7.

FIGURE 6.61
Normal distribution assumption
is not appropriate here.

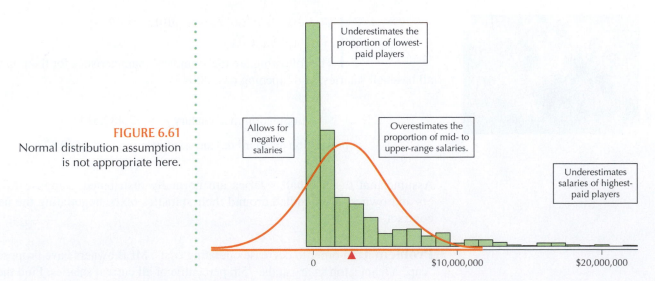

Underestimates the proportion of lowest-paid players

Allows for negative salaries

Overestimates the proportion of mid- to upper-range salaries.

Underestimates salaries of highest-paid players

0          $10,000,000 $20,000,000

STEP-BY-STEP TECHNOLOGY GUIDE: Finding Areas, Probabilities, and Percentiles for Any Normal Distribution

TI-83/84

Finding Areas or Probabilities for Any Normal Distribution

Step 1 Press **2ⁿᵈ**, then **DISTR** (the **VARS** key).
Step 2 Press **2** to choose **normalcdf(**.
Step 3 On the home screen, enter the smaller value of X, comma, the larger value of X, comma, the mean of X, comma, the standard deviation of X, then close parenthesis. See Figure 6.46 (page 302).
Step 4 Press **ENTER**.

Note: When finding the area to the right of a value of X, use **1E99** as the larger value. When finding the area to the left of a value of X, use **-1E99** as the smaller value. Also, the shortcut for using the standard normal distribution is to specify only

the lower and higher values of X. If you enter only two values, the calculator assumes you want the standard normal distribution.

Finding Percentiles for Any Normal Distribution
Step 1 Press **2ⁿᵈ**, then **DISTR** (the **VARS** key).
Step 2 Press **3** to choose **invNorm(**.
Step 3 On the home screen, enter the probability value or area, then the mean of X, then the standard deviation of X, then close parenthesis. See Figure 6.47 (see page 302).
Step 4 Press **ENTER**.

Note: A shortcut for finding standard normal percentiles is to enter only the value of X for this function, in which case the calculator assumes you want the standard normal distribution.

EXCEL

Finding Areas or Probabilities for Any Normal Distribution

Step 1 Select cell **A1** and click the **Insert Function** icon F_x.
Step 2 In the **Search for a function**, type **NORMDIST**, click **GO**, then **OK**.
Step 3 For **X**, enter the X-value that you want to find the probability for. For **Mean,** enter the value of μ. For **Standard_dev**, enter the value of σ. For **Cumulative**, always enter **true**. Click **OK**. See Figure 6.48 (page 303).
Step 4 Excel provides the cumulative probability, $P(X \leq X_1)$ (see Example 6.38). If you need to find $P(X > X_1)$, subtract the result from 1. If you need to find $P(X_1 \leq X \leq X_2)$, find the

two cumulative probabilities, and subtract the lesser from the greater, as in Example 6.38.

Finding Percentiles for Any Normal Distribution
Step 1 Select cell **A1** and click the **Insert Function** icon f_x .
Step 2 In the **Search for a function**, type **NORMINV**, click **GO**, then **OK**.
Step 3 For **Probability**, enter the desired percentile in decimal form (for example, 0.99). For **Mean**, enter the value of μ. For **Standard_dev**, enter the value of σ. Click **OK**. See Figure 6.50 (page 303).

MINITAB

Finding Areas or Probabilities for Any Normal Distribution

Step 1 Click **Calc > Probability Distributions > Normal**.
Step 2 Select **Cumulative Probability**, enter the mean μ and standard deviation σ.
Step 3 Select **Input Constant**, enter the X-value that you want to find the probability for.
Step 4 Minitab provides the cumulative probability, $P(X \leq X_1)$ (see Example 6.38 on page 303). If you need to find $P(X > X_1)$,

subtract the result from 1. If you need to find $P(X_1 \leq X \leq X_2)$, find the two cumulative probabilities, and subtract the lesser from the greater, as in Example 6.38).

Finding Percentiles for Any Normal Distribution
Step 1 Click **Calc > Probability Distributions > Normal**.
Step 2 Select **Inverse Cumulative Probability**, and enter the mean μ and standard deviation σ.
Step 3 Select **Input Constant**. For the constant, enter the desired percentile in decimal form (for example, 0.99).

Section 6.5 SUMMARY

1 Section 6.5 showed how to solve normal probability problems for any conceivable normal random variable by first standardizing and then using the methods of Section 6.4. Methods for finding probabilities for a given value of the normal random variable X were discussed.

2 For any normal probability distribution, values of X can be found for given probabilities using the "backward" methods of Section 6.4 and the formula $X_1 = Z\sigma + \mu$.

3 The normal distribution can be used to approximate binomial probabilities when $np \geq 5$ and $np(1 - p) \geq 5$.

Section 6.5 EXERCISES

CLARIFYING THE CONCEPTS

1. What does the word *standardize* mean? Explain how we use standardization in solving normal probability problems.
2. When finding a data value for a specified probability, explain why we can't just report the Z-value but must transform back to the original normal distribution.
3. Think of a data distribution that is clearly non-normal. What would happen if you assumed that the distribution was normal and reported probabilities based on that assumption?
4. How do we determine whether it is appropriate to use the normal approximation to the binomial probability distribution?

PRACTICING THE TECHNIQUES

For Exercises 5–16, assume that the random variable X is normally distributed with mean $\mu = 70$ and standard deviation $\sigma = 10$. Draw a graph of the normal curve with the desired probability and value of X indicated. Find the indicated probabilities by standardizing X to Z.

5. $P(X > 70)$
6. $P(X > 80)$
7. $P(X < 80)$
8. $P(X > 95)$
9. $P(X \geq 95)$
10. $P(X \geq 60)$
11. $P(X \geq 55)$
12. $P(60 < X < 100)$
13. $P(60 \leq X \leq 100)$
14. $P(90 \leq X \leq 100)$
15. $P(90 \leq X \leq 91)$
16. $P(60 \leq X \leq 70)$

For Exercises 17–22, assume that the random variable X is normally distributed with mean $\mu = 70$ and standard deviation $\sigma = 10$. Draw a graph of the normal curve with the desired probability and value of X indicated. Find the indicated values of X using the formula $X_1 = Z\sigma + \mu$.
17. The value of X larger than 95% of all X-values (that is, the 95th percentile)
18. The 10th percentile
19. The 90th percentile (note that the 10th and 90th percentiles are symmetric values of X that contain the central 80% of the area under the curve between them)
20. The two symmetric values of X that contain the central 90% of X between them
21. The two symmetric values of X that contain the central 95% of X between them
22. The two symmetric values of X that contain the central 99% of X between them

For Exercises 23–28, determine whether it is appropriate to use the normal approximation to the binomial probability distribution.
23. X is a binomial random variable with $n = 10$ and $p = 0.5$

24. X is a binomial random variable with $n = 8$ and $p = 0.5$
25. X is a binomial random variable with $n = 10$ and $p = 0.4$
26. X is a binomial random variable with $n = 13$ and $p = 0.4$
27. X is a binomial random variable with $n = 45$ and $p = 0.1$
28. X is a binomial random variable with $n = 50$ and $p = 0.1$

For Exercises 29–36, let X be a binomial random variable with $n = 40$ and $p = 0.5$. Use the normal approximation to find the following probabilities.
29. $P(X = 20)$
30. $P(X \geq 20)$
31. $P(X > 20)$
32. $P(X \leq 20)$
33. $P(X < 20)$
34. $P(18 \leq X \leq 22)$
35. $P(18 < X < 22)$
36. $P(18 \leq X < 22)$

For Exercises 37–44, let X be a binomial random variable with $n = 120$ and $p = 0.1$. Use the normal approximation to find the following probabilities.
37. $P(X = 10)$
38. $P(X \geq 10)$
39. $P(X > 10)$
40. $P(X \leq 8)$
41. $P(X < 8)$
42. $P(9 \leq X \leq 11)$
43. $P(9 < X < 11)$
44. $P(9 < X \leq 11)$

APPLYING THE CONCEPTS

Finding Probabilities and Data Values for Any Normal Distribution. The Applying the Concepts exercises for the normal approximation to the binomial distribution begin at Exercise 68.
45. Hungry Babies. Six-week-old babies consume a mean of $\mu = 15$ ounces of milk per day, with a standard deviation σ of 2 ounces. Assume that the distribution is normal.
 a. Find the probability that a randomly chosen baby consumes between 17 and 19 ounces of milk per day.
 b. Find the probability that a randomly selected baby consumes less than 11 ounces of milk per day.
46. Trading Volume. The Associated Press reports that the mean trading volume for equity and index options contracts was 3.6 million in July 2007. Assume that the distribution is normal with mean $\mu = 3.6$ (in millions) and standard deviation $\sigma = 0.5$ (in millions).
 a. Find the probability that a randomly selected trading volume exceeds 4.1 million.
 b. Find the trading volume greater than 90% of all trading volumes. (*Hint:* Use the formula $X_1 = Z\sigma + \mu$.)
47. Windy Frisco. The mean wind speed in San Francisco is 13.6 mph in July, according to the U.S. National Oceanic and Atmospheric Administration. Suppose that the distribution of

the wind speed in July in San Francisco is normal with mean $\mu = 13.6$ mph and standard deviation $\sigma = 6$ mph.
 a. Find the probability that a randomly selected wind speed in San Francisco in July is 7.2 mph or less.
 b. What proportion of days in July have wind speed greater than 20 mph?
 c. What percentage of days in July have wind speed between 15 and 20 mph?

48. Viewers of *60 Minutes*. Nielsen Media Research reported that, for the week ending July 11, 2004, 12 million viewers watched the television show *60 Minutes*. Suppose that the distribution of viewers of *60 Minutes* is normal with mean $\mu = 12$ million and standard deviation $\sigma = 4$ million.
 a. Find the probability that fewer than 10 million people will watch *60 Minutes*.
 b. What percentage of broadcasts are watched by more than 18 million viewers?
 c. What proportion of broadcasts are watched by between 10 million and 11 million viewers?

49. Windy Frisco. Refer to the information in Exercise 47.
 a. Tours to Alcatraz Island are canceled if the day is too windy, specifically if the wind speed is higher than 99% of all other wind speeds in July. Find the cutoff wind speed.
 b. Suppose that a particular day in July has no wind at all. Should this be considered unusual? On what do you base your answer?

50. Viewers of 60 *Minutes*. Refer to the information in Exercise 48.
 a. Find the number of viewers that represents the 75th percentile.
 b. On one particular night, 24 million people watched *60 Minutes*. Is this unusual? On what do you base your answer?

51. Math Scores. The National Center for Education Statistics reports that mean scores on the standardized math test for 12th-graders in 2000 declined significantly from previous years. The mean score in 2000 was $\mu = 147$; assume $\sigma = 10$.
 a. Find the probability that the test score of a randomly selected 12th-grader was greater than 150.
 b. What percentage of test scores were below 125?
 c. What proportion of test scores were between 155 and 160?

52. Long-Lived Tortoise. The Philadelphia Zoo reports that the mean life span of a Galapagos tortoise is 100 years. Assume that their life span follows a normal distribution with mean $\mu = 100$ years and standard deviation $\sigma = 15$ years.
 a. Find the probability that a randomly chosen tortoise will live longer than 70 years.
 b. What percentage of tortoises live longer than 125 years?
 c. What proportion of tortoises live for between 115 and 145 years?

53. Math Scores. Refer to the information in Exercise 51.
 a. Suppose students who scored at the 5th percentile or lower could not graduate. Find the 5th percentile test score.
 b. Suppose that you know someone who scored 118 on the test. Is this unusual? On what do you base your answer?

54. Long-Lived Tortoise. Refer to the information in Exercise 52.
 a. Suppose that a particular zoo claims to have a tortoise older than 99% of all such tortoises. Find the age of this tortoise.
 b. A particular tortoise is 150 years old. Is this unusual? On what do you base your answer?

55. Hospital Patient Length of Stays. A study of Pennsylvania hospitals showed that the mean patient length of stay in 2001 was 4.87 days with a standard deviation of 0.97 days. Assume that the distribution of patient length of stays is normal.
 a. Find the probability that a randomly selected patient has a length of stay of greater than 7 days.
 b. What percentage of patient lengths of stay are less than 4 days?
 c. What proportion of patient lengths of stay are between 3 and 5 days?

56. Tobacco-Related Deaths. The World Health Organization states that tobacco is the second leading cause of death in the world. Every year, a mean of 5 million people die of tobacco-related causes. Assume that the distribution is normal with $\mu = 5$ (in millions) and $\sigma = 2$ (in millions).
 a. Find the probability that more than 4 million people will die of tobacco-related causes in a particular year.
 b. Find the probability that fewer than 1 million people will die from tobacco-related causes in a particular year.
 c. What is the probability that more than 3 million people will die from tobacco-related causes in a particular year?

57. Hospital Patient Length of Stays. Refer to the information in Exercise 55.
 a. Find the 50th percentile of patient lengths of stay. What is the relationship between the mean and the median for normal distributions?
 b. A particular patient had a length of stay of 8 days. Is this unusual? On what do you base your answer?

58. Tobacco-Related Deaths. Refer to the information in Exercise 56.
 a. Find the 75th percentile of the distribution of tobacco-related deaths.
 b. In one particular year, 8 million people died from tobacco-related causes. Is this unusual? On what do you base your answer?

59. Grade Inflation. Many educators are concerned about grade inflation. One study showed that one low SAT-score high school (with mean combined SAT score = 750) had higher mean grade point average (mean GPA = 3.6) than a high SAT-score school (with mean combined SAT score = 1050 and mean GPA = 2.6).[8] Assume that GPA at the high-SAT-score school is normally distributed with mean $\mu = 2.6$ and standard deviation $\sigma = 0.5$.

a. What is the probability that a randomly selected student at the high-SAT-score school has a GPA of 3.6 or higher?

b. What percentage of students at the school have a GPA of 3.0 or higher?

c. What proportion of students have GPAs between 2.0 and 2.5?

60. Stock Shares Traded. The mean number of shares traded on the New York Stock Exchange in June 2004 was 2.4 billion per day. Assume that the distribution of shares traded is normal with $\mu = 2.4$ and $\sigma = 0.6$ (both in billions of shares).

a. Find the probability that the number of shares traded on a randomly selected day falls below 0.6 billion.

b. What percentage of days finds a volume of shares traded of more than 2.5 billion shares?

c. What proportion of days finds the volume of shares traded between 1 billion and 2 billion?

61. Grade Inflation. Refer to the information in Exercise 59.

a. Suppose that students with the lowest 5% of GPAs get put on academic probation. Find the cutoff grade.

b. One particular student had a GPA of 4.0. Is this unusual? Explain your answer.

62. Stock Shares Traded. Refer to the information in Exercise 60.

a. A slow trading day has fewer shares traded than 99% of all other days. Find the number of shares traded that represents this amount.

b. On one particular day, 3 billion shares were traded. Is this unusual? Explain your answer.

63. Sociology Salaries. Sociology graduates, upon entering the workforce, earn a mean salary of $30,000 with a standard deviation of $4000.

a. Jolene would like to earn between $27,000 and $38,000 when she graduates. What is the probability that a randomly chosen salary lies within this range?

b. Josefina didn't pull those all-nighters for nothing. She would like to make between $32,000 and $39,000. What is the probability that a randomly chosen salary lies within this range?

c. Jamal looks forward to spending that big salary after graduation. Anything over $27,000 would do just fine. What is the probability that a randomly chosen salary exceeds $27,000?

d. Jillian struggled through college, having to care for her children while holding down a full-time job. She hopes that her salary is at least the 10th percentile. Find the 10th percentile of sociology salaries.

e. Jung is not sure how much he will make upon graduation. He would be happy if his salary was in the middle 60% of salaries. Find the two salaries that are the cutoff values for the middle 60% of salaries.

64. Emotional Maturity. An emotional maturity test was administered at a local university hospital. The scores are normally distributed with a mean of 100 and a standard deviation of 10.

a. Find the probability that a randomly chosen subject scored lower than 72.

b. Find the probability that a randomly chosen subject scored between 95 and 101.

c. Find the scores that mark off the middle 95% of scores.

65. Calories per Gram. Figure 6.62 shows the number of calories per gram for 961 food items. Assume that the population mean calories per gram is 2.25 with a standard deviation of 2.

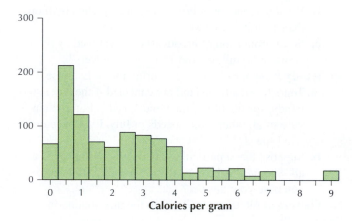

Figure 6.62

a. Assuming that the data follow a normal distribution, what is the 5th percentile of calories per gram?

b. Comment on whether your answer from (a) makes any sense.

c. The actual 5th percentile for this data set is 0.2 calories per gram. Looking at the histogram, does this make more sense than your answer from (a)?

d. Why is your answer in (a) wrong?

Use the *Normal Density Curve* applet for Exercises 66 and 67.

66. Use the applet to find the answers to the following exercises from this section.

a. Exercise 63a

b. Exercise 63b

67. Use the applet to find the answers to the following exercises from this section.

a. Exercise 64a

b. Exercise 64b

c. Exercise 64c

Normal Approximation to the Binomial Distribution. Use the normal approximation for each of Exercises 68–74.

68. Gas Tax. A *New York Times*/CBS News Poll conducted in April 2007 reported that 64% of Americans would favor an increased federal tax on gasoline if it would reduce dependence on foreign oil. For a sample of 200 Americans, approximate the following probabilities.

   a. Exactly 128 would favor such a tax.
   b. At least 128 would favor such a tax.
   c. More than 128 would favor such a tax.

69. Dress Casual. A survey conducted in 2007 by the Society for Human Resource Management found that the number of businesses allowing employees to "dress casually" every day has dropped from 53% in 2002 to 38% in 2007. Suppose we have a sample of 50 businesses from 2002 and a sample of 50 businesses from 2007. Approximate the following probabilities.

   a. At least 25 businesses allowed casual dress every day in 2002.
   b. At least 25 businesses allowed casual dress every day in 2007.
   c. Fewer than 15 businesses allowed casual dress every day in 2002.
   d. Fewer than 15 businesses allowed casual dress every day in 2007.

70. Hurricane Response. A survey found that 19% of respondents in New Orleans rated the overall response by government and volunteer agencies to major hurricanes in the past three years as good or excellent, while 57% of those living in other areas did so.[9] Suppose that we have a sample of 100 people living in New Orleans and 100 people living in other areas. Approximate the following probabilities.

   a. At least 30 of the respondents living in New Orleans rated the response as good or excellent.
   b. At least 30 of the respondents living in other areas rated the response as good or excellent.
   c. Fewer than 20 of the respondents living in New Orleans rated the response as good or excellent.
   d. Fewer than 20 of the respondents living in other areas rated the response as good or excellent.

71. Disease Outbreak. A survey found that only 9% of Americans were "very confident" that the U.S. government is prepared to handle a major outbreak of an infectious disease.[10] Suppose that we have a sample of 100 Americans. Approximate the following probabilities.

   a. Exactly 9 Americans are very confident.
   b. At least 9 Americans are very confident.
   c. More than 9 Americans are very confident.

72. Terrorism Concern. A September 2006 *New York Times*/CBS News Poll found that 69% of New York City residents were personally very concerned about a terrorist attack occurring in the area in which they live, compared with 22% of Americans as a whole. Suppose we have a sample of 100 New York City residents and 100 non–New Yorkers. Approximate the following probabilities.

   a. More than 50 New Yorkers are so concerned.
   b. More than 50 non–New Yorkers are so concerned.
   c. Fewer than 25 New Yorkers are so concerned.
   d. Fewer than 25 non–New Yorkers are so concerned.

Use the *Normal Approximation to the Binomial Distribution* applet for Exercises 73 and 74.

73. Select n (trials) = 10 and p (probability) = 0.2. The rectangles represent the binomial probabilities and the area under the curve represents the normal probabilities.

   a. For $n = 10$ and $p = 0.2$, is there a tight fit between the rectangles and the curve?
   b. What does this mean for whether the normal approximation should be used for a binomial distribution with $n = 10$ and $p = 0.2$?
   c. Verify whether the conditions are met for applying the normal approximation.

74. Select n (trials) = 63 and p (probability) = 0.2. The rectangles represent the binomial probabilities and the area under the curve represents the normal probabilities.

   a. For $n = 63$ and $p = 0.2$, is there a tight fit between the rectangles and the curve?
   b. What does this mean for whether the normal approximation should be used for a binomial distribution with $n = 63$ and $p = 0.2$?
   c. Verify whether the conditions are met for applying the normal approximation.

Try This in Class! In-class activities to enhance your understanding of statistics

PROBABILITY DISTRIBUTIONS AND A DECK OF CARDS

Bring in a deck of cards and perform the following experiment.

1. Take out the 12 face cards (kings, queens, and jacks).
2. Construct the probability distribution of the face value of the remaining cards (ace → X = 1, deuce → X = 2, and so on).
3. Find the population mean μ and standard deviation σ of this probability distribution.
4. Designate one student as shuffler, another student as selector. Have the selector draw one card at random from the abridged deck, observe it, record the value of X, and replace it.
5. The shuffler then shuffles the cards. The selector then selects another card at random, until 20 cards have been sampled with replacement.
6. Using the relative frequencies from Step 5, construct a relative frequency distribution of your observed face values. Is it very different from the population distribution from Step 2?
7. Calculate the mean and standard deviation of your relative frequency distribution. Are they very different from

the population values calculated in Step 3? How would you define "very different"?

A BINOMIAL PROBABILITY ACTIVITY

There is a famous problem in binomial probability. Which parents would wind up with the greater proportion of boy babies: the parents who simply stopped having babies after a certain number of babies or the parents who had babies until they had a boy? It might seem that the answer is that the parents who stopped after the first boy would have more boys, but the answer is that both sets of parents could expect to have the same proportion of boys. You will use simulation to try to confirm this. Assume that the probability that a randomly chosen

newborn is a boy is 50%. Then, clearly, the parents who simply stopped having babies after a certain number of babies could expect to have the same proportion (long-run probability!) of boys and girls. Now, take a single die and define the results of your rolls as follows: an odd-number result will represent a boy, and an even-number result will represent a girl.

1. Have your group roll the die. If you get a girl the first time, record it and roll again. Continue rolling and recording until you get a boy. That ends that particular trial.
2. Repeat until you have a large number (say, 30) of trials.
3. Then find the proportions of boys and girls in your 30 trials (representing 30 reproducing families). Your results shouldn't be too far away from a 50-50 split between boys and girls.

CHAPTER 6 FORMULAS AND VOCABULARY

SECTION 6.1
- **Continuous random variable** (p. 250)
- **Discrete random variable** (p. 250)
- **Expected value, or expectation, of a random variable** X (p. 257). Denoted $E(X)$
- **Mean μ of a discrete random variable** (p. 256). $\mu = \sum[X \cdot P(X)]$
- **Probability distribution of a discrete random variable** (p. 252)
- **Random variable** (p. 249)
- **Requirements for the probability distribution of a discrete random variable** (p. 252). $\sum P(X) = 1$ and $0 \leq P(X) \leq 1$
- **Variance and standard deviation of a discrete random variable** (p. 257).

  Definition formulas:

  $$\sigma^2 = \sum[(X - \mu)^2 \cdot P(X)]$$
  $$\sigma = \sqrt{\sum[(X - \mu)^2 \cdot P(X)]}$$

  Computational formulas:

  $$\sigma^2 = \sum[X^2 \cdot P(X)] - \mu^2$$
  $$\sigma = \sqrt{\sum[X^2 \cdot P(X)] - \mu^2}$$

SECTION 6.2
- **Binomial experiment** (p. 263)
- **Binomial probability distribution formula** (p. 266). $P(X) = (_nC_X)\, p^X\, (1 - p)^{n-X}$
- **Factorial (!)** (p. 265). $n! = n(n - 1)(n - 2) \ldots (2)(1)$

- **Mean, variance, and standard deviation of a binomial random variable X** (p. 269).

  Mean (or expected value): $\mu = n \cdot p$

  Variance: $\sigma^2 = n \cdot p \cdot (1 - p)$

  Standard deviation: $\sigma = \sqrt{n \cdot p \cdot (1 - p)}$

- **Number of combinations** (p. 265). $_nC_X = \dfrac{n!}{X!(n - X)!}$

SECTION 6.3
- **Continuous probability distribution** (p. 275)
- **Density curve** (p. 275)
- **Law of Total Probability for Continuous Random Variables** (p. 275)
- **Probability for continuous distributions** (p. 276)
- **Properties of the normal density curve (normal curve)** (p. 278)

SECTION 6.4
- **Standard normal (Z) distribution** (p. 284)
- **Z table for the standard normal random variable Z** (p. 284)

SECTION 6.5
- **Finding normal data values for specified probabilities** (p. 300)
- **Finding probabilities for any normal distribution** (p. 297)
- **Normal approximation to the binomial probability distribution** (p. 305)
- **Standardizing a normal random variable** (p. 296).

  $$Z = \frac{X - \mu}{\sigma}$$

CHAPTER 6 REVIEW EXERCISES

SECTION 6.1
For Exercises 1 and 2, use the given information to construct a probability distribution table and a probability distribution graph. Then answer the questions.

1. Take Me to the Ball Game. Andrew loves to go to the ballpark to watch Major League Baseball games. If this year is like most years, there is only a 5% chance that he will not catch a single game, 20% that he will see one game, 35%

that he will see two games, 25% that he will see three games, and 15% that he will see four games.

   a. What is the probability that Andrew will see at least one game this year?
   b. What is the most likely number of games that Andrew will see?
   c. What is the probability that Andrew will see at most two games this year?

2. **Early Lunch.** Chad has gotten to lunch early and is waiting for his friends to catch up. He figures that the probability that one friend shows up is 25%; two friends, 35%; three friends, 20%; and more than three friends, 5%.

   a. What is the probability that no friends show up?
   b. What is the probability that more than one friend shows up?
   c. What is the least likely outcome?

3. **Investment Profits.** Chang is looking forward to a profitable career as a day trader on Wall Street, after he has completed college. His present investments have the following probability distribution, with X = profit.

X	$-10,000$	5000	10,000
$P(X)$	0.3	0.5	0.2

   a. Find the mean, variance, and standard deviation of X.
   b. Draw a probability distribution graph.
   c. Plot the mean of the random variable on the graph from (b). Does this value for the mean make sense as the point where the distribution balances?

4. **Connecticut Lotto.** The Connecticut Lottery Corporation runs a game called Connecticut Classic Lotto. The player picks six different numbers from 1 to 44 and pays $1 to play.
   - If your picks match 3 of the 6 numbers chosen (probability 0.02381), you win $2.
   - If you match 4 out of 6 (probability 0.001495), you win $50.
   - If you match 5 out of 6 (probability 0.0000323), you win $2000.
   - If you match all 6 numbers (probability 0.0000001417), you win the jackpot. The jackpot on July 23, 2004, was $2,600,000.

   a. Construct the probability distribution of your winnings. Make sure to include the probability of not winning anything (in effect, losing $1), which equals 1 minus the four probabilities specified above.
   b. Find the expected winnings. Compare this with the price to play.

SECTION 6.2

5. **Gestational Diabetes.** Gestational diabetes occurs in 8% of pregnancies, according to the American Diabetes Association. A random sample of 20 pregnancies is taken.

   a. Find the probability that none of the pregnancies results in gestational diabetes.
   b. Find the probability that at least 1 of the pregnancies results in gestational diabetes.
   c. Find the probability that at most 2 of the pregnancies result in gestational diabetes.

6. **Female Students at DuPage.** According to the National Center for Education Statistics, 57% of the students at the College of DuPage in Glen Ellyn, Illinois, are female. A random sample of 10 students from the College of DuPage is taken.

   a. Find the probability that all 10 of the students are female.
   b. Find the probability that at most 9 of the students are female.
   c. Find the probability that at least 8 of the students are female.

7. **Watching the Price of Gas.** The Pew Research Center reports that, for the week of May 28, 2008, 35% of Americans said that the price of gasoline was the news story that they followed more closely than any other news item. A random sample of 15 Americans is taken.

   a. Find the probability that 5 Americans said that the price of gas was the news story they followed most closely.
   b. Find the probability that 4 Americans said that the price of gas was the news story they followed most closely.
   c. Find the probability that between 3 and 5 Americans said that the price of gas was the news story they followed most closely.

For Exercises 8–10, find the mean, variance, and standard deviation. Interpret the mean for each.

8. **Gestational Diabetes.** Gestational diabetes occurs in 8% of pregnancies, according to the American Diabetes Association. A random sample of 20 pregnancies is taken and the number of those with gestational diabetes is recorded.

9. **Female Students at DuPage.** According to the National Center for Education Statistics, 57% of the students at the College of DuPage in Glen Ellyn, Illinois, are female. A random sample of 10 students from the College of DuPage is taken and the number of females is recorded.

10. **Watching the Price of Gas.** The Pew Research Center reports that, for the week of May 28, 2008, 35% of Americans said that the price of gasoline was the news story that they followed more closely than any other news item. A random sample of 15 Americans is taken and the number who follow the price of gasoline most closely is recorded.

SECTION 6.3

Systolic Blood Pressure. Use the following information for Exercises 11–18. A study found that the mean systolic blood pressure was 106 mm Hg.[11] Assume that systolic blood pressure follows a normal distribution with mean $\mu = 106$ and standard deviation $\sigma = 8$.

11. What is the probability that a randomly chosen systolic blood pressure is equal to 106 mm Hg?
12. What is the probability that a randomly selected systolic blood pressure is more than 106 mm Hg?
13. What is the probability that a randomly chosen systolic blood pressure is at least 106 mm Hg?
14. Is the area to the right of $X = 110$ mm Hg greater than or less than 0.5? How do you know this?
15. Is the area to the left of $X = 110$ mm Hg greater than or less than 0.5? How do you know this?

16. What is the probability that a randomly selected systolic blood pressure is between 98 and 114 mm Hg?

17. What is the probability that a randomly chosen systolic blood pressure is between 90 and 122 mm Hg?

18. What is the probability that a randomly selected systolic blood pressure is between 82 and 130 mm Hg?

SECTION 6.4

For Exercises 19–24, **(a)** draw the graph, **(b)** estimate the area, **(c)** find the indicated area using the *Z* table or technology, and **(d)** check your estimate against your final answer.

Find the area under the standard normal curve that lies to the left of the following.

19. $Z = 2.1$

20. $Z = 2.9$

Find the area under the standard normal curve that lies to the right of the following.

21. $Z = -2.2$

22. $Z = -2.9$

Find the area under the standard normal curve that lies between the following.

23. $Z = -1.28$ and $Z = 1.28$

24. $Z = -1.04$ and $Z = 1.51$

25. Generation Worse Off? A *New York Times*/CBS News Poll conducted in 2007 found that 48% of young Americans (aged 17–29) think that their generation will be worse off than the previous generation. Suppose we take a random sample of 100 young Americans. Use the normal approximation to approximate the following probabilities.

   a. Exactly 50 think that their generation will be worse off.

   b. At least 50 think that their generation will be worse off.

   c. More than 50 think that their generation will be worse off.

   d. At most 50 think that their generation will be worse off.

   e. Fewer than 50 think that their generation will be worse off.

   f. Between 48 and 52 (inclusive) think that their generation will be worse off.

26. South Dakota Speeds. The National Motorists Association reports that, in South Dakota, the mean speed on interstate highways is 68.3 mph. Denote the mean to be $\mu = 68.3$ mph, and assume that the distribution is normal and $\sigma = 4$ mph.

   a. Find the probability that a randomly chosen vehicle is traveling faster than the 65 mph speed limit.

   b. What percentage of vehicles travel slower than 60 mph?

   c. What proportion of vehicles travel at speeds between 65 and 68.3 mph?

   d. The National Motorists Association asserts that speeding tickets should be issued only for drivers whose speeds exceed the 85th percentile. If the police in South Dakota followed this rule, then at what speed would they start handing out speeding tickets?

   e. Suppose that someone from South Dakota never drives faster than 55 mph on the interstate. Is this unusual? On what do you base your answer?

27. Reinstate the Draft? A *New York Times*/CBS News Poll conducted in 2007 found that 97% of young Americans (aged 17–29) oppose reinstating the military draft. Suppose we take a random sample of 400 young Americans. Use the normal approximation to approximate the following probabilities.

   a. Exactly 388 oppose reinstating the military draft.

   b. All 400 oppose reinstating the military draft.

   c. More than 388 oppose reinstating the military draft.

   d. At most 388 oppose reinstating the military draft.

   e. Fewer than 388 oppose reinstating the military draft.

   f. Between 385 and 390 (inclusive) oppose reinstating the military draft.

28. Drunk-Driving Deaths. In 2002 in the United States, a mean of 48 people *per day* were killed in vehicle accidents involving a drunk-driver. Assume that the distribution of drunk-driving accident deaths per day is normal, $\mu = 48$, and $\sigma = 12$.

   a. Find the probability that at most 12 people will be killed in drunk-driving accidents today.

   b. Find the probability that between 50 and 80 people will be killed in drunk-driving accidents today.

   c. Suppose that Mothers Against Drunk Driving holds special memorial services for the innocent people killed on days when the number of drunk-driving accident deaths reaches the 99.5th percentile. What is the cutoff number of deaths for the services to be held?

   d. Suppose on one particular day in the United States, 60 people are killed in drunk-driving accidents. Is this unusual? On what do you base your answer?

29. Hispanic Favorite: *Copa America*. According to Nielsen Media Research, the highest-rated television program among Hispanic Americans is *Copa America,* with a rating of 23.4 among Hispanic viewers for the week ending July 18, 2004. Assume that the ratings for this program are normally distributed with mean $\mu = 23.4$ and standard deviation $\sigma = 5$.

   a. Find the probability that the ratings for *Copa America* will be lower than 18.4.

   b. Find the probability that the ratings for *Copa America* will be between 15 and 20.

   c. Suppose that the show's producer must report to the UNI network executives any time the ratings fall below the 10th percentile. What is the ratings value that will trigger this report?

   d. Suppose that the ratings for *Copa America* in one particular week were 8.4. Is this unusual? On what do you base your answer? Will the producer have to file a report?

CHAPTER 6 Quiz

TRUE OR FALSE

1. True or false: The following is a continuous and not a discrete random variable: How much coffee there is in your next cup of coffee.

2. True or false: The following is an example of a binomial experiment: Rolling a pair of dice 3 times and observing the sum of the two dice.

3. True or false: Our distributions for continuous random variables are for samples and not for populations.

FILL IN THE BLANK

4. The probability that a randomly chosen value of a normally distributed random variable will be greater than the mean is _____.

5. The probability that a randomly chosen value of a normally distributed random variable will be equal to the mean is _____.

6. The standard deviation of a normal random variable can never take a value that is less than _____.

SHORT ANSWER

7. Is the following a discrete or continuous random variable: The number of goals your college soccer team will score in its next game.

8. Recording the gender of the next 20 babies born at City Hospital is an example of what kind of experiment?

9. What are the values for the mean and standard deviation of the standard normal distribution?

CALCULATIONS AND INTERPRETATIONS

10. Church Bazaar. Lenny has gone down to the church bazaar with his family. There is a game there where if you roll two dice and get a sum of at least 9 you will win $5; otherwise, you don't win anything.

    a. Construct the probability distribution for the amount you win playing this game.

    b. What are the expected winnings?

    c. What would be a fair (break-even) price for the church to ask you to pay to play this game?

11. CEOs Driving Luxury Cars. According to **CareerBuilder.com**, 19% of company CEOs drive luxury cars. Suppose a random sample is taken of 100 company CEOs.

    a. Find the probability that the sample contains 20 CEOs who drive luxury cars.

    b. What is the most likely number of CEOs who drive luxury cars?

    c. Find the mean, variance, and standard deviation. Interpret the mean.

    d. Suppose the sample contains 40 CEOs who drive luxury cars. Is this unusual? Explain how you determine this.

12. Gambling Losses. Treatment providers for problem gamblers report that men who approached them for intervention had lost a mean of $2849 in the preceding four weeks, according to a 2002 report.[12] Assume that the distribution of gambling losses is normally distributed with mean $\mu = \$2849$ and standard deviation $\sigma = \$900$.

    a. Find the probability that a randomly selected male had lost more than $4000.

    b. What percentage of males lost between $3000 and $4000?

    c. Suppose that a gambling support group is trying to identify those who lose the most, as measured by the 95th percentile. How much money in gambling losses does this represent?

    d. Suppose you know of a male problem gambler who lost $1000 in four weeks and then approached a treatment provider. Is this amount unusual? On what do you base your answer?

13. Cats and Dogs. The American Veterinary Medical Association reported that there were 61.1 million cats in America and 53.6 million dogs. It also reported that 31.4% of American households had a cat while 34.3% of American households had a dog. A random sample of 1000 households is taken. Use the normal approximation.

    a. Find the probability that the sample contains 320 households with a cat.

    b. Find the probability that the sample contains 320 households with a dog.

    c. Find the mean, variance, and standard deviation of the number of cats. Interpret the mean. Now do the same for dogs.

    d. Suppose that the sample contains 290 households with a cat. Is this unusual? Explain how you determine this.

14. Attendance at the Magic Kingdom. The Magic Kingdom at Walt Disney World in Orlando, Florida, has a mean of 38,000 customers per day. Assume that the distribution of attendance is normal, $\mu = 38,000$, and $\sigma = 10,000$.

    a. Find the probability that more than 40,000 people visit the Magic Kingdom in one day.

    b. Find the probability that between 35,000 and 41,000 people visit the Magic Kingdom in one day.

    c. Suppose that special overflow parking lots are opened for the days with the highest 10% attendance levels. What is the cutoff attendance number that triggers the opening of the overflow parking areas?

    d. Suppose that on one rainy day, only 15,000 people visit the Magic Kingdom. Is this unusual? On what do you base your answer?

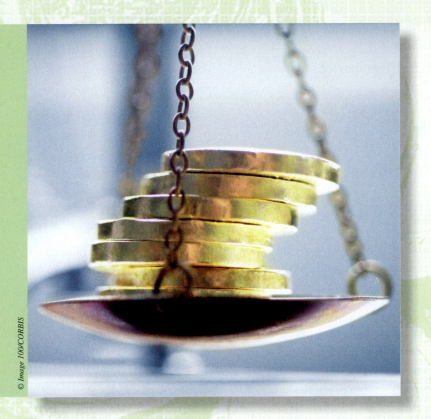

© Image 100/CORBIS

Sampling Distributions

Trial of the Pyx: How Much Gold Is in Your Gold Coins?

The kings of bygone England had a problem: How much gold should they put into their gold coins? After all, the very commerce of the kingdom depended on the purity of the currency. How did the lords of the realm ensure that the coins floating around the kingdom contained reliable amounts of gold?

From the year 1282, the Trial of the Pyx has been held annually in London to ensure that newly minted coins adhere to the standards of the realm. It is the responsibility of the presiding judge to ensure that the trial proceeds lawfully and to inform Her Majesty's Treasury of the verdict. Six members of the Company of Goldsmiths compose the jury, who are given two months to test the coins. It works like this: A ceremonial boxwood chest, called the Pyx, is brought forth, and a sample of 100 of the coins cast that year at the mint is put into it. The Pyx is then weighed. In times past, each gold coin, called a guinea, had an expected weight of 128 grams, so the total weight of the guineas in the Pyx was expected to be 12,800 grams.

If the weight of the coins in the Pyx was much less than 12,800 grams, the jury concluded that the Master of the Mint was cheating the crown by pocketing the excess gold, and he was severely punished. On the other hand, if the coins in the Pyx weighed much more than 12,800 grams, that wasn't good either, since it cut down on the profits produced by the kings' coin-minting monopoly.

By how much could the Master of the Mint debase the coinage before getting caught? We shall see in the chapter's Case Study, The Trial of the Pyx, which unfolds in Section 7.2.

The Big Picture

Where we are coming from, and where we are headed...

We have reached a major crossroads in our journey. Chapters 1–4 introduced us to the language of data analysis and methods for summarizing a data set using graphical or numerical methods. Next, to help us deal with the uncertainty implicit in generalizing from a sample to a population, we acquired the tools of probability in Chapters 5 and 6. We examined some of the rules of probability and learned how random variables can help us describe patterns in randomness.

Here, in Chapter 7, "Sampling Distributions," we approach a significant turning point. In the chapters on probability, we assumed that we knew the population mean and standard deviation. However, in the real world, such knowledge is rare. Uncovering the true values of population characteristics represents the very heart of data analysis: the notion of statistical inference. Statistical inference refers to learning about population characteristics by studying the same characteristics in a sample.

In this chapter, we discover that seemingly random statistics, such as the sample mean, have *predictable behaviors*. Taking repeated samples from the same population, we find that the sample means do not vary randomly but rather form a distinctive pattern or distribution. This special type of distribution is called the **sampling distribution** of the sample mean. We explore sampling distributions and uncover perhaps the most important result in statistical inference, the Central Limit Theorem.

Thus, the key to all the statistical inference in Chapters 8–11 is the set of patterns implicit in the sampling distributions that we learn about here in Chapter 7. When trying to infer the true value of a population parameter, data analysts often start with a point estimate. However, in Chapter 8, "Confidence Intervals," we will learn that point estimates, though widespread, have serious drawbacks. We need something better, confidence interval estimates, our first application of statistical inference. ■

7.1 Introduction to Sampling Distributions

Objectives: By the end of this section, I will be able to...

1 Compute point estimates and sampling error.

2 Explain the sampling distribution of the sample mean $\bar{x}$.

3 Describe the sampling distribution of the sample mean $\bar{x}$ when the population is normal.

4 Use normal probability plots to assess normality.

5 Find probabilities and percentiles for the sample mean when the population is normal.

✐ Point Estimates and Sampling Error

Recall from Chapter 1 that **statistical inference** consists of methods for estimating and drawing conclusions about population characteristics based on the information contained in a subset (sample) of that population. Sample characteristics, like the sample mean $\bar{x}$, are called **statistics**. Population characteristics, like the population mean μ, are called **parameters**. Table 7.1 summarizes the relationship between three of the most common statistics and their associated parameters. Using known statistics to estimate unknown parameters and reporting a single number as the estimate is called **point estimation**, and the value of the statistic is called the **point estimate**.

Table 7.1 Point estimation: Use statistics to estimate unknown population parameters

	Sample statistic	. . . estimates . . .	Population parameter
Mean	$\bar{x}$	. . . estimates . . .	μ
Standard deviation	s	. . . estimates . . .	σ
Proportion	$\hat{p}$	. . . estimates . . .	p

Example 7.1   Using a sample mean to estimate a population mean

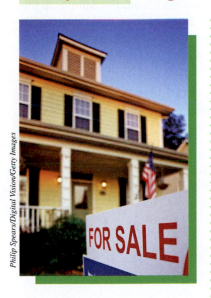

Philip Spears/Digital Vision/Getty Images

In Example 3.2 (page 81) we calculated the mean of a sample of 6 home sales prices for Broward County, Florida (Table 7.2). The mean price was

$$\bar{x} = \frac{\sum x}{n} = \frac{3,849,000}{6} = \$641,500$$

Find a point estimate for the mean price μ of all home sales prices for Broward County, Florida.

Table 7.2 Home sales prices, Broward County, Florida

Location	Price
Coral Springs	\$415,000
Weston	\$449,000
Coconut Creek	\$472,000
Davie	\$649,000
Hallandale	\$889,000
Fort Lauderdale	\$975,000

Solution

We use the sample mean $\bar{x} = \$641,500$ as the point estimate for μ. ■

It is not likely that the population mean home sales price μ for all houses in Broward County precisely equals the sample mean $\bar{x} = \$641,500$ because the sample is a subset of the population. Therefore, when using point estimates, we nearly always "miss" our target by a certain amount. This type of error is called **sampling error**. We define sampling error as *distance,* so it is never negative. Table 7.3 presents the sampling error for most common characteristics.

> **Sampling error** is the distance between the point estimate and its target parameter.

Table 7.3 Sampling error for common characteristics

Characteristic	Sample statistic	Population parameter	Sampling error
Mean	$\bar{x}$	μ	$|\bar{x} - \mu|$
Standard deviation	s	σ	$|s - \sigma|$
Proportion	$\hat{p}$	p	$|\hat{p} - p|$

Example 7.2 Sampling error for $\bar{x}$

Suppose that someone did an exhaustive search of the multiple listing service for all of Broward County and found the population mean home sales price μ to be \$650,000. Find the sampling error for $\bar{x}$ for the sample in Table 7.2.

Solution

From Example 7.1, we have $\bar{x} = \$641,500$, and from Table 7.3 we have the sampling error for $\bar{x}$ as $|\bar{x} - \mu|$. The sampling error is therefore

$$|\bar{x} - \mu| = |\$641,500 - \$650,000| = |-\$8,500| = \$8,500$$

Unfortunately, the true value of the sampling error usually is unknown because we rarely know the actual value of the population parameter. Since a sample is a subset (usually a tiny subset) of the population, generalizing from the sample to the population carries risks. Point estimates are not always useful because we cannot confidently tell how accurate the estimates are. That is, the sampling error may be small or large. In Chapter 8, we will learn another way to calculate estimates. But first we need to learn more about how statistics behave.

The *Law of Large Numbers for Means* applet allows you to simulate dice rolls and observe the behavior of the mean of the rolls as the number of rolls increases. As the sample size n increases, the value of $\bar{x}$ tends to settle down at μ. Thus, as n increases, the sampling error $|\bar{x} - \mu|$ tends to decrease.

The *Law of Large Numbers for Proportions* applet allows you to simulate coin tosses, and observe the proportion of heads as the number of tosses increases. As the sample size n increases, the value of $\hat{p}$ tends to settle down at p. Thus, as n increases, the sampling error $|\hat{p} - p|$ tends to decrease.

Sampling Distribution of the Sample Mean $\bar{x}$

We begin by taking some samples from a population and seeing what we can learn.

Example 7.3 Commuting times for student government members

We are interested in how long it takes the five members of the student government to commute to school. The times (in minutes) are given in Table 7.4. Since these five people are *all* the members of the student government, we can consider them to constitute a *population*.

Table 7.4 Commuting times for the five members of the student government

Amber	Brandon	Chantal	Dave	Emma
10	20	5	30	15

a. Calculate the population mean commuting time μ.

b. Take a sample of the following student government members: Amber, Brandon, and Chantal. Find the sample mean commuting time $\bar{x}$ and the sampling error $|\bar{x} - \mu|$.

Solution

a. The mean commuting time of this population is

$$\mu = \frac{\sum x}{N} = \frac{10 + 20 + 5 + 30 + 15}{5} = 16 \text{ minutes}$$

b. For Amber, Brandon, and Chantal, the sample mean commuting time is

$$\bar{x}_1 = \frac{\sum x}{n} = \frac{10 + 20 + 5}{3} \approx 11.67 \text{ minutes}$$

The sampling error of this sample mean is $|\bar{x}_1 - \mu| = |11.67 - 16| = 4.33$ minutes. ■

Developing Your Statistical Sense

Sampling Variability

Before we take a second sample, consider this question: When we take another sample, is the sample mean commuting time from the new sample likely to be exactly the same as the $\bar{x}_1 = 11.67$ minutes from the first sample? Probably not, since different samples provide different values for the sample statistics. This is called *sampling variability*. ■

Table 7.5 shows all possible samples of size 3 from the five students, along with the respective sample means and sampling errors. Note that, while the value for the sample mean varies, the population mean μ is constant: $\mu = 16$. A population parameter like μ is a constant (a fixed number). However, with each sample we take, we obtain a different estimate of μ. This is because the sample mean $\bar{x}$ is a random variable and exhibits sampling variability.

Table 7.5 All possible samples of size 3 from population of student government members

Sample	Amber Brandon Chantal	Amber Brandon Dave	Amber Brandon Emma	Amber Chantal Dave	Amber Chantal Emma	Amber Dave Emma	Brandon Chantal Dave	Brandon Chantal Emma	Brandon Dave Emma	Chantal Dave Emma
Data	10 20 5	10 20 30	10 20 15	10 5 30	10 5 15	10 30 15	20 5 15	20 5 15	20 30 15	5 30 15
$\bar{x}$	11.67	20	15	15	10	18.33	18.33	13.33	21.67	16.67
μ	16	16	16	16	16	16	16	16	16	16
$|\bar{x} - \mu|$	4.33	4	1	1	6	2.33	2.33	2.67	5.67	0.67

Fortunately, there are patterns in how the sample mean varies. These patterns depend on the sample size, on the shape of the population distribution, and on the underlying value of the population parameters, μ and σ. The highlighted row in Table 7.5 represents the sample means for all possible samples of size $n = 3$. This is called the **sampling distribution for the sample mean** $\bar{x}$ for this tiny population and sample size, obtained by examining every possible sample of size $n = 3$.

> The **sampling distribution of the sample mean** $\bar{x}$ for a given sample size n consists of the collection of the sample means of all possible samples of size n from the population.

Sampling distributions can tell us about the expected location and variability of a statistic. We are especially interested in the center and the spread of the sampling distribution. Like any distribution, the sampling distribution for the sample mean has a balance point and therefore a mean. Figure 7.1 provides a dotplot of the sample means in Table 7.5, along with the mean of these sample means, indicated at the balance point $\mu = 16$. Figure 7.1 represents the sampling distribution of the sample mean for this example.

FIGURE 7.1

Sampling distribution of the sample mean for Example 7.3.

Population mean = 16

The mean value of the ten sample means is

$$\frac{11.67 + 20 + 15 + 15 + 10 + 18.33 + 18.33 + 13.33 + 21.67 + 16.67}{10} = 16$$

Note that this value is exactly equal to the population mean $\mu = 16$. In fact, the sampling distribution of the sample mean $\bar{x}$ has a mean value of μ. That is, the sampling distribution of $\bar{x}$ is centered at μ. We generalize this result as follows.

Note: It is convenient to number a set of important *facts,* as we build toward the Central Limit Theorem for Means and the Central Limit Theorem for Proportions.

> **Fact 1: Mean of the Sampling Distribution of the Sample Mean $\bar{x}$**
>
> The mean of the sampling distribution of the sample mean $\bar{x}$ is the value of the population mean μ. It can be denoted as $\mu_{\bar{x}} = \mu$ and read as "the mean of the sampling distribution of $\bar{x}$ is μ."

Note: In this example, the precise relationship between the two standard deviations is

$$\sigma_{\bar{x}} = \sqrt{\frac{N-n}{N-1}} \cdot \frac{\sigma}{\sqrt{n}}$$

where N is the population size and n is the sample size. This gives

$$\sigma_{\bar{x}} = \sqrt{\frac{5-3}{5-1}} \cdot \frac{8.6023}{\sqrt{3}} \approx 3.5119$$

However, the coefficient

$$\sqrt{\frac{N-n}{N-1}}$$

called the *finite population correction factor,* is required only for special cases (like this textbook example) where the population is not much larger than the sample. This coefficient does not apply when sampling with replacement, and its value tends to zero as the sample size approaches the population size. However, for most real-world problems, and for the remainder of this book, we dispense with this coefficient and assume that the population size is very large compared to the sample size.

Next we would like to uncover information regarding the spread of the sampling distribution of $\bar{x}$. Now the population standard deviation of the original commute times in Table 7.4 is

$$\sigma = \sqrt{\frac{\sum(x-\mu)^2}{N}}$$

$$= \sqrt{\frac{[(10-16)^2 + (20-16)^2 + (5-16)^2 + (30-16)^2 + (15-16)^2]}{5}}$$

$$\approx 8.6023$$

And the population standard deviation of the sampling distribution of $\bar{x}$ is

$$\sigma_{\bar{x}} = \sqrt{\frac{\sum(\bar{x}-\mu)^2}{10}} = \sqrt{\frac{[(11.67-16)^2 + (20-16)^2 + \ldots + (16.67-16)^2]}{10}}$$

$$\approx 3.5119$$

Note that the standard deviation of the sample means is smaller than the original standard deviation. This is a good thing, since smaller variability is better.

> **Fact 2: Standard Deviation of the Sampling Distribution of the Sample Mean $\bar{x}$**
>
> The standard deviation of the sampling distribution of the sample mean $\bar{x}$ is $\sigma_{\bar{x}} = \sigma/\sqrt{n}$, where σ is the population standard deviation and n is the sample size.

Note the $\sqrt{n}$ in the denominator of the formula. Because of this factor, the larger the sample size, the tighter the resulting sampling distribution. Larger sample sizes result in more precise estimation.

Example 7.4   Finding the mean and standard deviation of the sampling distribution of $\bar{x}$

According to CanEquity Mortgage Company, the mean age of mortgage applicants in the city of Toronto is 37 years old. Assume that the standard deviation is 6 years. Find the mean and standard deviation for the sampling distribution of $\bar{x}$ for the following sample sizes: (a) 4, (b) 100, (c) 225.

Solution

We have $\mu_{\bar{x}} = \mu = 37$. Note that this value for $\mu_{\bar{x}}$ does not depend on the sample size, so the value is true for any sample size. We also have $\sigma = 6$.

a. $n = 4$. Then $\sigma_{\bar{x}} = \dfrac{\sigma}{\sqrt{n}} = \dfrac{6}{\sqrt{4}} = 3$.

b. $n = 100$. Then $\sigma_{\bar{x}} = \dfrac{\sigma}{\sqrt{n}} = \dfrac{6}{\sqrt{100}} = 0.6$.

c. $n = 225$. Then $\sigma_{\bar{x}} = \dfrac{\sigma}{\sqrt{n}} = \dfrac{6}{\sqrt{225}} = 0.4$.   ■

What Does This Number Mean?

Consider $\sigma_{\bar{x}} = 0.6$ for $n = 100$. This is a measure of the variability of the sampling distribution of $\bar{x}$ for this sample size. That is, if we take samples of size 100, our sampling error in estimating the population mean age μ of all mortgage applicants in Toronto will be within 0.6 year of the true population mean most of the time.   ■

3 Sampling Distribution of $\bar{x}$ for a Normal Population

To find out what form the sampling distribution of $\bar{x}$ takes when the population is normal, we consider the following example, in which we examine a small, normally distributed population and find the sample means for all possible samples of a certain size.

Example 7.5   Sampling distribution of $\bar{x}$ for a normal population

In Example 6.33 (page 296), the average statewide temperature in Georgia in the month of April was normally distributed with a mean of $\mu = 61.5°\text{F}$ and a standard deviation of $\sigma = 8°\text{F}$. Using Minitab, 1000 samples of size $n = 2$ were generated from this normal distribution, and the sample means $\bar{x}$ were calculated for each sample. Construct a histogram and observe the shape of the distribution.

Solution

Figure 7.2 shows the histogram of the means from the 1000 samples of size $n = 2$. As you may have expected, the histogram looks quite normal, even with this tiny sample size of $n = 2$.   ■

FIGURE 7.2 Sampling distribution of sample means of size $n = 2$ from a normal population.

Example 7.5 considered the smallest possible sample size, $n = 2$. In fact, this outcome is true for all normal populations. Using Facts 1 and 2, we can summarize this sampling distribution as follows.

Note: Let the notation

normal $(\mu, \sigma/\sqrt{n})$

denote a normal distribution with mean of μ and standard deviation of $\sigma/\sqrt{n}$.

> **Fact 3: Sampling Distribution of the Sample Mean for a Normal Population**
> For a normal population, the sampling distribution of the sample mean $\bar{x}$ is distributed as normal $(\mu, \sigma/\sqrt{n})$, where μ is the population mean and σ is the population standard deviation.

Once we know that the sample mean is normally distributed, we can standardize and produce Z, just as we would for any normal random variable.

> **Fact 4: Standardizing a Normal Sampling Distribution for Means**
> When the sampling distribution of $\bar{x}$ is normal, we may standardize to produce the standard normal random variable Z as follows:
> $$Z = \frac{\bar{x} - \mu_{\bar{x}}}{\sigma_{\bar{x}}} = \frac{\bar{x} - \mu}{\sigma/\sqrt{n}}$$
> where μ is the population mean, σ is the population standard deviation, and n is the sample size.

4 Assessing Normality Using Normal Probability Plots

Recall from Section 2.2 that histograms for the same data set can take different shapes, depending on the class width and number of classes. We therefore introduce a type of graph called the **normal probability plot**, which is a more reliable tool for assessing normality. A normal probability plot is a scatterplot of the estimated cumulative normal probabilities (expressed as percents) against the corresponding data values in the data set.

Figure 7.3 shows the normal probability plot for a sample of normally distributed data. The points are arrayed nicely along the straight line, and all the points lie within the curved bounds. Figure 7.4 shows the normal probability plot for a sample of right-skewed data. The points do not line up in a straight line, and many points lie outside the curved bounds, indicating that the data set is not normal.

> **Analyzing Normal Probability Plots**
>
> If the points in the normal probability plot either cluster around a straight line or nearly all fall within the curved bounds, then it is likely that the data set is normal. Systematic deviations off the straight line are evidence against the claim that the data set is normal.

FIGURE 7.3 Normal probability plot of normal data.

FIGURE 7.4 Normal probability plot of right-skewed data.

We illustrate how to construct normal probability plots using technology in the Step-by-Step Technology Guide at the end of this section (page 331).

5 Finding Probabilities and Percentiles Using a Sampling Distribution

Since we know that the sampling distribution of the sample mean $\bar{x}$ is normal when the population is normally distributed, we can use the techniques of Section 6.5 to answer questions about the means of samples taken from normal populations.

Example 7.6   Sampling distribution of sample mean statistics quiz score

Suppose that statistics quiz scores for a certain instructor are normally distributed with mean 70 and standard deviation 10.

a. Find the probability that a randomly chosen student's score will be above 80.

b. Suppose that, over the years, the instructor has had many sections of size $n = 25$. What kind of distribution is the sampling distribution of the sample mean?

Solution

a. This is a normal probability problem, which we learned how to do in Chapter 6.

$$P(X > 80) = P\left(\frac{X - 70}{10} > \frac{80 - 70}{10}\right)$$

$$= P(Z > 1) = 1 - 0.8413 = 0.1587$$

Example 7.8 **Finding a value of $\bar{x}$, given a probability or area**

Using the information in Example 7.7, find the 95th percentile of the class mean quiz scores.

Solution

Since we want the 95th percentile, we consider the right-hand tail area in Figure 7.6, which has area 0.05, leaving area 0.95 to the left of the 95th percentile. We then use the Z table *backward*. We seek 0.95 on the *inside* of the Z table. Since 0.95 is not in the Z table, we take the closest value. Since the two closest values, 0.9495 and 0.9505, are equally close, we split the difference. Working backward from 0.9495, we find $Z = 1.64$, and for 0.9505 we find $Z = 1.65$. Splitting the difference, we get $Z = 1.645$. This value of $Z = 1.645$ is the 95th percentile of the standard normal distribution.

Since we are looking for a sample mean quiz score, 1.645 cannot be the answer. We need to "unstandardize" by transforming this value of Z using a formula similar to the formula $x = Z \cdot \sigma + \mu$ that we used in Section 6.5:

$$\bar{x} = Z \cdot \sigma_{\bar{x}} + \mu = 1.645(2) + 70 = 73.29$$

Thus, the 95th percentile of the sample means for the statistics quizzes is 73.29. This is close to our earlier estimate of 73. ■

Example 7.9 **Finding probabilities and percentiles using sample means**

Use the information from Examples 7.6, 7.7, and 7.8.
a. Find the 5th percentile of the class mean quiz scores.
b. If a sample mean is chosen at random, what is the probability that it will lie between 66.71 and 73.29?
c. What two symmetric values for the sample mean quiz score contain 90% of all sample means between them?

Solution

a. Since the sampling distribution is normal, it is also symmetric. Thus, the 95th percentile and the 5th percentile are the same distance away from the mean. Since the 95th percentile is $(73.29 - 70) = 3.29$ above the mean, the 5th percentile must be 3.29 below the mean, or $(70 - 3.29) = 66.71$.
b. We seek $P(66.71 < \bar{x} < 73.29)$, as shown in Figure 7.7. Proceeding with the calculations, we have, as expected,

$$P(66.71 < \bar{x} < 73.29) = P\left(\frac{66.71 - 70}{2} < \frac{\bar{x} - 70}{2} < \frac{73.29 - 70}{2}\right)$$
$$= P(-1.645 < Z < 1.645) = 0.95 - 0.05 = 0.90$$

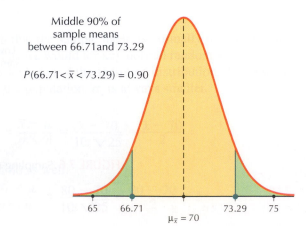

Middle 90% of sample means between 66.71 and 73.29

$P(66.71 < \bar{x} < 73.29) = 0.90$

FIGURE 7.7
Middle 90% of the sample means.

65   66.71   73.29   75
          $\mu_{\bar{x}} = 70$

c. This is just another way of asking for the 5th and 95th percentiles. The answer is 66.71 and 73.29. ∎

In Section 7.2, we tackle the more challenging problem of finding the sampling distribution of the sample mean for non-normal populations.

STEP-BY-STEP TECHNOLOGY GUIDE: Constructing Normal Probability Plots

TI-83/84
Assume that the data set is in list L1.
Step 1 Access STAT PLOTS by pressing **2nd Y**.
Step 2 Select **1:Plot1**. Press **ENTER**.
Step 3 Move the cursor over **On** and press **ENTER**.
Step 4 Select the normal probability plot type by moving the cursor to the lower-right plot among the choices for Type. Press **ENTER**.

Step 5 For **Data List**, enter **L1**.
Step 6 For **Data Axis**, choose **X**.
Step 7 Press **ZOOM**, then **9: ZoomStat**.

MINITAB
Assume that the data set is in column **C1**.
Step 1 From the menu, select **Graph**, then click **Probability Plot**.
Step 2 Select **Single** and click **OK**.

Step 3 In the Probability Plot dialog box, select **C1**, and click **OK**. The normal probability plot for the data set in **C1** is then generated.

Section 7.1 SUMMARY

1 We can use sample statistics as point estimates of the unknown population parameters. For each statistic, sampling error is the distance between the point estimate and its target parameter.

2 The sampling distribution of the sample mean $\bar{x}$ for a given sample size n consists of the collection of the sample means of all possible samples of size n from the population. The mean of the sampling distribution of the sample mean $\bar{x}$ is the value of the population mean μ (Fact 1). The standard deviation of the sampling distribution of the sample mean $\bar{x}$

is $\sigma_{\bar{x}} = \sigma/\sqrt{n}$, where σ is the population standard deviation (Fact 2).

3 For a normal population, the sampling distribution of the sample mean $\bar{x}$ is distributed as normal $(\mu, \sigma/\sqrt{n})$, where μ is the population mean and σ is the population standard deviation (Fact 3).

4 Normal probability plots are used to assess the normality of a data set.

5 We can use Fact 3 and Fact 4 to find probabilities and percentiles using sample means.

Section 7.1 EXERCISES

CLARIFYING THE CONCEPTS
1. Explain in your own words what sampling error means. In the real world, when would we know the actual value of sampling error?
2. Explain what a sampling distribution is. Why are sampling distributions so important?
3. For a normal population, what can we say about the sampling distribution of the sample mean?
4. Explain what we use a normal probability plot for. What should we look for in a normal probability plot?

PRACTICING THE TECHNIQUES
For Exercises 5–10, find the sampling error.
5. $\mu = 100, \bar{x} = 95$
6. $\mu = 0, \bar{x} = 1.5$
7. $\sigma = 20, s = 15$
8. $\sigma = 1, s = 0.9$
9. $p = 0.25, \hat{p} = 0.30$
10. $p = 0.75, \hat{p} = 0.70$

For Exercises 11–16, find $\mu_{\bar{x}}$ and $\sigma_{\bar{x}}$, the mean and standard deviation of the sampling distribution of $\bar{x}$.

11. $\mu = 100, \sigma = 20, n = 25$
12. $\mu = 100, \sigma = 20, n = 100$
13. $\mu = 0, \sigma = 10, n = 9$
14. $\mu = 0, \sigma = 10, n = 25$
15. $\mu = -10, \sigma = 5, n = 100$
16. $\mu = -10, \sigma = 5, n = 400$

For Exercises 17–24, assume that the random variable X is normally distributed, with mean $\mu = 10$ and standard deviation $\sigma = 4$. That is, X is normal (10,4). Let $n = 16$.

17. Describe the sampling distribution of $\bar{x}$ for $n = 16$.
18. Find the probability that $\bar{x}$ exceeds 11.
19. Without using your calculator, find the probability that $\bar{x}$ is less than 9.
20. Without using your calculator, find the probability that $\bar{x}$ lies between 9 and 11.
21. Find the value of $\bar{x}$ that is greater than 97.5% of all values of $\bar{x}$.
22. Find the value of $\bar{x}$ that is less than 97.5% of all values of $\bar{x}$.
23. What proportion of $\bar{x}$ lies between the values you found in Exercises 21 and 22?
24. What proportion of $\bar{x}$ lie outside the values you found in Exercises 21 and 22?

APPLYING THE CONCEPTS

Vehicle Weights. Use the following information for Exercises 25–28. The data set **Crash**, located on your data disk, contains the results of crash tests performed in 1992 for 352 different vehicles. The variable "weight" contains the weight in pounds of the vehicles. A random sample of 46 of these vehicles was selected and their weights noted. The following table contains descriptive statistics for the weights of the 46 vehicles in the sample and population parameter values for the weights of all 352 vehicles in the population of vehicles in the data set **Crash**.

	Sample statistic	Population parameter
Mean	$\bar{x} = 3021$	$\mu = 2930.34$
Standard deviation	$s = 607$	$\sigma = 627.13$

25. **a.** What is the point estimate for the population mean weight of all crash test vehicles?
    b. If we took another sample of vehicles, would you expect to get the same sample mean weight? Explain why or why not.
    c. Is it likely that the population mean weight of all crash test vehicles equals the sample mean weight from (**b**)? Why or why not?

26. **a.** What is the point estimate for the standard deviation weight of all crash test vehicles?

b. If we took another sample of vehicles, would you expect to get the same sample standard deviation weight? Explain why or why not.

27. **a.** What is the population mean weight for all the vehicles?
    b. Find the sampling error for the mean weight.

28. **a.** What is the population standard deviation of the weights?
    b. Find the sampling error for the standard deviation.

29. **Carbon Dioxide Concentration.** The Environmental Protection Agency reports (2008) that the mean concentration of the greenhouse gas carbon dioxide is 365 parts per million (ppm). Suppose the population is normally distributed, assume that the standard deviation is 100 ppm, and suppose we take atmospheric samples from 25 cities.
    a. Find the probability that the sample mean carbon dioxide concentration will be above 385 ppm.
    b. Find the probability that the sample mean carbon dioxide concentration will be between 305 and 425 ppm.
    c. Find the probability that the sample mean carbon dioxide concentration will be below 345.

30. **Short-Term Memory.** In a famous research paper in the psychology literature, George Miller found that the amount of information humans could process in short-term memory was 7 bits (pieces of information), plus or minus 2 bits.[1] Assume that the mean number of bits is 7 and the standard deviation is 2, and that the distribution is normal. Suppose we take a sample of 100 people and test their short-term memory skills.
    a. Find the probability that the sample mean number of bits retained in short-term memory is greater than 7.5.
    b. Find the probability that the sample mean number of bits retained in short-term memory is between 6.8 and 7.2.
    c. Find the probability that the sample mean number of bits retained in short-term memory is less than 6.75.

31. **Carbon Dioxide Concentration.** For the carbon dioxide concentrations in Exercise 29, do the following.
    a. Find the sample mean carbon dioxide concentration higher than 97.5% of all such sample means.
    b. Find the sample mean carbon dioxide concentration lower than 97.5% of all such sample means.
    c. Describe the area that lies between the values you found in (**a**) and (**b**).

32. **Short-Term Memory.** For the numbers of bits in Exercise 30:
    a. Find the 97.5th percentile of the sample mean number of bits retained in short-term memory.
    b. Find the 2.5th percentile of the sample mean number of bits retained in short-term memory.
    c. Between which two values do the middle 95% of sample mean number of bits retained in short-term memory lie?

2006 Olympic Figure Skating Scores. Use this information for Exercises 33–36. The table shows the Ladies' Short Program scores for the top five finishers in the 2006 Winter Olympic Games in Torino, Italy. *Consider these scores to be a population.*

Skater	Nation	Score
Sasha Cohen	United States	66.73
Irina Slutskaya	Russia	66.70
Shizuka Arakawa	Japan	66.02
Fumie Suguri	Japan	61.75
Kimmie Meissner	United States	59.40

33. a. How many samples of size $n = 2$ can we generate from this population of size 5?
   b. Compute the population mean μ. Do we usually know the value of the population mean in a typical real-world problem? Why or why not?
   c. Calculate the population standard deviation σ.
34. Take every possible sample of size $n = 2$ skaters from this tiny population.
   a. Find the mean score $\bar{x}$ of each sample.
   b. Construct a dotplot of the sample mean scores.
35. Refer to the dotplot in your answer to Exercise 34.
   a. Where would the balance point be located in the dotplot? Indicate it on the plot.
   b. Recall that the balance point is an estimate of the mean. What is your estimate of $\mu_{\bar{x}}$, using this balance point?
36. a. Find the mean of all the sample mean scores in Exercise 34.
   b. Is your estimate for Exercise 35b close to your answer to Exercise 36a?
   c. Does your answer to Exercise 36a agree with value of the population mean that you found in Exercise 33b?
37. A Fair Die. Consider a fair six-sided die. Suppose we take samples of size 16 and are interested in the population mean of the die rolls.
   a. Find $\mu_{\bar{x}}$.
   b. Find $\sigma_{\bar{x}}$. (*Hint:* First find the standard deviation of a fair die roll using a frequency distribution.)

SAT Math Scores. Use this information for Exercises 38–41. The College Board (www.collegeboard.com) reports that the nationwide mean SAT Math score was 515 in 2007. Assume that the standard deviation is 116 and that the scores are normally distributed.
38. What is the probability that a randomly selected SAT Math score will be less than 500?
39. As a researcher, you are looking at samples of SAT Math scores of size 16.
   a. Find $\mu_{\bar{x}}$.
   b. Find $\sigma_{\bar{x}}$.

c. What can you say about the sampling distribution of the sample mean? How do you know this?
40. Refer to Exercise 39.
   a. What is the probability that a sample of 16 students will have a sample mean SAT Math score below 500?
   b. Why is the probability so much lower for the sample mean than for a particular student?

41. *What if* the population standard deviation was greater than 116. Explain how this would affect the following, if at all.
   a. Probability that a sample of 16 SAT Math scores will have a mean less than 500
   b. 99.5th percentile of the sample mean SAT Math scores
   c. 0.5th percentile of the sample mean SAT Math scores

Law of Large Numbers for Means Applet. Use the *Law of Large Numbers for Means* applet for Exercises 42 and 43.
42. Set the number of **Rolls** to 10, and select **Show Mean.** Click **Roll dice.**
   a. Observe the behavior of the sample mean $\bar{x}$ compared to the population mean $\mu = 3.5$.
   b. Estimate the sampling error $|\bar{x} - \mu|$.
   c. Without pressing **Reset**, click **Roll dice** until the total number of dice equals $n = 20, 40, 60, 80$, and 100.
   d. In each case in (c), estimate the sampling error $|\bar{x} - \mu|$.
43. What can you conclude about the sampling error $|\bar{x} - \mu|$ as the sample size n increases?

Law of Large Numbers for Proportions Applet. Use the *Law of Large Numbers for Proportions* applet for Exercises 44 and 45.
44. Set the probability of heads to $p = 0.5$ and the number of tosses to $n = 40$. Click **Toss.**
   a. Record the sample proportion $\hat{p}$ of heads observed and the sampling error $|\hat{p} - p|$.
   b. Without pressing **Reset**, continue to click **Toss** until the total number of tosses is $n = 120$. Again record the sample proportion $\hat{p}$ of heads observed and the sampling error $|\hat{p} - p|$.
   c. Without pressing **Reset**, continue to click **Toss** until the total number of tosses is $n = 180$. Again record the sample proportion $\hat{p}$ of heads observed and the sampling error $|\hat{p} - p|$.
   d. Without pressing **Reset**, continue to click **Toss** until the total number of tosses is $n = 240$. Again record the sample proportion $\hat{p}$ of heads observed and the sampling error $|\hat{p} - p|$.
45. What can you conclude about the sampling error $|\hat{p} - p|$ as the sample size n increases?

7.2 Central Limit Theorem for Means

Objectives: **By the end of this section, I will be able to...**

1. Describe the sampling distribution of $\bar{x}$ for skewed and symmetric populations as the sample size increases.

2. Apply the Central Limit Theorem for Means to solve probability questions about the sample mean.

1 Sampling Distribution of $\bar{x}$ for Skewed and Symmetric Populations

In Section 7.1, we discovered that the sampling distribution for the sample mean for a normal population is also normal. What if the population is not normal? In this section, we use simulation studies to learn how the sampling distribution of the sample mean $\bar{x}$ for non-normal populations becomes approximately normal as the sample size increases. We start with small sample sizes and work our way up, keeping our eye on the sampling distribution of the sample mean $\bar{x}$.

Example 7.10   Simulation study: Sample means from strongly skewed population

The data set **Nutrition** on your CD and the companion Web site contains nutrition information on a population of 961 foods.
a. Construct a histogram of the potassium content of these 961 foods, and describe the shape of the population distribution.
b. Using Minitab, take 500 random samples of sizes $n = 10$, 20, and 30 from the population. Assess the normality of the resulting sampling distributions of $\bar{x}$ using histograms and normal probability plots.

Solution
a. A histogram of the potassium content of these foods is shown in Figure 7.8, revealing a strongly right-skewed, non-normal data set.

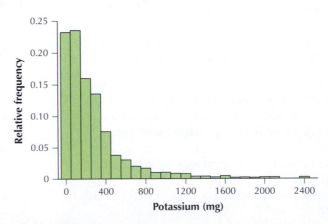

FIGURE 7.8 Potassium content is strongly right-skewed, not normal.

b. Using Minitab, we take 500 random samples of size $n = 10$ from the population. We find the means of the 500 samples shown in the graphs in Figure 7.9.

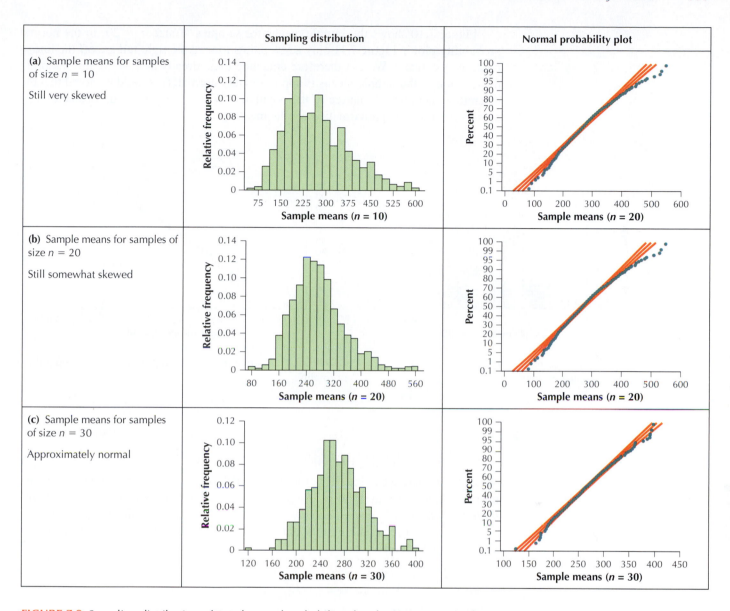

FIGURE 7.9 Sampling distribution of $\bar{x}$ and normal probability plots for $n = 10$, 20, and 30.

- $n = 10$: The sampling distribution of $\bar{x}$ is skewed (Figure 7.9a).
- $n = 20$: The sampling distribution of $\bar{x}$ is still somewhat skewed (Figure 7.9b).
- $n = 30$: Despite a few outliers, the sampling distribution of $\bar{x}$ is approximately normal (Figure 7.9c).

For a skewed population, we have seen that the sampling distribution of the sample mean becomes approximately normal as the sample size reaches 30. For a less skewed population, we can expect that the sampling distribution of $\bar{x}$ approximates a normal distribution for smaller sample sizes.

Recall that a distribution is **symmetric** if an axis of symmetry splits the image in half so that one side is the mirror image of the other. The population distribution of values from a single roll of a fair die is perfectly uniform and therefore symmetric since each number is equally likely to be rolled. Recall that the population mean, or the mean of all possible die rolls, is $\mu = 3.5$, and the population standard deviation of all possible die rolls is $\sigma = 1.7078$.

Figure 7.10 shows the sample means for samples of size $n = 20$. In the normal probability plot in Figure 7.11 only a few points toward the right tail are off the line—nothing systematic. We can therefore conclude that, already at $n = 20$, the sampling distribution of the sample means is approximately normal. (It should be noted that the discrete, noncontinuous nature of the population seems to be the main obstacle to an even better normal approximation for the means.)

FIGURE 7.10 Sample means of size $n = 20$: approximately normal.

FIGURE 7.11 Normal probability plot for $n = 20$: acceptable normality.

⚡ Using the Central Limit Theorem for Means to Solve Probability Problems About the Sample Mean

We can summarize our investigation into the behavior of the sample means for non-normal populations as follows: *Regardless of the population, the sampling distribution of the sample mean becomes approximately normal as the sample size gets larger.* We can then combine this statement with Fact 3 to form the **Central Limit Theorem for Means.**

Central Limit Theorem for Means

Given a population with mean μ and standard deviation σ, the sampling distribution of the sample mean $\bar{x}$ becomes approximately normal $(\mu, \sigma/\sqrt{n})$ as the sample size gets larger, regardless of the shape of the population.

The Central Limit Theorem (CLT) is one of the most important results in statistics. Worldwide, much statistical inference is based on the CLT. It actually makes fairly intuitive sense, doesn't it? If we find the mean of a sample of data values, in many cases the extreme values will tend to be blended away. However, remember that the mean is very sensitive to outliers. In a small sample, there may not be enough non-extreme values to balance the influence of the outliers. This is what was happening early in the potassium simulation (for example, Figure 7.9a). However, as the sample sizes increase, the influence of extreme values diminishes and the resulting sample means start to migrate toward the center.

How large does the sample size have to be before the Central Limit Theorem for Means takes effect? As we have seen, it depends on the degree of symmetry exhibited by the population. In general, however, we shall abide by the following rough rule of thumb.

Rule of Thumb for When Central Limit Theorem for Means Takes Effect

We consider $n \geq 30$ as large enough to apply the Central Limit Theorem for Means for any population.

The *Central Limit Theorem* applet allows you to experiment with various sample sizes and see how the Central Limit Theorem for Means behaves in action.

Combining Fact 3 and the Central Limit Theorem for Means, we can identify three cases for the sampling distribution of $\bar{x}$.

Three Cases for the Sampling Distribution of the Sample Mean $\bar{x}$

- **Case 1: The population is normal.** Therefore the sampling distribution of $\bar{x}$ is *normal* (Fact 3, page 326).

- **Case 2: The population is either non-normal or of unknown distribution *and* the sample size is at least 30.** Therefore the sampling distribution of $\bar{x}$ is *approximately normal* (Central Limit Theorem for Means).

- **Case 3: The population is either non-normal or of unknown distribution *and* the sample size is less than 30.** Therefore we have *insufficient information* to conclude that the sampling distribution of the sample mean $\bar{x}$ is either normal or approximately normal.

Of course, in the real world, no one will tell you which of the three cases applies. You need to investigate the assumptions of each of the cases to determine for yourself which case applies.

Example 7.11 Application of the Central Limit Theorem for the Mean

The U.S. Small Business Administration (SBA) provides information on the number of small businesses for each metropolitan area in the United States.[2] Figure 7.12 shows a histogram of our population for this example, the number of small businesses in each of the 328 cities nationwide. (For example, Austin, Texas, has 22,305 small businesses, while Pensacola, Florida, has 6020.) The mean is $\mu = 12,485$ and the standard deviation is $\sigma = 21,973$. Find the probability that a random sample of size $n = 36$ cities will have a mean number of small businesses greater than 17,000.

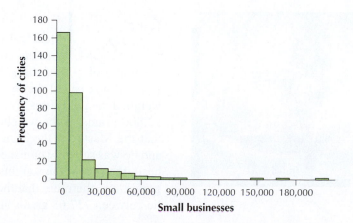

FIGURE 7.12

Solution

First we try to apply Case 1. Clearly, the population is not normal, so Case 1 does not apply. Next, we try to apply Case 2. Since the sample size $n = 36$ is large enough, the Central Limit Theorem applies. The sampling distribution of the sample mean $\bar{x}$ is approximately normal. Now that we know that the sampling distribution of $\bar{x}$ is approximately normal, we need to find $\mu_{\bar{x}}$ and $\sigma_{\bar{x}}$. Facts 1 and 2 tell us that

$$\mu_{\bar{x}} = \mu = 12{,}485 \qquad \text{and} \qquad \sigma_{\bar{x}} = \frac{\sigma}{\sqrt{n}} = \frac{21{,}973}{\sqrt{36}} \approx 3662.1667$$

Therefore, as the CLT indicates, the sampling distribution of $\bar{x}$ is approximately normal($\mu_{\bar{x}} = 12{,}485$, $\sigma_{\bar{x}} = 3662.1667$). We are then left to solve a normal probability problem, just as we did in Chapter 6. Figure 7.13 shows the sampling distribution of $\bar{x}$ and the probability we are interested in, $P(\bar{x} > 17{,}000)$. Using Fact 4, we standardize:

$$Z = \frac{17{,}000 - \mu_{\bar{x}}}{\sigma_{\bar{x}}} = \frac{17{,}000 - 12{,}485}{3662.1667} \approx 1.2329 \approx 1.23$$

Thus, $P(\bar{x} > 17{,}000) \approx P(Z > 1.23)$, as shown in Figure 7.14.

We therefore look up $Z = 1.23$ in the Z table and subtract this table area (0.8907) from 1 to get the desired tail area:

$$P(Z > 1.23) = 1 - 0.8907 = 0.1093$$

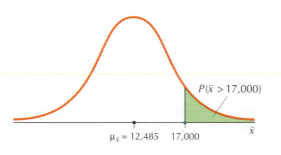

$P(\bar{x} > 17{,}000)$

$\mu_{\bar{x}} = 12{,}485$ $17{,}000$ $\bar{x}$

FIGURE 7.13

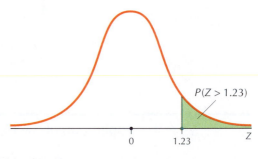

$P(Z > 1.23)$

0 1.23 Z

FIGURE 7.14

The probability is 0.1093 that a random sample of 36 cities will have a mean number of small businesses greater than 17,000. ■

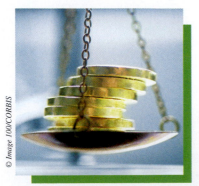

CASE STUDY | Trial of the Pyx: How Much Gold Is in Your Gold Coins?

Medieval English kings and queens devised a procedure to ensure that the coins of the realm contained the proper amount of gold. A sample of 100 of the gold coins that were cast each year was placed in a ceremonial box called the Pyx. At the chosen time, the Company of Goldsmiths jury weighed the gold coins. The weight of the entire sample of coins was supposed to be 12,800 grams. If the weight was much less than 12,800 grams, the jury concluded that the Master of the Mint was cheating the crown by pocketing the excess gold, and he was severely punished. We assume that, for the Master of the Mint to keep his head, the *sum* of the weights of the gold coins had to lie between 12,768 and 12,832 grams, and therefore that the average weight $\bar{x}$ of the sample of the 100 coins had to lie between 127.68 grams and 128.32 grams.

Note: The methods we have learned in this chapter have dealt with the sampling distribution of the sample *mean*, not the sum.

Since the sample mean $\bar{x} = \frac{\sum X}{n}$, we divide the total weight by $n = 100$. So the probability that the Master of the Mint passed the Trial of the Pyx is

$$P(12{,}768 < \textstyle\sum X < 12{,}832)$$

which, divided by $n = 100$, becomes

$$P\left(127.68 < \frac{\sum X}{n} < 128.32\right)$$

or $P(127.68 < \bar{x} < 128.32)$

That is, the jury would have accepted the guineas as pure if the mean weight of the guineas lay between 127.68 and 128.32 grams.

Further assume that "much less than" indicated a measurement that is 2 or more standard deviations below average. For the sampling distribution of $\bar{x}$, this would indicate a range of $0.32 = 2\,\sigma_{\bar{x}}$ between 127.68 and the mean 128. Therefore, $\sigma_{\bar{x}} = 0.16$ and, by the Empirical Rule, approximately 95% of the sample mean observations for the Trial of the Pyx would have been between 127.68 and 128.32. Since $\sigma_{\bar{x}} = \sigma/\sqrt{n}$, it follows that $\sigma = \sqrt{100} \cdot 0.16 = 1.6$ grams.

Problem 1. What were the chances that the Master of the Mint would have been caught and punished if he were in fact cheating? *Hint:* Note that σ equals 1.6 grams (see sidebar).

Solution

What if the Master of the Mint set the mean amount of gold in the population of all coins to be $\mu = 127.9$ grams instead of the required 128, shortchanging the crown by a tenth of a gram of gold per coin?

We've seen that the Master of the Mint would have passed the Trial of the Pyx if $127.68 < \bar{x} < 128.32$. Now, because 100 is a large sample size, the Central Limit Theorem tells us that the sampling distribution of $\bar{x}$ is approximately normal, with $\mu_{\bar{x}} = \mu = 127.9$ and $\sigma_{\bar{x}} = \dfrac{\sigma}{\sqrt{n}} = \dfrac{1.6}{\sqrt{100}} = 0.16.$

Standardize using Fact 4:

$$Z = \frac{127.68 - \mu_{\bar{x}}}{\sigma_{\bar{x}}} = \frac{127.68 - 127.9}{0.16} \approx -1.38 \quad\text{and}$$

$$Z = \frac{128.32 - \mu_{\bar{x}}}{\sigma_{\bar{x}}} = \frac{128.32 - 127.9}{0.16} \approx 2.63$$

Solving using Table 6.7 in Section 6.4 (page 385), we get

$$P(-1.38 < Z < 2.63) = 0.9957 - 0.0838 = 0.9119$$

That is, the chances of the Crown accepting the coins as pure, even if the Master of the Mint had been shortchanging by a tenth of a gram per coin, were over 91% (Figure 7.15).

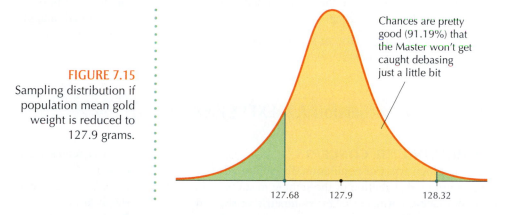

FIGURE 7.15
Sampling distribution if population mean gold weight is reduced to 127.9 grams.

Chances are pretty good (91.19%) that the Master won't get caught debasing just a little bit

127.68   127.9   128.32

Problem 2. Would the Master of the Mint have been satisfied with this small amount of debasement? Would he have quit while he was ahead?

Solution

No way! The following year the Master of the Mint decided to debase the currency even further, setting the mean amount of gold in the guineas to be $\mu = 127.3$ grams per coin.

We need to find the probability of the Master passing the Trial of the Pyx if the mean amount of gold in a coin was 127.3 grams instead of the required 128 grams per coin. We use the same calculations, with $\mu_{\bar{x}} = 127.3$ grams. Standardize:

$$Z = \frac{127.68 - \mu_{\bar{x}}}{\sigma_{\bar{x}}} = \frac{127.68 - 127.3}{0.16} \approx 2.38 \quad\text{and}$$

$$Z = \frac{128.32 - \mu_{\bar{x}}}{\sigma_{\bar{x}}} = \frac{128.32 - 127.3}{0.16} \approx 6.38$$

Then $P(2.38 < Z < 6.38) \approx 1 - 0.9913 = 0.0087$.

In other words, the Master of the Mint actually would have stood very little chance—less than 1% probability—of passing the Trial of the Pyx if he cheated by this much.

England is a great country for retaining fine old traditions. Today, England's Company of Goldsmiths still operates the London Assay Office, where the purity of the kingdom's coin is tested at the annual Trial of the Pyx.

Note: Sir William Sharington (1493–1553), Master of the Mint during the turbulent Tudor era in England, debased the currency and issued worthless coinage. He was arrested in 1548.

Section 7.2 SUMMARY

1 In this section, we examine the behavior of the sample mean when the population is not normal. The approximate normality of the sampling distribution of the sample mean kicks in much quicker when the original population is symmetric rather than skewed.

2 The Central Limit Theorem is one of the most important results in statistics and is stated as follows:

Given a population with mean μ and standard deviation σ, the sampling distribution of the sample mean $\bar{x}$ becomes approximately normal $(\mu, \sigma/\sqrt{n})$ as the sample size gets larger, regardless of the shape of the population. This approximation applies for smaller sample sizes when the original distribution is more symmetric.

Section 7.2 EXERCISES

CLARIFYING THE CONCEPTS

1. For which types of population distribution does the sampling distribution of the sample mean become approximately normal quicker: symmetric or skewed distributions? Explain in your own words why this happens.

2. Compare the x axes (the scale) of the histograms in Figure 7.9. What do you notice as the sample size increases? How does the scale change?

3. Refer to Exercise 2. What does this change indicate about the variability in the sampling distribution?

4. True or false: The Central Limit Theorem states that the sampling distribution of $\bar{x}$ is approximately normal, regardless of sample size.

PRACTICING THE TECHNIQUES

In Exercises 5–10, samples are taken. Determine which of the three cases for the sampling distribution of the sample mean applies.

5. SAT scores are normally distributed and the sample size is small.

6. SAT scores are not normally distributed and the sample size is large.

7. Accountant incomes are normally distributed and the sample size is small.

8. Accountant incomes are not normally distributed and the sample size is small.

9. Systolic blood pressure readings are not normally distributed and the sample size is large.

10. Systolic blood pressure readings are normally distributed and the sample size is small.

In Exercises 11–16, samples are taken. Provide (a) $\mu_{\bar{x}}$ and (b) $\sigma_{\bar{x}}$, and (c) determine whether the sampling distribution of $\bar{x}$ is normal, approximately normal, or unknown.

11. SAT scores are normally distributed, with $\mu = 516$ and $\sigma = 116$. A sample of size 9 is taken.

12. SAT scores are not normally distributed, with $\mu = 516$ and $\sigma = 116$. A sample of size 36 is taken.

13. Accountant incomes are normally distributed, with $\mu = \$60,000$ and $\sigma = \$10,000$. A sample of size 16 is taken.

14. Accountant incomes are not normally distributed, with $\mu = \$60,000$ and $\sigma = \$10,000$. A sample of size 16 is taken.

15. Systolic blood pressure readings are not normally distributed, with $\mu = 80$ and $\sigma = 8$. A sample of size 64 is taken.

16. Systolic blood pressure readings are normally distributed, with $\mu = 80$ and $\sigma = 8$. A sample of size 25 is taken.

For Exercises 17–22, if possible find the indicated probability. If not possible, explain why not.

17. The exchange rate for U.S. \$100 in euros for 2007 is not normally distributed, with $\mu = 74$ euros and $\sigma = 5$. A sample of size 9 is taken. Find $P(\bar{x} > 75)$.

18. The exchange rate for U.S.\$100 in euros for 2007 is not normally distributed, with $\mu = 74$ euros and $\sigma = 5$. A sample of size 36 is taken. Find $P(\bar{x} > 75)$.

19. 2007 prices for boned trout are normally distributed, with $\mu = \$3.10$ per pound and $\sigma = \$0.30$. A sample of size 16 is taken. Find $P(\bar{x} < \$3)$.

20. 2007 prices for boned trout are not normally distributed, with $\mu = \$3.10$ per pound and $\sigma = \$0.30$. A sample of size 16 is taken. Find $P(\bar{x} < \$3)$.

21. Systolic blood pressure readings are not normally distributed, with $\mu = 80$ and $\sigma = 8$. A sample of size 64 is taken. Find $P(78 < \bar{x} < 82)$.

22. Systolic blood pressure readings are normally distributed, with $\mu = 80$ and $\sigma = 8$. A sample of size 25 is taken. Find $P(78 < \bar{x} < 82)$.

For Exercises 23–28, find the indicated value of $\bar{x}$. If it is not possible, explain why not.

23. SAT scores are normally distributed, with $\mu = 516$ and $\sigma = 116$. A sample of size 9 is taken. Find the sample mean SAT score larger than 95% of all sample means.

24. SAT scores are not normally distributed, with $\mu = 516$ and $\sigma = 116$. A sample of size 36 is taken. Find the sample mean SAT score larger than 95% of all sample means.

25. Accountant incomes are normally distributed, with $\mu = \$60,000$ and $\sigma = \$10,000$. A sample of size 16 is taken. Find the sample mean accountant income that is smaller than 90% of all sample means.

26. Accountant incomes are not normally distributed, with $\mu = \$60,000$ and $\sigma = \$10,000$. A sample of size 16 is taken. Find the sample mean accountant income that is smaller than 90% of all sample means.

27. The pollen count distribution for Los Angeles in September is not normally distributed, with $\mu = 8.0$ and $\sigma = 1.0$. A sample of size 16 is taken. Find the 75th percentile of sample means.

28. The pollen count distribution for Los Angeles in September is not normally distributed, with $\mu = 8.0$ and $\sigma = 1.0$. A sample of size 64 is taken. Find the 75th percentile of sample means.

APPLYING THE CONCEPTS

29. Cholesterol Levels. The Centers for Disease Control reports that the mean serum cholesterol level in Americans in 2005 was 202. Assume that the standard deviation is 45. There is no information about the distribution. We take a sample of 36 Americans. Does Case 1 apply? If possible, find $P(\bar{x} > 212)$. If not possible, explain why not.

30. Tennessee Temperatures. According to the National Oceanic and Atmospheric Administration, the mean temperature for Nashville, Tennessee, in the month of January between 1872 and 2006 was 38.6°F. Assume that the standard deviation is 10 degrees and the distribution is normal. Suppose we take a sample of 25 days. If possible, find $P(\bar{x} > 40)$. If not possible, explain why not.

31. Computers per School. The National Center for Education Statistics (**http://nces.ed.gov**) reported that the mean number of instructional computers per public school nationwide was 124. Assume that the standard deviation is 49 computers and that there is no information about the shape of the distribution. Suppose we take a sample of size 75 public schools, and suppose we are interested in the probability that the sample mean number of instructional computers is less than 100.

    a. If possible, find $P(\bar{x} < 100)$. If not possible, explain why not.

    b. What is the probability that a sample of size 75 public schools will have more than 100 instructional computers? (*Hint:* Use your answer to (**a**) and take the easy way to a solution.)

 Use the *Central Limit Theorem* applet for Exercises 32 and 33.

32. Describe the shape of the sampling distribution of $\bar{x}$ for the following sample sizes.

    a. 2

    b. 5

    c. 30

33. At what sample size would you say the sampling distribution of $\bar{x}$ becomes approximately normal?

Coaching and SAT Scores. Use this information for Exercises 34–38. The College Board reports that the mean increase in SAT Math scores of students who attend review courses (coaching) is 18 points. Assume that the standard deviation of the change in score is 12 points and that changes in score are not normally distributed. We are interested in the probability that the sample mean score increase is negative, indicating a loss of points after coaching.

34. Suppose we take a sample of size 20. If possible, find $P(\bar{x} < 0)$, the probability that the sample mean score increase is negative, indicating a loss of points after coaching. If not possible, explain why not.

35. Suppose we take a sample of size 40, and we are interested in $P(\bar{x} < 0)$, indicating a loss of points after coaching. If possible find $P(\bar{x} < 0)$. If not possible, explain why not.

36. Suppose we take a sample of size 80, and we are interested in $P(\bar{x} < 0)$.

   a. *Before doing any calculations,* predict whether this probability will be greater or smaller than if we had used $n = 40$. Why?

   b. Verify your intuition by finding $P(\bar{x} < 0)$ for $n = 80$.

37. *What if* the sample size used was some unspecified value greater than 80. Describe how and why this change would have affected the following, if at all. Would the quantities increase, decrease, remain unchanged? Or is there insufficient information to tell what would happen? Explain your answers.

   a. $\mu_{\bar{x}}$

   b. $\sigma_{\bar{x}}$

   c. $Z = \dfrac{\bar{x} - \mu_{\bar{x}}}{\sigma_{\bar{x}}}$

   d. $P(\bar{x} < 0)$

38. Suppose we take a sample of 100 students.

   a. Find $P(\bar{x} > 0)$, the probability that the sample of 100 students had an increase in mean SAT score.

   b. Find the 95th percentile of SAT score sample mean increases after coaching.

   c. Find the 5th percentile of SAT score sample mean increases after coaching.

   d. Between which two symmetric values lie 90% of all sample mean SAT score increases?

   e. By hand, draw a normal curve with your answer to (d) indicated.

7.3   Central Limit Theorem for Proportions

Objectives: By the end of this section, I will be able to…

1 Explain the sampling distribution of the sample proportion $\hat{p}$.

2 Describe the sampling distribution of the sample proportion $\hat{p}$ for extreme and moderate values of p.

3 Apply the Central Limit Theorem for Proportions to solve probability questions about the sample proportion.

1 Sampling Distribution of the Sample Proportion $\hat{p}$

The sample mean is not the only statistic that can have a sampling distribution. Every sample statistic has a sampling distribution. One of the most important is the sampling distribution of the **sample proportion $\hat{p}$**.

Suppose each individual in a population either has or does not have a particular characteristic. If we take a sample of size n from this population, the **sample proportion $\hat{p}$** (read "p-hat") is

$$\hat{p} = \frac{x}{n}$$

where x represents the number of individuals in the sample that have the particular characteristic. We use $\hat{p}$ to estimate the unknown value of the population proportion p. In Section 6.2, we were introduced to $\hat{p}$ as the sample proportion of successes in a binomial experiment.

Just as there is a Central Limit Theorem for Means, there is a Central Limit Theorem for Proportions. In Section 7.2, we learned how the Central Limit Theorem for Means allows us to consider the sampling distribution of the sample mean as approximately normal for sufficiently large sample sizes. Here in Section 7.3, we develop a Central Limit Theorem for Proportions, where the **sampling distribution of the sample proportion** also becomes approximately normal if the right conditions are satisfied.

> The **sampling distribution of the sample proportion** $\hat{p}$ for a given sample size n consists of the collection of the sample proportions of all possible samples of size n from the population.
>
> In general, the **sampling distribution of any particular statistic** for a given sample size n consists of the collection of the values of that sample statistic across all possible samples of size n.

Example 7.12   Genders of the student government members

Photofusion Picture Library/Alamy

Suppose we are interested in the proportion of female members of the student government. Since 3 of the 5 members of the student government are female, the population proportion of females is $p = 3/5 = 0.6$.

Amber	Brandon	Chantal	Dave	Emma
Female	Male	Female	Male	Female

Suppose we take all possible samples of size 3. For example, the first sample of Amber, Brandon, and Chantal contains 2 females and 1 male, so the sample proportion of females is $\hat{p} = 2/3$. The second sample of Amber, Brandon, and Dave contains 1 female and 2 males, so $\hat{p} = 1/3$.

a. Make a table of the sample proportion $\hat{p}$ and the sampling error $|\hat{p} - p|$ for all possible samples of size $n = 3$.

b. Construct a dot plot of the values of $\hat{p}$ for all possible samples of size $n = 3$.

Solution

Table 7.6 shows the results from all possible samples of size 3. The highlighted row contains the values of the sampling distribution of the statistic $\hat{p}$. These values are plotted in Figure 7.16. ■

Table 7.6 All possible samples of size 3 from population of student government members

Sample	Amber Brandon Chantal	Amber Brandon Dave	Amber Brandon Emma	Amber Chantal Dave	Amber Chantal Emma	Amber Dave Emma	Brandon Chantal Dave	Brandon Chantal Emma	Brandon Dave Emma	Chantal Dave Emma
Data	F M F	F M M	F M F	F F M	F F F	F M F	M F M	M F F	M M F	F M F
$\hat{p}$	2/3	1/3	2/3	2/3	3/3	2/3	1/3	2/3	1/3	2/3
p	3/5	3/5	3/5	3/5	3/5	3/5	3/5	3/5	3/5	3/5
$|\hat{p} - p|$	1/15	4/15	1/15	1/15	6/15	1/15	4/15	1/15	4/15	1/15

FIGURE 7.16 Sampling distribution of the sample proportion for Example 7.12.

Just as the population parameter μ was a constant, so also is the target parameter p, the population proportion of females in the student government. However, with each sample we take, we obtain a different estimate of the population proportion p because the sample proportion $\hat{p}$ is a random variable. The sample proportion is a random variable because we do not know its value before the random sample of student government members is chosen.

Example 7.13   Mean of sample proportions

Calculate the mean of the ten sample proportions $\hat{p}$ from Table 7.6.

Solution
The mean of these ten sample proportions is

$$\frac{\left(\frac{2}{3} + \frac{1}{3} + \frac{2}{3} + \frac{2}{3} + \frac{3}{3} + \frac{2}{3} + \frac{1}{3} + \frac{2}{3} + \frac{1}{3} + \frac{2}{3}\right)}{10} = 0.6$$

It is no coincidence that the mean of all the sample proportions in the sampling distribution of $\hat{p}$ exactly equals the population proportion of females for the original population, $p = 3/5 = 0.6$. We generalize this result in Fact 5. ■

Fact 5: Mean of the Sampling Distribution of the Sample Proportion $\hat{p}$

The mean of the sampling distribution of the sample proportion $\hat{p}$ is the value of the population proportion p. This may be denoted as $\mu_{\hat{p}} = p$ and read as "the mean of the sampling distribution of $\hat{p}$ is p."

Fact 5 provides a measure of center for the sampling distribution of the sample proportion $\hat{p}$, and Fact 6 provides a measure of spread. We will use these and other results as we build toward the Central Limit Theorem for Proportions.

Note: Just as for $\sigma_{\bar{x}}$ (see page 324), the *finite population correction factor*

$$\sqrt{\frac{N - n}{N - 1}}$$

should be used when the population is not much larger than the sample.

Fact 6: Standard Deviation of the Sampling Distribution of the Sample Proportion p

The standard deviation of the sampling distribution of the sample proportion $\hat{p}$ is $\sigma_{\hat{p}} = \sqrt{\dfrac{p \cdot (1 - p)}{n}}$, where p is the population proportion and n is the sample size.

| Example 7.14 | Color blindness in men |

The National Institutes of Health reported that color blindness linked to the X chromosome afflicts 8% of men. Suppose we take a random sample of 100 men and let p denote the proportion of men in the population who have color blindness linked to the X chromosome. Find $\mu_{\hat{p}}$ and $\sigma_{\hat{p}}$.

Solution

First, we note that this is a binomial experiment with $p = 0.08$ and $n = 100$. Fact 5 tells us that $\mu_{\hat{p}} = p$, that is, the sampling distribution of the sample proportion $\hat{p}$ has a mean of $p = 0.08$. Fact 6 states that

$$\sigma_{\hat{p}} = \sqrt{\frac{p \cdot (1 - p)}{n}} = \sqrt{\frac{0.08 \cdot (1 - 0.08)}{100}} = \sqrt{0.000736} \approx 0.02713$$

What Do These Numbers Mean?

Imagine that we repeatedly draw random samples of 100 men and observe the proportion of men $\hat{p}$ in each sample who have color blindness linked to the X chromosome. Each sample provides us with a value for $\hat{p}$. Eventually, the values of $\hat{p}$, when graphed, form the sampling distribution shown in Figure 7.17.

FIGURE 7.17 Sampling distribution of sample proportion $\hat{p}$.

Note that $\mu_{\hat{p}} = p = 0.08$ is located at the balance point of this distribution, which we should expect since the mean proportion of these samples is $\mu_{\hat{p}} = p = 0.08$. Each arrow represents one standard deviation of this distribution, a measure of spread given by $\sigma_{\hat{p}} = 0.02713$. Note that nearly all the sample proportions lie within three standard deviations of the mean. ■

Sampling Distribution of $\hat{p}$ for Extreme and Moderate Values of p

Next, we investigate whether we can develop a Central Limit Theorem for Proportions using a simulation study as we did for the means.

Example 7.15   Simulation study: Proportion of high-calorie foods

We return to the **Nutrition** data set, this time examining the calorie content in the 961 foods. We define a high-calorie food item as one that contains at least 2000 calories. There are $X = 20$ high-calorie foods in the data set, shown in Table 7.7.

Defining "success" as a high-calorie food item, we find that the population proportion of successes is $p = \dfrac{20}{961} \approx 0.02081$. The very small proportion was chosen deliberately as an extreme case.

Table 7.7 Food items with 2000 calories or more

Food item	Serving size	Calories
Carrot cake with cream cheese frosting	1 cake	6175
Fruitcake, dark, homemade	1 cake	5185
White cake with white frosting, commercial	1 cake	4170
Sheet cake with white frosting, homemade	1 cake	4020
Yellow cake with chocolate frosting, commercial	1 cake	3895
Devil's food cake with chocolate frosting, mix	1 cake	3755
Yellow cake with chocolate frosting, mix	1 cake	3735
Pecan pie	1 pie	3450
Cheesecake	1 cake	3350
Sheet cake without frosting	1 cake	2830
Vanilla ice cream, 16% fat (rich)	½ gallon	2805
Cream pie	1 pie	2710
Cherry pie	1 pie	2465
Apple pie	1 pie	2420
Peach pie	1 pie	2410
Blueberry pie	1 pie	2285
Sherbet, 2% fat	½ gallon	2160
Vanilla ice cream, 11% fat	½ gallon	2155
Lemon meringue pie	1 pie	2140
Pound cake, homemade	1 loaf	2025

Generate 100 samples of size $n = 50$ and find the proportion $\hat{p}$ of each sample. Then, repeat the process for sample size $n = 250$. Assess the normality of each sampling distribution of $\hat{p}$.

Solution

* $n = 50$: The sampling distribution of $\hat{p}$ for these 100 samples is skewed (Figure 7.18a).
* $n = 250$: Generating 100 samples of size $n = 250$, we seem to have achieved approximate normality (Figure 7.18b).

(a) Sample proportions for sample size $n = 50$ still skewed

(b) Approximate normality achieved at $n = 250$

FIGURE 7.18 Sampling distribution of sample proportion for $n = 50$ and 250.

Approximate normality for the sampling distribution of the sample proportion $\hat{p}$ is achieved when the sample size has attained the rather dizzying size of $n = 250$. This is more than eight times larger than the sample size ($n = 30$) needed for the sampling distribution for sample means. Clearly, we need a new rule of thumb for proportions. Recall that the population proportion of high-calorie foods was very low, $p = 0.02081$, which is an extreme case. For values of p near 0.50, the sample size required for approximate normality is much smaller. ■

We thus have the following conditions for the approximate normality of the sampling distribution of $\hat{p}$.

Fact 7: Conditions for Approximate Normality of the Sampling Distribution of the Sample Proportion $\hat{p}$

The sampling distribution of the sample proportion $\hat{p}$ may be considered approximately normal only if both the following conditions hold:

$$(1)\ np \geq 5 \qquad \text{and} \qquad (2)\ n(1 - p) \geq 5$$

The *minimum sample size* required to produce approximate normality in the sampling distribution of $\hat{p}$ is the *larger* of either

$$n_1 = \frac{5}{p} \qquad \text{or} \qquad n_2 = \frac{5}{(1 - p)}$$

(rounded up to the next integer).

---

③ Central Limit Theorem for Proportions

We are finally ready to express the Central Limit Theorem for Proportions, using information from Facts 5, 6, and 7.

Central Limit Theorem for Proportions

The sampling distribution of the sample proportion $\hat{p}$ follows an approximately normal distribution with mean $\mu_{\hat{p}} = p$ and standard deviation $\sigma_{\hat{p}} = \sqrt{\dfrac{p \cdot (1 - p)}{n}}$ when both the following conditions are satisfied: (1) $np \geq 5$ *and* (2) $n(1 - p) \geq 5$.

Example 7.16 Color blindness in men

In Example 7.14, we learned that color blindness linked to the X chromosome afflicts 8% of men. Describe the sampling distribution of $\hat{p}$, the proportion of men who have color blindness linked to the X chromosome, for samples of size (a) 50 and (b) 100.

Solution

We need to check both conditions to find whether the sampling distribution of $\hat{p}$ is approximately normal.

a. We are given that $p = 0.08$ and $n = 50$.

$$np = 50 \cdot 0.08 = 4 \qquad \text{and} \qquad n(1 - p) = 50 \cdot (0.92) = 46$$

Since 4 is not ≥ 5, the first condition is not satisfied. The Central Limit Theorem for Proportions cannot be used. We cannot conclude that the sampling distribution of $\hat{p}$ is approximately normal.

b. Here $p = 0.08$ and $n = 100$.

$$np = 100 \cdot 0.08 = 8 \quad \text{and} \quad n(1 - p) = 100 \cdot (0.92) = 92$$

Since both 8 and 92 are ≥ 5, both conditions are satisfied. The Central Limit Theorem for Proportions takes effect, and we can conclude that the sampling distribution of $\hat{p}$ is approximately normal. From Example 7.14 we have $\mu_{\hat{p}} = 0.08$ and $\sigma_{\hat{p}} = 0.02713$. Thus, the sampling distribution of $\hat{p}$ is approximately normal with $\mu_{\hat{p}} = 0.08$ and $\sigma_{\hat{p}} = 0.02713$. ■

Example 7.17   Unemployment rate in Texas, March 2007

The Texas Workforce Commission reported that the state unemployment rate in March 2007 was 4.3%. Let $p = 0.043$ represent the population proportion of unemployed workers in Texas.

a. Find the minimum size of the samples that produces a sampling distribution of $\hat{p}$ that is approximately normal.

b. Describe the sampling distribution of $\hat{p}$ if we use this minimum sample size.

Solution

a.

***W**hat Results Might We Expect?*

The proportion of unemployed workers in Texas is only 0.043, compared to 0.957 for the proportion of employed workers. Since these proportions are relatively imbalanced, we would expect the minimum sample size required to be fairly large. ■

Using Fact 7, the minimum sample size required is the *larger* of either

$$n_1 = \frac{5}{p} \quad \text{or} \quad n_2 = \frac{5}{1 - p}$$

Here

$$n_1 = \frac{5}{p} = \frac{5}{0.043} \approx 116.3 \quad \text{and} \quad n_2 = \frac{5}{1 - p} = \frac{5}{0.957} \approx 5.2$$

The larger of n_1 and n_2 is $n_1 = 116.3$. However, it is unclear what "0.3" of a worker means. So we round up to the next integer: $n = 117$. Therefore, the minimum sample size required to produce a sampling distribution of $\hat{p}$ that is approximately normal is $n = 117$ Texas workers. We confirm that this satisfies our conditions:

$$np = (117)(0.043) = 5.031 \geq 5, \quad \text{and} \quad n(1 - p) = (117)(0.957) = 111.969 \geq 5$$

This sample size of $n = 117$ confirms our earlier expectation that this required sample size would be fairly large compared to the means case, where the required sample size is $n = 30$.

b. We have $\mu_{\hat{p}} = 0.043$ and

$$\sigma_{\hat{p}} = \sqrt{\frac{p(1 - p)}{n}} = \sqrt{\frac{0.043(0.957)}{117}} \approx \sqrt{0.00035172} \approx 0.01875$$

Since the conditions are met, we invoke the Central Limit Theorem for Proportions. The sampling distribution of the sample proportion $\hat{p}$ is an approximately normal $\left(p, \sqrt{\frac{p(1 - p)}{n}}\right)$ distribution. That is, normal $\left(0.043, \sqrt{\frac{0.043(0.957)}{117}}\right)$, or normal $(\mu_{\hat{p}} = 0.043, \sigma_{\hat{p}} = 0.01875)$. ■

In those cases where we determine that the sampling distribution of $\hat{p}$ is approximately normal, we can then proceed to determine probabilities or find percentiles using the normal distribution methods we have learned in Chapter 6. Fact 8 is similar to Fact 4.

Fact 8: Standardizing a Normal Sampling Distribution for Proportions

When the sampling distribution of $\hat{p}$ is normal (or approximately normal), we can standardize to produce the standard normal Z:

$$Z = \frac{\hat{p} - \mu_{\hat{p}}}{\sigma_{\hat{p}}} = \frac{\hat{p} - p}{\sqrt{\dfrac{p(1-p)}{n}}}$$

where p is the population proportion of successes and n is the sample size.

Example 7.18 Using the sampling distribution of $\hat{p}$ to find a probability

Using the information in Example 7.17, find the probability that a sample of Texas workers will have a proportion unemployed greater than 9% for samples of size (a) 30 respondents and (b) 117 respondents.

Solution

a. We found in Example 7.17a that this sample size of $n = 30$ does not meet the minimum sample size required for the sampling distribution of $\hat{p}$ to be approximately normal, so we cannot conclude that the sampling distribution of $\hat{p}$ is approximately normal. Thus, we cannot solve this problem.

b. From Example 7.17b, the sampling distribution of $\hat{p}$ is approximately normal with mean $\mu_{\hat{p}} = 0.043$ and standard deviation $\sigma_{\hat{p}} = 0.01875$. We are then faced with a normal probability problem similar to those in Section 6.5 (page 295). Figure 7.19 shows the sampling distribution of $\hat{p}$ and the probability we are interested in, $P(\hat{p} > 0.09)$. Using Fact 8, we standardize as follows:

$$Z = \frac{0.09 - \mu_{\hat{p}}}{\sigma_{\hat{p}}} = \frac{0.09 - 0.043}{0.01875} \approx 2.51$$

Thus, $P(\hat{p} > 0.09) = P(Z > 2.51)$, as shown in Figure 7.20.

FIGURE 7.19

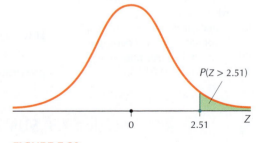

FIGURE 7.20

Following Table 6.7 (page 285), we look up $Z = 2.51$ in the Z table and subtract this table area (0.9940) from 1 to get the desired tail area. That is,

$$P(Z > 2.51) = 1 - 0.9940 = 0.0060$$

So the probability that the sample proportion of unemployed Texas workers will exceed 0.09 is 0.0060. ∎

Example 7.19   Using the sampling distribution of $\hat{p}$ to find a percentile

Using the information from Example 7.17, find the 99th percentile of the sampling distribution of $\hat{p}$ for $n = 117$.

Solution

The 99th percentile shown in Figure 7.21 separates the top 1% of sample proportions from the lower 99%. Thus, the area to the left of the 99th percentile is 0.99. We look up $Z = 0.99$ on the inside of the Z table, and the closest value we can find is 0.9901. The Z-value associated with 0.9901 is 2.33.

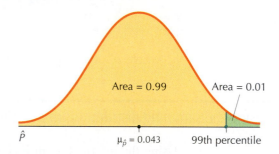

FIGURE 7.21

We need to transform this Z-value back to the scale of sample proportions. Use

$$\hat{p} = Z \cdot \sigma_{\hat{p}} + \mu_{\hat{p}} = (2.33)(0.01875) + 0.043 \approx 0.0867$$

The 99th percentile of the sampling distribution of $\hat{p}$ is 0.0867. ■

Example 7.20   Pitfalls of using an approximation

Using the information from Example 7.17, find the 1st percentile of the sampling distribution of $\hat{p}$ for $n = 117$.

Solution

Note: What can we do to estimate the 1st percentile? One way is to use simulation. Generate samples of size $n = 117$ from the population of the original survey respondents, record the sample proportion from each, and simply choose the 1st percentile. Proceeding in this manner, our estimate of the 1st percentile is 0.0128.

By symmetry, the 1st percentile will be the same distance below the mean that the 99th percentile is above the mean. The 99th percentile, 0.0867, lies $(0.0867 - 0.043) = 0.0437$ above the mean. Therefore, the 1st percentile lies 0.0437 below the mean:

$$\hat{p} = (0.043 - 0.0437) = -0.0007$$

However, this value of -0.0007 is negative and cannot represent a sample proportion of any kind. This is because the normality of the sampling distribution of $\hat{p}$ is only approximate and not exact. ■

Section 7.3 SUMMARY

1 The sampling distribution of the sample proportion $\hat{p}$ for a given sample size n consists of the collection of the sample proportions of all possible samples of size n from the population.

2 The approximate normality of the sampling distribution of the sample proportion kicks in much quicker when the population proportion is moderate rather than extreme.

3 According to the Central Limit Theorem for Proportions, the sampling distribution of the sample proportion $\hat{p}$ follows an approximately normal distribution with mean $\mu_{\hat{p}} = p$ and standard deviation $\sigma_{\hat{p}} = \sqrt{p \cdot (1 - p)/n}$ when both the following conditions are satisfied: (1) $np \geq 5$ and (2) $n(1 - p) \geq 5$. We can use Fact 8 to find probabilities and percentiles for sample proportions.

Section 7.3 EXERCISES

CLARIFYING THE CONCEPTS

1. For the rule of thumb governing sample size given in Fact 7, explain why we can't just use what we did in Section 7.2, n at least 30.

2. Explain in your own words the meaning of the Central Limit Theorem for Proportions.

PRACTICING THE TECHNIQUES

In Exercises 3–6, samples are taken. Find (a) $\mu_{\hat{p}}$ and (b) $\sigma_{\hat{p}}$, and (c) determine whether the sampling distribution of $\hat{p}$ is approximately normal or unknown.

3. $p = 0.5, n = 100$
4. $p = 0.5, n = 5$
5. $p = 0.01, n = 100$
6. $p = 0.01, n = 500$

In Exercises 7–12, find the minimum sample size that produces a sampling distribution of $\hat{p}$ that is approximately normal.

7. $p = 0.5$
8. $p = 0.25$
9. $p = 0.1$
10. $p = 0.05$
11. $p = 0.01$
12. $p = 0.001$

For Exercises 13–18, if possible find the indicated probability. If it is not possible, explain why not.

13. $p = 0.5, n = 100, P(\hat{p} > 0.55)$
14. $p = 0.5, n = 5, P(\hat{p} > 0.55)$
15. $p = 0.01, n = 100, P(\hat{p} > 0.011)$
16. $p = 0.01, n = 500, P(\hat{p} > 0.011)$
17. $p = 0.9, n = 40, P(0.88 < \hat{p} < 0.91)$
18. $p = 0.9, n = 50, P(0.88 < \hat{p} < 0.91)$

For Exercises 19–24, find the indicated value of $\hat{p}$. If it is not possible, explain why not.

19. $p = 0.5, n = 100$, value of $\hat{p}$ larger than 90% of all values of $\hat{p}$
20. $p = 0.5, n = 400$, value of $\hat{p}$ larger than 90% of all values of $\hat{p}$
21. $p = 0.9, n = 64$, 95th percentile of values of $\hat{p}$
22. $p = 0.9, n = 144$, 95th percentile of values of $\hat{p}$
23. $p = 0.1, n = 64$, 10th percentile of values of $\hat{p}$
24. $p = 0.1, n = 144$, 10th percentile of values of $\hat{p}$

APPLYING THE CONCEPTS

25. **The Green Party.** At a certain university, 25% of students support the Green Party. Suppose we take samples of size 100 students and are interested in the population proportion of students who support the Green Party.
    a. What is the mean of the sampling distribution of $\hat{p}$?
    b. What is the standard deviation of the sampling distribution of $\hat{p}$?
    c. Describe the sampling distribution of $\hat{p}$.

26. **Women's Radio Preferences.** In their 2001 study *What Women Want: Five Secrets to Better Ratings,* the Arbitron company reported that "Music I Like" is the biggest factor in deciding which radio station to tune to, chosen by 87% of women.
    a. Find the minimum sample size $n*$ that produces a sampling distribution of $\hat{p}$ that is approximately normal.
    b. Confirm that this sample size satisfies the conditions in Fact 7.
    c. Describe the sampling distribution of $\hat{p}$ if we use this minimum sample size. Which fact allows us to say this?
    d. Find $\mu_{\hat{p}}$ and $\sigma_{\hat{p}}$ for $n = 50$.
    e. Find the probability that, in a sample of 50 women, more than 45 chose "Music I Like" as the reason they decide which radio station to tune to.

27. **The Green Party.** Refer to Exercise 25.
    a. Find the probability that the sample mean proportion supporting the Green Party exceeds 0.30.
    b. Find the 95th percentile of sample proportions.

28. **Shaquille O'Neal.** In the 2003–2004 National Basketball Association regular season, Shaquille O'Neal led the league with a 58.4% field goal percentage.
    a. Find the minimum sample size that produces a sampling distribution of $\hat{p}$ that is approximately normal.
    b. Find $\mu_{\hat{p}}$ and $\sigma_{\hat{p}}$ for $n = 50$.
    c. Find the probability that, in a sample of 200 shots, Shaquille would score more than 120 baskets.

29. **Women's Access to the Internet.** The Pew Internet and American Life Project reported in 2003 that 61% of women had access to the Internet. Consider a sample of 100 women.
    a. Confirm that the Central Limit Theorem for Proportions applies.
    b. Find $\mu_{\hat{p}}$ and $\sigma_{\hat{p}}$.
    c. Find the probability that fewer than 60 of these 100 women had access to the Internet.

30. **High-Calorie Foods.** Refer to Example 7.15.
    a. Find the values of $\sigma_{\hat{p}}$ for the data in the graphs in Figure 7.18.
    b. Describe the relationship between $\sigma_{\hat{p}}$ and sample size.

Women in Management Courses. Use this information for Exercises 31–34. Suppose we are interested in the population proportion of women in management courses and we take samples of size $n = 20$ from a population of management courses. Assume that the population proportion of women in management courses is 0.40.

31. Follow steps (a)–(e).
    a. Find the minimum sample size n that produces a sampling distribution of $\hat{p}$ that is approximately normal.

b. Confirm that this sample size satisfies the conditions in Fact 7.

c. Describe the sampling distribution of $\hat{p}$ if we use this minimum sample size. What allows us to say this?

d. Find $\mu_{\hat{p}}$ and $\sigma_{\hat{p}}$ for $n = 20$.

e. Find, if possible, the probability that a sample of 20 people in management courses will have a proportion of women that is smaller than 38%. If it is not possible, clearly explain why not, and suggest what we should do.

32. Refer to Exercise 31.

a. Explain in your own words what we mean by the sampling distribution of the sample proportion of women in management courses.

b. Where will the sampling distribution of the sample proportion be centered?

c. Provide the value of a measure of spread of the sampling distribution.

d. Suppose a particular management course has 65% women. Is this unusual? How do we determine whether this is unusual?

e. Suppose a particular management course has 75% women. Is this unusual? How do we determine whether this is unusual?

33. Refer to Exercise 32.

a. Suppose we find the measure of spread in Exercise 32 to be too large and we would like a more precise measure of spread. How can we reduce the size of the measure of spread?

b. Provide a value of n that would result in a smaller standard deviation of this statistic.

34. Refer to Exercise 33b. Suppose a particular management course has 65% women.

a. Is this proportion to be considered unusual with our new value for the standard deviation for the statistic?

b. How does this proportion compare with the first time you checked whether it was unusual?

c. What has changed?

Try This in Class! In-class activities to enhance your understanding of statistics

SAMPLING DISTRIBUTIONS

Perform the following activity as a class. Use the *Random Sample* applet to generate the random samples requested.

1. Each student in the class should write down an estimate of the length of the following line segment. Use your eyes only! No rulers.

<hr style="border: 1px solid red;">

2. Consider your class to be a population. Find the mean μ and the standard deviation σ of all student estimates.

3. Choose one student at random from the class. Ask for this student's estimate.

4. Take a random sample of two students from the class, and find the mean $\bar{x}$ of their estimates. What is the mean

$\mu_{\bar{x}}$ and standard deviation $\sigma_{\bar{x}}$ of the sampling distribution of $\bar{x}$? Given these values for $\mu_{\bar{x}}$ and $\sigma_{\bar{x}}$, which values for the length of the line segment would be considered unusual? (*Hint:* Which would be outliers?)

5. Take a random sample of 10 students from the class, and find the mean $\bar{x}$ of their estimates. What are $\mu_{\bar{x}}$ and $\sigma_{\bar{x}}$ for $n = 10$? Now, which values for the length of the line segment would be considered unusual?

6. Take a random sample of 20 students and find their estimates. What are $\mu_{\bar{x}}$ and $\sigma_{\bar{x}}$? Now, which values for the length of the line segment would be considered unusual?

7. As the sample size increases, what happens to $\sigma_{\bar{x}}$, and what happens to the estimates that are considered unusual? How can you state this phenomenon in terms of sample size and precision of the estimate?

CHAPTER 7 FORMULAS AND VOCABULARY

SECTION 7.1

- **Mean of the sampling distribution of the sample mean** $\bar{x}$ (p. 324). Denoted as $\mu_{\bar{x}} = \mu$.
- **Normal probability plot** (p. 326)
- **Parameter** (p. 321)
- **Point estimation** (p. 321)
- **Sampling distribution of the sample mean** $\bar{x}$ (p. 323)
- **Sampling error** (p. 321)

- **Standard deviation of the sampling distribution of the sample mean** $\bar{x}$ (p. 324). $\sigma_{\bar{x}} = \sigma/\sqrt{n}$.
- **Standardizing a normal sampling distribution for means** (p. 326).

$$Z = \frac{\bar{x} - \mu_{\bar{x}}}{\sigma_{\bar{x}}} = \frac{\bar{x} - \mu}{\sigma/\sqrt{n}}$$

- **Statistic** (p. 321)

$$Z = \frac{p - \mu_{\hat{p}}}{\sigma_{\hat{p}}} = \frac{\hat{p} - p}{\sqrt{\dfrac{p(1 - p)}{n}}}$$

CHAPTER 7 REVIEW EXERCISES

SECTION 7.1
For Exercises 1–3, find the sampling error.
1. $\mu = 95, \bar{x} = 95$
2. $\sigma = 5, s = 7$
3. $p = 0.1, \hat{p} = 0$

For Exercises 4 and 5, find $\mu_{\bar{x}}$ and $\sigma_{\bar{x}}$, the mean and standard deviation of the sampling distribution of $\bar{x}$.
4. $\mu = 50, \sigma = 40, n = 4$
5. $\mu = 50, \sigma = 40, n = 16$

For Exercises 6–9, assume that X is normal(10, 4) and $n = 25$.
6. Find the sampling distribution of $\bar{x}$ for $n = 25$.
7. Find the probability that $\bar{x}$ exceeds 11.
8. Without using your calculator, find the probability that $\bar{x}$ is less than 9.
9. Without using your calculator, find the probability that $\bar{x}$ lies between 9 and 11.

SECTION 7.2
For Exercises 10 and 11, if possible find the indicated probability. If it is not possible, explain why not.
10. Scores on a psychological test are not normally distributed, with $\mu = 100$ and $\sigma = 15$. A sample of size 25 is taken. Find $P(94 < \bar{x} < 103)$.
11. Scores on a psychological test are normally distributed, with $\mu = 100$ and $\sigma = 15$. A sample of size 25 is taken. Find $P(94 < \bar{x} < 103)$.

For Exercises 12 and 13, find the indicated value of $\bar{x}$. If it is not possible, explain why not.
12. Scores on a psychological test are not normally distributed, with $\mu = 100$ and $\sigma = 15$. A sample of size 25 is taken. Find the 50th percentile of sample means.
13. Scores on a psychological test are normally distributed, with $\mu = 100$ and $\sigma = 15$. A sample of size 25 is taken. Find the 50th percentile of sample means.
14. **Cocaine and Heart Attacks.** The American Medical Association reported: "During the first hour after using cocaine, the user's risk of heart attack increases nearly 24 times. The average age of people in the study who suffered heart attacks soon after using cocaine was only 44. That's about 17 years younger than the average heart attack patient. Of the 38 cocaine users who had heart attacks, 29 had no prior symptoms of heart

disease."[3] Assume that the standard deviation of the age of people who suffered heart attacks soon after using cocaine was 10 years and we take a sample of size 38.
   a. Find the 97.5th percentile of the mean age at heart attack after using cocaine.
   b. Find the 2.5th percentile of the mean age at heart attack after using cocaine.
   c. Between which two sample mean ages that are symmetric about the population mean lie 95% of mean ages of all people who suffered heart attacks soon after using cocaine?
   d. By hand, sketch a plot of how this would look.

SECTION 7.3
For Exercises 15 and 16, if possible find the indicated probability. If it is not possible, explain why not.
15. $p = 0.1, n = 40, P(\hat{p} < 0.12)$
16. $p = 0.1, n = 50, P(\hat{p} < 0.12)$

For Exercises 17 and 18, find the indicated value of $\hat{p}$. If it is not possible, explain why not.
17. $p = 0.02, n = 400$, the value of $\hat{p}$ smaller than 75% of all p values
18. $p = 0.02, n = 625$, the value of $\hat{p}$ smaller than 75% of all p values
19. **Living with AIDS.** The Centers for Disease Control reported that, in 2004, 11% of white males living with AIDS contracted it through injection drug use. A sample of 120 white males living with AIDS is examined.
   a. Confirm that the Central Limit Theorem for Proportions applies.
   b. Find $\mu_{\hat{p}}$ and $\sigma_{\hat{p}}$.
   c. Find the probability that less than 10 of the 120 males contracted AIDS through injection drug use.
20. **Small Business Jobs.** According the the U.S. Small Business Administration, small businesses provide 75% of the net new jobs added to the economy. Consider a sample of 1000 new jobs.
   a. Confirm that the Central Limit Theorem for Proportions applies.
   b. Find $\mu_{\hat{p}}$ and $\sigma_{\hat{p}}$.
   c. Find the probability that more than 750 of the 1000 new jobs were provided by a small business.

The confidence level 95% is shown, along with the critical Z value, $Z_{\alpha/2} = 1.96$. The confidence interval is then shown:

lower bound = 22,478.96, upper bound = 25,556.43

This 99% confidence interval can also be expressed as (22,478.96, 25,556.43). ∎

4 Margin of Error

Suppose we wish to estimate the population mean μ to within a certain measure of accuracy. For example, when we estimate the mean age of when children learn to read, we might estimate it to within 1 year. This estimate of accuracy (1 year, in this case) is called the **margin of error**. In Example 8.1, we found that we are 95.44% confident that μ, the population mean number of runs per game for all MLB teams in 2006, lies within 1 run per game of our sample mean of $\bar{x} = 4.5$ runs per game:

$$\text{lower bound} = \bar{x} - Z_{\alpha/2}(\sigma/n) = 4.5 - 2(3/\sqrt{36}) = 4.5 - 1 = 3.5$$
$$\text{upper bound} = \bar{x} - Z_{\alpha/2}(\sigma/n) = 4.5 + 2(3/\sqrt{36}) = 4.5 + 1 = 5.5$$

In these calculations, note that we are adding (+) and subtracting (−) the same quantity $Z_{\alpha/2}(\sigma/\sqrt{n}) = 1$ to and from $\bar{x}$. The quantity $Z_{\alpha/2}(\sigma/\sqrt{n}) = 1$ is called the *margin of error E*, and the Z interval

$$(\bar{x} - Z_{\alpha/2}(\sigma/\sqrt{n}), \bar{x} + Z_{\alpha/2}(\sigma/\sqrt{n}))$$

can be written

$$\bar{x} \pm Z_{\alpha/2}(\sigma/\sqrt{n})$$

which consists of the point estimate $\bar{x}$ plus and minus the margin of error $Z_{\alpha/2}(\sigma/\sqrt{n})$.

> **CAUTION** Remember that the "±" notation *always* represents a pair of numbers.

Note the form that the confidence interval takes:

point estimate ± margin of error

Of course, we would like our confidence interval estimates to be as precise as possible. Therefore, we would like the margin of error to be as small as possible, which would in turn result in a shorter confidence interval. Shorter, tighter confidence intervals are better, since the likely maximum difference between the sample mean and the population mean is reduced.

> The **margin of error E** is a measure of the precision of the confidence interval estimate. For the Z interval, the margin of error takes the form
>
> $$E = Z_{\alpha/2}(\sigma/\sqrt{n})$$
>
> Interpret the margin of error E for a $(1 - \alpha)100\%$ confidence interval for μ as follows:
> "We can estimate μ to within E units with $(1 - \alpha)100\%$ confidence."

Note: When it comes to the margin of error E, smaller is better!

Example 8.5   Confidence interval as point estimate ± margin of error

In Example 8.3, the Z interval for the population mean lead contamination (in ppm) for all trout on the Spokane River is

$$\text{lower bound} = 1 - 1.96(0.5/\sqrt{100}) = 1 - 1.96(0.05) = 1 - 0.098 = 0.902$$
$$\text{upper bound} = 1 + 1.96(0.5/\sqrt{100}) = 1 + 1.96(0.05) = 1 + 0.098 = 1.098$$

a. Find the margin of error *E*.
b. Express the confidence interval in the form "point estimate ± margin of error."
c. Interpret the margin of error *E*.

Solution

a. We find the margin of error as follows:

$$E = Z_{\alpha/2}(\sigma/\sqrt{n}) = 1.96\,(0.5/\sqrt{100}) = 1.96\,(0.05) = 0.098$$

b. The point estimate is $\bar{x} = 1$. Thus, the 95% confidence interval for the population mean lead contamination (in ppm) for all trout on the Spokane River takes the following form:

$$\text{point estimate} \pm \text{margin of error}$$
$$= \bar{x} \pm Z_{\alpha/2}\,(\sigma/\sqrt{n})$$
$$= 1 \pm 0.098$$

c. We interpret *E* by saying that we can estimate the population mean lead contamination for all trout in the Spokane River to *within 0.098 ppm with 95% confidence.* ■

If a smaller margin of error is better, how do we make it smaller? Let's look at the margin of error for the *Z* interval:

$$E = Z_{\alpha/2}(\sigma/\sqrt{n})$$

Since the population standard deviation σ is fixed, only $Z_{\alpha/2}$ and *n* can vary. To make the margin of error smaller, we would need to make $Z_{\alpha/2}$ smaller, which is achieved by *decreasing the confidence level*. Next, notice that *n* is in the denominator of the formula. Therefore to make the margin of error smaller, we would need to make *n* larger, thereby *increasing the sample size*. Thus, there are two strategies for decreasing the margin of error:

1. Decrease the confidence level.
2. Increase the sample size.

Example 8.6   Decreasing the margin of error by decreasing the confidence level

For the confidence interval for the population mean lead contamination in Example 8.3, suppose we reduce the confidence level from 95% to 90% and leave everything else unchanged. Find the new margin of error. Describe how the margin of error has changed.

Solution

For confidence level 90%, $Z_{\alpha/2} = 1.645$, giving the following margin of error:

$$E = Z_{\alpha/2}(\sigma/\sqrt{n}) = 1.645\,(0.5/\sqrt{100}) \approx 0.082$$

Decreasing the confidence level from 95% to 90% decreases the margin of error from 0.098 to 0.082 ppm. ■

Developing
Your Statistical Sense

There's No Free Lunch

The margin of error in Example 8.6 is smaller than the one in Example 8.3. A smaller margin of error is good because it gives a more precise estimate of the unknown population mean. However, this smaller margin of error is due entirely to the decrease in the confidence level, which in turn decreased the value of $Z_{\alpha/2}$. And a decrease in the confidence level is not good. In statistical data analysis, there is rarely a free lunch. The trade-off here is that, while the margin of error went down, *so did the confidence level,* from 95% to 90%. For a given sample, the only difference in the formula between a 95% and a 90% confidence interval is the value of $Z_{\alpha/2}$, which is larger for higher confidence levels. The larger value of $Z_{\alpha/2}$ for the higher confidence level means that the margin of error $E = Z_{\alpha/2}(\sigma/\sqrt{n})$ is larger, which gives a wider confidence interval. Confidence intervals that are too wide are ineffectual and often useless. For example, we can be 99.9999% confident that the population mean age of college students in Florida lies between 15 and 75 years old. But, so what? The interval is too wide to be of practical use. More useful would be a 95% confidence interval that the population mean age of college students in Florida lies between 20 and 27. ■

This leads us to Strategy 2 for reducing the margin of error: increase the sample size. *The only way to have both high confidence and a tight interval is to boost the sample size.*

Example 8.7   **Decreasing the margin of error by increasing the sample size**

For the confidence interval for the population mean lead contamination in Example 8.3, suppose the results were based on a sample of size $n = 400$ rather than $n = 100$. Leaving everything else unchanged, find the new margin of error, and describe how the margin of error has changed.

Solution

For $n = 400$, the margin of error is

$$E = Z_{\alpha/2}(\sigma/\sqrt{n}) = 1.96\,(0.5/\sqrt{400}) = 0.049$$

Increasing the sample size from $n = 100$ to $n = 400$ has decreased the margin of error from 0.098 to 0.049 ppm. ■

"More data" is a familiar refrain in statistical analysis circles. Of course, increasing the sample size often raises pocketbook issues, since large samples can get very expensive ("We would like a large-sample estimate of the amount of damage sustained by Corvettes hitting a wall at 90 mph"). Sometimes obtaining large samples is simply impossible. Suppose an astronomer has developed a new technique for predicting corona effects during solar eclipses; she will have to wait a while (say, a few hundred years) to build up a large sample. So, take samples as large as realistically possible to keep the width of the confidence interval as narrow as possible.

5 Sample Size for Estimating the Population Mean

When samples are plentiful and cheap, *arbitrarily precise confidence intervals with arbitrarily high confidence are possible simply by taking sufficiently large samples.*

Therefore, the question arises: *How large a sample size do I need* to get a tight confidence interval with high confidence level?

Recall that the general form of the Z interval for the mean is

$$\text{point estimate} \pm \text{margin of error}$$

where the point estimate is the sample mean $\bar{x}$ and the margin of error $Z_{\alpha/2}(\sigma/\sqrt{n})$ measures the precision of the estimate. A smaller margin of error is associated with an estimate with higher precision. A larger margin of error has less precision and results in a wider confidence interval.

Example 8.8   Estimating the mean salary of business majors

Suppose we want to estimate to within $1000 the mean salary μ of all college graduates who were business majors. How many business majors would we sample to estimate the mean salary to within $1000 with 95% confidence?

Solution
"Within $1000" means that the margin of error E is $1000. Recall that the margin of error for 95% confidence is given by

$$E = 1.96\,(\sigma/\sqrt{n})$$

where 1.96 is the $Z_{\alpha/2}$ value associated with 95% confidence. Since the desired margin of error is 1000,

$$E = 1000 = 1.96\,(\sigma/\sqrt{n})$$

Note: We solve for n as follows:

$$1000 = 1.96\,(\sigma/\sqrt{n})$$

Multiply both sides by $\sqrt{n}$:

$$1000\sqrt{n} = 1.96\,\sigma$$

Divide both sides by 1000:

$$\sqrt{n} = \left(\frac{1.96\sigma}{1000}\right)$$

Square both sides to get the formula for n:

$$n = \left(\frac{1.96\sigma}{1000}\right)^2$$

To find the value of n that is the sample size that will produce an estimate within $1000 with 95% confidence, we solve for n, giving us:

$$n = \left(\frac{1.96\sigma}{1000}\right)^2$$

Suppose we know that $\sigma = \$5000$. Then:

$$n = \left(\frac{1.96 \cdot 5000}{1000}\right)^2 = 96.04$$

Thus, we need a sample size of $n = 97$ for a confidence level of 95%. ■

We generalize the result from Example 8.8 as follows.

> ### Sample Size for Estimating the Population Mean
> The sample size for a Z interval that estimates the population mean μ to within a margin of error E with confidence $100(1 - \alpha)\%$ is given by
>
> $$n = \left(\frac{(Z_{\alpha/2})\sigma}{E}\right)^2$$
>
> where $Z_{\alpha/2}$ is the value associated with the desired confidence level (Table 8.1), E is the desired margin of error, and σ is the population standard deviation. By convention, whenever this formula yields a sample size with a decimal, *always round up to the next whole number.*

Use the *Confidence Interval* applet for Exercises 49 and 50.

49. Set the confidence level to 90%. Click "Sample 50" to produce 50 simple random samples (SRSs) and display the resulting 90% confidence intervals for μ.

   a. What is the percent hit, that is, the proportion of the confidence intervals that actually contain the true value of μ?

   b. Keep clicking "Sample 50" until 1000 confidence intervals are generated. What is the percent hit?

   c. It is not likely (though it is possible) that the percent hit in **(b)** exactly equals 90%. Explain why the percent hit is not equal to 90% when we asked for a confidence level of 90%.

50. Set the confidence level to 99%. Click "Sample 50" to produce 50 99% confidence intervals for μ.

   a. Compare the widths of these intervals with the width of the 90% confidence intervals. Which set of intervals is wider?

   b. Explain why the intervals you identified in **(a)** are wider, using the concept of *margin of error*.

Use the *Normal Density Curve* applet for Exercises 51 and 52.

51. Use the applet to find $Z_{\alpha/2}$ critical values for unusual confidence levels. Select **2-Tail**, and click and drag the flags so that the central area and not the tail area is highlighted. Verify that the $Z_{\alpha/2}$ critical value for 95% confidence is 1.96.

52. Use the applet to find $Z_{\alpha/2}$ critical values for the following confidence levels.

   a. 80%

   b. 85%

   c. 98%

8.2 *t* Interval for the Mean

Objectives: **By the end of this section, I will be able to...**

1 Describe the characteristics of the *t* distribution.

2 Calculate and interpret a *t* interval for the mean, for either of two cases.

3 Compute and interpret the margin of error.

1 Introducing the *t* Distribution

In Section 8.1 we estimated the population mean assuming that the population standard deviation σ was known. This assumption may be valid for certain fields such as quality control. However, in many real-world problems, we do not know the value of σ. Therefore, we cannot use a Z interval to estimate the mean. In this section, we use the sample standard deviation s to estimate σ to construct a confidence interval that is likely to contain the population mean.

Suppose we do not know the value of the population standard deviation σ. What should a data analyst do? Estimate it! Recall that the sample standard deviation s may be used as an estimate of the population standard deviation σ. Fact 4 from Chapter 7 showed us that we could standardize $\bar{x}$ to derive the standard normal random variable:

$$Z = \frac{\bar{x} - \mu}{\sigma/\sqrt{n}}$$

Unfortunately, however, replacing the unknown σ in this equation with the known s, we can no longer obtain the standard normal Z because s is itself a random variable. Instead, $\dfrac{\bar{x} - \mu}{s/\sqrt{n}}$ follows an entirely new and different distribution, called the *t* **distribution**.

---

t Distribution

For a normal population, the distribution of

$$t = \frac{\bar{x} - \mu}{s/\sqrt{n}}$$

follows a *t* distribution, with $n - 1$ degrees of freedom, where $\bar{x}$ is the sample mean, μ is the unknown population mean, s is the sample standard deviation, and n is the sample size.

Developing Your Statistical Sense

Degrees of Freedom

Notice that the definition of the *t* distribution includes a new concept called *degrees of freedom*. Degrees of freedom is a measure that determines how the *t* distribution changes as the sample size changes. The idea of degrees of freedom is that, in a sum of *n* numbers, you need to know only the first $n - 1$ of these numbers to find the *n*th number because you already know the sum. For example, suppose you know that the sum of $n = 3$ numbers is 10 and are told that the first two numbers are 5 and 1. Then you can deduce that the last number is $10 - (5 + 1) = 4$. The first two numbers have the freedom to take on any values, but the third number must take a particular value. Thus, there are only $n - 1$ independent pieces of information. The concept is similar for the *t* distribution. Since we use the sample standard deviation *s* to estimate the unknown σ and since *s* is known, only $n - 1$ independent pieces of information are needed to find the value of *t*. Thus, we say that $t = \dfrac{\bar{x} - \mu}{s/\sqrt{n}}$ follows a *t* distribution with $n - 1$ degrees of freedom. ■

Figure 8.9 displays a comparison of some *t* curves with the *Z* curve. Note first of all that, although there is one and only one *Z* distribution (or curve), there are many different *t* curves. In fact, there is a different *t* curve for every different degrees of freedom (df), that is, for every different sample size. The degrees of freedom, $df = n - 1$, determines the shape of the *t* distribution, just as the mean and variance uniquely determine the shape of the normal distribution. All *t* curves have several characteristics in common.

FIGURE 8.9
Different *t* curve for different degrees of freedom ($df = n - 1$).

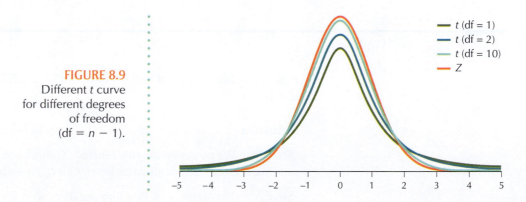

Characteristics of the t Distribution

- Centered at zero. The mean of t is zero, just as with Z.
- Symmetric about its mean zero, just as with Z.
- As df decreases, the t curve gets flatter, and the area under the t curve decreases in the center and increases in the tails. That is, the t curve has heavier tails than the Z curve.
- As df increases toward infinity, the t curve approaches the Z curve, and the area under the t curve increases in the center and decreases in the tails.

Similar to the definition of $Z_{\alpha/2}$ in Section 8.1, we can define $t_{\alpha/2}$ to be the value of the t distribution with area $\alpha/2$ to the right of it, as seen in Figure 8.10. Table 8.1 in Section 8.1 provides the $Z_{\alpha/2}$ values for certain common confidence levels. Unfortunately, because there is a different t curve for each sample size, there are many possible $t_{\alpha/2}$ values. You will need to use the t table in the Appendix (page T-11) to find the value of $t_{\alpha/2}$. Here is how to find $t_{\alpha/2}$ in the t table.

FIGURE 8.10
$t_{\alpha/2}$ has area $\alpha/2$ to the right of it.

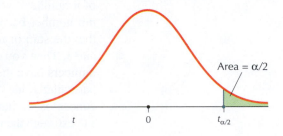

Area = $\alpha/2$

t 0 $t_{\alpha/2}$

Procedure for Finding $t_{\alpha/2}$

Step 1   Go across the row marked "Confidence level" in the t table (Table D in the Appendix, page T-11) until you find the column with the desired confidence level at the top. The $t_{\alpha/2}$ value is in this column somewhere.

Step 2   Go down the column until you see the correct number of degrees of freedom on the left. The number in that row and column is the desired value of $t_{\alpha/2}$.

Example 8.10   Finding $t_{\alpha/2}$

Find the value of $t_{\alpha/2}$ that will produce a 95% confidence interval for μ if the sample size is $n = 20$.

Solution

Step 1   We go across the row labeled "Confidence level" in the t table (Figure 8.11) until we see the 95% confidence level. Our $t_{\alpha/2}$ is somewhere in this column.

Step 2   The degrees of freedom are df $= n - 1 = 20 - 1 = 19$. We go down the column until we see 19 on the left. The number in that row is our $t_{\alpha/2}$, 2.093. ■

t-Distribution

df	80%	90%	95%	98%	99%
			Confidence level		
	80%	**90%**	**95%**	**98%**	**99%**
			Area in one tail		
	0.10	**0.05**	**0.025**	**0.01**	**0.005**
			Area in two tails		
	0.20	**0.10**	**0.05**	**0.02**	**0.01**
1	3.078	6.314	12.706	31.821	63.657
2	1.886	2.920	4.303	6.965	9.925
3	1.638	2.353	3.182	4.541	5.841
14	1.345	1.761	2.145	2.624	2.977
15	1.341	1.753	2.131	2.602	2.947
16	1.337	1.746	2.120	2.583	2.921
17	1.333	1.740	2.110	2.567	2.898
18	1.330	1.734	2.101	2.552	2.878
19	1.328	1.729	2.093	2.539	2.861
20	1.325	1.725	2.086	2.528	2.845
21	1.323	1.721	2.080	2.518	2.831

FIGURE 8.11 *t* table (excerpt).

t Interval for the Mean: Two Cases

The *t* distribution provides the following confidence interval for the unknown population mean μ, called the *t* **interval**.

---

t Interval for μ

A random sample of size n is taken from a population with unknown mean μ. A $100(1 - \alpha)\%$ confidence interval for μ is given by the interval

$$\text{lower bound} = \bar{x} - t_{\alpha/2}(s/\sqrt{n}), \text{ upper bound} = \bar{x} + t_{\alpha/2}(s/\sqrt{n})$$

where $\bar{x}$ is the sample mean, $t_{\alpha/2}$ is associated with the confidence level and $n - 1$ degrees of freedom, and s is the sample standard deviation. The *t* interval may also be written as

$$\bar{x} \pm t_{\alpha/2}(s/\sqrt{n})$$

and is denoted

$$\text{(lower bound, upper bound)}$$

The *t* interval applies whenever *either* of the following conditions is met:
- Case 1: The population is normal.
- Case 2: The sample size is large ($n \geq 30$).

Note: Suppose that σ is unknown, *and* the population is either non-normal or of unknown distribution, *and* the sample size is not large. Then we should not use the *t* interval. Rather, we need to turn to nonparametric methods, for example the *sign interval* or the *Wilcoxon interval*.

Example 8.11   Fourth-grade feet

Suppose a children's shoe manufacturer is interested in estimating the population mean length of fourth graders' feet. A random sample of 20 fourth graders' feet yielded the following foot lengths, in centimeters.[7]

| 22.4 | 23.4 | 22.5 | 23.2 | 23.1 | 23.7 | 24.1 | 21.0 | 21.6 | 20.9 |
| 25.5 | 22.8 | 24.1 | 25.0 | 24.0 | 21.7 | 22.0 | 22.7 | 24.7 | 23.5 |

Construct a 95% confidence interval for μ, the population mean length of all fourth-graders' feet.

Solution

We do not know the population standard deviation σ, so we cannot use the Z interval. We can construct a t interval whenever either the population is normal (Case 1) or the sample size is large (Case 2). The sample size here is 20, which is not large ($n \geq 30$), so Case 2 does not apply. The only remaining choice is to check for normality. Figure 8.12 shows the normal probability plot of the foot lengths. The points generally line up along the line, so the assumption of normality is validated for this data set. By Case 1, we can then proceed to construct the t interval for μ.

FIGURE 8.12 Fourth-grade foot lengths are normally distributed.

The TI-83/84 provides the summary statistics shown here, giving $n = 20, \bar{x} = 23.095$, and $s \approx 1.280$.

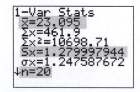

All that is left is to find $t_{\alpha/2}$. In Example 8.10, we found the value of $t_{\alpha/2}$ for confidence level = 95% and $n = 20$ to be $t_{\alpha/2} = 2.093$. The 95% confidence interval then becomes

$$\text{lower bound} = \bar{x} - t_{\alpha/2}(s/\sqrt{n})$$
$$= 23.095 - 2.093(1.280/\sqrt{20}) \approx 23.095 - 0.599 = 22.496$$

$$\text{upper bound} = \bar{x} + t_{\alpha/2}(s/\sqrt{n})$$
$$= 23.095 + 2.093(1.280/\sqrt{20}) \approx 23.095 + 0.599 = 23.694$$

We are 95% confident that the population mean length of fourth graders' feet lies between 22.496 and 23.694 cm. ■

*D*eveloping Your Statistical Sense

t Intervals May Offer More Peace of Mind than *Z* Intervals

In Example 8.11, if we had assumed that the population standard deviation σ was known ($\sigma = 1.280$), then the 95% *Z* interval for the population mean length of fourth-grade feet would have been

lower bound $= \bar{x} - Z_{\alpha/2}(\sigma/\sqrt{n})$
$$= 23.095 - 1.96(1.280/\sqrt{20}) \approx 23.095 - 0.561 = 22.534$$

upper bound $= \bar{x} + Z_{\alpha/2}(\sigma/\sqrt{n})$
$$= 23.095 + 1.96(1.280/\sqrt{20}) \approx 23.095 + 0.561 = 23.656$$

Note that this *Z* interval (22.534, 23.656) is only slightly more precise than the *t* interval (22.496, 23.694). However, the *Z* interval depends on prior knowledge of the value of σ. If the value of σ is inaccurate, then the *Z* interval will be misleading and overly optimistic. With even moderate sample sizes, reporting the *t* interval rather than the *Z* interval may offer peace of mind to the data analyst. ■

If the degrees of freedom needed to find $t_{\alpha/2}$ do not appear in the df column of the *t* table, a conservative solution is to take the next row with smaller df. Alternatively, we can use interpolation. Both methods are illustrated in Example 8.12.

Example 8.12 Water use in Israel: degrees of freedom not in the *t* table

© Ed Kashi/CORBIS

Water scarcity is a continuing serious problem for the countries in the Middle East, including Israel. The Jordan River (shown here) feeds the Sea of Galilee, which provides Israel with 40% of its potable water and 80% of the water used for agricultural purposes. The mean daily water consumption in Israel is reported to be 72 cubic meters per person.[8] Suppose a random sample of 49 Israelis showed a sample mean daily water consumption of 72 cubic meters, with a sample standard deviation of 28 cubic meters. Construct a 99% confidence interval for the population mean daily water consumption for all Israelis. Use (a) the conservative method and (b) interpolation.

Solution

Since the population standard deviation is unknown and the sample size is large, we proceed to construct the *t* interval for μ. Three of the four numbers we need to complete the *t* interval are sample size $n = 49$, sample mean $\bar{x} = 72$, and sample standard deviation $s = 28$. Now we must find $t_{\alpha/2}$. The confidence level is 99% and the degrees of freedom are $n - 1 = 49 - 1 = 48$. *Unfortunately, the value of 48 for the df does not appear in the df column.*

a. The next row with df smaller than 48 would be df = 40. Thus, the "conservative" $t_{\alpha/2}$ is 2.704. We then proceed to construct the 99% confidence interval:

$$\bar{x} \pm t_{\alpha/2}(s/\sqrt{n}) = 72 \pm 2.704(28/\sqrt{49}) = 72 \pm 10.816 = (61.184, 82.816)$$

b. Alternatively, you could interpolate, as follows. Since df = 48 is 8/10 of the distance between 40 and 50, we can estimate $t_{\alpha/2}$ by taking 8/10 of the distance between the associated *t* values for df = 40 and df = 50 and subtracting that result from the *t* value for df = 40:

$$\frac{8}{10}\left[(t_{\alpha/2} \text{ for df} = 40) - (t_{\alpha/2} \text{ for df} = 50)\right] = \frac{8}{10}(2.704 - 2.678) = 0.0208$$

Thus, $t_{\alpha/2}$ for df = 48 would be $2.704 - 0.0208 = 2.6832$, using interpolation. The 99% confidence interval using interpolation is thus

$$\bar{x} \pm t_{\alpha/2}(s/\sqrt{n}) = 72 \pm 2.6832(28/\sqrt{49}) = 72 \pm 10.7328 = (61.2672, 82.7328)$$

Note that the confidence interval using the conservative method is somewhat wider, reflecting the conservative choice of $t_{\alpha/2}$. ■

c. The 95% confidence interval is
$$\text{point estimate} \pm \text{margin of error}$$
$$= \hat{p} \pm Z_{\alpha/2}\sqrt{\frac{\hat{p}(1-\hat{p})}{n}}$$
$$= \hat{p} \pm E$$
$$\approx 0.63 \pm 0.03$$
$$= (\text{lower bound} = 0.60, \text{ upper bound} = 0.66)$$

Thus, we are 95% confident that the population proportion of all American adults who think that there should not be such a law lies between 60% and 66%. ■

Developing
Your Statistical Sense

Famous "Plus or Minus 3 Points"

Note that this confidence interval was obtained by adding and subtracting 3% from the 63% point estimate. That is, the poll has a margin of error of $E = 3$ percentage points $= 0.03$. This is the famous "plus or minus 3 percentage points" used in many news reports. However, newscasters rarely announce the confidence level of the poll. National pollsters almost always use 95% as their confidence level and usually try to select the sample size necessary to create a margin of error of about 3%. We learn how they do this next. ■

4 Sample Size for Estimating the Population Proportion

Next we consider the question: How large a sample size do I need to estimate the population proportion p to within margin of error E with $100(1 - \alpha)\%$ confidence? The margin of error of the confidence interval for proportions equals

$$E = Z_{\alpha/2} \cdot \sqrt{\frac{\hat{p}(1-\hat{p})}{n}}$$

To derive the formula for sample size, solve for the required sample size n:

$$E = Z_{\alpha/2}\sqrt{\frac{\hat{p}(1-\hat{p})}{n}}$$
$$\sqrt{n}\,E = Z_{\alpha/2}\sqrt{\hat{p}(1-\hat{p})}$$
$$\sqrt{n} = \sqrt{\hat{p}(1-\hat{p})}\left(\frac{Z_{\alpha/2}}{E}\right)$$
$$n = \hat{p}(1-\hat{p})\left(\frac{Z_{\alpha/2}}{E}\right)^2 \qquad \text{(Equation 8.1)}$$

Thus, we have the following formula.

> ### Sample Size for Estimating a Population Proportion When $\hat{p}$ Is Known
>
> The sample size needed to estimate the population proportion p to within a margin of error E with confidence $100(1 - \alpha)\%$ is given by
>
> $$n = \hat{p}(1-\hat{p})\left(\frac{Z_{\alpha/2}}{E}\right)^2$$
>
> where $Z_{\alpha/2}$ is the value associated with the desired confidence level, E is the desired margin of error, and $\hat{p}$ is the sample proportion of successes available from some earlier sample. Round up to the next integer.

This formula is illustrated using the following example.

Example 8.19 Sample size for estimating a sample proportion when $\hat{p}$ is known

Refer to Example 8.18. Suppose that the Gallup Organization now wanted to estimate the population proportion of those who think there should not be a law that would ban the possession of handguns to within a margin of error of $E = 0.01$ with 95% confidence. How large a sample size is needed?

Solution

From Example 8.18, we have the sample proportion $\hat{p} = 0.63$. The confidence level of 95% implies that our $Z_{\alpha/2} = 1.96$, and the desired margin of error is $E = 0.01$. Thus, the required sample size is

$$n = \hat{p}(1 - \hat{p})\left(\frac{Z_{\alpha/2}}{E}\right)^2 = 0.63(1 - 0.63)\left(\frac{1.96}{0.01}\right)^2 \approx 8954.77$$

Rounding up, this gives us a required sample size of 8955. The smaller margin of error requires a larger sample size. ■

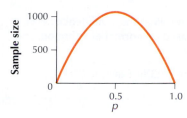

FIGURE 8.24 Sample size required for the range of values of *p*.

Equation 8.1 has $\hat{p}$ in it, which implies that we need to have information about $\hat{p}$ before we can find the required sample size *n*. However, what if we do not know the value of $\hat{p}$?

Figure 8.24 plots the sample size requirements for a 95% confidence interval for *p*, with a desired margin of error of 0.03, for values of $\hat{p}$ ranging from 0.01 to 0.99, representing all sample proportions from 1% to 99%. Note that the plot is symmetric, and therefore the largest required sample size occurs at the midpoint $\hat{p} = 0.5$. Thus, $\hat{p} = 0.5$ is the most conservative value for $\hat{p}$. When the actual value of $\hat{p}$ is not known, we use the following formula.

> **Sample Size for Estimating a Population Proportion When $\hat{p}$ Is Not Known**
>
> The sample size needed to estimate the population proportion *p* to within a margin of error *E* with confidence $100(1 - \alpha)\%$ is given by
>
> $$n = \left(\frac{0.5 \cdot Z_{\alpha/2}}{E}\right)^2$$
>
> where $Z_{\alpha/2}$ is the value associated with the desired confidence level, *E* is the desired margin of error, and 0.5 is a constant representing the most conservative estimate. Round up to the next integer.

Example 8.20 Required sample size for polls

Suppose the Dimes-Newspeak organization would like to take a poll on the proportion of Americans who will vote Republican in the next presidential election. How large a sample size does the Dimes-Newspeak organization need to estimate the proportion to within plus or minus three percentage points ($E = 0.03$) with 95% confidence?

Solution

The 95% confidence implies that the value for $Z_{\alpha/2}$ is 1.96. Since there is no information available about the value of the population proportion of all Americans who will vote Republican in the next election, we use 0.5 as our "worst-case scenario" value of *p*:

$$n = \left[\frac{0.5 \cdot Z_{\alpha/2}}{E}\right]^2 = \left[\frac{(0.5)(1.96)}{0.03}\right]^2 \approx 1067.11$$

So if the pollsters would like to estimate the population proportion of all American voters who will vote Republican in the upcoming election to within 3% with 95% confidence, they will need a sample of 1068 voters (don't forget to round up!). ■

STEP-BY-STEP TECHNOLOGY GUIDE: *Z* Confidence Intervals for *p*

We illustrate how to construct the *Z* confidence interval for *p* from Example 8.17 (page 388).

TI-83/84

Step 1   Press **STAT** and highlight **TESTS**.
Step 2   Scroll down to **A** (for **1-PropZInt,** see Figure 8.25), and press **ENTER**.
Step 3   For **x**, enter the number of success, **631**.
Step 4   For **n**, enter the sample size **830**.
Step 5   For **C-Level** (confidence level), enter the appropriate confidence level (e.g., **0.95**), and press **ENTER** (Figure 8.26).
Step 6   Highlight **Calculate** and press **ENTER**. The results are shown in Figure 8.22 in Example 8.17.

Figure 8.25   **Figure 8.26**

MINITAB

Step 1   Click **Stat > Basic Statistics > 1-Proportion**.
Step 2   Click **Summarized Data**.
Step 3   Enter the **Number of Trials** (n) **830** and the **Number of Events** (X) **631**.

Step 4   Click on **Options**, enter **95** as the **Confidence Level**, select **Use test and interval based on normal distribution**, and click **OK**. Then click **OK** again.

The results are shown in Figure 8.23 in Example 8.17.

Section 8.3 SUMMARY

1 The sample proportion of successes

$$\hat{p} = \frac{x}{n} = \frac{\text{number of successes}}{\text{sample size}}$$

is a point estimate of the population proportion *p*.

2 The $100(1 - \alpha)\%$ confidence interval for the population proportion *p*, is given by

$$\hat{p} \pm Z_{\alpha/2} \sqrt{\frac{\hat{p}(1 - \hat{p})}{n}}$$

where $\hat{p}$ is the sample proportion of successes, *n* is the sample size, and $Z_{\alpha/2}$ depends on the confidence level. The *Z* interval for *p* may be constructed only if *both* the following conditions apply: $n\hat{p} \geq 5$ and $n(1 - \hat{p}) \geq 5$.

3 Note that the confidence interval for *p* takes on the form

point estimate ± margin of error

where $\hat{p}$ is the point estimate of *p* and $E = Z_{\alpha/2}\sqrt{\hat{p}(1 - \hat{p})/n}$ is the margin of error.

4 Suppose we would like to estimate the population proportion *p* to within a margin of error *E* with confidence $100(1 - \alpha)\%$. If $\hat{p}$, is known, then the required sample size needed is given by

$$n = \hat{p}(1 - \hat{p})\left(\frac{Z_{\alpha/2}}{E}\right)^2$$

If $\hat{p}$ is not known, then the required sample size needed is given by

$$n = \left(\frac{0.5 \cdot Z_{\alpha/2}}{E}\right)^2$$

Section 8.3 EXERCISES

CLARIFYING THE CONCEPTS

1. Suppose the population proportion of successes *p* is known. Does it make sense to construct a confidence interval for *p*?
2. A news broadcast mentions that the sample size of a poll is about 1000 and that the margin of error is plus or minus three percentage points. How do we know that the pollsters are using a 95% confidence level?

PRACTICING THE TECHNIQUES

For Exercises 3–10, do the following.
    a. Construct a confidence interval for *p* with the indicated confidence level.
    b. Sketch the confidence interval on a number line.
3. Confidence level 95%, sample size 25, sample proportion 0.5

4. Confidence level 90%, sample size 81, sample proportion 0.1
5. Confidence level 99%, sample size 100, sample proportion 0.95
6. Confidence level 99%, sample size 400, sample proportion 0.02
7. Confidence level 95%, $n = 64, \hat{p} = 0.4$
8. Confidence level 99%, $n = 144, \hat{p} = 0.6$
9. Confidence level 90%, $n = 49, \hat{p} = 0.3$
10. Confidence level 95%, $n = 225, \hat{p} = 0.03$

For Exercises 11–18, do the following.
   a. Find $Z_{\alpha/2}$
   b. Determine whether the conditions are met.
   c. If the conditions are met, calculate the margin of error, $E = Z_{\alpha/2} \cdot \sqrt{\hat{p}(1 - \hat{p})/n}$.
   d. If the conditions are met, construct a confidence interval for p with the indicated confidence level, and sketch the confidence interval on a number line. If the conditions are not met, state why not.
11. Confidence level 95%, sample size 121, sample proportion 0.1
12. Confidence level 99%, sample size 121, sample proportion 0.04
13. Confidence level 90%, sample size 16, sample proportion 0.5
14. Confidence level 99%, sample size 16, sample proportion 0.2
15. Confidence level 95%, $n = 25, \hat{p} = 0.1$
16. Confidence level 99%, $n = 25, \hat{p} = 0.2$
17. Confidence level 90%, $n = 36, \hat{p} = 0.15$
18. Confidence level 95%, $n = 36, \hat{p} = 0.12$
19. For the following samples, find the margin of error E for a 95% confidence interval for p.
   a. 5 successes in 10 trials
   b. 50 successes in 100 trials
   c. 500 successes in 1000 trials
   d. 5000 successes in 10,000 trials
20. For the following samples, find the margin of error E for a 95% confidence interval for p.
   a. 10 successes in 100 trials
   b. 20 successes in 100 trials
   c. 30 successes in 100 trials
   d. 40 successes in 100 trials
   e. 50 successes in 100 trials
21. Refer to Exercise 19.
   a. Write a sentence describing what happens to the margin of error as the sample size increases while $\hat{p}$ remains constant.
   b. What effect will the behavior you observed in **(a)** have on the width of the confidence interval?
22. Refer to Exercise 20.
   a. Write a sentence describing what happens to the margin of error as the sample proportion approaches 0.5 while the sample size remains constant.

   b. What effect will the behavior you observed in **(a)** have on the width of the confidence interval?

For Exercises 23–28, we are estimating p and we know the value of $\hat{p}$. Find the required sample size.
23. Confidence level 95%, margin of error 0.03, $\hat{p} = 0.3$
24. Confidence level 95%, margin of error 0.03, $\hat{p} = 0.5$
25. Confidence level 95%, margin of error 0.03, $\hat{p} = 0.7$
26. Confidence level 95%, margin of error 0.03, $\hat{p} = 0.1$
27. Confidence level 95%, margin of error 0.03, $\hat{p} = 0.01$
28. Confidence level 95%, margin of error 0.03, $\hat{p} = 0.001$
29. Using Exercises 26–28, describe what happens to the required sample size when $\hat{p}$ gets very small.

For Exercises 30–35, we are estimating p and we do not know the value of $\hat{p}$. Find the required sample size.
30. Confidence level 90%, margin of error 0.03
31. Confidence level 95%, margin of error 0.03
32. Confidence level 99%, margin of error 0.03
33. Confidence level 95%, margin of error 0.015
34. Confidence level 95%, margin of error 0.0075
35. Confidence level 95%, margin of error 0.00375
36. Using Exercises 30–32, describe what happens to the required sample size as the confidence level increases.
37. Using Exercises 33–35, describe what happens to the required sample size when the margin of error is halved and the confidence level stays constant.

APPLYING THE CONCEPTS
For Exercises 38–41, do the following.
   a. Find $Z_{\alpha/2}$
   b. Determine whether the conditions are met for constructing a confidence interval for p.
   c. If the conditions are met, calculate the margin of error E and explain what this number means.
   d. If the conditions are met, construct and interpret a confidence interval for p with the indicated confidence level, and sketch the confidence interval on the number line. If the conditions are not met, state why not.
38. Global Assignments. On June 8, 2004, Cendant Mobility (now Cartus) reported the results of a survey it conducted asking randomly selected workers if they would be more likely, less likely, or just as likely to accept a global assignment rather than a domestic assignment in view of recent world events. Of the 548 subjects polled, 367 said that they would be more likely to accept a global assignment. Use a 99% confidence level.
39. Rather Be Fishing? A study found that Minnesota, at 38%, leads the nation in the proportion of people who go fishing.[12] Assume that the study sample size was 12 and use a 95% confidence level.
40. Spring Break and Drinking. A study released by the American Medical Association in 2006 found that 83% of college female respondents agreed that heavier drinking occurs on spring break trips than is typically found on campus. Assume that the sample size was 25 and use a 90% confidence level.

41. NASCAR Fans and Pickup Trucks. In April 2004, *American Demographics* magazine reported that 40% of a sample of NASCAR racing attendees said they owned a pickup truck. Suppose the sample size was 1000. Construct a 95% confidence interval for the population proportion of NASCAR racing attendees who own a pickup truck.

42. Politics of Gay Marriage. The Pew Research Center for the People and the Press (**http://people-press.org**) reported in 2004 that 322 out of 1149 randomly selected people surveyed would not vote for a political candidate who disagreed with their views on gay marriage.

   a. Is it appropriate to apply the Z interval for the population proportion?

   b. Find the margin of error. What does this number mean?

   c. Construct and interpret a 95% confidence interval for the population proportion of all people who would not vote for a political candidate who disagreed with their views on gay marriage.

43. Ecstasy Use in Twelfth Grade. According to the National Institute on Drug Abuse (**www.drugabuse.gov**), in 2005 there was a decline in the proportion of twelfth-graders who were using MDMA (Ecstasy).[13] Among whites, 3.9% reported use of the drug. Assume that the sample size is 200.

   a. Is it appropriate to use the Z interval to estimate the population proportion?

   b. Find the margin of error. What does this number mean?

   c. Construct and interpret a 99% confidence interval for the population proportion of all white twelfth-graders who were using Ecstasy.

44. Does Heavy Debt Lead to Ulcers? An AP–AOL poll reported on June 9, 2008, that 27% of respondents carrying heavy mortgage or credit card debt also said that they had stomach ulcers.[14] How large a sample size is needed to estimate the population proportion of respondents carrying heavy debt who also have stomach ulcers to within 1% with 99% confidence?

45. Male Binge Drinking. The American Medical Association is interested in commissioning a study on the population proportion of males who binge drink during spring break.

   a. How large a sample size is required to estimate p to within 0.05 with 90% confidence?

   b. How large a sample size is required to estimate p to within 0.01 with 90% confidence?

46. Global Assignment. On June 8, 2004, Cendant Mobility (now Cartus) reported the results of a survey it conducted asking workers if they would be more likely, less likely, or just as likely to accept a global assignment in view of recent world events. Of the 548 randomly chosen subjects polled, 367 said that they would be more likely to accept a global assignment.

   a. How large a sample size would have been required if Cendant wanted to estimate p to within 0.03 with 95% confidence?

   b. How large a sample size would have been required if Cendant wanted to estimate p to within 0.01 with 95% confidence?

47. Most Churchgoers Are Not Drinkers. According to Gallup.com, 60% of Americans who attend church weekly do not drink alcohol. How large a sample size do we need to estimate the population proportion to within 3% with 99% confidence?

48. Most Churchgoers Are Not Drinkers. According to Gallup.com, 60% of Americans who attend church weekly do not drink alcohol. The poll reported that the sample size was 1006 respondents. Assuming 95% confidence, what is the margin of error of this poll?

49. Objective News Source? A random sample of 1113 American adults found 240 who cited CNN as the media outlet that provides the most objective news.

   a. If appropriate, find the margin of error for confidence level 99%. What does this number mean?

   b. If appropriate, construct a 99% confidence interval for the population proportion of all American adults who cite CNN as the media outlet that provides the most objective news.

? 50. Most Churchgoers Are Not Drinkers. Refer to Exercise 48. Try to think about the relationship between the various statistics rather than calculating the statistics. *What if* the sample size is increased from 1006 and everything else stays the same.

   a. What will be the effect on the margin of error?

   b. What will be the effect on the reported sample proportion?

   c. What will be the effect on the confidence level?

51. Mozart Effect. Harvard University's Project Zero (**pzweb.harvard.edu**) found that listening to certain kinds of music, including Mozart, improved spatial-temporal reasoning abilities in children. Suppose that, in a sample of 100 randomly chosen fifth-graders, 65 performed better on a spatial-temporal achievement test after listening to a Mozart sonata. If appropriate, find a 95% confidence interval for the population proportion of all fifth-graders who performed better after listening to a Mozart sonata.

52. Taxes Too High? In a Gallup Poll of 1005 randomly selected Americans in April 2006, 482 responded "Yes" to the question "Do you consider the amount of federal income tax you have to pay as too high?"

   a. If appropriate, find a 90% confidence interval for the population proportion of all Americans who consider that the amount of federal income tax they pay is too high.

   b. If appropriate, find a 99% confidence interval for the population proportion of all Americans who consider that the amount of federal income tax they pay is too high.

? 53. Mozart Effect. Refer to Exercise 51. *What if* we increase the confidence level to 99% while changing nothing else. Explain what would happen to the following statistics and why.

   a. $Z_{\alpha/2}$

b. Margin of error

c. Width of the confidence interval

The Famous ± 3 Percentage Points. Use the information from Example 8.18 for Exercises 54 and 55.

54. *What if* the sample size is higher than 1012, but otherwise everything else is the same as the example. How would this affect the following?

  a. Margin of error

  b. $Z_{\alpha/2}$

  c. Width of the confidence interval

55. *What if* the confidence level is lower than 95%, but otherwise everything else is the same as the example. How would this affect the following?

  a. Margin of error

  b. $Z_{\alpha/2}$

  c. Width of the confidence interval

Drug Companies and Research Studies. Use this information for Exercises 56–58. The *Annals of Internal Medicine* reported that 39 of the 40 research studies with acknowledged sponsorship by a drug company had outcomes favoring the drug under investigation.[15]

56. If appropriate, construct and interpret a 90% confidence interval for the population proportion of all studies sponsored by drug companies that have outcomes favoring the drug. If not appropriate, clearly state why not.

57. The article in the *Annals of Internal Medicine* found that 89 of the 112 studies *without* acknowledged drug company support had outcomes favoring the drug. If appropriate, construct a 95% confidence interval for the population proportion of all studies without acknowledged drug company support which have outcomes favoring the drug. If not appropriate, clearly state why not.

58. Refer to Exercise 57. *What if* we decrease the confidence level to 90%, while changing nothing else. Explain precisely what would happen to the following statistics and why.

  a. $Z_{\alpha/2}$

  b. Margin of error

  c. Width of the confidence interval

8.4   Confidence Intervals for the Population Variance and Standard Deviation

Objectives: **By the end of this section, I will be able to…**

1 Describe the properties of the χ^2 (chi-square) distribution.

2 Find critical values for the χ^2 (chi-square) distribution.

3 Construct and interpret confidence intervals for the population variance and standard deviation.

We have seen how confidence intervals can be used to estimate the unknown value of a population mean or a population proportion. However, the variability of a population is also important. As we have learned, less variability is almost always better. For example, a tool manufacturer relies on a quality control technician (who has a strong background in statistics) to make sure that the tools the company is making do not vary appreciably from the required specifications. Otherwise, the tools may be too large or too small. Data analysts therefore construct confidence intervals to estimate the unknown value of the population parameters that measure variability: the population variance σ^2 and the population standard deviation σ.

The χ^2 **(chi-square) distribution** is used to construct these confidence intervals. So we first need to become acquainted with the χ^2 distribution.

1 Properties of the χ^2 (Chi-Square) Distribution

The χ^2 (pronounced *ky-square*, to rhyme with "my square") distribution was discovered in 1875 by the German physicist Friedrich Helmert and further developed in 1900 by the English statistician Karl Pearson.

The χ^2 random variable is continuous. Just as we did with the normal and *t* distributions, we can find probabilities associated with values of χ^2, and vice versa. Like any continuous distribution, probability is represented by area below the curve above an interval. We examine the properties of the χ^2 distribution and then learn how to use the χ^2 table to find the critical values of the χ^2 distribution.

Properties of the χ^2 Distribution
- Just as for any continuous random variable, the total area under the χ^2 curve equals 1.
- The value of the χ^2 random variable is never negative, so the χ^2 curve starts at 0. However, it extends indefinitely to the right, with no upper bound.
- Because of the characteristics just described, the χ^2 curve is right-skewed.
- There is a different curve for every different degrees of freedom, $n - 1$. As the number of degrees of freedom increases, the χ^2 curve begins to look more symmetric (Figure 8.27).

FIGURE 8.27
Shape of the χ^2 distribution for different degrees of freedom.

✌ Finding Critical Values for the χ^2 (Chi-Square) Distribution

To construct the confidence intervals in this section, we shall need to find the critical values of a χ^2 distribution for the given confidence level $100(1 - \alpha)\%$. To find these χ^2 critical values, we can use either the χ^2 table (Table E in the Appendix on page T-12) or technology. The χ^2 table is somewhat similar to the t table (Table D in the Appendix); both tables show the degrees of freedom in the left column. The area to the right of the χ^2 critical value is given across the top of the table.

Since the χ^2 distribution is not symmetric, we cannot construct the confidence interval for σ^2 using the "Point estimate $\pm$ Margin of error" method. Rather, the lower bound and upper bound for the confidence interval are determined using two χ^2 critical values:

$\chi^2_{1-\alpha/2}$ = the value of the χ^2 distribution with area $1 - \alpha/2$ to its right (Figure 8.28)

$\chi^2_{\alpha/2}$ = the value of the χ^2 distribution with area $\alpha/2$ to its right (Figure 8.28).

FIGURE 8.28
χ^2 critical values.

For instance, for a 95% confidence interval $(1 - \alpha) = 0.95$, $\alpha/2 = 0.025$ and $1 - \alpha/2 = 0.975$. Thus, $\chi^2_{0.975}$ represents the value of the χ^2 distribution with area $1 - \alpha/2 = 0.975$ to the right of the χ^2 critical value. The second critical value $\chi^2_{0.025}$ represents the value of the χ^2 distribution with area $\alpha/2 = 0.025$ to the right of the χ^2 critical value.

Example 8.21 Finding the χ^2 critical values

Find χ^2 critical values for a 90% confidence interval, where we have a sample size of size $n = 10$.

Solution

For a 90% confidence interval,

$$(1 - \alpha) = 0.90 \qquad \frac{\alpha}{2} = \frac{0.10}{2} = 0.05 \qquad 1 - \frac{\alpha}{2} = 1 - 0.05 = 0.95$$

Note: If the appropriate degrees of freedom are not given in the χ^2 table, the conservative solution is to take the next row with the smaller df.

So we are seeking (1) $\chi^2_{0.95}$, the critical value with area $1 - \alpha/2 = 0.95$ to the right of it, and (2) $\chi^2_{0.05}$, the critical value with area $\alpha/2 = 0.05$ to the right of it.

Since $n = 10$, the degrees of freedom is df $= n - 1 = 10 - 1 = 9$. To find $\chi^2_{0.95}$ for df $= 9$, go across the top of the χ^2 table (Table E in the Appendix) until you see 0.95 (Figure 8.29). $\chi^2_{0.95}$ is somewhere in that column. Now go down that column until you see your number of degrees of freedom df $= 9$. Thus, for df $= 9$, $\chi^2_{0.95} = 3.325$. For a χ^2 distribution with 9 degrees of freedom, there is area $= 0.95$ to the right of 3.325. Similarly, $\chi^2_{0.05}$ is found in the column labeled "0.05" and the row corresponding to df $= 9$. We find that $\chi^2_{0.05} = 16.919$, as shown in Figure 8.30. ■

Chi-Square (χ^2) Distribution
Area to the Right of Critical Value

Degrees of Freedom	0.995	0.99	0.975	0.95	0.90	0.10	0.05	0.025	0.01	0.005
1	—	—	0.001	0.004	0.016	2.706	3.841	5.024	6.635	7.879
2	0.010	0.020	0.051	0.103	0.211	4.605	5.991	7.378	9.210	10.597
3	0.072	0.115	0.216	0.352	0.584	6.251	7.815	9.348	11.345	12.838
4	0.207	0.297	0.484	0.711	1.064	7.779	9.488	11.143	13.277	14.860
5	0.412	0.554	0.831	1.145	1.610	9.236	11.071	12.833	15.086	16.750
6	0.676	0.872	1.237	1.635	2.204	10.645	12.592	14.449	16.812	18.548
7	0.989	1.239	1.690	2.167	2.833	12.017	14.067	16.013	18.475	20.278
8	1.344	1.646	2.180	2.733	3.490	13.362	15.507	17.535	20.090	21.955
9	1.735	2.088	2.700	3.325	4.168	14.684	16.919	19.023	21.666	23.589
10	2.156	2.558	3.247	3.940	4.865	15.987	18.307	20.483	23.209	25.188

FIGURE 8.29 Finding $\chi^2_{0.95}$ and $\chi^2_{0.05}$ using the χ^2 table.

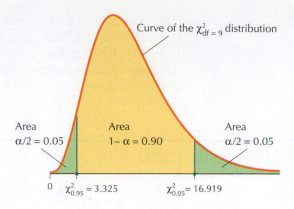

FIGURE 8.30

3 Constructing Confidence Intervals for the Population Variance and Standard Deviation

We derive the formula for a $100(1 - \alpha)\%$ confidence interval for the population variance σ^2. Suppose we take a random sample of size n from a normal population with mean μ and standard deviation σ. Then the statistic

$$\chi^2 = \frac{(n-1)s^2}{\sigma^2}$$

follows a χ^2 distribution with $n - 1$ degrees of freedom, where s^2 represents the sample variance. From Figure 8.28, we see that $100(1 - \alpha)\%$ of the values of χ^2 lie between $\chi^2_{1-\alpha/2}$ and $\chi^2_{\alpha/2}$. These values are described as

$$\chi^2_{1-\alpha/2} < \frac{(n-1)s^2}{\sigma^2} < \chi^2_{\alpha/2}$$

Rearranging this inequality so that σ^2 is in the numerator gives us the formula for the $100(1 - \alpha)\%$ confidence interval for σ^2:

$$\frac{(n-1)s^2}{\chi^2_{\alpha/2}} < \sigma^2 < \frac{(n-1)s^2}{\chi^2_{1-\alpha/2}}$$

Thus the lower bound of the confidence interval for σ^2 is $\dfrac{(n-1)s^2}{\chi^2_{\alpha/2}}$, and the upper bound is $\dfrac{(n-1)s^2}{\chi^2_{1-\alpha/2}}$.

Confidence Interval for the Population Variance σ^2

Suppose we take a sample of size n from a normal population with mean μ and standard deviation σ. Then a $100(1 - \alpha)\%$ confidence interval for the population variance σ^2 is given by

$$\text{lower bound} = \frac{(n-1)s^2}{\chi^2_{\alpha/2}}, \quad \text{upper bound} = \frac{(n-1)s^2}{\chi^2_{1-\alpha/2}}$$

where s^2 represents the sample variance and $\chi^2_{1-\alpha/2}$ and $\chi^2_{\alpha/2}$ are the critical values for a χ^2 distribution with $n - 1$ degrees of freedom.

Confidence Interval for the Population Standard Deviation σ

A $100(1 - \alpha)\%$ confidence interval for the population standard deviation σ is then given by

$$\text{lower bound} = \sqrt{\frac{(n-1)s^2}{\chi^2_{\alpha/2}}}, \quad \text{upper bound} = \sqrt{\frac{(n-1)s^2}{\chi^2_{1-\alpha/2}}}$$

Example 8.22 Constructing confidence intervals for the population variance σ^2 and population standard deviation σ

The accompanying table shows the city gas mileage for 6 hybrid cars, as reported by the Environmental Protection Agency and **www.hybridcars.com** in 2007. The normal probability plot in Figure 8.31 indicates that the data are normally distributed.

Vehicle	Mileage (mpg)
Honda Accord	30
Ford Escape (2 wd)	36
Toyota Highlander	33
Saturn VUE Green Line	27
Lexus RX 400h	31
Lexus GS 450h	25

a. Find the critical values $\chi^2_{1-\alpha/2}$ and $\chi^2_{\alpha/2}$ for a confidence interval with a 95% confidence level.
b. Construct and interpret a 95% confidence interval for the population variance of hybrid gas mileage.
c. Construct and interpret a 95% confidence interval for the population standard deviation of hybrid gas mileage.

FIGURE 8.31 Normal probability plot of mileage.

Solution

a. There are $n = 6$ hybrid cars in our sample, so the degrees of freedom equal $n - 1 = 5$. For a 95% confidence interval,

$$(1 - \alpha) = 0.95 \qquad \alpha/2 = 0.025 \qquad 1 - \alpha/2 = 0.975$$

From the χ^2 table (Table E in the Appendix), therefore,

$$\chi^2_{1-\alpha/2} = \chi^2_{0.975} = 0.831 \qquad \chi^2_{\alpha/2} = \chi^2_{0.025} = 12.833$$

Figures 8.32 and 8.33 show these results using Excel and Minitab.

A1				f_x	=CHIINV(0.975,5)
	A	B	C	D	E
1	0.831212				
2	12.8325				

(a)

A2				f_x	=CHIINV(0.025, 5)
	A	B	C	D	E
1	0.831212				
2	12.8325				

(b)

FIGURE 8.32 Excel results.

```
Inverse Cumulative Distribution Function
Chi-Square with 5 DF

P( X <= x )            x
      0.025   0.831212
```

(a)

```
Inverse Cumulative Distribution Function
Chi-Square with 5 DF

P( X <= x )            x
      0.975   12.8325
```

(b)

FIGURE 8.33 Minitab results.

b. Figure 8.34 shows the descriptive statistics for the hybrid car gas mileages, as obtained by the TI-83/84. The sample standard deviation is $s = 3.983298466$.

FIGURE 8.34 TI-83/84 results.

Thus, our 95% confidence interval for σ^2 is given by

$$\text{lower bound} = \frac{(n-1)s^2}{\chi^2_{\alpha/2}} = \frac{(5)3.983298466^2}{12.833} \approx 6.181978754 \approx 6.18$$

$$\text{upper bound} = \frac{(n-1)s^2}{\chi^2_{1-\alpha/2}} = \frac{(5)3.983298466^2}{0.831} \approx 95.46730848 \approx 95.47$$

We are 95% confident that the population variance σ^2 lies between 6.18 and 95.47 miles per gallon squared, that is, (mpg)^2. (Recall that the variance is measured in *units squared*.) Since it is unclear what miles per gallon squared means, we prefer to construct a confidence interval for the population standard deviation σ.

c. Using the results from **(b)**.

$$\text{lower bound} = \sqrt{\frac{(n-1)s^2}{\chi^2_{\alpha/2}}} = \sqrt{6.181978754} \approx 2.486358533 \approx 2.49$$

$$\text{upper bound} = \sqrt{\frac{(n-1)s^2}{\chi^2_{1-\alpha/2}}} = \sqrt{95.46730848} \approx 9.770737356 \approx 9.77$$

We are 95% confident that the population standard deviation σ lies between 2.49 and 9.77 miles per gallon. Figure 8.35 shows the two confidence intervals obtained using Minitab. ∎

```
                                  CI for        CI for
Variable   Method       StDev        Variance
Mileage    Standard   (2.49, 9.77)   (6.2, 95.4)
```

FIGURE 8.35 Minitab results showing the confidence intervals (excerpt).

STEP-BY-STEP TECHNOLOGY GUIDE: χ^2 Distribution

EXCEL

Finding the Critical Values $\chi^2_{1-\alpha/2}$ and $\chi^2_{\alpha/2}$

Step 1   Select cell **A1**. Click the **Insert Function** icon f_x.
Step 2   For **Search for a Function**, type **chiinv**, click **GO**, then click **OK**.
Step 3   To find $\chi^2_{1-\alpha/2}$: For **Probability**, enter $1 - \alpha/2$ (such as **0.975** for a 95% confidence interval), and for **Deg_freedom**

enter the degrees of freedom. Excel displays the value of $\chi^2_{1-\alpha/2}$ in the cell.

Step 4   To find $\chi^2_{\alpha/2}$: Repeat Steps 1 – 2. For **Probability**, enter $\alpha/2$ (such as **0.025** for a 95% confidence interval), and for **Deg_freedom** enter the degrees of freedom. Excel displays the value of $\chi^2_{\alpha/2}$ in the cell.

MINITAB

Finding the Critical Values $\chi^2_{1-\alpha/2}$ and $\chi^2_{\alpha/2}$

Step 1 Click **Calc > Probability Distributions > Chi-Square**.

Step 2 Select **Inverse cumulative probability**, and enter the **Degrees of freedom**.

Step 3 To find $\chi^2_{1-\alpha/2}$: For **Input constant**, enter the area to the *left* of the desired critical value. For $\chi^2_{1-\alpha/2}$, this will be $\alpha/2$ (such as **0.025**). Click **OK**.

Step 4 To find $\chi^2_{\alpha/2}$: Repeat Steps 1 – 2. For **Input constant**, enter the area to the *left* of the desired critical value. For $\chi^2_{\alpha/2}$, this will be $1 - \alpha/2$ (such as **0.975**). Click **OK**.

Step 5 Minitab displays the values of $\chi^2_{1-\alpha/2}$ and $\chi^2_{\alpha/2}$ in the session window.

Finding a $100(1 - \alpha)\%$ Confidence Interval for σ

Step 1 Enter the data into column **C1**.

Step 2 Select **Stat > Basic Statistics > 1 Variance ...**

Step 3 For **Samples in columns**, select **C1**.

Step 4 Click **Options**, choose the confidence level, and click **OK**. The confidence interval for σ is reported in the output, as shown in Figure 8.35.

Section 8.4 SUMMARY

1 The χ^2 continuous random variable takes values that are never negative, so the χ^2 distribution curve starts at 0 and extends indefinitely to the right. Thus, the χ^2 curve is right-skewed and not symmetric. There is a different curve for every different degrees of freedom, $n - 1$.

2 To find χ^2 critical values, we can use either the χ^2 table or technology.

3 If the population is normally distributed, we use the χ^2 distribution to construct a $100(1 - \alpha)\%$ confidence interval for the population variance σ^2, which is given by

$$\text{lower bound} = \frac{(n-1)s^2}{\chi^2_{\alpha/2}}, \text{ upper bound} = \frac{(n-1)s^2}{\chi^2_{1-\alpha/2}}$$

where s^2 represents the sample variance and $\chi^2_{1-\alpha/2}$ and $\chi^2_{\alpha/2}$ are the critical values for a χ^2 distribution with $n - 1$ degrees of freedom. A $100(1 - \alpha)\%$ confidence interval for the population standard deviation σ is then given by

$$\text{lower bound} = \sqrt{\frac{(n-1)s^2}{\chi^2_{\alpha/2}}}$$

$$\text{upper bound} = \sqrt{\frac{(n-1)s^2}{\chi^2_{1-\alpha/2}}}$$

Section 8.4 EXERCISES

CLARIFYING THE CONCEPTS

Determine whether each proposition in Exercises 1–4 is true or false. If it is false, restate the proposition correctly.

1. The χ^2 curve is symmetric.
2. The value of the χ^2 random variable is never negative.
3. The χ^2 curve is right-skewed.
4. The total area under the χ^2 curve equals 1.

PRACTICING THE TECHNIQUES

For Exercises 5–10, find the critical values $\chi^2_{1-\alpha/2}$ and $\chi^2_{\alpha/2}$ for the given confidence level and sample size.

5. Confidence level 90%, $n = 25$
6. Confidence level 95%, $n = 25$
7. Confidence level 99%, $n = 25$
8. Confidence level 95%, $n = 10$
9. Confidence level 95%, $n = 15$
10. Confidence level 95%, $n = 20$
11. Consider the critical values you calculated in Exercises 5–7. Describe what happens to the critical values for a given sample size as the confidence level increases.

12. Consider the critical values you calculated in Exercises 8–10. Describe what happens to the critical values for a given confidence level as the sample size increases.

In Exercises 13–18, a random sample is drawn from a normal population. The sample of size $n = 25$ has a sample variance of $s^2 = 10$. Construct the specified confidence interval.

13. 90% confidence interval for the population variance σ^2
14. 95% confidence interval for the population variance σ^2
15. 99% confidence interval for the population variance σ^2
16. 90% confidence interval for the population standard deviation σ
17. 95% confidence interval for the population standard deviation σ
18. 99% confidence interval for the population standard deviation σ
19. Consider the confidence intervals you constructed in Exercises 13–15. Describe what happens to the lower bound and upper bound of a confidence interval for σ^2 as the confidence level increases but the sample size stays the same.

20. Consider the confidence intervals you constructed in Exercises 16–18. Describe what happens to the lower bound and upper bound of a confidence interval for σ as the confidence level increases but the sample size stays the same.

In Exercises 21–26, a random sample is drawn from a normal population. The sample variance is $s^2 = 10$. Construct the specified confidence interval.

21. 95% confidence interval for the population variance σ^2 for a sample of size $n = 10$
22. 95% confidence interval for the population variance σ^2 for a sample of size $n = 15$
23. 95% confidence interval for the population variance σ^2 for a sample of size $n = 20$
24. 95% confidence interval for the population standard deviation σ for a sample of size $n = 10$
25. 95% confidence interval for the population standard deviation σ for a sample of size $n = 15$
26. 95% confidence interval for the population standard deviation σ for a sample of size $n = 20$
27. Consider the confidence intervals you constructed in Exercises 21–23. Describe what happens to the lower bound and upper bound of a confidence interval for σ^2 as the sample size increases but the confidence level stays the same.
28. Consider the confidence intervals you constructed in Exercises 24–26. Describe what happens to the lower bound and upper bound of a confidence interval for σ as the sample size increases but the confidence level stays the same.

APPLYING THE CONCEPTS
29. Prisoner Deaths. The table contains the numbers of prisoners who died in state custody in 2005 for a random sample of 5 states. Assume the data are normally distributed.

State	Prisoner deaths
New York	171
Pennsylvania	149
Michigan	140
Ohio	121
Georgia	122

Source: U.S. Bureau of Justice Statistics.

a. Find the critical values $\chi^2_{1-\alpha/2}$ and $\chi^2_{\alpha/2}$ for a 95% confidence interval for σ^2.
b. Construct and interpret a 95% confidence interval for the population variance σ^2 of the number of prisoners who died in state custody.
c. Construct and interpret a 95% confidence interval for the population standard deviation σ of the number of prisoners who died in state custody.
30. Most Active Stocks. Table 8.8 shows the ten most traded stocks on the New York Stock Exchange on October 3, 2007, together with their closing prices and net change in price, in dollars. Use only the net change data for this analysis. Assume that the net change data are normally distributed.

TABLE 8.8   New York Stock Exchange, October 3, 2007

Stock	Closing price	Net change
Micron Technology	$10.74	−1.05
Ford Motor Company	$ 8.43	−0.14
Citigroup	$47.89	0.03
Advanced Micro Devices	$13.23	0.03
EMC Corporation	$21.13	−0.24
Commerce Bancorp	$38.84	−0.63
General Electric Company	$41.55	−0.57
Avaya	$16.95	−0.07
Sprint Nextel Corporation	$18.76	−0.24
iShares:Taiwan	$17.18	−0.18

Source: USA Today. http://markets.usatoday.com/custom/usatoday=com/html=mktscreener.asp.

a. Find the critical values $\chi^2_{1-\alpha/2}$ and $\chi^2_{\alpha/2}$ for a 95% confidence interval for σ^2.
b. Construct and interpret a 95% confidence interval for the population variance σ^2 of net price changes
31. Prisoner Deaths. Refer to Exercise 29.
a. What are the units you used to interpret your confidence interval in (b)?
b. Do you think that those units would be easily understood by most people?
c. What are the units you used to interpret your confidence interval in (c)?
d. Do you think that these units are more easily understood by most people?
32. Most Active Stocks. Refer to Exercise 30.
a. What are the units you used to interpret your confidence interval in (b)?
b. Do you think that those units would be easily understood by most people?
c. What would the units be for a confidence interval for the population standard deviation σ?
d. Proceed to construct and interpret a 95% confidence interval for σ.
33. Fourth-Grade Feet. Suppose a children's shoe manufacturer is interested in estimating the variability of fourth graders' feet. A random sample of 20 fourth graders' feet yielded the following foot lengths, in centimeters.[16] The normality of the data was verified in Example 8.11 in Section 8.2.

22.4	23.4	22.5	23.2	23.1	23.7	24.1	21.0	21.6	20.9
25.5	22.8	24.1	25.0	24.0	21.7	22.0	22.7	24.7	23.5

a. Find the critical values $\chi^2_{1-\alpha/2}$ and $\chi^2_{\alpha/2}$ for a 95% confidence interval for σ^2.
b. Construct and interpret a 95% confidence interval for σ^2.
c. Construct and interpret a 95% confidence interval for σ.
d. Is the interpretation in (b) or (c) easier to understand?

34. Sodium in Breakfast Cereal. In Table 8.9 is a list of 23 breakfast cereals, along with their sodium content in milligrams per serving. Assume the data are normally distributed.

    a. Find the critical values $\chi^2_{1-\alpha/2}$ and $\chi^2_{\alpha/2}$ for a 95% confidence interval for σ^2.

    b. Construct and interpret a 95% confidence interval for σ^2.

    c. Construct and interpret a 95% confidence interval for σ.

    d. Is the interpretation in (**b**) or (**c**) easier to understand? Explain.

TABLE 8.9   Sodium content of 23 breakfast cereals

Cereals	Sodium (mg)
Apple Jacks	125
Cap'n Crunch	220
Cheerios	290
Cinnamon Toast Crunch	210
Cocoa Puffs	180
Corn Flakes	290
Count Chocula	180
Cream of Wheat	80
Froot Loops	125
Frosted Flakes	200
Fruity Pebbles	135
Grape Nuts Flakes	140
Kix	260
Life	150
Lucky Charms	180
Mueslix Crispy	150
Raisin Bran	210
Rice Chex	240
Rice Krispies	290
Special K	230
Total Whole Grain	200
Trix	140
Wheaties	200

35. Biomass Power Plants. Power plants around the country are retooling in order to consume biomass instead of or in addition to coal. Table 8.10 contains a random sample of 10 such power plants and the amount of biomass they consumed in 2006 in trillions of BTUs (British thermal units). Assume the data are normally distributed.

    a. Find the critical values for a 99% confidence interval for σ.

    b. Construct and interpret a 99% confidence interval for σ.

TABLE 8.10   Ten power plants that consume biomass instead of fuel

Power plant	Location	Biomass consumed (trillions of BTUs)
Georgia Pacific	Choctaw, AL	13.4
Jefferson Smurfit	Nassau, FL	12.9
International Paper	Richmond, GA	17.8
Gaylord Container	Washington, LA	15.1
Escanaba Paper	Delta, MI	19.5
Weyerhaeuser	Martin, NC	18.6
International Paper	Georgetown, SC	13.8
Bowater Newsprint	McMinn, TN	10.6
Covington Facility	Covington, VA	12.7
Mosinee Paper	Marathon, WI	17.6

Sources: Energy Information Administration, Form EIA-860, "Annual Electric Generator Report"; and Form EIA-906, "Power Plant Report."

Try This in Class! In-class activities to enhance your understanding of statistics

CONFIDENCE INTERVALS FOR THE POPULATION MEAN:
Estimating the Mean Shoe Size for All Female and Male Students at Your School

1. Guess what the population mean shoe size is for all female students at your school. Do the same for the males.

2. Collect data on the shoe sizes of the female students in your class. Find the mean of this sample.

3. What would you say is the probability that your guessed value for the sample mean is exactly equal to the mean foot size of all female students at your school?

4. Use technology to determine if the female shoe sizes are normally distributed. If not, and if the number of females is not at least 30, discuss whether it is appropriate to construct a *t* confidence interval for the population mean shoe size of all females at your school, and go to Step 7.

5. Construct a 95% confidence interval for the population mean shoe size for all female students at your school.

6. Does your guess for the population mean shoe size fall within the confidence interval you constructed?

7. Repeat Steps 2–6 for the males in your class.

CHAPTER 8 FORMULAS AND VOCABULARY

SECTION 8.1
- **Confidence interval** (p. 358)
- **Margin of error E for the Z interval for μ** (p. 364).

$$E = Z_{\alpha/2}(\sigma/\sqrt{n})$$

- **Sample size for estimating the population mean** (p. 367).

$$n = \left[\frac{(Z_{\alpha/2})\sigma}{E}\right]^2$$

- **Z interval for μ** (p. 359).

$$\text{lower bound} = \bar{x} - Z_{\alpha/2}(\sigma/\sqrt{n})$$
$$\text{upper bound} = \bar{x} + Z_{\alpha/2}(\sigma/\sqrt{n})$$

SECTION 8.2
- **Margin of error E for the t interval for μ** (p. 378).

$$E = t_{\alpha/2}(s/\sqrt{n})$$

- **t Distribution** (p. 373)
- **t interval for μ** (p. 375).

$$\text{lower bound} = \bar{x} - t_{\alpha/2}(s/\sqrt{n})$$
$$\text{upper bound} = \bar{x} + t_{\alpha/2}(s/\sqrt{n})$$

SECTION 8.3
- **Central limit theorem for proportions** (p. 386)
- **Margin of error E for the Z interval for p** (p. 389).

$$E = Z_{\alpha/2}\sqrt{\frac{\hat{p}(1-\hat{p})}{n}}$$

- **Sample size for estimating a population proportion when $\hat{p}$ is not known** (p. 391).

$$n = \left[\frac{(0.5)(Z_{\alpha/2})}{E}\right]^2$$

- **Sample size for estimating a population proportion when $\hat{p}$ is known** (p. 390).

$$n = \hat{p}(1-\hat{p})\left(\frac{Z_{\alpha/2}}{E}\right)^2$$

- **Z interval for p** (p. 386).

$$\text{lower bound} = \hat{p} - Z_{\alpha/2}\sqrt{\frac{\hat{p}(1-\hat{p})}{n}}$$
$$\text{upper bound} = \hat{p} + Z_{\alpha/2}\sqrt{\frac{\hat{p}(1-\hat{p})}{n}}$$

SECTION 8.4
χ^2 (chi square) distribution properties (p. 396)
- **Confidence interval for the population variance σ^2** (p. 398).

$$\text{lower bound} = \frac{(n-1)s^2}{\chi^2_{\alpha/2}} \qquad \text{upper bound} = \frac{(n-1)s^2}{\chi^2_{1-\alpha/2}}$$

- **Confidence interval for the population standard deviation σ** (p. 398).

$$\text{lower bound} = \sqrt{\frac{(n-1)s^2}{\chi^2_{\alpha/2}}} \qquad \text{upper bound} = \sqrt{\frac{(n-1)s^2}{\chi^2_{1-\alpha/2}}}$$

CHAPTER 8 REVIEW EXERCISES

SECTION 8.1
For Exercises 1 and 2, answer the following questions.
  a. Calculate $\sigma/\sqrt{n}$.
  b. Find $Z_{\alpha/2}$ for a confidence interval for μ with 95% confidence.
  c. Compute and interpret E, the margin of error for a confidence interval μ with 95% confidence.
  d. Construct and interpret a 95% confidence interval for μ.
1. A sample of $n = 25$ with sample mean $\bar{x} = 50$ is drawn from a normal population in which $\sigma = 10$.
2. A sample of $n = 100$ with sample mean $\bar{x} = 50$ is drawn from a population in which $\sigma = 10$.
3. The Mozart Effect. A random sample of 45 children showed a mean increase of 7 IQ points after listening to a Mozart piano sonata for about 10 minutes. The distribution of such increases is unknown, but the standard deviation is assumed to be 2 IQ points.

  a. Find the point estimate of the increase in IQ points for all children after listening to Mozart.
  b. Calculate $\sigma/\sqrt{n}$.
  c. Find $Z_{\alpha/2}$ for a confidence interval with 90% confidence.
  d. Compute and interpret the margin of error for a confidence interval with 90% confidence.
  e. Construct and interpret a 90% confidence interval for the mean increase in IQ points for all children after listening to a Mozart piano sonata for about 10 minutes.
4. Working Women's Salaries. A random sample of 25 working women in California yielded a sample mean income of $45,000. Assume that the population standard deviation is $6000 and that the distribution of such incomes is normal.
  a. Find the point estimate of the income of all working women.

b. Find $\sigma/\sqrt{n}$.

c. Find $Z_{\alpha/2}$ for a confidence interval with 99% confidence.

d. Compute and interpret the margin of error for a confidence interval with 99% confidence.

e. Construct and interpret a 99% confidence interval for the mean income of the population of all working women in California.

Suppose we are estimating μ. For Exercises 5–7, find the required sample size.

5. $\sigma = 50$, confidence level 95%, margin of error 10

6. $\sigma = 30$, confidence level 95%, margin of error 10

7. $\sigma = 10$, confidence level 95%, margin of error 10

8. Clinical Psychology. A clinical psychologist would like to estimate the population mean number of episodes her patients have suffered in the past year. Assume that the standard deviation is 10 episodes. How many patients will she have to examine if she wants her estimate to be within 2 episodes with 90% confidence?

SECTION 8.2

For Exercises 9–11, construct the indicated confidence interval if appropriate. If it is not appropriate, explain why not.

9. Confidence level 90%, $n = 25$, $\bar{x} = 22$, $s = 5$, non-normal population

10. Confidence level 90%, $n = 25$, $\bar{x} = 22$, $s = 5$, normal population

11. Confidence level 90%, $n = 100$, $\bar{x} = 22$, $s = 5$, non-normal population

12. Cigarette Consumption. Health officials are interested in estimating the population mean number of cigarettes smoked per capita in order to evaluate the efficacy of the antismoking campaign. A random sample of eight U.S. counties yielded the following numbers of cigarettes smoked per capita: 2206, 2391, 2540, 2116, 2010, 2791, 2392, 2692. Assure the data are normally distributed.

a. Construct a 95% confidence interval for the population mean per capita number of cigarettes smoked in all U.S. counties.

b. Construct a 99% confidence interval for the population mean per capita number of cigarettes smoked in all U.S. counties.

c. Compare the intervals from **(a)** and **(b)**. Name one strength and one weakness of each interval.

13. Parking Meters. A random sample of 75 parking meters yielded a mean of $120 per meter with a standard deviation of $30.

a. Find $t_{\alpha/2}$ for a confidence interval with 90% confidence.

b. Compute and interpret the margin of error E for a confidence interval with 90% confidence.

c. Construct and interpret a 90% t confidence interval for the population mean amount of money collected by all parking meters.

SECTION 8.3

For Exercises 14 and 15, follow steps **(a)–(d)**.

a. Find $Z_{\alpha/2}$.

b. Determine whether the conditions are met.

c. Calculate and interpret the margin of error, $E = Z_{\alpha/2} \sqrt{\hat{p}(1 - \hat{p})/n}$

d. Construct a confidence interval for p with the indicated confidence level, and sketch the confidence interval on the number line.

14. Confidence level 95%, $n = 100$, $\hat{p} = 0.1$

15. Confidence level 95%, $n = 500$, $\hat{p} = 0.99$

16. Ecstasy and Emergency Room Visits. According to the National Institute on Drug Abuse (**www.drugabuse.gov**), in 2001, 77% of the emergency room patients who mentioned MDMA (Ecstasy) as a factor in their admission were age 25 and under. Assume that the sample size is 200.

a. Calculate and interpret the margin of error for confidence level 95%.

b. Construct and interpret a 95% confidence interval for the population proportion of all emergency room patients mentioning MDMA (Ecstasy) as a factor in their admission who are age 25 and under.

17. Too Many Erasures? The Iowa Test of Basic Skills is a comprehensive national test given annually to many third-graders and fifth-graders. At one school, investigators found that 3 to 5 times as many erasures were made to the tests, compared to other schools. They also found that 89% of the erasures resulted in a correct answer, compared to less than 70% for other schools. Assume that the sample size is 1000. Let's investigate whether there is evidence of tampering.

a. Calculate and interpret the margin of error for confidence level 99%.

b. Construct a 99% confidence interval for p.

c. Compare this confidence interval to the proportion for other such schools. Do you think there is evidence of tampering?

For Exercises 18–20, we are estimating p and we know the value of $\hat{p}$. Find the required sample size.

18. Confidence level 99%, margin of error 0.03, $\hat{p} = 0.9$

19. Confidence level 95%, margin of error 0.03, $\hat{p} = 0.99$

20. Confidence level 95%, margin of error 0.03, $\hat{p} = 0.999$

For Exercises 21–23, we are estimating p and we do not know the value of $\hat{p}$. Find the required sample size.

21. Confidence level 90%, margin of error 0.05

22. Confidence level 90%, margin of error 0.03

23. Confidence level 90%, margin of error 0.01

SECTION 8.4

For Exercises 24–27, a random sample is drawn from a normal population. The sample of size $n = 36$ has a sample variance of $s^2 = 100$. Construct the specified confidence interval.

24. 90% confidence interval for the population variance σ^2

25. 95% confidence interval for the population variance σ^2

26. 90% confidence interval for the population standard deviation σ

27. 95% confidence interval for the population standard deviation σ

28. Union Membership. The table contains the total union membership for seven randomly selected states in 2006. Construct and interpret a 95% confidence interval for σ. Assume the data are normally distributed.

State	Union membership (thousands)
Florida	397
Indiana	334
Maryland	342
Massachusetts	414
Minnesota	395
Texas	476
Wisconsin	386

Source: U.S. Bureau of Labor Statistics.

CHAPTER 8 *Quiz*

TRUE OR FALSE

1. True or false: In Table 8.2 (page 361), since the confidence level is 95%, then 95% of the intervals must contain μ. Explain your answer.

2. True or false: The t curve is symmetric about 0, just like the Z curve was. Therefore we can use all our symmetry techniques with the t curve as well.

FILL IN THE BLANK

3. Suppose we cut a margin of error in half. The sample size requirement then becomes _____ times larger.

4. Our estimate of μ is _____ precise using the t curve rather than the Z curve.

SHORT ANSWER

5. α is used to find the value of $Z_{\alpha/2}$. Is α a probability or a value of x or a value of Z?

6. What are the conditions for constructing a t interval?

CALCULATIONS AND INTERPRETATIONS

7. College Education Costs. A random sample of 49 colleges yielded a mean cost of college education of $30,500 per year. Assume that the population standard deviation is $3000.

    a. Compute and interpret the margin of error for a confidence interval with 90% confidence.

    b. Construct and interpret a 90% confidence interval for the population mean cost of college education.

8. Crash Test Data. The National Highway Traffic Safety Administration collects data on crash tests for new motor vehicles. They reported that the mean femur load (force applied to the femur) in a frontal crash for the passenger in a 2005 Ford Equinox SUV was 1003 pounds. Assume that the population standard deviation was 210 pounds and the sample size was 49.

    a. Compute and interpret the margin of error for a confidence interval with 90% confidence.

    b. Construct and interpret a 90% confidence interval for the population mean femur load in a frontal crash for the passenger in a 2005 Ford Equinox SUV.

9. Wolves Returning to Isle Royale? A random sample of 81 day-counts of wolves from Isle Royale in the middle of Lake Superior resulted in a mean of 57 wolves and a standard deviation of 10 wolves. Assume that the data are normally distributed. Calculate and interpret a 90% confidence interval for the population mean day-count of wolves in Isle Royale.

10. The Fog Index. The Fog Index is a scale that measures the readability of a particular English passage. For example, a Fog Index of 11.5 indicates that the material is about eleventh-grade level. A random sample of 64 passages by a well-known horror author yielded a mean Fog Index of 9.5 with a standard deviation of 0.8. Construct and interpret a 95% confidence interval for the mean Fog Index for all the books by this horror author.

11. 9/11 and Religious Attendance. The Pew Research Center reported that, in a survey of 3733 randomly selected respondents, 991 had attended a religious service in response to the attacks on the World Trade Center and the Pentagon.

    a. If appropriate, find the margin of error for confidence level 95%. What does this number mean?

    b. Construct, if appropriate, a 95% confidence interval for the population proportion of Americans who attended a religious service in response to the attacks on the World Trade Center and the Pentagon.

12. Independence for Quebec? A January 2006 poll by the newspaper *La Presse* reported that 340 of 1000 randomly chosen Quebec adults surveyed would vote "Yes" in a referendum for independence from Canada.

    a. If appropriate, find the margin of error for confidence level 99%. What does this number mean?

    b. If appropriate, find a 99% confidence interval for the population proportion of all Quebec residents who favor independence for the Province of Quebec.

13. Delivery Times. Recall from Section 3.2 that Amanda and Bethany have contracted with a local document dispatch service to deliver documents in midtown Manhattan. One particularly important client kept careful track of the delivery times (in minutes) for documents delivered by Amanda and Bethany, shown in the accompanying table. Assume both data sets are normally distributed.

Amanda	15	17	12	15	10	16	14	20	16	15
Bethany	5	22	15	25	4	18	10	29	17	5

   a. Construct and interpret a 95% confidence interval for the population standard deviation of Amanda's delivery times.
   b. Construct and interpret a 95% confidence interval for the population standard deviation of Bethany's delivery times.

14. Rather Be Fishing? A study found that Minnesota, at 38%, leads the nation in the proportion of people who go fishing. How large a sample size is required to estimate p to within 0.05 with 95% confidence?

15. Genetic Analysis. A biotechnician would like to estimate the mean number of genes that she will have to analyze in her research. Assume that the standard deviation is 120 genes. How many genes does she have to sample if she would like the estimate to be accurate to within 10 genes with 95% confidence?

16. Quality of Education in America. The National Assessment of Educational Progress (NAEP) administers exams to a nationwide sampling of students to assess the quality of education in America. Suppose NAEP would like to estimate the population proportion of American schoolchildren who would answer a given question correctly. Find a sample size which would give a margin of error of 0.03 with 90% confidence.

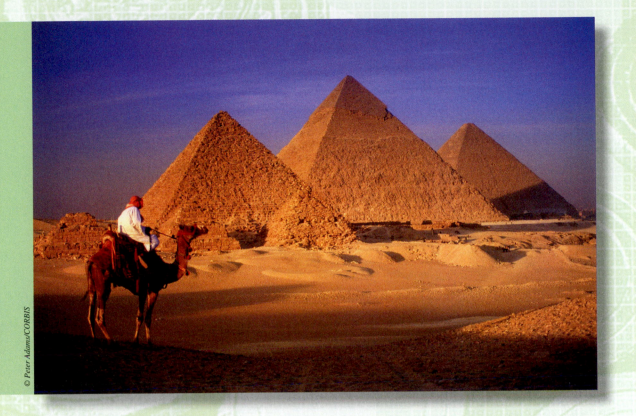

© Peter Adams/CORBIS

Hypothesis Testing

CASE STUDY | **The Golden Ratio**

What do Euclid's *Elements*, the Parthenon of ancient Greece, the *Mona Lisa*, and the beadwork of the Shoshone tribe of Native Americans have in common? An appreciation for the *golden ratio*. Suppose we have two quantities A and B, with $A > B > 0$. Then, A/B is called the golden ratio if

$$\frac{A + B}{A} = \frac{A}{B}$$

that is, if the ratio of the sum of the quantities to the larger quantity equals the ratio of the larger to the smaller.

The golden ratio permeates ancient, medieval, Renaissance, and modern art and architecture. For example, the Egyptians constructed their great pyramids using the golden ratio. (Specifically, in Figure 9.1, if $A = \overline{XY}$ is the height from the top vertex to the base, and $B = \overline{YZ}$ is the distance from the center of the base to the edge, then $(A + B)/A = A/B$.) Some mathematicians have said that the golden ratio may be intrinsically pleasing to the human species. Support for this conjecture would

FIGURE 9.1

be especially strong if evidence was found for the use of the golden ratio in non-Western artistic traditions. In the Case Study on pages 450–451, we use hypothesis testing to determine whether the decorative beaded rectangles sewn by the Shoshone tribe of Native Americans follow the golden ratio.

OVERVIEW

The Big Picture

Where we are coming from, and where we are headed…

Hypothesis testing is the most widely used method for statistical inference in the world. Hypothesis testing forms the bedrock of the scientific method, and it touches nearly every field of scientific endeavor, from biology and genetics, through physics and chemistry, to sociology and psychology. It also forms the basis for decision science, and thus for all business-oriented decision-making methods. In this chapter we will discover what hypothesis testing is all about.

In Chapter 8, we learned about confidence interval estimation. But confidence intervals represent only the first of a large family of topics in inferential statistics. In fact, each of Chapters 8–11 covers a fresh and distinct facet of statistical inference. In Chapters 7 and 8, we learned that, since the value of the target population parameter is unknown, we need to estimate it using point and interval estimates provided by sample data. However, because the target population parameter is unknown, people may disagree about its value. In this chapter, we introduce the statistical method for resolving conflicting claims about the value of a population parameter, hypothesis testing.

Researchers are interested in investigating many different types of questions, such as the following:

- An accountant may wish to examine whether evidence exists for corporate tax fraud.

- A Department of Homeland Security executive may want to test whether a new surveillance method will uncover terrorist activity.

- A sociologist may want to examine whether the mayor's economic policy is increasing poverty in the city.

Questions such as these can be tackled using statistical **hypothesis testing,** which is a process for rendering a decision about the unknown value of a population parameter. We will learn how to make decisions about the values of a population mean, a population proportion, and a population standard deviation.

In Chapter 10, we will consider confidence intervals and hypothesis tests for two-sample problems. ■

9.1 Introduction to Hypothesis Testing

Objectives: By the end of this section, I will be able to...

1 Construct the null hypothesis and the alternative hypothesis from the statement of the problem.

2 State the two types of errors made in hypothesis tests: the Type I error, made with probability α, and the Type II error, made with probability β.

1 Constructing the Hypotheses

Recall that inferential statistics refers to methods for learning about the unknown value of a population parameter by studying the information in a sample. The basic idea of hypothesis testing is the following:

1. We need to make a *decision* about the value of a population parameter (such as μ or p).
2. Unfortunately, the true value of that parameter is *unknown*.
3. Therefore, there may be different *hypotheses* about the true value of this parameter.

Statistical hypothesis testing is a way of formalizing the decision-making process so that a decision can be rendered about the value of the parameter. **Hypothesis testing** is a procedure where claims about the value of a population parameter may be investigated using the sample evidence. We craft *two competing statements (hypotheses)* about the value of the population parameter (either μ, p, or σ). We must find evidence to conclude that one of the hypotheses is likely to be true.

The Hypotheses
- The status quo hypothesis represents what has been tentatively assumed about the value of the parameter and is called the **null hypothesis,** denoted as H_0.
- The **alternative hypothesis,** or **research hypothesis,** denoted as H_a, represents an alternative claim about the value of the parameter.

Hypothesis testing is like conducting a criminal trial. In a trial in the United States, the defendant is innocent until proven guilty, and the jury must evaluate the truth of two competing hypotheses:

H_0: defendant is not guilty   versus H_a: defendant is guilty

The not-guilty hypothesis is considered the **null hypothesis H_0** because the jurors must assume it is true until proven otherwise. The **alternative hypothesis H_a,** that the defendant is guilty, *must be demonstrated* to be true, beyond a reasonable doubt. How does a court of law determine whether the defendant is convicted or acquitted? This judgment is based upon the *evidence,* the hard facts heard in court. Similarly, in hypothesis testing, the researcher draws a conclusion based on the evidence provided by the sample data.

In Sections 9.1–9.4, we will examine hypotheses for the unknown mean μ. The null hypothesis will be a claim about certain values for μ, and the alternative hypothesis will be a claim about other values for μ. We will begin by considering hypotheses about the population mean μ for some fixed-value μ_0 (that is, μ_0 is just a specified value for the unknown mean μ). The hypotheses have one of the three possible forms shown in Table 9.1. The right-tailed test and the left-tailed test are called one-tailed tests. In Section 9.3 we will find out why we use this terminology.

Table 9.1 The three possible forms for the hypotheses for a test for μ

Form	Null and alternative hypotheses
Right-tailed test	$H_0: \mu = \mu_0$ versus $H_a: \mu > \mu_0$
Left-tailed test	$H_0: \mu = \mu_0$ versus $H_a: \mu < \mu_0$
Two-tailed test	$H_0: \mu = \mu_0$ versus $H_a: \mu \neq \mu_0$

Example 9.1   Invasive species in the Great Lakes

Edward Kinsman/Photo Researchers

Invasive species discharged from the ballast tanks of transoceanic ships are a growing problem in the Great Lakes. The round goby is an invasive species that carries the parasite *Neochasmus umbellus.* A 2007 study reported that the mean intensity (mean number of parasites per infected host) was 7.2 parasites.[1] Scientists worry that the population mean number of parasites in the round goby has *increased* since 2007 and want to test that hypothesis.

Solution

We can write this situation as two hypotheses about the mean population number of parasites. Note that the term *increased* means *greater than.* Thus, we write a null hypothesis and an alternative hypothesis for a right-tailed test:

$$H_0: \mu = 7.2 \quad \text{versus} \quad H_a: \mu > 7.2$$

The null hypothesis H_0 states that the population mean μ equals 7.2 parasites per fish. The alternative hypothesis H_a states that the population mean number of parasites is *more than* 7.2. Here, $\mu_0 = 7.2$, which is the possible value of μ specified in the example. ■

Recall that each hypothesis is a claim that the statement is true. We must find evidence that is consistent with one of the two claims H_0 or H_a. The first task in hypothesis testing is to form hypotheses. To convert a word problem into two hypotheses, look for certain key words that can be expressed mathematically. Table 9.2 shows how to convert words typically found in word problems into symbols.

Table 9.2 Key English words, with mathematical symbols and synonyms

English words	Symbols	Synonyms
Equal	$=$	Is; is the same as
Not equal	$\neq$	Is different from; has changed from; differs from
Greater than	$>$	Is more than; is larger than; exceeds
Less than	$<$	Is below; is smaller than

Once you have identified the key words, use the associated mathematical symbol to write two hypotheses. The following strategies can be used to write the hypotheses.

Strategy for Constructing the Hypotheses About μ

Step 1   Search the word problem for certain key English words and select the associated symbol from Table 9.2.

Step 2   Determine the form of the hypotheses listed in Table 9.1 that uses this symbol.

Step 3   Find the value of μ_0 (the number that answers the question: "greater than what?" or "less than what?") and write your hypotheses in the appropriate forms.

Example 9.2   Has mean gas mileage continued to fall?

According to the U.S. Department of Transportation, the mean gas mileage of U.S. passenger cars (exclusive of SUVs, trucks, and vans) decreased from 21.2 miles per gallon (mpg) in 2001 to 20.8 mpg in 2004. Has this downward trend continued until today? Economists are interested in testing whether the population mean gas mileage for U.S. passenger cars has decreased from the mean of 20.8 mpg in 2004. Write a null hypothesis and an alternative hypothesis that describe this situation.

Solution

Let's use our strategy to construct the hypotheses needed to test this claim.

Step 1   **Search the word problem for certain key English words and select the appropriate symbol.** The problem uses the words "decreased from," which means "less than." Thus, we will write a hypothesis that contains the $<$ symbol.

Step 2   **Determine the form of the hypotheses.** From Table 9.1, we see that the symbol $<$ means that we use a left-tailed test:

$$H_0: \mu = \mu_0 \quad \text{versus} \quad H_a: \mu < \mu_0$$

Step 3   **Find the value for μ_0 and write your hypotheses.** The alternative hypothesis H_a states that the mean gas mileage μ is less than some value μ_0. Less than what? The 20.8 mpg from 2004. Write the two hypotheses with $\mu_0 = 20.8$.

$$H_0: \mu = 20.8 \quad \text{versus} \quad H_a: \mu < 20.8$$

 CAUTION

Do not blindly apply this strategy without thinking about what you are doing. Rather, use the strategy to help formulate your own hypotheses. *There is no substitute for thinking through the problem!*

Note: The test statistic is called Z_{data} because its value is derived largely from the sample data and the population data. The sample mean $\bar{x}$ and the sample size n are characteristics of the sample data, and the population standard deviation σ is a characteristic of the population data.

Example 9.3   Finding Z_{da}

Unusual and extreme values of $\bar{x}$, and therefore of Z_{data}, will have a small p-value, while values of $\bar{x}$ and Z_{data} nearer to the center of the distribution will have a large p-value.

> **Assuming H_0 is true:**
>
> Unusual and extreme values of $\bar{x}$ and Z_{data} ↔ Small p-value (close to 0)
>
> Values of $\bar{x}$ and Z_{data} near center ↔ Large p-value (greater than, say, 0.15)

A small p-value indicates a conflict between your sample data and the null hypothesis. Therefore, since the data are real and the hypothesis is a conjecture, this *small p-value will lead us to reject H_0.*

However, *how small is small?* We have already met another fellow in this chapter, a type of probability called α, the probability of a Type I error. (Recall that a Type I error is rejecting H_0 when H_0 is true.) The researcher chooses α to be some small constant like 0.01, 0.05, or 0.10 and then compares the p-value with α. A p-value is small if it is smaller than α. This leads us to the **rejection rule** that tells us the conditions under which we may reject the null hypothesis. In fact, this rejection rule can be applied to any type of p-value hypothesis test we perform.

> The rejection rule for performing a hypothesis test using the p-value method is
> **Reject H_0 when the p-value is less than α. Otherwise do not reject H_0.**

Note that this rejection rule calls for a new role for α.

> **Level of Significance, α**
> The value of α represents the boundary between results that are statistically significant (where we reject H_0) and results that are not statistically significant (where we do not reject H_0). Thus, α is called the level of significance of the hypothesis test.

𝒲hat Does the Conclusion Mean?

Interpreting Your Conclusion for Nonspecialists

Recall that a data analyst needs to interpret the results so that nonspecialists can understand them. You can use the following generic interpretation for the two possible conclusions. Just remember that generic interpretations are no substitute for thinking clearly about the problem and the implications of the conclusion. ■

> **Interpreting the Conclusion**
> * If you reject H_0, the interpretation is *"There is evidence that [whatever H_a says]."*
> * If you do not reject H_0, the interpretation is *"There is insufficient evidence that [whatever H_a says]".*

Example 9.4	*p*-value for the price of CDs

For the data from Example 9.3:
a. Find the *p*-value.
b. Compare the *p*-value with our level of significance $\alpha = 0.05$, and draw a conclusion.
c. Interpret the conclusion.

Solution

a. Since we have a right-tailed test,

$$H_0: \mu = 10 \quad \text{versus} \quad H_a: \mu > 10$$

and $Z_{data} = 3.33$, we have

$$p\text{-value} = P(Z > Z_{data}) = P(Z > 3.33)$$

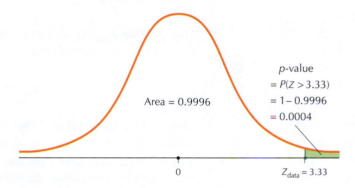

Area = 0.9996

p-value
$= P(Z > 3.33)$
$= 1 - 0.9996$
$= 0.0004$

$Z_{data} = 3.33$

0

FIGURE 9.6

as shown in Figure 9.6. The Z table tells us that the area to the left of $Z_{data} = 3.33$ is 0.9996. Thus, using Case 2 from Table 6.7 in Section 6.4 (page 285), we find the area to the right of $Z_{data} = 3.33$ to be

$$p\text{-value} = 1 - 0.9996 = 0.0004$$

What Results Might We Expect?

The value of the test statistic $Z_{data} = 3.33$ was shown to lie deep in the right-hand tail of Figure 9.6. This is evidence that our observed sample mean $\bar{x} = \$15$ is unusual or extreme if we assume that H_0 is correct. Thus, by the "essential idea about hypothesis testing for the mean," we expect our conclusion to be "Reject H_0." ■

b. Since $\alpha = 0.05$, we will reject H_0 if *p*-value < 0.05. Since our *p*-value 0.0004 is less than our level of significance $\alpha = 0.05$, our conclusion is therefore to reject H_0, as expected. The difference between $\bar{x} = \$15$ and $\mu_0 = \$10$ is statistically significant.
c. Since the conclusion for the hypothesis test is to reject H_0, the generic interpretation is: "There is evidence that [whatever H_a says]." Here, H_a states that $\mu > 10$; that is, the population mean cost of a CD is greater than $10. Hence, we interpret our conclusion as follows: "There is evidence that the population mean cost of a CD is greater than \$10." ■

3 Performing the *Z* test for the Population Mean Using the *p*-Value Method

We summarize the methods used in Examples 9.3 and 9.4 in the following step-by-step method for performing the *Z* test for μ using the *p*-value method.

Z Test for the Population Mean μ: *p*-Value Method

When a random sample of size *n* is taken from a population where the population standard deviation σ is known, you can use the *Z* test if either of the following conditions is satisfied:

- Case 1: the population is normal, or

- Case 2: the sample size is large ($n \geq 30$).

Step 1   State the hypotheses and the rejection rule. Use one of the forms from Table 9.1 to write H_0 and H_a. Define μ (that is, "μ represents the population mean ___"). The rejection rule is "Reject H_0 if the *p*-value is less than α."

Step 2   Find Z_{data}. Either use technology to find the value of the test statistic Z_{data}, or calculate the value of Z_{data} as follows:

$$Z_{\text{data}} = \frac{\bar{x} - \mu_0}{\sigma_{\bar{x}}} = \frac{\bar{x} - \mu_0}{\sigma/\sqrt{n}}$$

Step 3   Find the *p*-value. Either use technology to find the *p*-value, or calculate it using the form in Table 9.4 that corresponds to your hypotheses.

Step 4   State the conclusion and interpretation. If the *p*-value is less than α, then reject H_0. Otherwise, do not reject H_0. Interpret your conclusion so that a nonspecialist can understand.

Example 9.5   The *Z* test for the mean using the *p*-value method: left-tailed test

The technology Web site **www.cnet.com** publishes user reviews of computers, software, and other electronic gadgetry. The mean user rating, on a scale of 1–10, for the Dell XPS 410 desktop computer as of September 10, 2007, was 7.2. Assume that the population standard deviation of user ratings is known to be $\sigma = 0.9$. A random sample taken this year of $n = 81$ user ratings for the Dell XPS 410 showed a mean of $\bar{x} = 7.05$. Using level of significance $\alpha = 0.05$, test whether the population mean user rating for this computer has fallen since 2007.

Solution
The sample size $n = 81$ is large, and the population standard deviation is known. We may therefore perform the *Z* test for the mean.

Step 1   State the hypotheses and the rejection rule. The key words here are "has fallen," which means "is less than." The answer to the question "Less than what?" gives us $\mu_0 = 7.2$. Thus, our hypotheses are

$$H_0: \mu = 7.2 \quad \text{versus} \quad H_a: \mu < 7.2$$

where μ refers to the population mean user rating for the Dell XPS 410 computer. We will reject H_0 if the *p*-value is less than $\alpha = 0.05$.

Step 2   **Find Z_{data}.** We have $\bar{x} = 7.05$, $\mu_0 = 7.2$, $n = 81$, and $\sigma = 0.9$. Thus, our test statistic is

$$Z_{\text{data}} = \frac{\bar{x} - \mu_0}{\sigma_{\bar{x}}} = \frac{\bar{x} - \mu_0}{\sigma/\sqrt{n}} = \frac{7.05 - 7.2}{0.9/\sqrt{81}} = \frac{-0.15}{0.1} = -1.5$$

Step 3   **Find the *p*-value.** Our hypotheses represent a left-tailed test from Table 9.4. Thus,

$$p\text{-value} = P(Z < Z_{\text{data}}) = P(Z < -1.5)$$

This is a Case 1 problem from Table 6.7 (page 285). The Z table provides us with the area to the left of $Z = -1.5$ (Figure 9.7):

$$P(Z < -1.5) = 0.0668$$

Thus, the *p*-value is 0.0668.

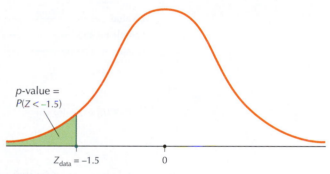

FIGURE 9.7 The *p*-value for a left-tailed test.

Step 4   **State the conclusion and interpretation.** Our level of significance is $\alpha = 0.05$. (Step 1) Since the *p*-value = 0.0668 is *not* < 0.05, we do *not* reject H_0. There is insufficient evidence that the population mean user rating for a Dell XPS 410 computer is less than 7.2. ■

Example 9.6   The *p*-value method using technology: two-tailed test

The birth weights, in grams (1000 grams = 1 kilogram ≈ 2.2 pounds), of a random sample of 44 babies from Brisbane, Australia, have a sample mean weight is $\bar{x} = 3276$ grams. Formerly, the mean birth weight of babies in Brisbane was 3200 grams. Assume that the population standard deviation $\sigma = 528$ grams. Is there evidence that the population mean birth weight of Brisbane babies now differs from 3200 grams? Use technology to perform the appropriate hypothesis test, with level of significance $\alpha = 0.10$.

Solution

Since the sample size $n = 44$ is large and $\sigma = 528$ is known, we may proceed with the hypothesis test, using Case 2 ($n \geq 30$).

Step 1   **State the hypotheses and the rejection rule.** The key words *differs from* means that we have a two-tailed test:

$$H_0: \mu = 3200 \quad \text{versus} \quad H_a: \mu \neq 3200$$

where μ refers to the population mean birth weight of Brisbane babies. We will reject H_0 if the p-value is less than $\alpha = 0.10$.

Step 2   **Find Z_{data}.** We will use the instructions provided in the Step-by-Step Technology Guide at the end of this section (page 429). Figure 9.8 shows the TI-83/84 results from the Z test for μ:

$$Z_{\text{data}} = \frac{\bar{x} - \mu_0}{\sigma/\sqrt{n}} = \frac{3276 - 3200}{528/\sqrt{44}} = 0.9547859245 \approx 0.9548$$

Form of H_a:

Z_{data}

p-value

Sample mean $\bar{x}$

Sample size n

FIGURE 9.8 TI-83/84 results.

Figure 9.9 shows the Minitab results, where

- **Test of mu = 3200 vs. not 3200** refers to the hypotheses being tested, $H_0 : \mu = 3200$ versus $H_a : \mu \neq 3200$.
- **The assumed standard deviation = 528** refers to our assumption that $\sigma = 528$.
- **SE Mean** refers to the standard deviation of the sampling distribution of $\bar{x}$, that is, $\sigma/\sqrt{n}$. You can see that $528/\sqrt{44} \approx 79.60$.
- **90% CI** represents a 90% Z confidence interval for μ.
- **Z** refers to our test statistic:

$$Z_{\text{data}} = \frac{\bar{x} - \mu_0}{\sigma/\sqrt{n}} = (3276 - 3200)/(528/\sqrt{44}) \approx 0.95$$

- **P** represents our p-value of 0.340.

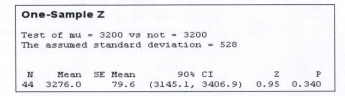

One-Sample Z

Test of mu = 3200 vs not = 3200
The assumed standard deviation = 528

N	Mean	SE Mean	90% CI	Z	P
44	3276.0	79.6	(3145.1, 3406.9)	0.95	0.340

FIGURE 9.9 Minitab results.

SUMMARY STATISTICS		
Sample Mean	**Sample Size**	**Standard Deviation**
3276	44	528

ONE SAMPLE Z TEST - CONFIDENCE INTERVAL		
Confidence Level	**Z Value**	**Critical Z Value**
90 %	0.954786	1.645

Confidence Interval
3276	+/-	130.9403
3145.06	to	3406.94

Population Mean (Null Hypothesis Value)
3200

Alternative Hypothesis	**P-Value**	
> 3200	0.169843	
		1-Sided
< 3200	0.830157	
Not = 3200	0.339686	**2-Sided**

FIGURE 9.10 WHFStat Add-ins for Excel results.

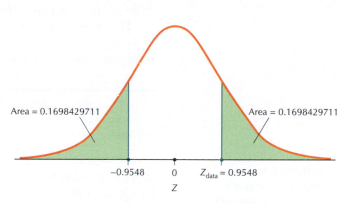

FIGURE 9.11 *P*-value is sum of two tail areas.

Figure 9.10 shows output from WHFStat Add-ins for Excel.

Step 3 **Find the *p*-value.** We have a two-tailed test from Step 1, so that from Table 9.4 our *p*-value is (Figure 9.11)

$$p\text{-value} = 2 \cdot P(Z > |Z_{\text{data}}|) = 2 \cdot P(Z > 0.9548) \approx 2 \cdot (0.1698429711)$$
$$= 0.3396859422 \approx 0.3397.$$

Step 4 **State the conclusion and interpretation.** Since 0.340 is *not* less than 0.10, we do *not* reject H_0. There is insufficient evidence that the population mean birth weight differs from 3200 grams. ■

The *p-Value* applet allows you to experiment with various hypotheses, means, standard deviations, and sample sizes in order to see how changes in these values affect the *p*-value.

4 Assessing the Strength of Evidence Against the Null Hypothesis

The hypothesis-testing methods we have shown so far deliver a simple "yes-or-no" conclusion: either "Reject H_0" or "Do not reject H_0." There is no indication of how strong the evidence is for rejecting the null hypothesis. Was the decision close? Was it a no-brainer? On the other hand, the *p*-value itself implicitly represents the *strength of evidence against the null hypothesis*. There is extra information here, which we should not ignore.

For instance, we can directly compare the results of hypothesis tests. Suppose that we have two hypothesis tests that both result in not rejecting the null hypothesis, with level of significance $\alpha = 0.05$. However, Test A has a *p*-value of 0.06, while Test B has a *p*-value of 0.57. Clearly, Test A came very close to rejecting the null hypothesis and shows a fair amount of evidence against the null hypothesis, while Test B shows no evidence at all against the null hypothesis. A simple statement of the "yes-or-no" conclusion misses the clear distinction between these two hypothesis tests.

The *p*-value provides us with the smallest level of significance at which the null hypothesis would be rejected, that is, the smallest value of α at which the results would be considered significant. Of course, we are free to determine whether the results are significant using whatever α level we wish. For example, Test A above would have rejected H_0 for any α level higher than 0.06. Some data analysts in fact do not think in terms of rejecting or not rejecting the null hypothesis. Rather, they think completely in terms of *assessing the strength of evidence against the null hypothesis.*

For many (though not all) data domains, Table 9.5 provides a thumbnail impression of the strength of evidence against the null hypothesis for various *p*-values. For certain domains (such as the physical sciences), however alternative interpretations are appropriate.

Table 9.5 Strength of evidence against the null hypothesis for various levels of *p*-value

p-Value	Strength of evidence against H_0
p-value ≤ 0.001	Extremely strong evidence
$0.001 < p$-value ≤ 0.01	Very strong evidence
$0.01 < p$-value ≤ 0.05	Solid evidence
$0.05 < p$-value ≤ 0.10	Mild evidence
$0.10 < p$-value ≤ 0.15	Slight evidence
$0.15 < p$-value	No evidence

Note: Use Table 9.5 for all exercises that ask for an assessment of the strength of evidence against the null hypothesis.

Example 9.7 Assessing the strength of evidence against H_0

Assess the strength of evidence against H_0 shown by the *p*-values in (a) Example 9.4, (b) Example 9.5, and (c) Example 9.6.

Solution

a. In Example 9.4 we tested $H_0: \mu = 10$ versus $H_a: \mu > 10$, where μ is the population mean price of CDs. Our *p*-value of 0.0004 implies that there is *extremely strong* evidence against the null hypothesis that the population mean price of CDs equals \$10.

b. In Example 9.5 we tested $H_0: \mu = 7.2$ versus $H_a: \mu < 7.2$ where μ refers to the population mean user rating for the Dell XPS 410 computer. Our *p*-value of 0.0668 implies that there is *mild* evidence against the null hypothesis that the population mean user rating for the Dell XPS 410 computer is 7.2 or higher.

c. In Example 9.6 we tested $H_0: \mu = 3200$ versus $H_a: \mu \neq 3200$ where μ refers to the population mean birth weight of Brisbane babies (in grams). Our *p*-value of 0.3397 implies that there is *no* evidence against the null hypothesis that the population mean birth weight of Brisbane babies equals 3200 grams. ∎

Developing Your Statistical Sense

The Role of the Level of Significance α

Suppose that in Example 9.5, our level of significance α was 0.10 rather than 0.05. Would this have changed anything? Certainly. Since our *p*-value of 0.0668 is less than the new $\alpha = 0.10$, we would now reject H_0. Think about that for a moment. *The data haven't changed at all, but our conclusion is reversed simply by changing α.* What is a data analyst to make of a situation like this? There are two alternatives.

1. Since we don't want the choice of α to dictate our conclusion, then perhaps we should turn to a direct assessment of the strength of evidence against the null hypothesis, as provided in Table 9.5. In this case, the *p*-value of about 0.0668 would offer mild evidence against the null hypothesis, *regardless of the value of α.*

2. Obtain more data, perhaps through a call for further research. ∎

STEP-BY-STEP GUIDE TO TECHNOLOGY: Z test for μ

We will use the weight data from Example 9.6 (page 425).

TI-83/84

If you have the data values:
Step 1 Enter the data into list **L1**.
Step 2 Press **STAT**, highlight **TESTS**, and press **ENTER**.
Step 3 Press **1** (for **Z-Test**; see Figure 9.12).
Step 4 For input (**Inpt**), highlight **Data** and press **ENTER** (Figure 9.13).
a. For μ_0, enter the value of μ_0, **3200**.
b. For σ, enter the value of σ, **528**.
c. For **List**, press **2nd**, then **L1**.
d. For **Freq**, enter **1**.
e. For μ, select the form of H_a. Here we have a two-tailed test, so highlight $> \mu_0$ and press **ENTER**.
f. Highlight **Calculate** and press **ENTER**. The results are shown in Figure 9.8 in Example 9.6.

If you have the summary statistics:
Step 1 Press **STAT**, highlight **TESTS**, and press **ENTER**.
Step 2 Press **1** (for **Z-Test**; see Figure 9.12).
Step 3 For input (**Inpt**), highlight **Stats** and press **ENTER** (Figure 9.14).
a. For μ_0, enter the value of μ_0, **3200**.
b. For σ, enter the value of σ, **528**.
c. For $\bar{x}$, enter the sample mean **3276**.
d. For **n**, enter the sample size **44**.
e. For μ, select the form of H_a. Here we have a two-tailed test, so highlight $> \mu_0$ and press **ENTER**.
f. Highlight **Calculate** and press **ENTER**. The results are shown in Figure 9.8 in Example 9.6.

Figure 9.12

Figure 9.13

Figure 9.14

EXCEL

WHFStat Add-ins
Step 1 Enter the data into column A. (If you have only the summary statistics, go to Step 2.)
Step 2 Load the **WHFStat Add-ins**.
Step 3 Select **Add-Ins > Macros > Testing a Mean > Z Test − Confidence Interval − One Sample**.

Step 4 Select cells A1 to A44 as the **Dataset Range**. (Alternatively, you may enter the summary statistics.)
Step 5 Select your **Confidence level**, which should be $1 - \alpha$. Here, because $\alpha = 0.10$, we select **90%**.
Step 6 Enter the **Population Standard Deviation**, $\sigma = 528$.
Step 7 Enter the **Null Hypothesis Value**, $\mu_0 = 3200$, and click **OK**. The results are shown in Figure 9.10 in Example 9.6.

MINITAB

If you have the data values:
Step 1 Enter the data into column C1.
Step 2 Click **Stat > Basic Statistics > 1-Sample Z**.
Step 3 Click **Samples in Columns** and select **C1**.
Step 4 Enter **528** as **Standard Deviation**.
Step 5 For **Test Mean**, enter **3200**.
Step 6 Click **Options**.
a. Choose your **Confidence Level** as $100(1 - \alpha)$. Our level of significance α here is 0.10, so the confidence level is 90.0.
b. Select **not equal** to symbolize the two-tailed test.

Step 7 Click **OK** and click **OK** again. The results are shown in Figure 9.9 in Example 9.6.

If you have the summary statistics:
Step 1 Click **Stat > Basic Statistics > 1-Sample Z**.
Step 2 Click **Summarized Data**.
Step 3 Enter the **Sample Size 44** and the **Sample Mean 3276**.
Step 4 Click **Options**.
a. Choose your **Confidence Level** as $100(1 - \alpha)$. Our level of significance α here is 0.10, so the confidence level is 90.0.
b. Select **not equal** to symbolize the two-tailed test.
Step 5 Click **OK** and click **OK** again. The results are shown in Figure 9.9 in Example 9.6.

Section 9.2 SUMMARY

The essential idea about hypothesis testing for the mean may be described as follows. When the observed value of $\bar{x}$ is unusual or extreme in the sampling distribution of $\bar{x}$ that is based on the assumption that H_0 is correct, we should reject H_0. Otherwise, there is insufficient evidence against H_0, and we should not reject H_0. All of the remaining steps and calculations are simply explanations of how to implement this essential idea about hypothesis testing for the mean.

2 The *p*-value can be thought of as the probability of observing a sample statistic at least as extreme as the statistic in your sample if we assume that the null hypothesis is true. The rejection rule for performing a hypothesis test using the *p*-value method is to reject H_0 when the *p*-value is less than the level of significance α.

3 Use the Z test only when the population standard deviation σ is known and when either (Case 1) the population is normal or (Case 2) the sample size is large ($n \geq 30$).

4 The *p*-value can be used to assess the strength of evidence against the null hypothesis.

Section 9.2 EXERCISES

CLARIFYING THE CONCEPTS

1. Explain in your own words how a researcher determines if a particular value of the sample mean is extreme.
2. What do we mean by the term *test statistic*? On what does the value of the test statistic depend?
3. How do we tell when a *p*-value is small? That is, to what do we compare the *p*-value to determine if it is small?
4. Does α have anything to do with the sample data?

PRACTICING THE TECHNIQUES

For Exercises 5–8, Suppose that we are testing the hypotheses $H_0 : \mu = \mu_0$ versus $H_a : \mu < \mu_0$ using level of significance $\alpha = 0.05$, and that the conditions for performing the Z test are satisfied. For each exercise, provide
   a. the conclusion and
   b. the interpretation of the conclusion.
5. *p*-value = 0.01
6. *p*-value = 0.06
7. *p*-value = 0.5
8. *p*-value = 0.05

For Exercises 9–12, suppose that we are testing the hypotheses $H_0 : \mu = \mu_0$ versus $H_a : \mu > \mu_0$, and that the conditions for performing the Z test are satisfied. For each exercise,
   a. draw a sketch of the Z distribution, showing Z_{data} as a point on the number line and the *p*-value as an area under the curve, and
   b. calculate the *p*-value.
9. $Z_{data} = 1.0$
10. $Z_{data} = 1.5$
11. $Z_{data} = 2.0$
12. $Z_{data} = 2.5$
13. What pattern do you observe in the *p*-value for right-tailed tests as the value of Z_{data} increases (gets more extreme)?

For Exercises 14–17:
   a. State the hypotheses and the rejection rule.
   b. Find Z_{data}, and round it to two decimal places.
   c. Find the *p*-value, using Table 9.4. Draw the standard normal curve, with Z_{data} and the *p*-value indicated on it.
   d. Formulate and interpret your conclusion.
14. We are interested in testing at level of significance $\alpha = 0.05$ whether the population mean is greater than 100. A

random sample of size 64 is taken, with a mean of 102. Assume $\sigma = 8$.
15. We would like to test at level of significance $\alpha = 0.10$ whether the population mean is less than 3.0. A random sample of size 36 is taken, with a mean of 2.9. Assume $\sigma = 0.5$.
16. We want to test at level of significance $\alpha = 0.01$ whether the population mean differs from 20. A random sample of size 49 is taken, with a mean of 27. Assume $\sigma = 5$.
17. We are interested in testing at level of significance $\alpha = 0.05$ whether the population mean is greater than 0. A random sample of size 16 is taken, with a mean of 2. Assume $\sigma = 4$ and that the population is normally distributed.

For the hypothesis tests in Exercises 18–23, assess the strength of evidence against the null hypothesis using Table 9.5.

18. Exercise 6	19. Exercise 7
20. Exercise 14	21. Exercise 15
22. Exercise 16	23. Exercise 17

APPLYING THE CONCEPTS
For Exercises 24–30, do the following.
   a. State the hypotheses and the rejection rule.
   b. Find Z_{data}.
   c. Find the *p*-value.
   d. State the conclusion and the interpretation.
24. **Stock Market.** The *Statistical Abstract of the United States* reports that the mean daily number of shares traded on the New York Stock Exchange in 2005 was 1.6 billion. Let this value represent the hypothesized population mean, and assume that the population standard deviation equals 0.5 billion shares. Suppose that, in a random sample of 36 days from the present year, the mean daily number of shares traded equals 1.5 billion. We are interested in testing whether the population mean daily number of shares traded has increased since 2005, using level of significance $\alpha = 0.05$.
25. **Child Abuse.** The U.S. Administration for Children and Families reported that the national rate for child abuse referrals was 43.9 per 1000 children in 2005. Suppose that a random sample of 1000 children taken this year shows 47 child abuse referrals. Assume $\sigma = 5$. Test whether the population mean referral rate has increased this year from the 2005 level, using level of significance $\alpha = 0.10$.
26. **California Warming.** A 2007 report found that the mean temperature in California has increased from 1950 to

2000 by 2 degrees Fahrenheit (°F). Suppose that a random sample of 36 California locations showed a mean increase of 4°F over 1950 levels. Assume $\sigma = 0.5$. Test whether the population mean temperature increase in California is greater than 2°F, at level of significance $\alpha = 0.05$.

27. Eating Trends. According to an NPD Group report the mean number of meals prepared and eaten at home is less than 700 per year. Suppose that a random sample of 100 households showed a sample mean number of meals prepared and eaten at home of 650. Assume $\sigma = 25$. Test whether the population mean number of such meals is less than 700, using level of significance $\alpha = 0.10$.

28. DDT in Breast Milk. Researchers compared the amount of DDT in the breast milk of 12 Hispanic women in the Yakima Valley of Washington State with the amount of DDT in breast milk in the general U.S. population.[5] They measured the mean DDT level in the general population to be 47.2 parts per billion (ppb) and the mean DDT level in the 12 Hispanic women to be 219.7 ppb. Assume $\sigma = 36$ and a normally distributed population. Test whether the population mean DDT level in the breast milk of Hispanic women in the Yakima Valley is greater than that of the general population, using level of significance $\alpha = 0.01$.

29. Tree Rings. Do trees grow more quickly when they are young? The International Tree Ring Data Base collected data on a particular 440-year-old Douglas fir tree.[6] The mean annual ring growth in the tree's first 80 years of life was 1.4261 millimeters (mm). A random sample of size 100 taken from the tree's later years showed a sample mean growth of 0.56 mm per year. Assume $\sigma = 0.5$ mm and a normally distributed population. Test whether the population mean annual ring growth in the tree's later years is less than 1.4261 mm, using level of significance $\alpha = 0.05$.

30. Hybrid Vehicles. A 2006 study by **Edmunds.com** examined the time it takes for owners of hybrid vehicles to recoup their additional initial cost through reduced fuel consumption. Suppose that a random sample of 9 hybrid cars showed a sample mean time of 2.1 years. Assume that the population is normal with $\sigma = 0.2$. Test using level of significance $\alpha = 0.01$ whether the population mean time it takes owners of hybrid cars to recoup their initial cost is less than three years.

Sodium in Breakfast Cereal. Use the following information for Exercises 31–33. A random sample of 23 breakfast cereals containing sodium had a mean sodium content per serving of 192.39 grams. Assume that the population standard deviation equals 50 grams. We are interested in whether the population mean sodium content per serving is less than 210 grams.

31. a. The normal probability plot of the sodium content is shown in Figure 9.15. Should we proceed to apply the Z test? Why or why not?
   b. Test whether the population mean sodium content per serving is less than 210 grams, using level of significance $\alpha = 0.01$.

Figure 9.15 Normal probability plot for Exercises 31–33.

32. *What if* the population standard deviation of 50 grams had been a typo, and the actual population standard deviation was smaller. How would this have affected the following?
   a. The standard deviation of the sampling distribution
   b. Z_{data}
   c. *p*-value
   d. The conclusion

33. Assess the strength of the evidence against the null hypothesis.

Online Shopping Privacy Protection. Use the *p-Value* applet and the following information for Exercises 34–37. A 2007 Carnegie Mellon University study reports that online shoppers were willing to pay an average of an extra 60 cents on a \$15 purchase in order to have better online privacy protection.[7] Assume that the population standard deviation is 20 cents. Suppose that a study conducted this year found that the mean amount online shoppers were willing to pay for privacy was $\bar{x} = 65$ cents. Does this represent evidence that the mean amount online shoppers are willing to pay for privacy has increased from 60 cents? As it turns out, the answer depends on the sample size. If this $\bar{x} = 65$ cents comes from a large sample, then the evidence against H_0 will be stronger than if it comes from a small sample. Use the *P-Value* applet to see this, using level of significance $\alpha = 0.05$.

34. Enter $H_0 : \mu = 60$, $H_a : \mu > 60$, $\sigma = 20$, and $\bar{x} = 65$.
   a. Enter $n = 30$. Click **Show P** to get the *p*-value. What is the *p*-value? Sketch the normal curve, showing the *p*-value.
   b. Would we reject the null hypothesis based on these values of $\bar{x}$ and n?

35. Repeat Exercise 34 for $n = 40$. Compare the *p*-value with that from Exercise 34.

36. Repeat Exercise 34 for $n = 50$. Compare the *p*-value with those from Exercises 34 and 35.

37. Note that $\bar{x}$ has not changed in Exercises 34–36; only the sample size has changed. Write a sentence to describe the pattern you see in the *p*-values as the sample size increases.

Table 9.6 shows values of Z_{crit} for the most commonly used values of the level of significance α : 0.01, 0.05, and 0.10. In Example 9.8, the critical region consists of the values of the test statistic Z_{data} that are *greater than* Z_{crit} because the alternative hypothesis states that the population mean is *greater than* μ_0. This is why this form of the test is called a right-tailed test—because the critical region is in the right (upper) tail.

Example 9.9   Finding the critical region for a left-tailed test

According to a report in the *New England Journal of Medicine,* a regimen of estrogen plus progestin therapy can help reduce cholesterol levels in postmenopausal women with high cholesterol.[9] The study examined whether the population mean cholesterol level could be lowered by this regimen to below 305 milligrams per deciliter (mg/dl). Find the critical region for this hypothesis test, using level of significance $\alpha = 0.05$.

Solution

Our hypotheses are H_0: $\mu = 305$ versus H_a: $\mu < 305$ where μ is the population mean cholesterol level in milligrams per deciliter. Since this is a left-tailed test, and $\alpha = 0.05$, we obtain from Table 9.6 that the critical value Z_{crit} is -1.645. Because the alternative hypothesis states that the population mean is *less than* μ_0, we will reject the null hypothesis for values of the test statistic that are less than -1.645. The critical region consists of the values in the left tail that are the values of Z_{data} less than Z_{crit} (Figure 9.18). ■

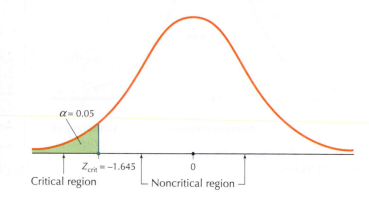

FIGURE 9.18
Critical region for a left-tailed test lies in the left (lower) tail.

Finally, for a two-tailed hypothesis test, H_0: $\mu = \mu_0$ versus H_a: $\mu \neq \mu_0$, the alternative hypothesis states that the population mean is *not equal to* μ_0. This can happen in either of two ways: (1) the population mean may be greater than μ_0, or (2) the population mean may be less than μ_0. So the critical region must contain values of Z_{data} that are either extremely large positively (in the right tail) or extremely large negatively (in the left tail), as shown in the following example. ■

Example 9.10   Finding the critical region for a two-tailed test

When the level of hemoglobin in the blood is too low, a person is anemic. Unusually high levels of hemoglobin are undesirable as well and can be associated with dehydration. We are interested in testing whether the population mean hemoglobin level differs from 13.8 grams per deciliter (g/dl), using level of significance $\alpha = 0.05$. Find the critical region for this test.

Solution

The key words "differs from" give us the folllowing hypotheses:

$$H_0: \mu = 13.8 \quad \text{versus} \quad H_a: \mu \neq 13.8$$

where μ is the population mean hemoglobin level. Then, from Table 9.6, the critical value Z_{crit} is 1.96. Because it is a two-tailed hypothesis test, we reject H_0 for values of Z_{data} that are either larger than 1.96 or smaller than -1.96. That is, the critical region consists of the values of Z_{data} that are greater than Z_{crit} and the values of Z_{data} that are less than $-Z_{\text{crit}}$ (Figure 9.19). ■

FIGURE 9.19
Critical region for a two-tailed test.

Table 9.7 summarizes the rejection rules for each form of the hypothesis test. Notice that you are essentially comparing the test statistic Z_{data} with the critical value Z_{crit} and basing your decision on their relative values.

Table 9.7 Rejection rules for *Z* test for the mean

Form of test		Rejection rules: "Reject H_0 if…"
Right-tailed	$H_0: \mu = \mu_0$ vs. $H_a: \mu > \mu_0$	$Z_{\text{data}} > Z_{\text{crit}}$
Left-tailed	$H_0: \mu = \mu_0$ vs. $H_a: \mu < \mu_0$	$Z_{\text{data}} < Z_{\text{crit}}$
Two-tailed	$H_0: \mu = \mu_0$ vs. $H_a: \mu \neq \mu_0$	$Z_{\text{data}} > Z_{\text{crit}}$ or $Z_{\text{data}} < -Z_{\text{crit}}$

Performing the *Z* Test for the Population Mean Using the Critical-Value Method

Now that we know how to find the critical value and the critical region, we are ready to carry out an actual hypothesis test. *It is important to understand that the critical-value method for the Z test is equivalent to the p-value method for the Z test.* Therefore, the conditions under which we may use the Z test critical-value method are the same as the conditions for using the Z test *p*-value method.

Z **Test for the Population Mean** μ**: Critical-Value Method**

When a random sample of size *n* is taken from a population where the population standard deviation σ is known, you can use the Z test if either of the following conditions is satisfied:

- Case 1: the population is normal, or
- Case 2: the sample size is large ($n \geq 30$).

Step 1 **State the hypotheses.** Use one of the forms from Table 9.1. Define μ.

Step 2 **Find Z_{crit} and state the rejection rule.** Use Tables 9.6 and 9.7.

Step 3   **Find Z_{data}.** Either use technology to find the value of the test statistic Z_{data} or calculate the value of Z_{data} as follows:

$$Z_{data} = \frac{\bar{x} - \mu_0}{\sigma_{\bar{x}}} = \frac{\bar{x} - \mu_0}{\sigma/\sqrt{n}}$$

Step 4   **State the conclusion and the interpretation.** If Z_{data} falls in the critical region, then reject H_0. Otherwise, do not reject H_0. Interpret your conclusion so that a nonspecialist can understand.

Example 9.11   **Critical-value method for the Z test for μ**

Let's use the critical-value method to test, using level of significance $\alpha = 0.01$, whether people use debit cards on average more than 11 times per month (Example 9.8). Suppose a random sample of 36 people used debit cards last month an average of $\bar{x} = 11.5$ times. Assume that the population standard deviation equals 3.

Solution

Step 1   **State the hypotheses.** In Example 9.8 our hypotheses were

$$H_0: \mu = 11 \quad \text{versus} \quad H_a: \mu > 11$$

where μ is the population mean number of times people use their debit cards per month.

Step 2   **Find Z_{crit} and state the rejection rule.** We have a right-tailed test and level of significance $\alpha = 0.01$, which, from Table 9.6, tell us that $Z_{crit} = 2.33$. Using Table 9.7, because we have a right-tailed test, the rejection rule will be "Reject H_0 if Z_{data} is greater then Z_{crit}"; that is, "Reject H_0 if Z_{data} is greater than 2.33" (see Figure 9.20).

Step 3   **Find Z_{data}.** We have a sample of $n = 36$ people who used their debit cards an average of $\bar{x} = 11.5$ times, and we assume $\sigma = 3$. Also, μ_0 is the hypothesized value of μ, in this case 11. Using these values, we get

$$Z_{data} = \frac{\bar{x} - \mu_0}{\sigma_{\bar{x}}} = \frac{\bar{x} - \mu_0}{\sigma/\sqrt{n}} = \frac{11.5 - 11.0}{3/\sqrt{36}} = 1.0$$

Step 4   **State the conclusion and interpretation.** Our rejection rule states that we will reject H_0 if Z_{data} is greater than 2.33. Since Z_{data} equals 1.0, which is *not* greater than 2.33, the conclusion is to *not* reject H_0 (Figure 9.20). Even though the sample mean of 11.5 exceeds 11.0, it does not do so by a wide enough margin to dispel the reasonable doubt that the difference between this sample mean and μ may have been due to chance. We interpret our conclusion as follows: "There is insufficient evidence that the population mean monthly debit card use is greater than 11 times per month." ■

FIGURE 9.20 Critical region for a right-tailed test.

3 Using Confidence Intervals for μ to Perform Two-Tailed Hypothesis Tests About μ

Since our Z interval for μ from Chapter 8 and our Z test here in Chapter 9 are both founded on the behavior of the Z statistic, namely

$$Z = \frac{\bar{x} - \mu_0}{\sigma_{\bar{x}}} = \frac{\bar{x} - \mu_0}{\sigma/\sqrt{n}}$$

these two inference methods are equivalent. Here we show how to use a single confidence interval to test as many possible values for μ_0 as one wants. If a certain value for μ_0 is rejected in a two-tailed Z test with level of significance α, then μ_0 will lie *outside* the $100(1 - \alpha)\%$ Z confidence interval for μ. Thus, μ_0 is not a plausible value for μ (see Figure 9.21).

FIGURE 9.21
Reject H_0 for values of μ_0 that lie outside confidence interval.

Alternatively, if a certain value for μ_0 is not rejected in a two-tailed Z test with level of significance α, then μ_0 will lie *inside* the $100(1 - \alpha)\%$ Z confidence interval for μ and is a plausible value for μ. Table 9.8 shows the confidence levels and associated α levels of significance that will produce the equivalent inference.

Table 9.8 Confidence levels for equivalent α levels of significance

Confidence level for $100(1 - \alpha)\%$ confidence interval	Level of significance α for two-tailed test H_0: $\mu = \mu_0$ versus H_a: $\mu \neq \mu_0$
90%	0.10
95%	0.05
99%	0.01

Example 9.12 Lead contamination in trout, revisited

Recall Example 8.3 from Section 8.1 (page 362), where we were 95% confident using a Z interval that the population mean lead contamination for all trout in the Spokane River lies between 0.902 and 1.098 ppm. Once we have constructed the 95% confidence interval, we may test as many possible values for μ_0 as we like. If any prospective values of μ_0 lie inside the confidence interval, that is, between 0.902 and 1.098, we will not reject H_0 for this value of μ_0. If any prospective values of μ_0 lie outside the confidence interval, that is, either to the left of 0.902 or to the right of 1.098, we will reject H_0, as shown in Figure 9.22.

FIGURE 9.22 Reject H_0 for values of μ_0 that lie outside (0.902, 1.098).

Test using level of significance $\alpha = 0.01$ whether the population mean sodium content per serving differs from these values: **(a)** 0.900, **(b)** 0.910, **(c)** 1.100.

Solution

We set up the three two-tailed hypothesis tests as follows:

a. $H_0 : \mu = 0.900$ vs. $H_a : \mu \neq 0.900$
b. $H_0 : \mu = 0.910$ vs. $H_a : \mu \neq 0.910$
c. $H_0 : \mu = 1.100$ vs. $H_a : \mu \neq 1.100$

To perform each hypothesis test, simply observe where each value of μ_0 falls on the number line shown in Figure 9.22. For example, in the first hypothesis test, the hypothesized value $\mu_0 = 0.900$ lies outside the interval (0.902, 1.098). Thus, we reject H_0. The three hypothesis tests are summarized here. ■

Value of μ_0	Form of hypothesis test, with $\alpha = 0.01$	Where μ_0 lies in relation to 95% confidence interval	Conclusion of hypothesis test
a. 0.900	$H_0 : \mu = 0.900$ vs. $H_a : \mu \neq 0.900$	Outside	Reject H_0
b. 0.910	$H_0 : \mu = 0.910$ vs. $H_a : \mu \neq 0.910$	Inside	Do not reject H_0
c. 1.100	$H_0 : \mu = 1.100$ vs. $H_a : \mu \neq 1.100$	Outside	Reject H_0

4   Probability of a Type II Error and the Power of a Hypothesis Test

In Section 9.1 we defined a Type II error as follows:

Type II error: not rejecting H_0 when H_0 is false

In this section we learn how to calculate the probability of making a Type II error, called β (beta), and to use the value of β to compute the power of a hypothesis test.

Calculating β, the Probability of a Type II Error

Use the following steps to calculate β, the probability of a Type II error.

Step 1 Let $\bar{x}_{crit}$ be the value of the sample mean $\bar{x}$ associated with Z_{crit}. The following table shows how to calculate $\bar{x}_{crit}$ for the three forms of the hypothesis test.

Form of test		Value of $\bar{x}_{crit}$
Right-tailed	$H_0: \mu = \mu_0$ vs. $H_a: \mu > \mu_0$	$\bar{x}_{crit} = \mu_0 + Z_{crit} \cdot \dfrac{\sigma}{\sqrt{n}}$
Left-tailed	$H_0: \mu = \mu_0$ vs. $H_a: \mu < \mu_0$	$\bar{x}_{crit} = \mu_0 - Z_{crit} \cdot \dfrac{\sigma}{\sqrt{n}}$
Two-tailed	$H_0: \mu = \mu_0$ vs. $H_a: \mu \neq \mu_0$	$\bar{x}_{crit, lower} = \mu_0 - Z_{crit} \cdot \dfrac{\sigma}{\sqrt{n}}$ $\bar{x}_{crit, upper} = \mu_0 + Z_{crit} \cdot \dfrac{\sigma}{\sqrt{n}}$

Here, μ_0 is the hypothesized value of the population mean, σ is the population standard deviation, and n is the sample size.

Step 2 Let μ_a represent a particular value for the population mean μ chosen from the values indicated in the alternative hypothesis H_a. Draw a normal curve centered at μ_a, with the value or values of $\bar{x}_{crit}$ from Step 1 indicated (see Example 9.13).

Step 3 Calculate β for the particular μ_a chosen using the following table.

Form of test		β = probability of Type II error
Right-tailed	$H_0: \mu = \mu_0$ vs. $H_a: \mu > \mu_0$	The area under the normal curve drawn in Step 2 to the left of $\bar{x}_{crit}$.
Left-tailed	$H_0: \mu = \mu_0$ vs. $H_a: \mu < \mu_0$	The area under the normal curve drawn in Step 2 to the right of $\bar{x}_{crit}$.
Two-tailed	$H_0: \mu = \mu_0$ vs. $H_a: \mu \neq \mu_0$	The area under the normal curve drawn in Step 2 between $\bar{x}_{crit,lower}$ and $\bar{x}_{crit,upper}$.

It is a correct decision to reject the null hypothesis when the null hypothesis is false. The probability of making this type of correct decision is called the **power of the test**.

> **Power of a Hypothesis Test**
>
> The **power of a hypothesis test** is the probability of rejecting the null hypothesis when the null hypothesis is false. Power is calculated as
>
> $$\text{power} = 1 - \beta$$

Let us illustrate the steps for calculating β, the probability of a Type II error, using an example.

Example 9.13 Calculating β (the probability of a Type II error) and the power of a test

In Example 9.11, we tested whether people use debit cards on average more than 11 times per month. The hypotheses are

$$H_0: \mu = 11 \quad \text{versus} \quad H_a: \mu > 11$$

where μ represents the population mean debit card usage per month. From Example 9.11 we have $n = 36$, $\bar{x} = 11.5$, $Z_{crit} = 2.33$, and $\sigma = 3$.
a. State what a Type II error would be in this case.
b. Let $\mu_a = 13$. That is, suppose the population mean debit card usage is actually 13 times per month. Calculate β, the probability of making a Type II error when $\mu_a = 13$.
c. Calculate the power of the test.

Solution
a. We make a Type II error when we do not reject H_0 when H_0 is false. In this case, a Type II error would be to conclude that the population mean debit card usage was 11 times per month when in actuality it was more than 11 times per month.
b. We follow the steps for calculating β.

Step 1 We have a right-tailed test, so that

$$\bar{x}_{crit} = \mu_0 + Z_{crit} \cdot \frac{\sigma}{\sqrt{n}} = 11 + 2.33 \cdot \frac{3}{\sqrt{36}} = 12.165$$

Step 2   Figure 9.23 shows the normal curve centered at $\mu_a = 13$, with $\bar{x}_{crit} = 12.165$ labeled.

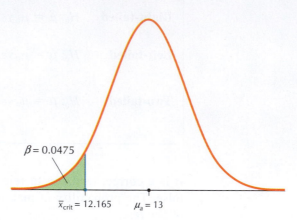

$\beta = 0.0475$

$\bar{x}_{crit} = 12.165$ $\mu_a = 13$

FIGURE 9.23 β, probability of a Type II error.

Step 3   The right-tailed test tells us that β equals the area under the normal curve drawn in Step 2 to the left of $\bar{x}_{crit} = 12.165$. This is the shaded area in Figure 9.23. Since area represents probability, we have

$$\beta = P(\bar{x} < 12.165) \text{ when } \mu_a = 13$$

Standardizing with $\mu_a = 13$, $\sigma = 3$, and $n = 36$:

$$\beta = P(\bar{x} < 12.165) = P\left(Z < \frac{12.165 - 13}{3/\sqrt{36}}\right) = P(Z < -1.67) = 0.0475$$

Thus, $\beta = 0.0475$. This represents the probability of making a Type II error, that is, of not rejecting the hypothesis that the population mean debit card usage is 11 times per month when in actuality it is 13 times per month.

c.   From part (b) we have $\beta = 0.0475$. Thus, the power of the hypothesis test is

$$\text{power} = 1 - \beta = 1 - 0.0475 = 0.9525$$

The probability of correctly rejecting the null hypothesis is 0.9525. ■

Section 9.3 SUMMARY

1 The critical region consists of the range of values of the test statistic Z_{data} for which we reject the null hypothesis. The value of Z that separates the critical region from the noncritical region is called the critical value Z_{crit}.

2 The critical-value method of performing the Z test for the mean is equivalent to the p-value method in that the same conclusion would be drawn using either method. In the critical-value method, we compare one Z-value (Z_{data}) with another Z-value (Z_{crit}).

3 We can use a single $100(1 - \alpha)\%$ confidence interval for μ to help us perform any number of two-tailed hypothesis tests about μ.

4 We can calculate β, the probability of a Type II error, for a given value of μ_a. The power of the hypothesis test is the probability of rejecting the null hypothesis when the null hypothesis is false. Power is calculated as power $= 1 - \beta$.

Section 9.3 EXERCISES

CLARIFYING THE CONCEPTS

For Exercises 1–3, indicate whether the quantity represents sample information or a threshold for significance.

1. Z_{crit} **2.** p-value   **3.** α

For Exercises 4–6, indicate whether or not the quantity represents a probability.

4. Z_{data} **5.** p-value   **6.** α

PRACTICING THE TECHNIQUES

For Exercises 7–10, assume that the conditions for performing the Z test are met. Do the following.

  a. Find the value of Z_{crit}.
  b. Find the critical-value rejection rule.
  c. Draw a standard normal curve and indicate the critical region.

7. $H_0 : \mu = \mu_0$ vs. $H_a : \mu < \mu_0$, $\alpha = 0.10$
8. $H_0 : \mu = \mu_0$ vs. $H_a : \mu > \mu_0$, $\alpha = 0.05$
9. $H_0 : \mu = \mu_0$ vs. $H_a : \mu \neq \mu_0$, $\alpha = 0.05$
10. $H_0 : \mu = \mu_0$ vs. $H_a : \mu \neq \mu_0$, $\alpha = 0.01$

For each of the following hypothesis tests in Exercises 11–14, assume that the conditions for performing the Z test are met. Do the following.

  a. Find the value of Z_{crit}.
  b. Find the critical-value rejection rule.
  c. Draw a standard normal curve and indicate the critical region.
  d. State the conclusion.
  e. State the interpretation.

11. $H_0 : \mu = \mu_0$ vs. $H_a : \mu > \mu_0$, $\alpha = 0.05$, $Z_{data} = 2.0$
12. $H_0 : \mu = \mu_0$ vs. $H_a : \mu > \mu_0$, $\alpha = 0.05$, $Z_{data} = 1.5$
13. $H_0 : \mu = \mu_0$ vs. $H_a : \mu < \mu_0$, $\alpha = 0.10$, $Z_{data} = -1.3$
14. $H_0 : \mu = \mu_0$ vs. $H_a : \mu \neq \mu_0$, $\alpha = 0.01$, $Z_{data} = 2.0$

For Exercises 15 and 16, do the following.

  a. State the hypotheses.
  b. Find Z_{crit} and the rejection rule.
  c. Find Z_{data}. Also, draw a standard normal curve showing Z_{crit}, the critical region, and Z_{data}.
  d. Formulate and interpret the conclusion.

15. We are interested in testing at level of significance $\alpha = 0.05$ whether the population mean is greater than 100. A random sample of size 64 is taken, with a mean of 102. Assume $\sigma = 8$.

16. We want to test at level of significance $\alpha = 0.01$ whether the population mean differs from 20. A random sample of size 49 is taken, with a mean of 27. Assume $\sigma = 5$.

17. A 95% Z confidence interval for the population mean is $(-2.7, 6.9)$. Using the equivalence between Z intervals and Z tests, test using level of significance $\alpha = 0.05$ whether the population mean differs from each of the following hypothesized values.

  a. -3 **b.** -2 **c.** 0 **d.** 5 **e.** 7

18. A 99% Z confidence interval for the population mean is $(45, 55)$. Using the equivalence between Z intervals and Z tests, test using level of significance $\alpha = 0.01$ whether the population mean differs from each of the following hypothesized values.

  a. 0 **b.** 44 **c.** 50 **d.** 54 **e.** 56

APPLYING THE CONCEPTS

For Exercises 19–22, do the following.

  a. State the hypotheses.
  b. Find Z_{crit} and the critical region. Also, draw a standard normal curve showing the critical region.
  c. Find Z_{data}. Also, draw a standard normal curve showing Z_{crit}, the critical region, and Z_{data}.
  d. State the conclusion and the interpretation.

19. Price of Milk. The Bureau of Labor Statistics reports that the mean price for a gallon of milk in 2005 was $3.24. A random sample of 100 retail establishments this year provides a mean price of $3.40. Assume $\sigma = \$0.25$. Perform a hypothesis test using level of significance $\alpha = 0.05$ to investigate whether the population mean price of milk this year has increased from the 2005 value.

20. Household Size. The U.S. Census Bureau reported that the mean household size in 2002 was 2.58 persons. A random sample of 900 households this year provides a mean size of 2.5 persons. Assume $\sigma = 0.1$. Conduct a hypothesis test using level of significance $\alpha = 0.10$ to determine whether the population mean household size this year has decreased from the 2002 level.

21. Salaries of Assistant Professors. Suppose that the mean salary for assistant professors in science in 2005 was $50,000. A random sample of 81 assistant professors in science taken this year shows a mean salary of $51,750. Assume $\sigma = \$1000$. Perform a hypothesis test using level of significance $\alpha = 0.01$ to determine whether the population mean salary of assistant professors in science has increased since 2005.

22. Americans' Height. A random sample of 400 Americans yields a mean height of 176 centimeters. Assume $\sigma = 2.5$ centimeters. Conduct a hypothesis test to investigate whether the population mean height of Americans this year has changed from 175 centimeters, using level of significance $\alpha = 0.05$.

23. Cost of Education. The College Board reports that the mean annual cost of education at a private four-year college was $22,218 for the 2006–2007 school year. Suppose that a random sample of 49 private four-year colleges this year gives a mean cost of $24,000 per year. Assume the population standard deviation is $3000.

  a. Construct a 95% confidence interval for the population mean annual cost.
  b. Use the confidence interval to test at level of significance $\alpha = 0.05$ whether the population mean annual cost differs from the following amounts.

    i. $24,000   **ii.** $23,000
    iii. $23,200   **iv.** $25,000

Health Care Premiums. Use the following information for Exercises 24–26. According to the National Coalition on Health Care, the mean annual premium for an employer health plan covering a family of four cost $11,500 in 2007. A random sample of 100 families of four taken this year showed a mean annual premium of $11,750. Assume $\sigma = \$1500$.

24. Test whether the population mean annual premium has increased, using level of significance $\alpha = 0.05$.

25. *What if* the sample mean premium equaled some value larger than $11,750, while everything else stayed the same. Explain how this change would affect the following, if at all.

    a. The hypotheses   **d.** Z_{data}
    b. Z_{crit} **e.** The conclusion
    c. The critical region

26. Test whether the population mean annual premium has increased, using level of significance $\alpha = 0.01$. Comment.

27. Credit Score in Florida. According to CreditReport.com, the mean credit score in Florida in 2006 was 673. Suppose we have a random sample of 900 credit scores in Florida, and assume that the population standard deviation is 150. We are interested in testing using level of significance $\alpha = 0.05$ whether the population mean credit score in Florida has decreased since that time.

    a. Describe what a Type II error would mean in the context of this problem.
    b. What is the probability of making a Type II error, when the actual mean credit score is 670?
    c. What is the probability of making a Type II error, when the actual mean credit score is 665?
    d. Calculate the power of the hypothesis test for the values of μ_a given in **(b)** and **(c)**.

9.4 *t* Test for the Population Mean μ

Objectives: By the end of this section, I will be able to...

1 Perform the *t* test for the mean using the *p*-value method.

2 Perform the *t* test for the mean using the critical-value method.

In many real-world scenarios, the value of the population standard deviation σ is unknown. When this occurs, we should use *neither* the Z interval nor the Z test. Recall from Chapter 8 that, when the population standard deviation is unknown, and we have either a normal population or a large sample, we use a *t* interval for the mean. The situation is similar for hypothesis testing: when the value of σ is unknown, we should see whether we may apply the *t* test instead.

1 *t* Test for μ Using the *p*-Value Method

Recall that in Section 8.2 we used the *t* distribution to find a confidence interval for the mean when the population standard deviation was not known.

Let $\bar{x}$ be the sample mean, μ be the unknown population mean, s be the sample standard deviation, and n be the sample size. The *t* statistic

$$t = \frac{\bar{x} - \mu}{s/\sqrt{n}}$$

with $n - 1$ degrees of freedom may be used when either (Case 1) the population is normal or (Case 2) the sample size is large. We call this *t* statistic t_{data} because its value depends largely on the sample data and the population data.

> The test statistic used for the *t* test for the mean is
>
> $$t_{data} = \frac{\bar{x} - \mu_0}{s/\sqrt{n}}$$

Extreme values of $\bar{x}$, that is, values of $\bar{x}$ that are significantly far from the hypothesized μ, will translate into extreme values of t_{data}. In other words, just as with Z_{data}, when $\bar{x}$ is far from μ_0, t_{data} will be far from 0.

---

t Test for the Population Mean μ: p-Value Method

When a random sample of size n is taken from a population, you can use the t test if either of the following conditions is satisfied:

- Case 1: the population is normal, or
- Case 2: the sample size is large ($n \geq 30$).

Step 1   State the hypotheses and the rejection rule. Use one of the forms from Table 9.1. Define μ. The rejection rule is "Reject H_0 if the p-value is less than α."

Step 2   Find t_{data}. Either use technology to find the value of the test statistic t_{data} or calculate the value of t_{data} as follows:

$$t_{data} = \frac{\bar{x} - \mu_0}{s/\sqrt{n}}$$

Step 3   Find the p-value. Either use technology to find the p-value or estimate the p-value using Table D, t-Distribution, in the Appendix.

Step 4   State the conclusion and the interpretation. If the p-value is less than α, then reject H_0. Otherwise, do not reject H_0. Interpret your conclusion so that a nonspecialist can understand.

The definition of p-values for t tests is the same as that for the Z test for the mean. Unusual and extreme values of $\bar{x}$, and therefore of t_{data}, will have a small p-value, while values of $\bar{x}$ and t_{data} nearer to the center of the distribution will have a large p-value. Table 9.9 summarizes the definition of the p-value for t tests. Note that we will not be finding these p-values manually but will either (a) use a computer or calculator or (b) estimate them using the t table.

Table 9.9 *p*-Values for *t* tests

Form of test	The *p*-value equals…	
Right-tailed test $H_0: \mu = \mu_0$ $H_a: \mu > \mu_0$	$P(t > t_{data})$ Area to the right of t_{data}	
Left-tailed test $H_0: \mu = \mu_0$ $H_a: \mu < \mu_0$	$P(t < t_{data})$ Area to the left of t_{data}	
Two-tailed test $H_0: \mu = \mu_0$ $H_a: \mu \neq \mu_0$	$P(t > |t_{data}|) + P(t < -|t_{data}|)$ $= 2 \cdot P(t > |t_{data}|)$ Sum of the two tail areas	

Example 9.14   A *t* test for the mean number of home runs

Suppose you have a collection of 14 baseball cards from the American League players who had at least 100 at-bats in the 2007 season (Table 9.10). You are trying to resolve a dispute with a friend about the population mean number of home runs hit per player in the American League. Use Minitab to test whether the population mean number of home runs is less than 16, with level of significance $\alpha = 0.10$.

Table 9.10   Home runs from collection of 14 baseball cards

Player	Home runs
Jermaine Dye	28
Carl Crawford	11
John McDonald	1
Jason Michaels	7
Melvin Mora	14
Jason Varitek	17
Orlando Cabrera	8
Tony Pena	2
Jason Kubel	13
Mark Ellis	19
Jose Vidro	6
Brad Wilkerson	20
Curtis Granderson	23
Hideki Matsui	25

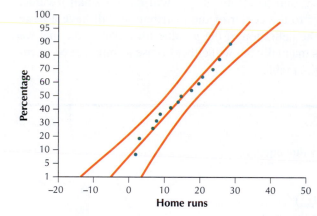

FIGURE 9.24 Normal probability plot of home runs.

Solution

Since we do not know the value of σ, we cannot perform a *Z* test. To use a *t* test, either the population must be normal or the sample size must be large. Since $n = 14$ is not a large sample size, we check whether the distribution of home runs is normal. Figure 9.24 shows the normal probability plot of the number of home runs per player for the baseball card collection. The normality is acceptable, allowing us to proceed with the *t* test. We use the Minitab instructions provided in the Step-by-Step Technology Guide at the end of the section.

Step 1   **State the hypotheses and the rejection rule.** The hypotheses are

$$H_0: \mu = 16 \quad \text{versus} \quad H_a: \mu < 16$$

where μ represents the population mean number of home runs. We will reject H_0 if we obtain a *p*-value less than 0.10.

Step 2   **Find t_{data}.** In Figure 9.25, Minitab tells us that $n = 14$, $\bar{x} = 13.8571$, $s = 8.5021$, and

$$t_{\text{data}} = \frac{\bar{x} - \mu_0}{s/\sqrt{n}} = \frac{13.8571 - 16}{8.5021/\sqrt{14}} \approx -0.94$$

Variable	N	Mean	StDev	SE Mean	T	P
Home Runs	14	13.8571	8.5021	2.2723	-0.94	0.181
	n	$\bar{x}$	s	$s/\sqrt{n}$	t_{data}	p-value

FIGURE 9.25 Minitab results for *t* test.

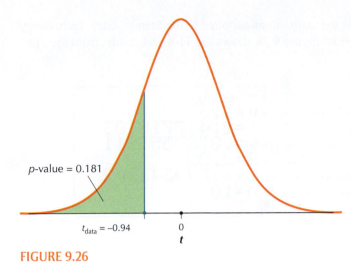

FIGURE 9.26

Step 3 **Find the *p*-value.** Also from Figure 9.25, we have the *p*-value, which is

$$P(t < t_{\text{data}}) = P(t < -0.94) = 0.181$$

Note that we could not have found this *p*-value without a computer or calculator. Using the *t* table, the *p*-value may only be estimated.

Step 4 **State the conclusion and interpretation.** Since the *p*-value 0.181 is not less than $\alpha = 0.10$, we do not reject H_0. There is insufficient evidence that the population mean number of home runs per player is less than 16. Although $\bar{x} = 13.8571$ is less than 16, the difference is not statistically significant. ■

Example 9.15 Performing the *t* test for μ using technology: nurse-to-patient ratios

The table below contains a random sample of ten highly rated cancer care facilities, along with their nursing index (nurse-to-patient ratio), in 2007.[10] Suppose that the population mean nursing index in 2005 was 1.6 nurses per cancer patient. Test whether the population mean index has changed using level of significance $\alpha = 0.05$.

Hospital	Index
Memorial Sloan Kettering Cancer Center	1.5
M. D. Anderson Cancer Center	2.0
Johns Hopkins Hospital	2.3
Mayo Clinic	2.8
Dana Farber Cancer Institute	0.8
Univ. of Washington Medical Center	2.2
Duke University Medical Center	1.8
Univ. of Chicago Hospitals	2.3
UCLA Medical Center	2.2
UC San Francisco Medical Center	2.3

Solution

Since the sample size is small, we check normality. The normal probability plot (Figure 9.27) is not perfectly linear, but there are no points outside the bounds, and it is difficult to determine normality for such small sample sizes. We proceed to perform the *t* test using Case 1, with the caveat that the normality assumption could be better supported and that more data would be helpful.

Step 1 **State the hypotheses and the rejection rule.** The key words "has changed" means that we have a two-tailed test:

$$H_0: \mu = 1.6 \quad \text{versus} \quad H_a: \mu \neq 1.6$$

where μ represents the population mean nursing index. We will reject H_0 if the *p*-value is less than $\alpha = 0.05$.

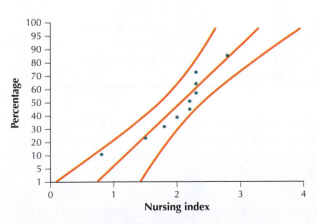

FIGURE 9.27 Normal probability plot of nursing index.

Step 2 **Find t_{data}.** We use the instructions supplied in the Step-by-Step Technology Guide at the end of this section. Figure 9.28 shows the TI-83/84 results from the t test for μ.

Form of H_a: ──────────────→ μ≠1.6
t_{data} ──────────────→ t=2.417718103
p-value ──────────────→ P=.0387553911
Sample mean $\bar{x}$ ──────────→ x̄=2.02
Sample standard deviation s ───→ Sx=.549343042
Sample size n ──────────────→ n=10

T-Test

FIGURE 9.28 TI-83/84 results.

Using the statistics from Figure 9.28, we have the test statistic

$$t_{data} = \frac{\bar{x} - \mu_0}{s/\sqrt{n}} = \frac{2.02 - 1.6}{0.549343042/\sqrt{10}} = 2.417718103$$

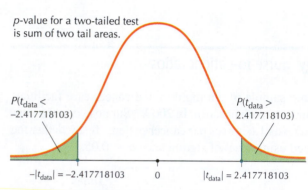

p-value for a two-tailed test is sum of two tail areas.

$P(t_{data} < -2.417718103)$

$P(t_{data} > 2.417718103)$

$-|t_{data}| = -2.417718103$ 0 $|t_{data}| = 2.417718103$

FIGURE 9.29 The *p*-value for a two-tailed test.

Step 3 **Find the *p*-value.** From Figures 9.28 and 9.29, we have

$$p\text{-value} = P(t > |2.417718103|) + P(t < -|2.417718103|)$$
$$= 0.0387553911$$

Step 4 **State the conclusion and interpretation.** The *p*-value of 0.0387553911 is less than $\alpha = 0.05$. We therefore reject H_0. There is evidence that the population mean nurse-to-patient ratio differs from 1.6. ■

Example 9.16   The t test for μ using the estimated *p*-value method

The table below lists the prices in dollars for a sample of 17 mathematics journals in 2005, as reported by the American Mathematical Society.

Journal	Price	Journal	Price	Journal	Price
Bull. Amer. Math.	$402	Ann. Statist.	$250	SIAM J. Math. Anal.	$717
J. Amer. Math. Soc.	276	Statistical Science	90	SIAM J. Matrix Anal.	497
Math. Comp.	467	SIAM J. Appl. Math.	527	SIAM J. Numer. Anal.	567
Proc. Amer. Math.	1022	SIAM J. Comput.	536	SIAM J. Optim.	438
Ann. Appl. Probab.	850	SIAM J. Control Optim.	633	SIAM J. Sci. Comput.	644
Ann. Probab.	250	SIAM J. Discrete Math.	450		

Suppose that the mean cost of all mathematics journals in 2001 was $400, and we are interested in whether the population mean cost is increasing. Evaluate the normality assumption, and perform the appropriate hypothesis test by *estimating the p-value* and comparing it with level of significance $\alpha = 0.05$.

Solution

Though not perfect, the normal probability plot indicates an acceptable level of normality of the math journal prices. We may therefore proceed to apply the t test.

Normal probability plot for price of math journals.

MINITAB Descriptive Statistics

Variable	N	Mean	SE Mean	StDev
Prices	17	506.82	55.8	230.0

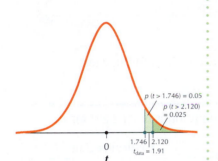

p (*t* > 1.746) = 0.05

p (*t* > 2.120) = 0.025

Step 1 **State the hypotheses and the rejection rule.** The hypotheses are

$$H_0: \mu = 400 \quad \text{versus} \quad H_a: \mu > 400$$

where μ represents the population mean math journal price. We will reject the null hypothesis if the *p*-value is less than $\alpha = 0.05$.

Step 2 **Find t_{data}.** Our sample statistics are $\bar{x} = \$506.82$, $s = \$230$, and $n = 17$, so our test statistic is

$$t_{data} = \frac{506.82 - 400}{230/\sqrt{17}} \approx 1.91$$

Step 3 **Find the *p*-value.** We estimate the *p*-value as follows. We follow the row in the *t* table for df = $n - 1 = 16$, looking for where $t_{data} = 1.91$ would fit:

Area in one tail	0.10	**0.05**	**0.025**	0.01	0.005
Value of *t*	1.337	**1.746**	**2.120**	2.583	23921

$t_{data} = 1.91$ would fall here between 1.746 and 2.120.

We see that our value of $t_{data} = 1.91$ would fit between $t = 1.746$ and $t = 2.120$. Therefore, the tail area associated with $t_{data} = 1.91$ is between area 0.025 and area 0.05. Therefore, since this is a right-tailed test, the estimated *p*-value also lies between 0.025 and 0.05.

Step 4 **State the conclusion and interpretation.** Since the *p*-value is less than $\alpha = 0.05$, we reject H_0. There is evidence that the population mean price of math journals is higher than $400. ■

t Test for μ Using the Critical-Value Method

So far we have learned how to perform a *t* test using the *p*-value method, which compares one probability (the *p*-value) with another probability (α). Now we will learn the critical-value method of performing the *t* test. We know that extreme values of $\bar{x}$ are associated with extreme values of t_{data} and will thus lead us to reject the null hypothesis. We answer the question "How extreme is extreme?" using the critical-value method by finding a *critical value* of *t*, called t_{crit}. This threshold value t_{crit} separates the values of t_{data} for which we reject H_0 (the *critical region*) from the values of t_{data} for which we will not reject H_0 (the *noncritical region*). Because there is a different *t* curve for every different sample size, you need to know the following to find the value of t_{crit}:

- the form of the hypothesis test (one-tailed or two-tailed)

- the degrees of freedom (df = $n - 1$)

- the level of significance α

t **Test for the Population Mean μ: Critical-Value Method**

When a random sample of size n is taken from a population, you can use the *t* test if either of the following conditions is satisfied:

- Case 1: the population is normal, or
- Case 2: the sample size is large ($n \geq 30$).

Step 1 **State the hypotheses.** Use one of the forms from Table 9.11. Define μ.

Step 2 **Find t_{crit} and state the rejection rule.** Use Table 9.11.

Step 3 **Find t_{data}.** Either use technology to find the value of the test statistic t_{data} or calculate the value of t_{data} as follows:

$$t_{\text{data}} = \frac{\bar{x} - \mu_0}{s/\sqrt{n}}$$

Step 4 **State the conclusion and the interpretation.** If t_{data} falls within the critical region, then reject H_0. Otherwise, do not reject H_0. Interpret your conclusion so that a nonspecialist can understand.

Table 9.11 contains the rejection rules and critical regions for the *t* test.

Table 9.11 Critical regions and rejection rules for various forms of the *t* test for μ

Form of test	Rejection rule	Critical region
Right-tailed test $H_0: \mu = \mu_0$ $H_a: \mu > \mu_0$	Reject H_0 if $t_{\text{data}} > t_{\text{crit}}$	
Left-tailed test $H_0: \mu = \mu_0$ $H_a: \mu < \mu_0$	Reject H_0 if $t_{\text{data}} < -t_{\text{crit}}$	
Two-tailed test $H_0: \mu = \mu_0$ $H_a: \mu \neq \mu_0$	Reject H_0 if $t_{\text{data}} > t_{\text{crit}}$ or $t_{\text{data}} < -t_{\text{crit}}$	

Example 9.17   Has the mean age at onset of anorexia nervosa been decreasing?

We are interested in testing, using level of significance $\alpha = 0.05$, whether the mean age at onset of anorexia nervosa in young women has been decreasing. Assume that the previous mean age at onset was 15 years old. Data were gathered for a study of the onset age for this disease.[11] From these data, a random sample was taken of 20 young women who were admitted under this diagnosis to the Toronto Hospital for Sick Children. The Minitab descriptive statistics shown here indicate a sample mean age of 14.251 years and a sample standard deviation of 1.512 years. If appropriate, perform the *t* test.

Variable	N	Mean	StDev
Patient Age	20	14.251	1.512

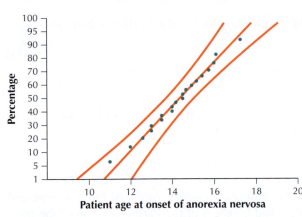

Normal probability plot for age at onset of anorexia nervosa.

Solution

Since the sample size $n = 20$ is not large, we need to verify normality. The normal probability plot of the ages at onset, shown here, indicates that the ages in the sample are normally distributed. We may proceed to perform the *t* test for the mean.

Step 1 **State the hypotheses.** The key word "decreasing" guides us to state our hypotheses as follows:

$$H_0: \mu = 15 \quad \text{versus} \quad H_a: \mu < 15$$

where μ refers to the population mean age at onset.

Step 2 **Find t_{crit} and state the rejection rule.** We determine that the hypothesis test is a left-tailed test and the degrees of freedom are df $= 20 - 1 = 19$, and we choose level of significance $\alpha = 0.05$. Referring to the *t* table with df $= 19$ and $\alpha = 0.05$ using a one-tailed test, we get $t_{crit} = 1.729$. Because we have a left-tailed test, the rejection rule from Table 9.11 is "Reject H_0 if $t_{data} < -t_{crit}$"; that is, we will reject H_0 if $t_{data} < -1.729$.

Step 3 **Find t_{data}.** The sample data show that our sample of $n = 20$ patients has a mean age at onset of $\bar{x} = 14.251$ years with a standard deviation of $s = 1.512$ years. Also, $\mu_0 = 15$, since this is the hypothesized value of μ stated in H_0. Therefore, our test statistic is

$$t_{data} = \frac{\bar{x} - \mu_0}{s/\sqrt{n}} = \frac{14.251 - 15}{1.512/\sqrt{20}} \approx -2.2154$$

Step 4 **State the conclusion and interpretation.** The rejection rule from Step 2 says to reject H_0 if $t_{data} < -1.729$. From Step 3, we have $t_{data} = -2.2154$. Since -2.2154 is less than -1.729, our conclusion is to reject H_0. If you prefer the graphical approach, consider Figure 9.30, which shows where t_{data} falls in relation to the critical region. Since $t_{data} = -2.2154$ falls within the critical region, our conclusion is to reject H_0. There is evidence that the population mean age of onset has decreased from its previous level of 15 years. ∎

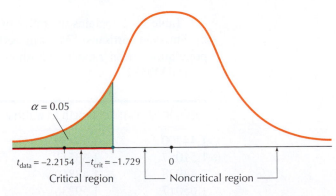

FIGURE 9.30 Our $t_{data} = -2.2154$ falls in the critical region.

FIGURE 9.31

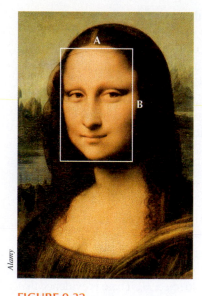

FIGURE 9.32

CASE STUDY

The Golden Ratio

Euclid's *Elements,* the Parthenon, the *Mona Lisa,* and the beadwork of the Shoshone tribe all have in common an appreciation for the **golden ratio.**

Suppose we have two quantities A and B, with $A > B > 0$. Then A/B is called the **golden ratio** if

$$\frac{A + B}{A} = \frac{A}{B}$$

that is, if the ratio of the sum of the quantities to the larger quantity equals the ratio of the larger to the smaller (see Figure 9.31).

Euclid wrote about the golden ratio in his *Elements,* calling it the "extreme and mean ratio." The ratio of the width A and height B of the Parthenon, one of the most famous temples in ancient Greece, equals the golden ratio (Figure 9.31). If you enclose the face of Leonardo da Vinci's *Mona Lisa* in a rectangle, the resulting ratio of the long side to the short side follows the golden ratio (Figure 9.32). The golden ratio has a value of approximately 1.6180339887.

Now we will test whether there is evidence for the use of the golden ratio in the artistic traditions of the Shoshone, a Native American tribe from the American West. Figure 9.33 shows a detail of a nineteenth-century Shoshone beaded dress that belonged to Nahtoma, the daughter of Chief Washakie of the Eastern Shoshone.[12] It is intriguing to consider whether Shoshone beaded rectangles such as those on this dress follow the golden ratio.

FIGURE 9.33

Table 9.12 contains the ratios of lengths to widths of 18 beaded rectangles made by Shoshone artisans.[13] We will perform a hypothesis test to determine whether the population mean ratio of Shoshone beaded rectangles equals the golden ratio of 1.6180339887.

Table 9.12 Ratio of length to width of a sample of Shoshone beaded rectangles

1.44300	1.75439	1.64204	1.66389	1.63666
1.51057	1.33511	1.52905	1.73611	1.80832
1.44928	1.48810	1.62602	1.49254	
1.65017	1.59236	1.49701	1.65017	

Since the population standard deviation for such rectangles is unknown, we must use a *t* test rather than a *Z* test. Our sample size *n* = 18 is not large, so we must assess whether the data are normally distributed. Figure 9.34 shows the normal probability plot indicating acceptable support for the normality assumption. We proceed with the *t* test, using level of significance $\alpha = 0.05$.

Shoshone beaded rectangle ratios (*n* = 18)

FIGURE 9.34 Normal probability plot (*n* = 18).

Solution

We use the TI-83/84 to perform this hypothesis test, using the instructions provided at the end of this section.

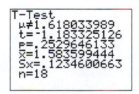

FIGURE 9.35

Step 1 **State the hypotheses and the rejection rule.** Since we are interested in whether the population mean length-to-width ratio of Shoshone beaded rectangles *equals* the golden ratio of 1.6180339887, we perform a two-tailed test:

$$H_0: \mu = 1.6180339887 \quad \text{versus} \quad H_a: \mu \neq 1.6180339887$$

where μ represents the population mean length-to-width ratio of Shoshone beaded rectangles. We will reject H_0 if the *p*-value is less than 0.05.

Step 2 **Find t_{data}.** From Figure 9.35 we have

$$t_{data} = \frac{\bar{x} - \mu_0}{s/\sqrt{n}} = \frac{1.583599444 - 1.6180339887}{0.1234600663/\sqrt{18}} = -1.183325126$$

(Note that the TI-83/84 output rounds μ_0 to 1.618033989.)

Step 3 **Find the *p*-value.** From Figures 9.35 and 9.36, we have

$$p\text{-value} = P(t > |1.183325126|) + P(t < -|1.183325126|)$$
$$= 0.2529646133$$

p-value for a two-tailed test is sum of two tail areas.

$P(t_{data} < -1.183325126)$

$P(t_{data} > 1.183325126)$

$-|t_{data}| = -1.183325126$ 0 $|t_{data}| = 1.183325126$

FIGURE 9.36

Step 4 **State the conclusion and interpretation.** Since *p*-value = 0.2529646133 is *not* less than $\alpha = 0.05$, we do *not* reject H_0. Thus, there is insufficient evidence that the population mean ratio differs from 1.6180339887. In other words, the data do not reject the claim that Shoshone beaded rectangles follow the same golden ratio exhibited by the Parthenon and the *Mona Lisa*.

STEP-BY-STEP TECHNOLOGY GUIDE: *t* test for *μ*

We will use the nurse-to-patient ratio data from Example 9.15 (page 445).

TI-83/84

If you have the data values:
Step 1 Enter the data into list **L1**.
Step 2 Press **STAT**, highlight **TESTS**, and press **ENTER**.
Step 3 Press **2** (for **T-Test**; see Figure 9.37).
Step 4 For input (**Inpt**), highlight **Data** and press **ENTER** (Figure 9.38).
a. For μ_0, enter the value of μ_0, **1.6**.
b. For **List**, press **2ⁿᵈ**, then **L1**.
c. For **Freq**, enter **1**.
d. For *μ*, select the form of H_a. Here we have a two-tailed test, so highlight ≠ μ_0 and press **ENTER** (Figure 9.38).
e. Highlight **Calculate** and press **ENTER**. The results are shown in Figure 9.28 in Example 9.15.

If you have the summary statistics:
Step 1 Press **STAT**, highlight **TESTS**, and press **ENTER**.
Step 2 Press **2** (for **T-Test**; see Figure 9.37).
Step 3 For input (**Inpt**), highlight **Stats** and press **ENTER** (Figure 9.39).
a. For μ_0, enter the value of μ_0, **1.6**.
b. For **Sx**, enter the value of *s*, **0.549343042**.
c. For $\bar{x}$, enter the sample mean **2.02**.
d. For **n**, enter the sample size **10**.
e. For *μ*, select the form of H_a. Here we have a two-tailed test, so highlight ≠ μ_0 and press **ENTER** (Figure 9.39).
f. Highlight **Calculate** and press **ENTER**. The results are shown in Figure 9.28 in Example 9.15.

Figure 9.37

Figure 9.38

Figure 9.39

EXCEL

WHFStat Add-ins
Step 1 Enter the data into column A. (If you have only the summary statistics, go to Step 2.)
Step 2 Load the **WHFStat Add-ins**.
Step 3 Select **Add-Ins > Macros > Testing a Mean > t Test – Confidence Interval – One Sample**.

Step 4 Select cells A1 to A10 as the **Dataset Range**. (Alternatively, you may enter the summary statistics.)
Step 5 Select your **Confidence level**, which should be $1 - \alpha$. Here, because $\alpha = 0.05$, we select **95%**.
Step 6 Enter the **Null Hypothesis Value**, $\mu_0 = 1.6$, and click **OK**.

MINITAB

If you have the data values:
Step 1 Enter the data into column C1.
Step 2 Click **Stat > Basic Statistics > 1-Sample t**.
Step 3 Click **Samples in Columns** and select **C1**.
Step 4 For **Test Mean**, enter 1.6.
Step 5 Click **Options**.
a. Choose your **Confidence Level** as $100(1 - \alpha)$. Our level of significance α here is 0.05, so the confidence level is 95.0.
b. Select **not Equal** for the **Alternative**.
Step 6 Click **OK** and click **OK** again.

If you have the summary statistics:
Step 1 Click **Stat > Basic Statistics > 1-Sample t**.
Step 2 Click **Summarized Data**,
Step 3 Enter the **Sample Size 10**, the **Sample Mean 2.02**, and the **Sample Standard Deviation 0.549343042**.
Step 4 Click **Options**.
a. Choose your **Confidence Level** as $100(1 - \alpha)$. Our level of significance α here is 0.05, so the confidence level is 95.0.
b. Select **not Equal** for the two-tailed test.
Step 5 Click **OK** and click **OK** again.

Section 9.4 SUMMARY

1 The test statistic used for the *t* test for the mean is

$$t_{\text{data}} = \frac{\bar{x} - \mu_0}{s/\sqrt{n}}$$

with $n - 1$ degrees of freedom. The *t* test may be used under either of the following conditions: (Case 1) the population is normal, or (Case 2) the sample size is large

($n \geq 30$). We perform hypothesis testing for the mean with the *t* test, using either the *p*-value or the critical-value method. For the *p*-value method, we reject H_0 if the *p*-value is less than α.

2 For the critical-value method, we compare the values of t_{data} and t_{crit}. If t_{data} falls in the critical region, we reject H_0.

Section 9.4 EXERCISES

CLARIFYING THE CONCEPTS

1. What assumption is required for performing the Z test that is not required for the t test?

2. What do we use to estimate the unknown population standard deviation σ?

3. What are the two conditions for using the t test for the mean?

4. What three elements are needed for finding the t critical value?

PRACTICING THE TECHNIQUES

For each of the hypothesis tests in Exercises 5–7, find the critical value t_{crit}, and sketch the critical region under the t curve. Assume normality.

5. $H_0: \mu = 40$ vs. $H_a: \mu < 40$, $n = 12$, $\alpha = 0.10$

6. $H_0: \mu = 40$ vs. $H_a: \mu < 40$, $n = 12$, $\alpha = 0.05$

7. $H_0: \mu = 40$ vs. $H_a: \mu < 40$, $n = 12$, $\alpha = 0.01$

8. Describe what happens to the t critical value t_{crit} for the left-tailed tests in Exercises 5–7 as the level of significance α decreases.

For Exercises 9–12, find the value of t_{data}.

9. $\bar{x} = 20$, $s = 4$, $n = 36$, $\mu_0 = 22$

10. $\bar{x} = 2$, $s = 1$, $n = 49$, $\mu_0 = 3$

11. $\bar{x} = 10$, $s = 3$, $n = 64$, $\mu_0 = 11$

12. $\bar{x} = 80$, $s = 5$, $n = 64$, $\mu_0 = 82$

For Exercises 13–16, do the following.
  a. State the hypotheses and the rejection rule using the p-value method.
  b. Calculate the test statistic t_{data}.
  c. Find the p-value. (Use technology or estimate the p-value.)
  d. State the conclusion and the interpretation.

13. A random sample of size 9 from a normal population yields $\bar{x} = 1$ and $s = 0.5$. Researchers are interested in finding whether the population mean differs from 0, using level of significance $\alpha = 0.05$.

14. A random sample of size 400 from a population with an unknown distribution yields a sample mean of 230 and a sample standard deviation of 5. Researchers are interested in finding whether the population mean is greater than 200, using level of significance $\alpha = 0.05$.

15. A random sample of size 100 from a population with an unknown distribution yields $\bar{x} = 27$ and $s = 10$. Researchers are interested in finding whether the population mean is less than 28, using level of significance $\alpha = 0.05$.

16. A random sample of size 16 from a normal population yields $\bar{x} = 2.2$ and $s = 0.3$. Researchers are interested in finding whether the population mean differs from 2.0, using level of significance $\alpha = 0.01$.

For Exercises 17–20, do the following.
  a. State the hypotheses.
  b. Calculate the t critical value t_{crit} and state the rejection rule. Also, sketch the critical region.
  c. Find the test statistic t_{data}.
  d. State the conclusion and the interpretation.

17. A random sample of size 25 from a normal population yields $\bar{x} = 104$ and $s = 10$. Researchers are interested in finding whether the population mean exceeds 100, using level of significance $\alpha = 0.01$.

18. A random sample of size 100 from a population with an unknown distribution yields a sample mean of -5 and a sample standard deviation of 5. Researchers are interested in finding whether the population mean is less than -4, using level of significance $\alpha = 0.05$.

19. A random sample of size 36 from a population with an unknown distribution yields $\bar{x} = 10$ and $s = 3$. Researchers are interested in finding whether the population mean differs from 9, using level of significance $\alpha = 0.10$.

20. A random sample of size 16 from a normal population yields $\bar{x} = 995$ and $s = 15$. Researchers are interested in finding whether the population mean is less than 1000, using level of significance $\alpha = 0.01$.

APPLYING THE CONCEPTS

Internet Response Times. Use the following information for Exercises 21–23. The Web site **www.Internettrafficreport.com** monitors Internet traffic worldwide and reports on the response times of randomly selected servers.

21. On July 13, 2004, the Web site reported the following response times to Asia, in milliseconds:

> 441 271 283 244 789 211 205
> 148 330 886 238 197 187

We are interested in testing whether the population mean response time is slower than 200 milliseconds, using a t test and level of significance $\alpha = 0.05$. Here is a boxplot of the data. (*Hint:* The boxplot is right-skewed and the normal distribution is symmetric.) Can we proceed with the t test? Explain.

22. On July 13, 2004, the Web site reported the following response times to North America, in milliseconds:

> 100 53 67 37 59 75 87
> 88 78 70 58 71 77 69

We are interested in testing whether the population mean response time is slower than 60 milliseconds, using a t test and $\alpha = 0.05$. Here is a normal probability plot of the data. Can we proceed with the t test? Explain.

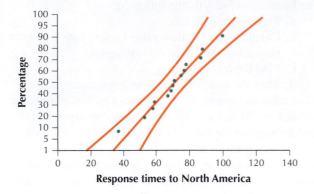

23. Refer to the previous exercise.
    a. Construct the hypotheses.
    b. Find the t critical value and state the rejection rule.
    c. Find t_{data}.
    d. State the conclusion and interpretation.

24. Health Care Costs. The U.S. Agency for Healthcare Research and Quality (**www.ahrq.gov**) reports that, in 2002, the mean cost of a stay in the hospital for American women aged 18–44 was $12,300. A random sample taken this year of 400 hospital stays of women aged 18–44 showed a mean cost of $15,000, with a standard deviation of $5000. Test whether the mean cost has increased since then, using $\alpha = 0.05$.

Obsession. Use the following information for Exercises 25 and 26. **Perfume.com** listed Obsession by Calvin Klein as their fifth best-selling fragrance for the 2006 holiday season. The mean price per bottle in 1996 was $46.42, according to the NPD Group. Suppose a random sample of 15 retail establishments yielded a mean retail price for Obsession this year of $48.92, with a standard deviation of $2.50. Assume these prices are normally distributed.

25. We are interested in whether the population mean price for Obsession perfume has increased.
    a. Is it appropriate to apply the t test for the mean? Why or why not?
    b. We have a sample mean of $48.92, which is greater than the mean of $46.42 in the null hypothesis. Isn't this enough by itself to reject the null hypothesis? Explain why or why not.
    c. Find t_{data}.
    d. Has there been a change in the population mean price since 1996? Test using the estimated p-value method at level of significance $\alpha = 0.01$.

26. *What if* the sample size was not 15 but some number greater than 15. Otherwise, everything else remains unchanged. How does this affect the following?
    a. t_{data}

b. t_{crit}
c. The p value
d. The conclusion

Top Gas Mileage. Use the following information for Exercises 27–28. The top ten vehicles for city gas mileage in 2007, as reported by the Environmental Protection Agency, are shown in the following table, along with the normal probability plot (Figure 9.40).

Car	Mileage	Car	Mileage
Toyota Yaris	39	Honda Fit	38
Chevrolet Aveo	37	Nissan Versa	34
Pontiac G5	34	Dodge Caliber	32
VW Eos	32	Ford Escape	31
Saturn Sky	30	BMW 525	30

Figure 9.40 Normal probability plot of gas mileage.

27. Test, using the estimated p-value method at level of significance $\alpha = 0.01$, whether the population mean city mileage exceeds 30 mpg.
28. Answer the following.
    a. Repeat your test from the previous exercise, this time using level of significance $\alpha = 0.001$.
    b. How do you think we should resolve the apparent contradiction in the preceding two results?
    c. Assess the strength of the evidence against the null hypothesis. Does this change depend on which level of α you use?

Community College Tuition. Use the following information for Exercises 29 and 30. The College Board reported that the mean tuition and fees at community colleges nationwide in 2006 was $2272. Data were gathered on the total tuition and fees for a random sample of ten community colleges this year. The normal probability plot (Figure 9.41) and Minitab t test output (Figure 9.42) are shown on the following page.

Figure 9.41 Normal probability plot of tuition and fees.

```
Test of mu = 2272 vs not = 2272

Variable   N    Mean    StDev  SE Mean        95% CI         T      P
tuition    10  2538.92  404.75  127.99  (2249.38, 2828.46)  2.09  0.067
```

Figure 9.42 Minitab *t* test output.

29. Analysts are interested in whether the population mean tuition and fees this year have increased.
   a. Is it appropriate to apply the *t* test for the mean? Why or why not?
   b. What are the sample size, sample mean cost, and sample standard deviation of the cost?
   c. Find t_{data} in the Minitab output.
   d. It appears that the data analyst who produced the Minitab printout asked for the wrong hypothesis test. How can we tell?

30. Refer to your work in the previous exercise.
   a. Test whether the population mean tuition and fees have increased using the *p*-value method and level of significance $\alpha = 0.05$. How can we use the *p*-value on the Minitab printout to find the *p*-value needed for this right-tailed hypothesis test?
   b. Compare the conclusion from (**a**) with the conclusion we would have gotten had we not noticed that the data analyst performed the wrong hypothesis test. What are some of the possible consequences of making an error of this sort?
   c. Based on your experiences in these exercises, write a sentence about the importance of understanding the statistical modeling behind the "point and click" power of statistical software.

Marriage Age. Use the following information for Exercises 31–33. The U.S. Census Bureau reported that the median age at first marriage for females in 2003 was 25.3. Let's assume that the mean age at first marriage among females in 2003 was 26 years old. A random sample of 157 females this year yielded a mean age at first marriage of 26.5 with a standard deviation of 4 years. A histogram of the ages was decidedly not normal.

31. Answer the following.
   a. Is it appropriate to apply the *t* test for the mean? Why or why not?
   b. Is there evidence that the population mean age at first marriage for females has changed since 2003? Test using the critical-value method at level of significance $\alpha = 0.10$.

32. Test using the critical-value method at level of significance $\alpha = 0.10$ whether the population mean age at first marriage for females has increased since 2003.

33. Challenger Exercise.
   a. Compare your conclusions in the previous two exercises. Note that we have concluded that there is insufficient evidence that the population mean marriage age has changed, but that there is evidence that the population mean marriage age has increased. How can the mean marriage age have increased without changing? Explain what is going on here, in terms of either critical regions or *p*-values.
   b. For which of the two hypothesis tests is the test statistic closer to the critical value?

34. Texas Towns. Work with the **Texas** data set for the following.
   a. How many observations are in the data set? How many variables?
   b. Use technology to explore the variable *tot_occ,* which lists the total occupied housing units for each county in Texas. Generate numerical summary statistics and graphs for the total occupied housing units. What is the sample mean? The sample standard deviation? Comment on the symmetry or skewness of the data set.
   c. Suppose we are using the data in this data set as a sample of the total occupied housing units of all the counties in the southwestern United States. Use technology to test at level of significance $\alpha = 0.05$ whether the population mean total occupied housing units for these counties differs from 40,000.

9.5   *Z* Test for the Population Proportion *p*

Objectives: **By the end of this section, I will be able to...**

1   Explain the essential idea about hypothesis testing for the proportion.

2   Perform and interpret the *Z* test for the proportion using the *p*-value method.

3 Perform and interpret the Z test for the proportion using the critical-value method.

4 Use confidence intervals for p to perform two-tailed hypothesis tests about p.

1 The Essential Idea About Hypothesis Testing for the Proportion

Thus far in Chapter 9 we have dealt with testing hypotheses about the population mean μ only. Here in Section 9.5, we will learn how to perform the Z test for the population proportion p.

For our point estimate of the unknown population proportion p, we use the sample proportion $\hat{p} = x/n$, where x equals the number of successes. Recall from Section 7.3 that the mean of the sampling distribution is p, and the standard deviation is

$$\sigma_{\hat{p}} = \sqrt{\frac{p\,(1-p)}{n}}$$

We learned that this sampling distribution of $\hat{p}$ is approximately normal whenever both of the following conditions are met: $np \geq 5$ and $n(1-p) \geq 5$. Just as with the Z test for the mean, in the Z test for the proportion the null hypothesis will include a certain hypothesized value for the unknown parameter, which we call p_0. For example, the hypotheses for the two-tailed test have the following form:

$$H_0: p = p_0 \quad \text{versus} \quad H_a: p \neq p_0$$

where p_0 represents a particular hypothesized value of the unknown population proportion p. For instance, if a researcher is interested in determining whether the population proportion of Americans who support increased funding for higher education differs from 50%, then $p_0 = 0.50$.

If we assume H_0 is correct, then the population proportion of successes is p_0, and so the sampling distribution of p is centered at p_0. The standard deviation of the sampling distribution of $\hat{p}$ takes the form

$$\sigma_{\hat{p}} = \sqrt{\frac{p(1-p)}{n}} = \sqrt{\frac{p_0\,(1-p_0)}{n}}$$

since we claim in H_0 that $p = p_0$. This leads us to the following statement of the essential idea about **hypothesis testing for the proportion.**

> **The Essential Idea About Hypothesis Testing for the Proportion**
> When the sample proportion $\hat{p}$ is unusual or extreme in the sampling distribution of $\hat{p}$ that is based on the assumption that H_0 is correct, we reject H_0. Otherwise, there is insufficient evidence against H_0, and we should not reject H_0.

2 Z Test for p Using the p-Value Method

The remainder of Section 9.5 explains the details of implementing hypothesis testing for the proportion. Recall that, in the Z test for the mean, Z_{data} represents the number of standard deviations ($\sigma_{\bar{x}}$) the observed sample mean $\bar{x}$ lies above or below the hypothesized mean μ_0. In this section, Z_{data} will represent the number of standard deviations ($\sigma_{\hat{p}}$) the sample proportion $\hat{p}$ lies above or below the hypothesized proportion p_0.

> The test statistic used for the Z test for the proportion is
>
> $$Z_{data} = \frac{\hat{p} - p_0}{\sigma_{\hat{p}}} = \frac{\hat{p} - p_0}{\sqrt{\dfrac{p_0(1 - p_0)}{n}}}$$
>
> where $\hat{p}$ is the observed sample proportion of successes, p_0 is the value of p hypothesized in H_0, and n is the sample size.

Just as for hypothesis tests for the mean, there are three possible forms of hypothesis tests for the proportion, as shown in Table 9.13.

Table 9.13 The three possible forms for the hypotheses for a test for p

Form	Null and alternative hypotheses
Right-tailed test, one-tailed test	$H_0: p = p_0$ versus $H_a: p > p_0$
Left-tailed test, one-tailed test	$H_0: p = p_0$ versus $H_a: p < p_0$
Two-tailed test	$H_0: p = p_0$ versus $H_a: p \neq p_0$

> **Z Test for the Population Proportion p: p-Value Method**
>
> When a random sample of size n is taken from a population, you can use the Z test for the proportion if both of the normality conditions are satisfied:
>
> $$np_0 \geq 5 \quad \text{and} \quad n(1 - p_0) \geq 5$$
>
> **Step 1   State the hypotheses and the rejection rule.** Use one of the forms from Table 9.13. Be sure to explicitly state the meaning of p. State the rejection rule as "Reject H_0 when the p-value is less than α."
>
> **Step 2   Find Z_{data}.** Either use technology to find the value of Z_{data} or calculate Z_{data} as follows:
>
> $$Z_{data} = \frac{\hat{p} - p_0}{\sigma_{\hat{p}}} = \frac{\hat{p} - p_0}{\sqrt{\dfrac{p_0(1 - p_0)}{n}}}$$
>
> **Step 3   Find the p-value.** Either use technology to find the p-value, or calculate it using the form in Table 9.4 that corresponds to your hypotheses.
>
> **Step 4   State the conclusion and the interpretation.** If the p-value is less than α, then reject H_0. Otherwise do not reject H_0. Interpret your conclusion so that a nonspecialist can understand.

Note that the p-value has precisely the same definition and behavior as in the Z test for the mean. That is, the p-value is roughly a measure of how extreme your value of Z_{data} is and takes values between 0 and 1, with small values indicating extreme values of Z_{data}.

eveloping
Your Statistical Sense

The Difference Between the *p*-Value and the Population Proportion *p*

Be careful to distinguish between the *p*-value and the population proportion *p*. The latter represents the population proportion of successes for a binomial experiment and is a population parameter. The *p*-value is the probability of observing a value of Z_{data} at least as extreme as the Z_{data} actually observed. The *p*-value depends on the sample data, but the population proportion *p* does not depend on the sample data. ■

Example 9.18   Are car accidents among young drivers increasing?

Getty Images/Stockbyte Platinum

The National Transportation Safety Board publishes statistics on the number of automobile crashes that people in various age groups have. In 2003, young people aged 18–24 had an accident rate of 12%, meaning that on average 12 out of every 100 young drivers per year had an accident. A more recent study examined 1000 young drivers aged 18–24 and found that 134 had an accident this year. We are interested in whether the population proportion of young drivers having accidents has increased since 2003, using the *p*-value method with level of significance $\alpha = 0.05$.

Solution

First we check that both of our normality conditions are met. Since we are interested in whether the proportion has increased from 12%, we have $p_0 = 0.12$.

$$np_0 = (1000)(0.12) = 120 \geq 5 \quad \text{and} \quad n(1 - p_0) = (1000)(0.88) = 880 \geq 5$$

The normality conditions are met and we may proceed with the hypothesis test.

Step 1   State the hypotheses and the rejection rule. Our hypotheses are

$$H_0 : p = 0.12 \quad \text{versus} \quad H_a : p > 0.12$$

where *p* represents the population proportion of young people aged 18–24 who had an accident. We reject the null hypothesis if the *p*-value is less than $\alpha = 0.05$.

Step 2   Find Z_{data}. Our sample proportion is $\hat{p} = 134/1000 = 0.134$. Since $p_0 = 0.12$, the standard deviation of the sampling distribution of $\hat{p}$ is

$$\sigma_{\hat{p}} = \sqrt{\frac{p_0(1 - p_0)}{n}} = \sqrt{\frac{(0.12)(0.88)}{1000}} \approx 0.0103$$

Thus, our test statistic is

$$Z_{\text{data}} = \frac{\hat{p} - p_0}{\sigma_{\hat{p}}} = \frac{\hat{p} - p_0}{\sqrt{\frac{p_0(1 - p_0)}{n}}} = \frac{0.134 - 12}{\sqrt{\frac{(0.12)(0.88)}{1000}}} \approx 1.36$$

That is, the sample proportion $\hat{p} = 0.134$ lies approximately 1.36 standard deviations above the hypothesized proportion $p_0 = 0.12$.

Step 3   Find the *p*-value. Since we have a right-tailed test, our *p*-value from Table 9.4 is $P(Z > Z_{\text{data}})$. This is a Case 2 problem from Table 6.7 (page 285), where we find the tail area by subtracting the Z table area from 1 (Figure 9.43):

$$P(Z > Z_{\text{data}}) = P(Z > 1.36) = 1 - 0.9131 = 0.0869$$

FIGURE 9.43

Note: If α were 0.10, then our conclusion would have been to reject H_0

Step 4 **State the conclusion and the interpretation.** Since the *p*-value is not less than $\alpha = 0.05$, we do not reject H_0. There is insufficient evidence that the population proportion of young people aged 18–24 who had an accident has increased. ∎

Example 9.19   Performing the *Z* test for *p* using technology

A 2006 study reported that 1% of American Internet users who are married or in a long-term relationship met on a blind date or through a dating service.[14] A survey taken this year of 500 American Internet users who are married or in a long-term relationship found 8 who met on a blind date or through a dating service. If appropriate, test whether the population proportion has increased since 2006. Use the *p*-value method with level of significance $\alpha = 0.05$.

Solution

We have $p_0 = 0.01$ and $n = 500$. Checking the normality conditions, we have

$$np_0 = (500)(0.01) = 5 \geq 5 \quad \text{and} \quad n(1 - p_0) = (500)(0.99) = 495 \geq 5$$

The normality conditions are met and we may proceed with the hypothesis test.

Step 1 **State the hypotheses and the rejection rule.** Our hypotheses are

$$H_0: p = 0.01 \quad \text{versus} \quad H_a: p > 0.01$$

where *p* represents the population proportion of American Internet users who are married or in a long-term relationship and who met on a blind date or through a dating service. We will reject H_0 if *p*-value < 0.05.

Step 2 **Find Z_{data}.** We use the instructions supplied in the Step-by-Step Technology Guide at the end of this section. Figure 9.44 shows the TI-83/84 results from the *Z* test for *p*, and Figure 9.45 shows the results from Minitab.

Form of H_a:
Z_{data}
p-value
Sample proportion $\hat{p}$
Sample size *n*

```
1-PropZTest
 prop>.01
 z=1.348399725
 p=.0887649866
 p̂=.016
 n=500
■
```

FIGURE 9.44 TI-83/84 results.

```
Test of p = 0.01 vs p > 0.01
                                 95%
                               Lower
    Sample   X    N   Sample p  Bound   Z-Value  P-Value
    1        8  500  0.016000 0.006770    1.35    0.089
             X    n      p̂    (not used)  Z_data  p-value
```

FIGURE 9.45 Minitab results.

From Figure 9.44, we have

$$Z_{data} = \frac{\hat{p} - p_0}{\sigma_{\hat{p}}} = \frac{\hat{p} - p_0}{\sqrt{\dfrac{p_0(1 - p_0)}{n}}} = \frac{0.016 - 0.01}{\sqrt{\dfrac{(0.01)(0.99)}{500}}} \approx 1.348399725$$

Step 3 **Find the *p*-value.** From Figures 9.44, 9.45, and 9.46, we have

$$p\text{-value} = P(Z > 1.348399725) = 0.0887649866$$

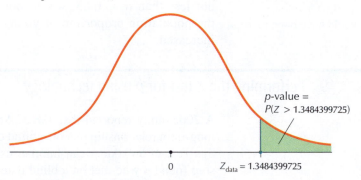

p-value =
$P(Z > 1.3484399725)$

0 $Z_{data} = 1.3484399725$

FIGURE 9.46 *p*-value for a right-tailed test.

Step 4 **State the conclusion and interpretation.** Since *p*-value ≈ 0.08876 is *not* less than our level of significance $\alpha = 0.05$, we do *not* reject H_0. There is insufficient evidence that the population proportion of American Internet users who are married or in a long-term relationship and who met on a blind date or through a dating service has increased since 2006. ▪

3 Z Test for *p* Using the Critical-Value Method

The critical-value method for the Z test for the proportion is equivalent to the *p*-value method for the Z test for the proportion. Therefore, the normality conditions for the Z test critical-value method and the Z test *p*-value method are the same, namely, $np_0 \geq 5$ and $n(1 - p_0) \geq 5$. In the critical-value method, we compare one Z-value (Z_{data}) with another Z-value (Z_{crit}) rather than comparing the *p*-value with α. You can use Table 9.6 to find the critical values Z_{crit}. To find the rejection rules or the critical regions, you can use Table 9.14 (which is the same as Table 9.7 except that the μ's have been changed to *p*'s).

Table 9.14 Rejection rules for *Z* test for a proportion

Form of test		Rejection rules: "Reject H_0 if…"
Right-tailed	$H_0: p = p_0$ vs. $H_a: p > p_0$	$Z_{data} > Z_{crit}$
Left-tailed	$H_0: p = p_0$ vs. $H_a: p < p_0$	$Z_{data} < Z_{crit}$
Two-tailed	$H_0: p = p_0$ vs. $H_a: p \neq p_0$	$Z_{data} > Z_{crit}$ or $Z_{data} < -Z_{crit}$

> ### Z Test for the Population Proportion *p*: Critical-Value Method
>
> When a random sample of size n is taken from a population, you can use the Z test for the proportion if both of the normality conditions are satisfied:
>
> $$np_0 \geq 5 \quad \text{and} \quad n(1 - p_0) \geq 5$$
>
> **Step 1   State the hypotheses.** Use one of the forms from Table 9.14. Clearly describe the meaning of *p*.
>
> **Step 2   Find Z_{crit} and state the rejection rule.** Use Tables 9.6 and 9.14.
>
> **Step 3   Find Z_{data}.** Either use technology to find the value of Z_{data} or calculate Z_{data} as follows:
>
> $$Z_{data} = \frac{\hat{p} - p_0}{\sigma_{\hat{p}}} = \frac{\hat{p} - p_0}{\sqrt{\dfrac{p_0(1 - p_0)}{n}}}$$
>
> **Step 4   State the conclusion and the interpretation.** If Z_{data} falls in the critical region, then reject H_0. Otherwise, do not reject H_0. Interpret the conclusion so that a nonspecialist can understand.

Example 9.20   Proportion of Americans who smoke

The National Center for Health Statistics reports on the proportion of Americans who smoke tobacco. Although the proportion decreased throughout the 1990s, from 2004 to 2006 there was little change in the proportion of Americans who smoke, 21%. A random sample taken this year of 400 Americans found 78 who smoked. Test whether the population proportion of Americans who smoke has changed since 2006, using the critical-value method with level of significance $\alpha = 0.10$.

Solution

First we check that both of our normality conditions are met. Since we are interested in whether the proportion has changed from 21%, we have $p_0 = 0.21$.

$$np_0 = (400)(0.21) = 84 \geq 5 \quad \text{and} \quad n(1 - p_0) = (400)(0.79) = 316 \geq 5$$

The normality conditions are met and we may proceed with the hypothesis test.

Step 1   State the hypotheses. Our hypotheses are

$$H_0: p = 0.21 \quad \text{versus} \quad H_a: p \neq 0.21$$

where *p* represents the population proportion of Americans who smoke tobacco.

Step 2   Find Z_{crit} and state the rejection rule. We have a two-tailed test, with $\alpha = 0.10$. This gives us our critical value $Z_{crit} = 1.645$ from Table 9.6 and the rejection rule from Table 9.14: Reject H_0 if $Z_{data} > 1.645$ or $Z_{data} < -1.645$ (Figure 9.47).

Step 3   Find Z_{data}. Our sample proportion is $\hat{p} = 78/400 = 0.195$. Since $p_0 = 0.21$, the standard deviation of the sampling distribution of $\hat{p}$ is

$$\sigma_{\hat{p}} = \sqrt{\frac{p_0(1 - p_0)}{n}} = \sqrt{\frac{(0.21)(0.79)}{400}} \approx 0.020365$$

Thus, our test statistic is

$$Z_{data} = \frac{\hat{p} - p_0}{\sigma_{\hat{p}}} = \frac{\hat{p} - p_0}{\sqrt{\dfrac{p_0(1 - p_0)}{n}}} = \frac{0.195 - 0.21}{\sqrt{\dfrac{(0.21)(0.79)}{400}}} \approx -0.74$$

Critical region   Critical region

$-1.645 \quad z_{data} = -0.74 \quad 0 \qquad 1.645$

z

FIGURE 9.47 The test statistic does not lie within the critical range.

Step 4 **State the conclusion and the interpretation.** The test statistic $Z_{data} = -0.74$ is not greater than 1.645 and not less than -1.645. Thus, as expected, we do not reject H_0. There is insufficient evidence that the population proportion of Americans who smoke tobacco has changed since 2006. ■

4 Using Confidence Intervals for p to Perform Two-Tailed Hypothesis Tests About p

Just as for μ, we can use a $100(1 - \alpha)\%$ confidence interval for the population proportion p in order to perform a set of two-tailed hypothesis tests for p.

Example 9.21 Using confidence intervals for p to determine significance

In 2007, the Pew Internet and American Life Project reported that 91% of Americans who have completed a bachelor's degree currently use the Internet. Pew also reports that the margin of error for this survey (confidence level = 95%) was $\pm3\%$. The 95% confidence interval for the population proportion of Americans with a bachelor's degree who currently use the Internet is therefore

$$0.91 \pm 0.03 = (0.88, 0.94)$$

Use the confidence interval to test, using level of significance $\alpha = 0.05$, whether the population proportion differs from
a. 0.85
b. 0.90
c. 0.95

Solution
There is equivalence between a $100(1 - \alpha)\%$ confidence interval for p and a two-tailed test for p with level of significance α. Values of p_0 that lie outside the confidence interval lead to rejection of the null hypothesis, while values of p_0 within the confidence interval lead to not rejecting the null hypothesis. Figure 9.48 illustrates the 95% confidence interval for p.

FIGURE 9.48 Reject H_0 for values p_0 that lie outside the interval (0.88, 0.94).

We would like to perform the following two-tail hypothesis tests:
a. $H_0: p = 0.85$ versus $H_a: p \neq 0.85$
b. $H_0: p = 0.90$ versus $H_a: p \neq 0.90$
c. $H_0: p = 0.95$ versus $H_a: p \neq 0.95$
To perform each hypothesis test, simply observe where each value of p_0 falls on the number line. For example, in the first hypothesis test, the hypothesized value $p_0 = 0.85$ lies outside the interval (0.88, 0.94). Thus, we reject H_0. The three hypothesis tests are summarized here. ■

Value of p_0	Form of hypothesis test, with $\alpha = 0.05$	Where p_0 lies in relation to 95% confidence interval	Conclusion of hypothesis test
a. 0.85	$H_0: p = 0.85$ $H_a: p \neq 0.85$	Outside	Reject H_0
b. 0.90	$H_0: p = 0.90$ $H_a: p \neq 0.90$	Inside	Do not reject H_0
c. 0.95	$H_0: p = 0.95$ $H_a: p \neq 0.95$	Outside	Reject H_0

STEP-BY-STEP TECHNOLOGY GUIDE: *Z* test for *p*

We will use the information from Example 9.19 (page 459).

TI-83/84

Step 1   Press **STAT**, highlight **TESTS**, and press **ENTER**.
Step 2   Press 5 (for **1-PropZTest**; see Figure 9.49).
Step 3   For p_0, enter the value of p_0, **0.01**.
Step 4   For **x**, enter the number of successes, **8**.
Step 5   For **n**, enter the number of trials **500**.
Step 6   For **prop**, enter the form of H_a. Here we have a right-tailed test, so highlight **>p_0** and press **ENTER** (see Figure 9.50).
Step 7   Highlight **Calculate** and press **ENTER**. The results are shown in Figure 9.44 in Example 9.19.

```
EDIT CALC TESTS
1:Z-Test…
2:T-Test…
3:2-SampZTest…
4:2-SampTTest…
5⬛1-PropZTest…
6:2-PropZTest…
7↓ZInterval…
```

Figure 9.49

```
1-PropZTest
 p0:.01
 x:8
 n:500
prop≠p0 <p0 >p0
Calculate Draw
```

Figure 9.50

EXCEL

WHFStat Add-ins

Step 1   Enter the data into column A. (If you have only the summary statistics, go to Step 2.)
Step 2   Load the **WHFStat Add-ins**.
Step 3   Select **Add-Ins > Macros > Testing a Proportion > One Sample**.

Step 4   Enter the **Number of successes 8**.
Step 5   Enter the **Sample size 500**.
Step 6   Enter the **Testing Proportion, p_0= 0.01**.
Step 7   Select your **Confidence level**, which should be $1 - \alpha$. Here, because $\alpha = 0.05$, we select **95%**.
Step 8   Click **OK**.

MINITAB

If you have the summary statistics:

Step 1   Click **Stat > Basic Statistics > 1 Proportion**.
Step 2   Click **Summarized Data**.
Step 3   Enter the **Number of trials 500** and the **Number of Events 8**.
Step 4   Click **Options**.

a.   Choose your **Confidence Level** as $100(1 - \alpha)$. Our level of significance α here is 0.05, so the confidence level is 95.0.
b.   Enter **0.01** for the **Test Proportion**.
c.   Select **Greater than** for the **Alternative**.
d.   Check **Use test and interval based on normal distribution**.
Step 5   Click **OK** and click **OK** again. The results are shown in Figure 9.45 in Example 9.19.

Section 9.5 SUMMARY

1 The essential idea about hypothesis testing for the proportion is as follows. When the sample proportion $\hat{p}$ is unusual or extreme in the sampling distribution of $\hat{p}$ that is based on the assumption that H_0 is correct, we should reject H_0. Otherwise, we should not reject H_0.

2 The test statistic used for the *Z* test for the proportion is

$$Z_{\text{data}} = \frac{\hat{p} - p_0}{\sigma_{\hat{p}}} = \frac{\hat{p} - p_0}{\sqrt{\dfrac{p_0(1 - p_0)}{n}}}$$

where $\hat{p}$ is the observed sample proportion of successes, p_0 is the value of p hypothesized in H_0, and n is the sample size. Z_{data} represents the number of standard deviations ($\sigma_{\hat{p}}$) the

sample proportion $\hat{p}$ lies above or below the hypothesized proportion p_0. Extreme values of $\hat{p}$ will be associated with extreme values of Z_{data}. The *Z* test for the proportion may be performed using either the *p*-value method or the critical-value method. For the *p*-value method, we reject H_0 if the *p*-value is less than α.

3 For the critical-value method, we compare the values of Z_{data} and Z_{crit}. If Z_{data} falls in the critical region, we reject H_0.

4 We can use a single $100(1 - \alpha)\%$ confidence interval for p to help us perform any number of two-tailed hypothesis tests about p.

Florida	397
Indiana	334
Maryland	342
Massachusetts	414
Minnesota	395
Texas	476
Wisconsin	386

a. State the hypotheses and the rejection rule.
b. The sample variance is $s^2 = 2245.67$. Calculate χ^2_{data}.
c. Calculate the p-value.
d. Compare the p-value with α. State your conclusion.
e. Interpret your conclusion.

24. Fourth-Grade Feet. Suppose a children's shoe manufacturer is interested in estimating the variability of fourth-graders' feet. A random sample of 20 fourth-graders' feet yielded the following foot lengths, in centimeters.[26] The normality of the data was verified in Example 8.11 (page 376). We are interested in whether the population standard deviation of foot lengths σ is less than 1 centimeter.

22.4	23.4	22.5	23.2	23.1	23.7	24.1	21.0	21.6	20.9
25.5	22.8	24.1	25.0	24.0	21.7	22.0	22.7	24.7	23.5

a. State the hypotheses.
b. Find the critical value $\chi^2_{1-\alpha}$ for level of significance $\alpha = 0.05$.
c. The sample variance is $s^2 = 1.638$. Calculate χ^2_{data}.
d. Compare χ^2_{data} with $\chi^2_{1-\alpha}$. State your conclusion.
e. Interpret your conclusion.

25. Does Score Variability Differ by Gender? Recently, researchers have been examining the evidence for whether there is greater variability in boys' scores than girls' scores on cognitive abilities tests. For example, one study found that boys were overrepresented at both the top and the bottom of nonverbal reasoning tests and quantitative reasoning tests.[27] Suppose that the standard deviation for girls' scores is known to be 50 points for a particular test and that the population of all scores is normal. A random sample of 101 boys has a sample variance of 2600. Test whether the population standard deviation for boys exceeds 50 points, using level of significance $\alpha = 0.05$. Use either the p-value method or the critical-value method.

26. Heart Rate Variability. A reduction in heart rate variability is associated with elevated levels of stress, since the body continues to pump adrenaline after high-stress situations, even when at rest.[28] Suppose the standard deviation of heartbeats in the general population is 20 beats per minute, and that the population of heart rates is normal. A random sample of 50 individuals leading high-stress lives has a sample variance of 200 beats per minute. Test using level of significance $\alpha = 0.05$ whether the population standard deviation for those leading high-stress lives is lower than that in the general population.

Try This in Class! In-class activities to enhance your understanding of statistics

TESTING FOR THE POPULATION MEAN OF THE SUM OF TWO FAIR DICE

Is the population mean μ of the sum of two fair dice equal to 7? The population standard deviation is $\sigma = 2.4152$. Each student will generate sample data to perform the following hypothesis test with level of significance $\alpha = 0.10$:

$$H_0: \mu = 7 \quad \text{versus} \quad H_a: \mu \neq 7$$

where μ represents the population mean sum of two fair dice. Assume that the distribution is normal. The instructor should prepare a number line on the board to display the collection of test statistic Z_{data} values. For $\alpha = 0.10$ and a two-tailed test, the value of Z_{crit} is 1.645. Thus, the instructor should draw lines separating the critical regions $Z_{data} < -1.645$ and $Z_{data} > 1.645$ from the noncritical region $-1.645 < Z_{data} < 1.645$.

1. Each student should toss two dice 10 times, recording the sum each time, and then calculate the mean sum $\bar{x}$ of this sample of size 10.

2. Using $n = 10$, $\sigma = 2.4152$, and $\mu_0 = 7$, each student should calculate his or her own value for the test statistic Z_{data} using his or her own value of $\bar{x}$ generated above.

3. Give each student a Post-It Note. Each student should record the following on the note: his or her name and the sample mean $\bar{x}$ and test statistic Z_{data} for his or her sample.

4. Each student should place the Post-It Note on the number line that the instructor prepared earlier.

5. Identify any Post-It Notes that indicate evidence against the null hypothesis that the population mean is 7. What is the value of Z_{data}? Of the sample mean $\bar{x}$? Do you agree that these sample means are unusual?

6. Approximately what percentage of sample means indicate evidence against $H_0: \mu_0 = 7$? Approximately what percentage of sample means indicate evidence consistent with $H_0: \mu_0 = 7$?

7. Overall, what is your personal conclusion about whether or not the population mean of the dice equals 7? What evidence can you point to in support of your conclusion?

CHAPTER *9* FORMULAS AND VOCABULARY

SECTION 9.1
- **α (alpha)** (p. 416)
- **Alternative hypothesis** (p. 411)
- **β (beta)** (p. 416)
- **Null hypothesis** (p. 411)
- **Hypothesis testing** (p. 411)
- **Statistical significance** (p. 414)
- **Type I error** (p. 415)
- **Type II error** (p. 415)

SECTION 9.2
- **Essential idea about hypothesis testing for the mean** (p. 419)
- **Level of significance α** (p. 422)
- **p-value** (p. 421)
- **Rejection rule for performing a hypothesis test using the p-value method** (p. 422)
- **Test statistic** (p. 420)
- **Z_{data}** (p. 420)

$$Z_{\text{data}} = \frac{\bar{x} - \mu_0}{\sigma_{\bar{x}}} = \frac{\bar{x} - \mu_0}{\sigma/\sqrt{n}}$$

SECTION 9.3
- **Critical region** (p. 432).
- **Critical value, Z_{crit}** (p. 432).
- **Noncritical region** (p. 432).
- **Power of a hypothesis test** (p. 439).

SECTION 9.4
- **Critical value, t_{crit}** (p. 447)
- **t_{data}** (p. 442)

$$t_{\text{data}} = \frac{\bar{x} - \mu_0}{s_{\bar{x}}} = \frac{\bar{x} - \mu_0}{s/\sqrt{n}}$$

SECTION 9.5
- **Essential idea about hypothesis testing for the proportion** (p. 456)
- **Z_{data} for the hypothesis test for the population proportion** (p. 457).

$$Z_{\text{data}} = \frac{\hat{p} - p_0}{\sigma_{\hat{p}}} = \frac{\hat{p} - p_0}{\sqrt{\dfrac{p_0(1 - p_0)}{n}}}$$

SECTION 9.6
- **χ^2_{data}** (p. 467).

$$\chi^2_{\text{data}} = \frac{(n - 1)s^2}{\sigma_0^2}$$

- **Essential idea about hypothesis testing for the standard deviation** (p. 466)

CHAPTER *9* REVIEW EXERCISES

SECTION 9.1

For Exercises 1 and 2, do the following.
  a. Provide the null and alternative hypotheses.
  b. Describe the two ways a correct decision could be made.
  c. Describe what a Type I error would mean in the context of the problem.
  d. Describe what a Type II error would mean in the context of the problem.
1. **Household Size.** The U.S. Census Bureau reported that the mean household size in 2002 was 2.58 persons. We conduct a hypothesis test to determine whether the population mean household size this year has changed.
2. **Salaries of Assistant Professors. Salaries.com** reports that the median salary for assistant professors in science as of 2005 was $49,934. We use this median salary to estimate that the mean salary in 2005 was $50,000. A hypothesis test was conducted to determine if the population mean salary of assistant professors in science has increased.

SECTION 9.2
For Exercises 3–5, find the value of Z_{data}.
3. $\bar{x} = 59$, $\sigma = 10$, $n = 100$, $\mu_0 = 60$
4. $\bar{x} = 59$, $\sigma = 5$, $n = 100$, $\mu_0 = 60$
5. $\bar{x} = 59$, $\sigma = 1$, $n = 100$, $\mu_0 = 60$

For Exercises 6 and 7, perform the following steps.
  a. State the hypotheses and the rejection rule for the p-value method.
  b. Find Z_{data}.
  c. Find the p-value. Draw the standard normal curve, with Z_{data} and the p-value indicated on it.
  d. State the conclusion and the interpretation.

Step 3   **Find t_{data}.** We need to calculate $\bar{x}_d$ and s_d.

$$\bar{x}_d = \frac{\sum x}{n} = \frac{0.00 + 0.70 + 0.20 + 0.80 + 0.40 - 0.50}{6} \approx \$0.267$$

$$s_d = \sqrt{\frac{\sum (x - \bar{x})^2}{n - 1}}$$

$$= \sqrt{\frac{(0.00 - 0.267)^2 + (0.70 - 0.267)^2 + (0.20 - 0.267)^2 + (0.80 - 0.267)^2 + (0.40 - 0.267)^2 + (-0.50 - 0.267)^2}{5}}$$

$$\approx \$0.48$$

This gives

$$t_{\text{data}} = \frac{\bar{x}_d}{s_d / \sqrt{n}} = \frac{0.267}{0.48 / \sqrt{6}} \approx 1.360$$

Step 4   **State the conclusion and the interpretation.** Since 1.36 is not greater than 2.015 (Figure 10.3), do not reject H_0. There is insufficient evidence that brand-name grocery items cost more on average than store-brand items. It appears that the brand-name milk was on sale; otherwise the conclusion may very well have been different. ■

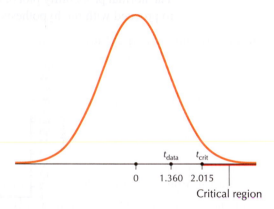

t_{data} t_{crit}

0   1.360   2.015

Critical region

FIGURE 10.3 $t_{\text{data}} = 1.360$ does not fall within the critical region.

STEP-BY-STEP TECHNOLOGY GUIDE: Confidence Intervals and Hypothesis Tests for μ_d

TI-83/84

Confidence Interval

(Example 10.2 is used to illustrate the procedure.)

Step 1   Enter samples 1 and 2 in lists **L1** and **L2**.

Step 2   Type **(L1 – L2) STO→ L3** and press **ENTER** (Figure 10.4).

Step 3   Press **STAT** and highlight **TESTS**.

Step 4   Press **8** (for the **TInterval**).

Step 5   For input (**Inpt**), highlight **Data** and press **ENTER**. (If given the summary statistics for the differences, choose **STATS**.)

Step 6   For **List**, press **2nd** then **L3**. For **Freq**, enter **1**. Enter the **C-Level** (confidence level, such as **0.95** for 95%), and press **ENTER** (Figure 10.5).

Step 7   Highlight **Calculate** and press **ENTER**. The results are shown in Figure 10.6.

Figure 10.4   Figure 10.5   Figure 10.6

Hypothesis Test

(Example 10.4 is used to illustrate the procedure.)
Step 1   Enter samples 1 and 2 in lists L1 and L2.
Step 2   Type **(L1 – L2) STO→ L3** and press **ENTER**.
Step 3   Press **STAT** and highlight **TESTS**.
Step 4   For the hypothesis test, press **2** (for the **T-Test**). The T-Test menu appears.
Step 5   For input (**Inpt**), highlight **Data** and press **ENTER**. (If given the summary statistics for the differences, choose **STATS**.)
Step 6   For μ_0, enter the hypothesized value. For **List**, press **2nd** then **L3**. For **Freq**, enter **1**. Choose the form of the hypothesis test, and press **ENTER** (Figure 10.7).

Step 7   When the cursor is over **Calculate**, make sure all your entries are correct, and press **ENTER**. The results are shown in Figure 10.8.

Figure 10.7

Figure 10.8

EXCEL

Hypothesis Test

Step 1   Enter samples 1 and 2 in columns A and B.
Step 2   Click **Data > Data Analysis > t-Test: Paired Two Sample for Means**, and click **OK**.

Step 3   For **Variable 1 Range**, highlight the cells for sample 1 in column A, and for **Variable 2 Range**, highlight the cells for sample 2 in column B.
Step 4   Enter the **Hypothesized Mean Difference** (usually 0), and enter a value for **alpha**. Then click **OK**.

MINITAB

Confidence Interval and Hypothesis Test

Step 1   Enter samples 1 and 2 in columns C1 and C2.
Step 2   Click **Stat > Basic Statistics > Paired t**.
Step 3   For **First Sample**, enter **C1**, and for **Second Sample**, enter **C2**.

Step 4   Click **Options**.
a.   For the confidence interval, specify the **Confidence Level**, then click **OK** twice.
b.   For the hypothesis test, specify the form of the alternative hypothesis, then click **OK** twice.

Section 10.1 SUMMARY

1 Two samples are independent when the subjects selected for the first sample do not determine the subjects in the second sample. Two samples are dependent when the subjects in the first sample determine the subjects in the second sample. The data from dependent samples are called matched-pair or paired samples. The key concept in this section is that we consider the differences of matched pair data as a sample, and perform inference on this sample of differences.

2 A $100(1 - \alpha)\%$ confidence interval for μ_d, the population mean of the differences, is given by $\bar{x}_d \pm t_{\alpha/2}(s_d/\sqrt{n})$, where

$\bar{x}_d$ and s_d represent the sample mean and sample standard deviation of the differences, respectively, of the set of n paired differences, $d_1, d_2, d_3, \ldots, d_n$, and where $t_{\alpha/2}$ is based on $n - 1$ degrees of freedom.

3 The paired sample t test for the population mean of the differences μ_d can be used under either of the following conditions:
- Case 1: the population is normal, or
- Case 2: the sample size is large ($n \geq 30$).

The test may be carried out using either the p-value method or the critical-value method.

Section 10.1 EXERCISES

CLARIFYING THE CONCEPTS

1. When are two samples considered independent?
2. When are two samples considered dependent?
3. What do we call the data obtained from dependent sampling?
4. How do we interpret the meaning of μ_d?

PRACTICING THE TECHNIQUES
Determine whether the experiments in Exercises 5–8 represent an independent sampling method or a dependent sampling method. Explain your answer.

5. The Jacksonville Jaguars are interested in comparing the performance of their first-year players. For each player, a sample is taken of their games from their last year in college and compared to a sample of games taken from their first year in the pros.

6. For her senior project, an exercise science major takes a sample of females majoring in exercise science, and a sample of females from her college not majoring in exercise science. She records the body mass index for each subject.

7. Before the first lecture, an algebra instructor gives a pre-test to his students to determine the students' algebra readiness. At the end of the course, the instructor gives a post-test to the same students and compares the results with the pre-test.

8. The sheriff's department takes a sample of vehicle speeds on a certain stretch of road and compares the results to a sample of vehicle speeds on a certain stretch of a different road. Both roads have the same posted speed limit.

In Exercises 9–12, assume that samples of differences are obtained through dependent sampling and follow a normal distribution.

   a. Calculate $\bar{x}_d$ and s_d.
   b. Construct a 95% confidence interval for μ_d.

9.

Subject	1	2	3	4	5
Sample 1	3.0	2.5	3.5	3.0	4.0
Sample 2	2.5	2.5	2.0	2.0	1.5

10.

Subject	1	2	3	4	5	6
Sample 1	10	12	9	14	15	8
Sample 2	8	11	10	12	14	9

11.

Subject	1	2	3	4	5	6	7
Sample 1	20	25	15	10	20	30	15
Sample 2	30	30	20	20	25	35	25

12.

Subject	1	2	3	4	5	6	7
Sample 1	1.5	1.8	2.0	2.5	3.0	3.2	4.0
Sample 2	1.0	1.7	2.1	2.0	2.7	2.9	3.3

13. For the data in Exercise 9, test whether $\mu_d > 0$, using the p-value method and level of significance $\alpha = 0.05$.

14. For the data in Exercise 9, test whether $\mu_d > 0$, using the p-value method and level of significance $\alpha = 0.01$.

15. For the data in Exercise 10, test whether $\mu_d > 0$, using the p-value method and level of significance $\alpha = 0.05$.

16. For the data in Exercise 11, test whether $\mu_d < 0$, using the critical-value method and level of significance $\alpha = 0.01$.

17. For the data in Exercise 11, test whether $\mu_d \neq 0$, using the critical-value method and level of significance $\alpha = 0.10$.

18. For the data in Exercise 11, test whether $\mu_d < 0$, using the critical-value method and level of significance $\alpha = 0.05$.

APPLYING THE CONCEPTS
19. New Car Prices. Kelley's Blue Book (**kbb.com**) publishes data on new and used cars. The table contains the manufacturer's suggested retail price for four vehicles, model years 2006 and 2007. We are interested in the difference in price between the 2006 models and the 2007 models. Assume that the population of price differences is normally distributed.
   a. Find the mean of the differences, $\bar{x}_d$, and the standard deviation of the differences, s_d.
   b. Construct a 95% confidence interval for μ_d, the population mean difference in price.

	Subaru Forester	Honda CR-V	Toyota RAV-4	Nissan Sentra
2006	$22,420	$20,990	$22,980	$13,815
2007	$21,820	$22,395	$23,630	$15,375

20. High and Low Temperatures. The University of Waterloo Weather Station tracks the daily low and high temperatures in Waterloo, Ontario, Canada. Table 10.7 contains a random sample of the daily high (sample 1) and low (sample 2) temperatures for May 1–May 10, 2006, in degrees centigrade. Assume that the temperature differences are normally distributed.
   a. Find the mean of the differences, $\bar{x}_d$, and the standard deviation of the differences, s_d.
   b. Construct a 99% confidence interval for the population mean difference between high and low temperatures.

21. New Car Prices. For the data in Exercise 19, test whether 2007 models are on average more expensive, using level of significance $\alpha = 0.05$.

22. High and Low Temperatures. Use the confidence interval from Exercise 20 to test using level of significance $\alpha = 0.01$ whether the population mean difference between high and low temperatures differs from zero.

23. Falling Home Sales Prices. A credit crunch gripped the nation in 2007–2008, leading to record numbers of mortgage foreclosures and declines in home sales prices. The following table provides the median home sales prices for four regions of the country in the first quarter (January–March) of 2007 and the first quarter of 2008. Assume that the differences are normally distributed.
   a. Find the mean of the differences $\bar{x}_d$ and the standard deviation of the differences s_d.

TABLE 10.7   High and low temperatures

May date	1	2	3	4	5	6	7	8	9	10
High temp	19.0	19.8	23.3	21.1	15.2	9.9	17.2	21.7	21.2	23.9
Low temp	7.4	3.0	3.9	7.9	4.4	0.7	−1.1	2.3	6.6	5.8

b. Construct a 90% confidence interval for μ_d, the population mean difference in price.

	Northeast	Midwest	South	West
Jan.–Mar. 2007	$370,300	$212,800	$222,900	$341,500
Jan.–Mar. 2008	$326,600	$201,900	$204,800	$298,900

Source: U.S. Census Bureau.

24. Mozart Effect? A researcher claims that listening to Mozart improves scores on math quizzes. A random sample of five students took math quizzes, first before and then after listening to Mozart. Perform the appropriate hypothesis test for determining whether the results support the researcher's claim, using level of significance $\alpha = 0.01$. Assume normality.

Student	1	2	3	4	5
Before	75	50	80	85	95
After	85	45	85	95	95

25. Falling Home Sales Prices. Use the confidence interval you constructed in Exercise 23 and level of significance $\alpha = 0.01$ to test whether the population mean difference μ_d between the first quarter 2007 median price and the first quarter 2008 median price differs from 0.

26. Spell-Checking. A software manufacturer claims that its spell-checker can reduce spelling errors in text documents. A random sample of nine documents were written and the number of spelling errors counted before and after the spell-checker was used. Assume that the differences are normally distributed.

Document	1	2	3	4	5	6	7	8	9
Before	5	3	10	2	0	1	5	0	3
After	2	1	2	0	0	1	0	0	0

Is there evidence for the manufacturer's claim? Perform the appropriate hypothesis test, using level of significance $\alpha = 0.10$.

27. Science Scores Worldwide. The National Center for Educational Statistics publishes the results from the Trends in International Math and Science Study (TIMSS). Table 10.8 contains the 1995 and 2003 mean science scores for fourth graders from various countries. Assume that the population of score differences is normally distributed.

a. Construct a 90% confidence interval for μ_d, the population mean difference in score.
b. Using level of significance $\alpha = 0.10$, test whether the 2003 scores are higher than the 1995 scores, on average.

TABLE 10.8 Fourth-grade science scores

Country	1995	2003
Singapore	523	565
Japan	553	543
Hong Kong	508	542
England	528	540
United States	542	536
Hungary	508	530
Latvia	486	530
Netherlands	530	525
New Zealand	505	523
Australia	521	521
Scotland	514	502
Slovenia	464	490
Cyprus	450	480
Norway	504	466
Iran	380	414

28. Reading Levels. An advertisement for a phonetics game claims that it will improve reading scores by at least 10 points. The reading levels of eight randomly selected children were tested without the phonetics game and retested after exposure to the game. Assume that the differences are normally distributed.

Child	1	2	3	4	5	6	7	8
Pre-test	70	65	85	50	75	80	90	60
Post-test	75	70	90	70	90	80	95	70

a. Find a 95% confidence interval for the population mean of the differences, μ_d.
b. Should we use the confidence interval in (a) to test the manufacturer's claim? Why or why not?
c. Using level of significance $\alpha = 0.05$, perform the appropriate hypothesis test.

29. NHL Goals Scored. The entire 2004–2005 National Hockey League season was canceled due to a lockout by the owners looking to introduce a player salary cap. During this time, the league also introduced rule changes designed to increase the number of goals scored per game. Table 10.9 shows the goals scored per game for a random sample of six NHL teams in the season before (2003–2004) and the season after (2005–2006) the lockout. Assume that the differences are normally distributed.

a. Construct a 95% confidence interval for μ_d.
b. Using level of significance $\alpha = 0.05$, test whether the population mean number of home sales differs from 2006 to 2007.

30. NHL Goals Scored. Refer to Exercise 29. *What if* we added a positive fixed amount to every entry in the 2005–2006 season results. How would this change affect the conclusion?

TABLE 10.9 NHL goals scored

	Detroit Red Wings	Tampa Bay Lightning	Phoenix Coyotes	Atlanta Thrashers	Colorado Avalanche	Dallas Stars
2003–2004 season	3.11	2.99	2.29	2.61	2.88	2.37
2005–2006 season	3.67	3.00	2.95	3.37	3.42	3.08

10.2 Inference for Two Independent Means

Objectives: **By the end of this section, I will be able to...**

1 Describe the sampling distribution of $\bar{x}_1 - \bar{x}_2$.

2 Compute and interpret t intervals for $\mu_1 - \mu_2$.

3 Perform and interpret t tests about $\mu_1 - \mu_2$.

4 Use confidence intervals for $\mu_1 - \mu_2$ to perform two-tailed hypothesis tests about $\mu_1 - \mu_2$.

1 Sampling Distribution of $\bar{x}_1 - \bar{x}_2$

In Chapter 3 we used boxplots to find evidence of a difference between male and female body temperature for a sample of 65 women and a sample of 65 men.[5] However, since the female subjects did not determine the male subjects, and vice versa, the 65 women and 65 men represent independent samples, and we cannot use the dependent sampling methods we learned in Section 10.1. We need to develop new sampling methods to determine beyond a reasonable doubt whether a difference exists between male and female mean body temperatures. The summary statistics are shown in Table 10.10.

Table 10.10 Summary statistics for female versus male body temperatures in °F

Gender	Sample size	Sample mean body temperature	Sample standard deviation	Population mean body temperature
Females (sample 1)	$n_1 = 65$	$\bar{x}_1 = 98.394$	$s_1 = 0.743$	$\mu_1 = ?$
Males (sample 2)	$n_2 = 65$	$\bar{x}_2 = 98.105$	$s_2 = 0.699$	$\mu_2 = ?$

Since we are considering two independent samples, we now have two sample sizes, n_1 and n_2, two sample means, $\bar{x}_1$ and $\bar{x}_2$, and two sample standard deviations, s_1 and s_2. These samples are taken from different populations, which have unknown population means μ_1 and μ_2. Since we are interested in the difference in the population means, we consider the quantity

$$\mu_1 - \mu_2$$

Since both μ_1 and μ_2 are population parameters, so also is $\mu_1 - \mu_2$, whose value is unknown.

Developing Your Statistical Sense

The Difference Difference

There is a difference in interpretation between the quantity $\mu_1 - \mu_2$ and the quantity μ_d from Section 10.1. Here, $\mu_1 - \mu_2$ refers to the difference in population means, whereas μ_d represents the population mean of the paired differences. ■

In previous chapters, to estimate μ, we examined the behavior of the statistic $\bar{x}$. For the difference in population means $\mu_1 - \mu_2$, we shall examine the behavior of the statistic $\bar{x}_1 - \bar{x}_2$. Note from Table 10.10 that the value of $\bar{x}_1 - \bar{x}_2$ for these samples is

$$\bar{x}_1 - \bar{x}_2 = 98.394 - 98.105 = 0.289$$

What if we repeat the experiment with two different samples and get $\bar{x}_1 = 98.4$ and $\bar{x}_2 = 97.9$? Then

$$\bar{x}_1 - \bar{x}_2 = 98.4 - 97.9 = 0.5$$

If we repeat the experiment an infinite number of times, then the values of $\bar{x}_1 - \bar{x}_2$ will form a distribution, called the **sampling distribution of $\bar{x}_1 - \bar{x}_2$**. Just as we used the sampling distribution of $\bar{x}$ to estimate μ, we use the sampling distribution of $\bar{x}_1 - \bar{x}_2$ to estimate $\mu_1 - \mu_2$.

It is unlikely that the experimenter will have knowledge of both population standard deviations σ_1 and σ_2. Therefore, we use the estimates of σ_1 and σ_2 provided by the sample standard deviations s_1 and s_2. Recall from Chapter 8 that, when the population standard deviation σ is unknown, and if either the population is normal or the sample size is large, the quantity

$$t = \frac{\bar{x} - \mu}{s_{\bar{x}}} = \frac{\bar{x} - \mu}{s/\sqrt{n}}$$

has a t distribution with $n - 1$ degrees of freedom. By analogy, we have the following sampling distribution.

Sampling Distribution of $\bar{x}_1 - \bar{x}_2$

When random samples are drawn independently from two populations with population means μ_1 and μ_2, and either

- Case 1: the two populations are normally distributed, or
- Case 2: the two sample sizes are large (at least 30), then the quantity

$$t = \frac{(\bar{x}_1 - \bar{x}_2) - (\mu_1 - \mu_2)}{s_{\bar{x}_1 - \bar{x}_2}} = \frac{(\bar{x}_1 - \bar{x}_2) - (\mu_1 - \mu_2)}{\sqrt{\dfrac{s_1^2}{n_1} + \dfrac{s_2^2}{n_2}}}$$

approximately follows a t distribution with degrees of freedom equal to the smaller of $n_1 - 1$ and $n_2 - 1$, where $\bar{x}_1$ and s_1 represent the mean and standard deviation of the sample taken from population 1, and $\bar{x}_2$ and s_2 represent the mean and standard deviation of the sample taken from population 2.

This t statistic is called *Welch's approximate t,* after the twentieth-century English statistician Bernard Lewis Welch. Although there are other distributions that statisticians use to estimate the difference between two population means, we use this approximation because it is conservative and easy to calculate.

Confidence Intervals for $\mu_1 - \mu_2$

Recall from Chapter 8 that to estimate the unknown population mean μ, we can use a t confidence interval:

$$\bar{x} \pm E = \bar{x} \pm t_{\alpha/2}(s_{\bar{x}}) = \bar{x} \pm t_{\alpha/2}(s/\sqrt{n})$$

where E is the **margin of error**. By analogy, here the t interval for $\mu_1 - \mu_2$ takes the following form.

> **Confidence Interval for $\mu_1 - \mu_2$**
>
> For two independent random samples taken from two populations with population means μ_1 and μ_2, a **100(1 − α)% confidence interval for $\mu_1 - \mu_2$** is given by
>
> $$(\bar{x}_1 - \bar{x}_2) \pm t_{\alpha/2}\sqrt{\frac{s_1^2}{n_1} + \frac{s_2^2}{n_2}}$$
>
> where $\bar{x}_1$, s_1, and n_1 represent the mean, standard deviation, and sample size of the sample taken from population 1 and $\bar{x}_2$, s_2, and n_2 represent the mean, standard deviation, and sample size of the sample taken from population 2, and $t_{\alpha/2}$ is associated with the confidence level and degrees of freedom of the smaller of $n_1 - 1$ and $n_2 - 1$.
>
> The *t* interval applies whenever *either* of the following conditions is met:
>
> - Case 1: both populations are normally distributed, or
> - Case 2: both sample sizes are large.
>
> **Margin of Error *E***
>
> The **margin of error** for a 100(1 − α)% confidence interval for $\mu_1 - \mu_2$ is given by
>
> $$E = t_{\alpha/2} \cdot \sqrt{\frac{s_1^2}{n_1} + \frac{s_2^2}{n_2}}$$

Thus, the confidence interval for $\mu_1 - \mu_2$ takes the form $(\bar{x}_1 - \bar{x}_2) \pm E$.

This is a confidence interval for the difference in two population means, which is not the same as in Section 10.1, which was for the population mean of the differences of matched pairs.

Example 10.5   Confidence interval for difference in women's and men's mean body temperatures

Find a 95% confidence interval for the difference in women's and men's population mean body temperatures, using the data in Table 10.10.

Solution

Both sample sizes are large, so the sampling distribution of $\bar{x}_1 - \bar{x}_2$ has a *t* distribution. Now,

$$\sqrt{\frac{s_1^2}{n_1} + \frac{s_2^2}{n_2}} = \sqrt{\frac{(0.743)^2}{65} + \frac{(0.699)^2}{65}} \approx 0.1265$$

For $t_{\alpha/2}$, the required degrees of freedom is the smaller of $n_1 - 1$ and $n_2 - 1$, which are both equal to $65 - 1 = 64$, so the degrees of freedom is also 64. This df = 64 is not listed in the *t* table, so we choose the next-lowest df listed, 60. For 95% confidence, then, $t_{\alpha/2} = 2.00$. The margin of error is

$$E = t_{\alpha/2} \cdot \sqrt{\frac{s_1^2}{n_1} + \frac{s_2^2}{n_2}} \approx (2.00) \cdot (0.1265) = 0.253$$

The 95% confidence interval is then

$$(\bar{x}_1 - \bar{x}_2) \pm E = (98.394 - 98.105) \pm 0.253 = 0.289 \pm 0.253 = (0.036, 0.542).$$

We are 95% confident that the difference in population means $\mu_1 - \mu_2$ lies between 0.036°F and 0.542°F. ∎

What Do These
Numbers Mean?

Remember that we are estimating the *difference* in population means $\mu_1 - \mu_2$ between women's and men's body temperatures. The margin of error $E \approx 0.253$ means that, when we repeatedly take samples from these two populations, 95% of the time our estimate $\bar{x}_1 - \bar{x}_2$ will be within 0.253°F of the unknown value of $\mu_1 - \mu_2$. ∎

3 Independent Sample *t* Test for $\mu_1 - \mu_2$

Researchers are often interested in testing whether the mean of one population is greater than, less than, or different from the mean of another population. Thus, we next learn how to perform hypothesis tests for the difference in population means $\mu_1 - \mu_2$. Usually the most important hypothesized value for $\mu_1 - \mu_2$ is 0. Consider the two-tailed hypothesis test

$$H_0 : \mu_1 - \mu_2 = 0 \quad \text{versus} \quad H_a : \mu_1 - \mu_2 \neq 0$$

If $\mu_1 - \mu_2 = 0$, that is, if there is no difference in the population means, then it follows that $\mu_1 = \mu_2$. Thus, the foregoing hypothesis test is equivalent to the hypothesis test

$$H_0 : \mu_1 = \mu_2 \quad \text{versus} \quad H_a : \mu_1 \neq \mu_2$$

Hypothesis tests about $\mu_1 - \mu_2$ take one of three forms, shown in Table 10.11.

Table 10.11 Three possible forms for the hypotheses for a test about $\mu_1 - \mu_2$

Form	Null hypothesis	Alternative hypothesis	Type of test
1	$H_0 : \mu_1 = \mu_2$	$H_a : \mu_1 > \mu_2$	Right-tailed test (one-tailed test)
2	$H_0 : \mu_1 = \mu_2$	$H_a : \mu_1 < \mu_2$	Left-tailed test (one-tailed test)
3	$H_0 : \mu_1 = \mu_2$	$H_a : \mu_1 \neq \mu_2$	Two-tailed test

The null hypothesis states that $\mu_1 = \mu_2$, that is, $\mu_1 - \mu_2 = 0$. The hypothesis test is based on the same approximate *t* distribution as the *t* interval on page 494, as long as the two samples are independent. In practice, the hypothesized difference between the two population means is nearly always $(\mu_1 - \mu_2)_0 = 0$. Thus, the test statistic takes the following form.

$$t_{\text{data}} = \frac{(\bar{x}_1 - \bar{x}_2) - 0}{\sqrt{\dfrac{s_1^2}{n_1} + \dfrac{s_2^2}{n_2}}} = \frac{(\bar{x}_1 - \bar{x}_2)}{\sqrt{\dfrac{s_1^2}{n_1} + \dfrac{s_2^2}{n_2}}}$$

Just as in Chapter 9, if t_{data} is extreme, then it represents evidence against the null hypothesis.

The hypothesis test applies whenever *either* of the following conditions is met:

- Case 1: both populations are normally distributed, or

- Case 2: both sample sizes are large.

The hypothesis test may be carried out using either the *p*-value method or the critical value method.

> **Hypothesis Test for the Difference in Two Population Means: *p*-Value Method**
>
> **Step 1   State the hypotheses and the rejection rule.** Use one of the forms from Table 10.11. Clearly state the meaning of μ_1 and μ_2. The rejection rule is *Reject H_0 if the p-value is less than α.*
>
> **Step 2   Find t_{data}.**
>
> $$t_{data} = \frac{(\bar{x}_1 - \bar{x}_2)}{\sqrt{\dfrac{s_1^2}{n_1} + \dfrac{s_2^2}{n_2}}}$$
>
> which follows an approximate *t* distribution with degrees of freedom the smaller of $n_1 - 1$ and $n_2 - 1$.
>
> **Step 3   Find the *p*-value.** If you have access to technology, use it to find the *p*-value. Otherwise, calculate the *p*-value using one of the forms in Table 10.12.
>
> **Step 4   State the conclusion and the interpretation.** Compare the *p*-value with α.

Table 10.12 *p*-Value

Form of test	The *p*-value equals…	Graphic
Form 1 Right-tailed test $H_0 : \mu_1 = \mu_2$ $H_a : \mu_1 > \mu_2$	*p*-value = $P(t > t_{data})$ Area to the right of t_{data}	
Form 2 Left-tailed test $H_0 : \mu_1 = \mu_2$ $H_a : \mu_1 < \mu_2$	*p*-value = $P(t < t_{data})$ Area to the left of t_{data}	
Form 3 Two-tailed test $H_0 : \mu_1 = \mu_2$ $H_a : \mu_1 \neq \mu_2$	*p*-value = $P(t > |t_{data}|) + P(t < -|t_{data}|)$ $= 2 \cdot P(t > |t_{data}|)$	

Example 10.6   Finding the estimated *p*-value from the *t* table

Recall the independent samples of men's and women's body temperatures from Table 10.10. Test whether women's population mean body temperature differs from that of men using the estimated *p*-value method and level of significance $\alpha = 0.05$.

Solution

Both sample sizes are large ($n_1 = n_2 = 65 \geq 30$), so we can perform the hypothesis test.

Step 1   **State the hypotheses and the rejection rule.** The key words "differs from" indicate that we will use the two-tailed test from Table 10.11:

$$H_0 : \mu_1 = \mu_2 \quad \text{versus} \quad H_a : \mu_1 \neq \mu_2$$

where μ_1 and μ_2 represent the population mean body temperature for women and men, respectively. We will reject H_0 if the p-value is less than 0.05.

Step 2   **Find t_{data}.** In Example 10.5 we found that $\sqrt{\dfrac{s_1^2}{n_1} + \dfrac{s_2^2}{n_2}} \approx 0.1265$. Thus,

$$t_{\text{data}} = \frac{(\bar{x}_1 - \bar{x}_2)}{\sqrt{\dfrac{s_1^2}{n_1} + \dfrac{s_2^2}{n_2}}} = \frac{(98.394 - 98.105)}{0.1265} = \frac{0.289}{0.1265} \approx 2.28$$

Step 3   **Find the p-value.** We estimate the p-value using Table 10.12:

$$\begin{aligned} p\text{-value} &= P(t > |t_{\text{data}}|) + P(t < -|t_{\text{data}}|) \\ &= P(t > |2.28|) + P(t < -|2.28|) \\ &= 2 \cdot P(t > 2.28) \end{aligned}$$

The p-value for a two-tailed test is the sum of the areas in the two tails, as shown in Figure 10.9.

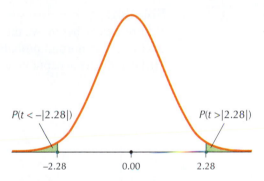

FIGURE 10.9 The p-value is the sum of two tail areas.

Using only the t table, we cannot find the exact value for $2 \cdot P(t > 2.28)$. But we can estimate it. The required degrees of freedom is the smaller of $n_1 - 1$ and $n_2 - 1$, which are both equal to $65 - 1 = 64$. Again, this df $= 64$ is not listed in the t table, so we choose the next lowest df listed, 60. To estimate $2 \cdot P(t > 2.28)$, consider this excerpt from the t table:

Two-tailed α	0.20	0.10	0.05	0.02	0.01
60		1.296	1.671	2.000	2.390	2.660

Since $t_{\text{data}} = 2.28$ lies between 2.000 and 2.390 in the row with df $= 60$, this p-value must be between 0.02 and 0.05. Note that we use the row labeled "two-tailed α" since this is a two-tailed test.

Step 4   **State the conclusion and the interpretation.** In Step 1 we said that we would reject H_0 if the p-value were less than 0.05. From Step 3, we know the p-value is between 0.02 and 0.05, thus less than 0.05. We therefore reject H_0. There is evidence that the population mean body temperatures differs for women and men. ■

Example 10.7   Finding the *p*-value using technology

Many baseball fans hold that, because of the designated hitter rule, there are more runs scored in the American League than in the National League. Perform an independent samples *t* test to find out whether that was indeed the case in 2006. Use the TI-83/84 or Excel, the *p*-value method, and level of significance $\alpha = 0.05$. Table 10.13 contains the mean runs per game (RPG) for a random sample of six teams from each league.

Table 10.13 Major League Baseball runs scored per game, 2006 regular season

American League: Sample 1		National League: Sample 2	
Team	**RPG**	**Team**	**RPG**
New York Yankees	5.74	Philadelphia Phillies	5.34
Chicago White Sox	5.36	Atlanta Braves	5.24
Texas Rangers	5.15	Colorado Rockies	5.02
Detroit Tigers	5.07	Arizona Diamondbacks	4.77
Boston Red Sox	5.06	Florida Marlins	4.68
Los Angeles Angels	4.73	Houston Astros	4.54

Solution

Before we can begin, we must determine whether both populations are normally distributed. The normal probability plots for RPG for each league, Figures 10.10 and 10.11, indicate acceptable normality, so we may perform the hypothesis test.

FIGURE 10.10

FIGURE 10.11

Step 1 **State the hypotheses and the rejection rule.** Since the American League represents sample 1 and we are interested in whether the American League has scored *more* runs than the National League, we have the following hypotheses:

$$H_0 : \mu_1 = \mu_2 \quad \text{versus} \quad H_a : \mu_1 > \mu_2$$

where μ_1 and μ_2 represent the population mean runs per game for the American League and National League, respectively. Since we are using the *p*-value method, the rejection rule is to reject H_0 if *p*-value < 0.05.

Note: Our degrees of freedom, the smaller of $n_1 - 1$ and $n_2 - 1$, is $6 - 1 = 5$. However, the TI-83/84 shows df = 9.966314697, and the Excel output rounds this to 10. Why does the technology use different degrees of freedom than we do? Recall that we are using Welch's approximation to the *t* distribution. The TI-83/84, Excel, Minitab, and other technology calculate the degrees of freedom as follows:[6]

$$df = \frac{\left(\dfrac{s_1^2}{n_1} + \dfrac{s_2^2}{n_2}\right)^2}{\dfrac{\left(\dfrac{s_1^2}{n_1}\right)^2}{n_1 - 1} + \dfrac{\left(\dfrac{s_2^2}{n_2}\right)^2}{n_2 - 1}}$$

This provides a more accurate determination of the degrees of freedom than our method. However, our method is a conservative estimate that is easier to calculate, and it is recommended for hand calculations.

Step 2   **Find t_{data}.** We use the instructions provided in the Step-by-Step Technology Guide at the end of this section. From either Figure 10.12 or Figure 10.13,

$$t_{\text{data}} = \frac{(\bar{x}_1 - \bar{x}_2)}{\sqrt{\dfrac{s_1^2}{n_1} + \dfrac{s_2^2}{n_2}}} \approx \frac{(5.185 - 4.932)}{\sqrt{\dfrac{0.339^2}{6} + \dfrac{0.320^2}{6}}} \approx 1.3301$$

```
2-SampTTest
μ1>μ2
t=1.330123458
p=.1065564474
df=9.966314697
x̄1=5.185
↓x̄2=4.931666667
```

	American	National
Mean	5.185	4.9317
Variance	0.11515	0.1025
Observations	6	6
Hypothesized Mean Difference	0	
df	10	
t Stat	**1.3301**	
P(T<=t) one-tail	**0.1065**	
t Critical one-tail	1.8125	
P(T<=t) two-tail	0.2130	
t Critical two-tail	2.2281	

FIGURE 10.12 TI-83/84 output.

FIGURE 10.13 Excel output.

Step 3   **Find the *p*-value.** From either Figure 10.12 or Figure 10.13,

$$p\text{-value} = P(t > t_{\text{data}}) = P(t > 1.3301) = 0.1065$$

This *p*-value is illustrated in Figure 10.14.

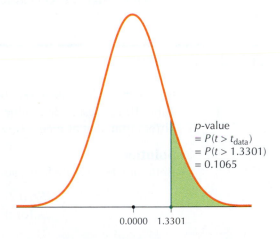

p-value
$= P(t > t_{\text{data}})$
$= P(t > 1.3301)$
$= 0.1065$

0.0000   1.3301

FIGURE 10.14 The *p*-value for the right-tailed test.

Step 4   **State the conclusion and the interpretation.** In Step 1, we said we would reject H_0 if the *p*-value was less than 0.05. Our *p*-value of 0.1065 is not less than 0.05. Therefore, we do not reject H_0. There is insufficient evidence that the population mean runs scored per game is greater in the American League than in the National League. ■

The independent sample *t* test for the difference in population means $\mu_1 - \mu_2$ may also be performed using the critical-value method.

Hypothesis Test for the Difference in Two Population Means: Critical-Value Method

Step 1   State the hypotheses. Use one of the forms from Table 10.11. Clearly state the meaning of μ_1 and μ_2.

Step 2   Find t_{crit} and state the rejection rule. To find t_{crit}, use the t table and degrees of freedom the smaller of $n_1 - 1$ and $n_2 - 1$. To find the rejection rule, use Table 10.14.

Table 10.14 Rejection rules for the t test for $\mu_1 - \mu_2$

Type of hypothesis test	Rejection rules: Reject H_0 if…
Right-tailed test $H_0 : \mu_1 = \mu_2$ versus $H_a : \mu_1 > \mu_2$	$t_{data} > t_{crit}$
Left-tailed test $H_0 : \mu_1 = \mu_2$ versus $H_a : \mu_1 < \mu_2$	$t_{data} < -t_{crit}$
Two-tailed test $H_0 : \mu_1 = \mu_2$ versus $H_a : \mu_1 \neq \mu_2$	$t_{data} > t_{crit}$ or $t_{data} < -t_{crit}$

Step 3   Find t_{data}.

$$t_{data} = \frac{(\bar{x}_1 - \bar{x}_2)}{\sqrt{\dfrac{s_1^2}{n_1} + \dfrac{s_2^2}{n_2}}}$$

which follows an approximate t distribution with degrees of freedom the smaller of $n_1 - 1$ and $n_2 - 1$.

Step 4   State the conclusion and the interpretation. Compare t_{data} with t_{crit}.

Example 10.8   Critical-value method

Recall the independent samples of men's and women's body temperatures from Table 10.10. Test to determine whether women's population mean body temperature differs from that of men, using the critical-value method and $\alpha = 0.05$.

Solution

Both sample sizes are large, so we can perform the hypothesis test.

Step 1   State the hypotheses. The hypotheses are the same as for the p-value method:

$$H_0 : \mu_1 = \mu_2 \quad \text{versus} \quad H_a : \mu_1 \neq \mu_2$$

where μ_1 and μ_2 represent the population mean body temperature for women and men, respectively.

Step 2   Find t_{crit} and state the rejection rule. Again df = 64, but we use the conservative df = 60 in the t table in the Appendix (page T-11). For $\alpha = 0.05$, this gives $t_{crit} = 2.000$. We have a two-tailed test, so Table 10.14 gives us the following rejection rule:

Reject H_0 if $t_{data} > 2.000$ or $t_{data} < -2.000$

Step 3   Find t_{data}. From Example 10.6 we have $t_{data} \approx 2.28$.

Step 4   State the conclusion and the interpretation. The test statistic $t_{data} = 2.28$ is greater than $t_{crit} = 2.000$ (see Figure 10.15).

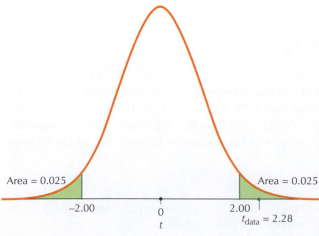

Area = 0.025   Area = 0.025

−2.00   0   2.00
t $t_{data} = 2.28$

FIGURE 10.15 $t_{data} = 2.28$ falls within the critical region.

We therefore reject H_0. There is evidence that the difference in population mean body temperatures differs for women and men. Note that this result matches the conclusion from Example 10.6. ■

4 Using Confidence Intervals for $\mu_1 - \mu_2$ to Perform Two-Tailed Hypothesis Tests About $\mu_1 - \mu_2$

Just as we did in previous chapters, we use the confidence interval to assess significance. Recall that a $100(1 - \alpha)\%$ confidence interval contains all the plausible values for $\mu_1 - \mu_2$ and that any hypothesized values that lie outside this interval are implausible, with confidence $100(1 - \alpha)\%$. Thus, if the $100(1 - \alpha)\%$ confidence interval for $\mu_1 - \mu_2$ contains 0, then it is plausible that $\mu_1 - \mu_2 = 0$ and therefore that $\mu_1 = \mu_2$. But if the interval does not contain 0, then the hypothesis that $\mu_1 - \mu_2 = 0$ is implausible, and we reject this hypothesis.

> **Using Confidence Intervals to Assess Significance**
> If a $100(1 - \alpha)\%$ confidence interval for $\mu_1 - \mu_2$ contains 0, then with $100(1 - \alpha)\%$ confidence we do not reject the hypothesis that there is no difference in the population means. If this confidence interval does not contain 0, then reject the null hypothesis with $100(1 - \alpha)\%$ confidence.

Example 10.9   Using a confidence interval to assess significance

a. Construct a 95% confidence interval for the difference in runs scored per game in the American League and National League, using the data from Example 10.7.
b. Determine with 95% confidence whether the population mean number of runs scored per game in the American League differs significantly from the population mean number of runs scored in the National League.

Solution

a. Figure 10.16 shows the 95% confidence interval for $\mu_1 - \mu_2$, where μ_1 and μ_2 represent the population mean runs per game in the American and National Leagues, respectively:

$$(-0.1712, 0.6779)$$

b. The confidence interval in (**a**) does contain 0. That is, 0 lies between –0.1712 and 0.6779. Therefore, with 95% confidence, we do not reject the hypothesis that there is no difference between population mean runs scored per game in the American and National leagues. ■

```
2-SampTInt
 (-.1712,.6779)
 df=9.966314697
 x̄1=5.185
 x̄2=4.931666667
 Sx1=.33933759
↓Sx2=.320151006
```

FIGURE 10.16

CASE STUDY | Do Prior Student Evaluations Influence Students' Ratings of Professors?

Recall that in this case study the students in one sample were shown positive evaluations of an instructor and the students in a second sample were shown negative evaluations of the instructor. Then all subjects were shown the same 20-minute lecture video given by the same instructor. They were then asked to rate the instructor using three questions, and a summary rating score was calculated. Were students' ratings influenced by the prior student evaluations?

We investigate this question by constructing a 95% confidence interval for the difference in population mean ratings $\mu_1 - \mu_2$. Assume that both populations are normally distributed and that the samples are drawn independently.

Getty Images/Punchstock

Example 10.10 Teen privacy in online social networks: Confidence interval

Punchstock/Banana Stock

The Pew Internet and American Life Project (www.pewinternet.org) tracks the behavior of Americans on the Internet. In 2007, they published a report that described some of the behaviors of American teenagers in online social networks, such as Facebook.[8] Teenagers who had online profiles were asked: "We'd like to know if your last name is posted to your profile or not." The results are shown in Table 10.15. Assume that the samples are independent.

Table 10.15 Proportions of teenage boys and girls who post their last names in online profiles

	Boys	Girls
Number responding "Yes"	$x_1 = 195$	$x_2 = 93$
Sample size	$n_1 = 487$	$n_2 = 487$
Sample proportion	$\hat{p}_1 = x_1/n_1 = 195/487 \approx 0.4004$	$\hat{p}_2 = x_2/n_2 = 93/487 \approx 0.1910$

a. Find the point estimate $\hat{p}_1 - \hat{p}_2$ for the difference in population proportions $p_1 - p_2$.

b. For a 95% confidence level, calculate the margin of error.

$$E = Z_{\alpha/2} \cdot \sqrt{\frac{\hat{p}_1 \cdot \hat{q}_1}{n_1} + \frac{\hat{p}_2 \cdot \hat{q}_2}{n_2}}$$

c. Construct and interpret a 95% confidence interval for the difference in population proportions of girls and boys whose last name is posted to their online profile.

Solution

We first check whether the conditions for the Z interval are valid. We have $x_1 = 195 \geq 5$, $(n_1 - x_1) = (487 - 195) = 292 \geq 5$, $x_2 = 93 \geq 5$, and $(n_2 - x_2) = (487 - 93) = 394 \geq 5$. We can therefore proceed.

a. The point estimate of the difference in population proportions $p_1 - p_2$ is

$$\hat{p}_1 - \hat{p}_2 \approx 0.4004 - 0.1910 = 0.2094$$

b. $\hat{q}_1 = 1 - \hat{p}_1 = 1 - 0.4004 = 0.5996$ $\qquad$ $\hat{q}_2 = 1 - \hat{p}_2 = 1 - 0.1910 = 0.8090$

The $Z_{\alpha/2}$ value for a 95% confidence level is 1.96. Therefore, the margin of error is

$$E = Z_{\alpha/2} \cdot \sqrt{\frac{\hat{p}_1 \cdot \hat{q}_1}{n_1} + \frac{\hat{p}_2 \cdot \hat{q}_2}{n_2}} = (1.96) \sqrt{\frac{(0.4004)(0.5996)}{487} + \frac{(0.1910)(0.8090)}{487}}$$

$$\approx (1.96)(0.0285) \approx 0.0559$$

c. The 95% confidence interval is therefore

$$(\hat{p}_1 - \hat{p}_2) \pm E = (0.4004 - 0.1910) \pm 0.0559$$
$$= 0.2094 \pm 0.0559$$
$$= (0.1535, 0.2653)$$

We are 95% confident that the difference in population proportions of teenage boys and girls who put their last names on their online profiles lies between 0.1535 and 0.2653. ∎

3 Independent Sample *Z* Test for $p_1 - p_2$

Researchers may also be interested in testing hypotheses regarding the difference in population proportions $p_1 - p_2$. The three possible forms for such a hypothesis test are given in Table 10.16.

Table 10.16 Three possible forms for the hypotheses for a test about $p_1 - p_2$

Form	Null hypothesis	Alternative hypothesis	Type of test
1	$H_0 : p_1 = p_2$	$H_a : p_1 > p_2$	Right-tailed test (one-tailed test)
2	$H_0 : p_1 = p_2$	$H_a : p_1 < p_2$	Left-tailed test (one-tailed test)
3	$H_0 : p_1 = p_2$	$H_a : p_1 \neq p_2$	Two-tailed test

In each form of the hypotheses, the null hypothesis asserts that $p_1 = p_2$. We denote this *common population proportion* as p. Since the null hypothesis is assumed true, the test statistic takes the following form.

$$Z_{\text{data}} = \frac{(\hat{p}_1 - \hat{p}_2) - (p_1 - p_2)}{\sqrt{\dfrac{p_1(1 - p_1)}{n_1} + \dfrac{p_2(1 - p_2)}{n_2}}} = \frac{(\hat{p}_1 - \hat{p}_2) - 0}{\sqrt{\dfrac{p_1(1 - p_1)}{n_1} + \dfrac{p_2(1 - p_2)}{n_2}}}$$

$$= \frac{\hat{p}_1 - \hat{p}_2}{\sqrt{\dfrac{p(1 - p)}{n_1} + \dfrac{p(1 - p)}{n_2}}} = \frac{\hat{p}_1 - \hat{p}_2}{\sqrt{p(1 - p)\left(\dfrac{1}{n_1} + \dfrac{1}{n_2}\right)}}$$

Since the common population proportion p is unknown, we estimate it using the following **pooled estimate of p**.

$$\hat{p}_{\text{pooled}} = \frac{x_1 + x_2}{n_1 + n_2}$$

Substituting this into the formula for the test statistic gives

$$Z_{\text{data}} = \frac{\hat{p}_1 - \hat{p}_2}{\sqrt{\hat{p}_{\text{pooled}} \cdot (1 - \hat{p}_{\text{pooled}})\left(\dfrac{1}{n_1} + \dfrac{1}{n_2}\right)}}$$

Z_{data} measures the distance between the sample proportions. Extreme values of Z_{data} indicate evidence against the null hypothesis.

The Big Picture

Where we are coming from, and where we are headed...

In Chapters 8–10, we learned how to perform inference for continuous variables. However, not all variables are continuous. In this chapter, we are introduced to the *multinomial* random variable, an extension of the binomial random variable. Multinomial data are not continuous but *categorical* (qualitative). In Sections 11.1 and 11.2, we will learn methods for performing hypothesis tests for this type of data. These methods rely on the χ^2 (chi-square) distribution, which we learned in Chapters 8 and 9. Section 11.3 introduces us to *analysis of variance,* in which we compare the population means of several different groups and determine whether significant differences exist between these means. Finally, in Section 11.4 we will use inference methods in regression to examine whether there is evidence for a relationship between two continuous variables.

11.1 χ^2 Goodness of Fit Test

Objectives: By the end of this section, I will be able to...

1. Explain what a multinomial random variable is and how to calculate expected frequencies.

2. Describe how a χ^2 goodness of fit test works.

3. Perform and interpret the results from the χ^2 goodness of fit test using the critical-value method, the *p*-value method, and the estimated *p*-value method.

According to **NetApplications.com**, the market share for the leading Internet browsers in March 2007 was as follows: Microsoft Internet Explorer, 78.5%; Firefox, 15%; Safari, 4.5%; others, 2%. Change is rapid in the online environment. Have these market shares changed since 2007? How would we go about performing a hypothesis test to determine whether market shares have changed significantly? In Section 11.1, we examine this question using a new type of random variable which follows a multinomial distribution, together with a new type of hypothesis test called a χ^2 *goodness of fit test*. We begin by first considering a new type of random variable that is used to represent categorical data.

1 The Multinomial Random Variable

Recall from Chapter 1 that *categorical* (qualitative) variables are not numeric but take values that can be classified into categories. In Chapter 6, we considered binomial random variables, for which there are only two possible outcomes. Now let's consider the following type of random variable, which can have more than two possible values. Suppose we choose a card from a deck of 52 playing cards at random and with replacement, and we define our random variable to be

$$X = \text{suit of the card}$$

The random variable X has 4 possible outcomes (clubs, spades, hearts, diamonds). We say that X is a **multinomial random variable** because it satisfies the following properties:

Multinomial Random Variable

A random variable is *multinomial* if it satisfies each of the following conditions:

- Each independent trial of the experiment has k possible outcomes, $k = 2, 3, 4, \ldots$
- The ith outcome (category) occurs with probability p_i, where $i = 1, 2, \ldots, k$ (that is, p_i is the population proportion for category i)
- $\sum_{i=1}^{k} p_i = 1$ (Law of Total Probability)

Data from a **multinomial random variable** are said to follow a *multinomial distribution*.

Note: The binomial distribution may be considered a special case of the multinomial distribution, with $k = 2$.

The random variable $X = suit$ of the card has 4 outcomes, and the probability of each outcome is

$$p_{\text{clubs}} = p_{\text{spades}} = p_{\text{hearts}} = p_{\text{diamonds}} = 1/4$$

Successive draws are independent since they are made at random and with replacement. Thus, since

$$\sum_{i=1}^{4} p_i = \frac{1}{4} + \frac{1}{4} + \frac{1}{4} + \frac{1}{4} = 1$$

the random variable $X = suit$ is a multinomial random variable.

Next, recall from Chapter 6 that the formula for finding the expected value (mean) of a binomial random variable having n trials and probability of success p is

$$\text{expected value} = n \cdot p$$

For a multinomial random variable, the **expected frequency** of the ith category is

$$\text{expected frequency}_i = E_i = n \cdot p_i$$

where n represents the number of trials, and p_i represents the population proportion for the ith category.

Example 11.1   Browser market share

According to **NetApplications.com**, the market share for the leading Internet browsers in March 2007 was as shown in Table 11.1. Let $X = browser$ of a randomly selected Internet user.

a.   Verify that $X = browser$ is a valid multinomial random variable.

b.   Find the expected frequency for each category in a series of 200 trials.

Table 11.1 Distribution of browser market share

Browser	Relative frequency
Microsoft Internet Explorer	0.785
Firefox	0.15
Safari	0.045
Other	0.02

Solution

a. There are $k = 4$ possible outcomes: Microsoft Internet Explorer, Firefox, Safari, and Other. Assigning probabilities using the relative frequency method, we have the following hypothesized proportions for each browser:

$$p_{\text{MS IE}} = 0.785, \quad p_{\text{Firefox}} = 0.15, \quad p_{\text{Safari}} = 0.045, \quad p_{\text{Other}} = 0.02$$

And

$$\sum_{i=1}^{4} p_i = 0.785 + 0.15 + 0.045 + 0.02 = 1$$

Therefore, $X = browser$ is a valid multinomial random variable.

b. We have $n = 200$ trials (sample size = 200), so the expected frequencies are as provided in Table 11.2.

Table 11.2 Expected frequencies for browser preference in sample of size 200

Category	Expected frequency$_i = E_i = n \cdot p_i$
Microsoft Internet Explorer	$E_{\text{MS IE}} = 200 \cdot 0.785 = 157$
Firefox	$E_{\text{Firefox}} = 200 \cdot 0.15 = 30$
Safari	$E_{\text{Safari}} = 200 \cdot 0.045 = 9$
Other	$E_{\text{Other}} = 200 \cdot 0.02 = 4$

As a check on the calculations, we have $\sum E_i = n$. In this case,

$$\sum E_i = 157 + 30 + 9 + 4 = 200 = n$$

What Do These Expected Frequencies Mean?

Recall that the expected value of a random variable refers to the long-run mean of that random variable after an arbitrarily large number of trials. For example, if we repeatedly took samples of 200 Internet users and asked about browser preference, the mean number of persons who preferred Firefox would approach 30 as the number of trials increased, *if the proportions given in Table 11.1 are correct*. Similarly, since 15% of the entire population of Internet users prefer Firefox, we would *expect* about 15% of any given sample of 200 Internet users to prefer Firefox, since the sample is a subset of the population. This of course begs the question: are the proportions in Table 11.1 still true? That is the type of question we will learn how to address here in Section 11.1. ■

🖉 What Is a χ^2 Goodness of Fit Test?

Market shares often shift, especially in the world of the Internet. Do the 2007 market shares still hold true today? In other words, has the distribution of the multinomial random variable *browser* given in Table 11.1 changed since 2007? To determine this, we introduce a new type of hypothesis test, called a goodness of fit test. Specifically, it is called a **χ^2 goodness of fit test** because we use the χ^2 distribution to perform the hypothesis test.

χ^2 Goodness of Fit Test

A **χ^2 goodness of fit test** is a hypothesis test used to determine whether a random variable follows a particular distribution. In a goodness of fit test, the hypotheses are

H_0 : The random variable follows a particular distribution.
H_a : The random variable does not follow the distribution specified in H_0.

For Example 11.1, the null hypothesis completely specifies each of the probabilities in the relative frequency distribution, as follows:

$$H_0 : p_{\text{MS IE}} = 0.785, p_{\text{Firefox}} = 0.15, p_{\text{Safari}} = 0.045, p_{\text{Other}} = 0.02$$

The alternative hypothesis simply denies the claim made by the null hypothesis:

$$H_a : \text{The random variable does not follow the distribution specified in } H_0.$$

In other words, H_a claims that the browser market shares have changed since 2007.

Developing Your Statistical Sense

Fitting the Model to the Data

Now, a goodness of fit test sounds like something you do in a clothing store dressing room. Actually, the analogy to clothes is rather appropriate. Suppose winter is coming and you are in the market for a new pair of gloves. You find one pair that is especially attractive, but the gloves don't fit your hands. What do you do? You reject the ill-fitting gloves and search for a new pair. In statistics, the gloves represent the models and your hands represent the actual "hard data" observed in the sample.

The null hypothesis H_0 represents what is called a *model,* a working theory of how the proportions are distributed. Our working model of how the market shares are distributed is stated in the null hypothesis:

Model 1. $H_0 : p_{\text{MS IE}} = 0.785, p_{\text{Firefox}} = 0.15, p_{\text{Safari}} = 0.045, p_{\text{Other}} = 0.02$

Of course, we could also try other models if we think the market has changed, such as the following:

Model 2. $H_0 : p_{\text{MS IE}} = 0.80, p_{\text{Firefox}} = 0.12, p_{\text{Safari}} = 0.05, p_{\text{Other}} = 0.03$
Model 3. $H_0 : p_{\text{MS IE}} = 0.75, p_{\text{Firefox}} = 0.20, p_{\text{Safari}} = 0.04, p_{\text{Other}} = 0.01$

In hypothesis testing, we "try on" only one model at a time. ■

In statistics, a goodness of fit test determines if the actual "hard data" observed in the sample are consistent with the null hypothesis. Hard data are needed if we are to determine whether or not the market shares have changed. Market researchers would collect data on the actual preferences of a sample of 200 real Internet users. They would come back with a set of *observed frequencies* of Internet users who prefer the various browsers and would compare them with the expected frequencies found in Example 11.1.

> **How a Goodness of Fit Test Works**
>
> The goodness of fit test is based on a comparison of the *observed frequencies* (actual data from the field) with the *expected frequencies* when H_0 is true. That is, we compare what we actually see with what we would expect to see if H_0 were true. If the difference between the observed and expected frequencies is large, we reject H_0.

As usual, it comes down to how large a difference is large. The hypothesis test we conduct to answer this question relies on the χ^2 distribution.

> **Test Statistic for the χ^2 Goodness of Fit Test**
>
> For a multinomial random variable with k categories and n trials, let O_i represent the observed frequency for category i, and let E_i represent the expected frequency for category i. Then the **test statistic for a goodness of fit test**
>
> $$\chi^2_{\text{data}} = \sum \frac{(O_i - E_i)^2}{E_i}$$
>
> approximately follows a χ^2 (chi-square) distribution with $k - 1$ degrees of freedom, if the following conditions are satisfied:
>   **a.** None of the expected frequencies is less than 1.
>   **b.** At most 20% of the expected frequencies are less than 5.

If the conditions are not satisfied, then it may be possible to combine two or more categories so that the conditions may then be fulfilled.

Example 11.2 Combining low-frequency categories

Determine if the conditions are met for performing the goodness of fit test for the data in Example 11.1. If necessary, combine the two low-frequency categories to fulfill the conditions.

Solution
From Table 11.2, the expected frequencies are

$$E_{\text{MS IE}} = 157 \qquad E_{\text{Firefox}} = 30 \qquad E_{\text{Safari}} = 9 \qquad E_{\text{Other}} = 4$$

Since none of these expected frequencies is less than 1, condition (a) is satisfied. However, one of the four expected frequencies (that is, 25%) is less than 5. This violates condition (b), which states that at most 20% of the expected frequencies be less than 5. We therefore combine the two categories with the smallest expected frequencies, Safari and Other, into a single category, Safari/Other. The relative frequency for this new category is, from Table 11.1, $0.045 + 0.02 = 0.065$. The probabilities may be added because the two categories are mutually exclusive. The expected frequency of the combined category in a sample of size 200 is $E_{\text{Safari/Other}} = 200 \cdot 0.065 = 13$. Now, none of the expected frequencies are less than 5. Therefore, condition (b) is now satisfied, and we may proceed to perform the goodness of fit test. However, we also need to update our hypotheses, as follows:

$$H_0 : p_{\text{MS IE}} = 0.785, p_{\text{Firefox}} = 0.15, p_{\text{Safari/Other}} = 0.065$$

$$H_a : \text{The random variable does not follow the distribution specified in } H_0.$$

3 Performing the χ^2 Goodness of Fit Test

The χ^2 goodness of fit test may be performed using (a) the critical-value method, (b) the *p*-value method, or (c) the estimated *p*-value method.

χ^2 Goodness of Fit Test: Critical-Value Method

Step 1   State the hypotheses and check the conditions.

- The null hypothesis states that the multinomial random variable follows a particular distribution.

- The alternative hypothesis states that the random variable does not follow that distribution.

The following conditions must be met:

a.   None of the expected frequencies is less than 1.

b.   At most, 20% of the expected frequencies are less than 5.

The expected frequency for the *i*th category is $E_i = n \cdot p_i$ where n represents the number of trials and p_i represents the population proportion for the *i*th category.

Step 2   Find the χ^2 critical value χ^2_{crit} and state the rejection rule. Reject H_0 if $\chi^2_{data} > \chi^2_{crit}$.

Step 3   Find the test statistic χ^2_{data}.

$$\chi^2_{data} = \sum \frac{(O_i - E_i)^2}{E_i}$$

where O_i = observed frequency, and E_i = expected frequency.

Step 4   State the conclusion and the interpretation. Compare χ^2_{data} with χ^2_{crit}.

Example 11.3   Critical-value method for the χ^2 goodness of fit test

Perform the hypothesis test shown in Example 11.2, using the observed frequencies of browser preference from a survey of 200 Internet users in Table 11.3. Use level of significance $\alpha = 0.05$.

Table 11.3 Observed frequencies of browser preference in sample of 200 internet users

Browser	Observed frequency
Microsoft Internet Explorer	140
Firefox	40
Safari/Other	20

Solution

Step 1   **State the hypotheses and check the conditions.** The hypotheses are as given in Example 11.2:

$H_0 : p_{MS\ IE} = 0.785, p_{Firefox} = 0.15, p_{Safari/Other} = 0.065$

$H_a :$ The random variable does not follow the distribution specified in H_0.

Example 11.2 shows that the conditions for applying the χ^2 goodness of fit test are satisfied.

Step 2 **Find the χ^2 critical value χ^2_{crit} and state the rejection rule.** We have degrees of freedom $k - 1 = 3 - 1 = 2$ and $\alpha = 0.05$. Turning to the χ^2 table (Table E in the Appendix, page T-12) in the column labeled $\chi^2_{0.05}$ and the row containing df $= 2$, we find $\chi^2_{\text{crit}} = \chi^2_{0.05} = 5.991$. The rejection rule is "Reject H_0 if $\chi^2_{\text{data}} > 5.991$."

Step 3 **Find the test statistic χ^2_{data}.** The observed frequencies O_i are found in Table 11.3 and the expected frequencies are given in Table 11.2. Then

$$\chi^2_{\text{data}} = \sum \frac{(O_i - E_i)^2}{E_i} = \frac{(140 - 157)^2}{157} + \frac{(40 - 30)^2}{30} + \frac{(20 - 13)^2}{13}$$
$$\approx 1.8408 + 3.3333 + 3.7692 = 8.9433$$

Table 11.4 gives the quantities needed to calculate χ^2_{data}.

Table 11.4 Calculating χ^2_{data}

Category	p_i	O_i	E_i	$O_i - E_i$	$(O_i - E_i)^2$	$\dfrac{(O_i - E_i)^2}{E_i}$
MS IE	0.785	140	157	-17	289	1.8408
Firefox	0.15	40	30	10	100	3.3333
Safari/Other	0.065	20	13	7	49	3.7692
Total	1.0	200	200	0		$\chi^2_{\text{data}} = 8.9433$

Step 4 **State the conclusion and the interpretation.** Compare χ^2_{data} with χ^2_{crit}. $\chi^2_{\text{data}} = 8.9433$ is greater than $\chi^2_{\text{crit}} = 5.991$, as shown in Figure 11.1. Therefore, we reject H_0.

FIGURE 11.1 Reject H_0 when $\chi^2_{\text{data}} > \chi^2_{\text{crit}}$.

There is evidence that the random variable *browser* does not follow the distribution specified in H_0. In other words, there is evidence that the market shares for Internet browsers have changed. ■

Developing Your Statistical Sense

Be Careful How You Interpret the Conclusion

Note carefully what this conclusion says and what it doesn't say. The χ^2 goodness of fit test shows that there is evidence that the random variable does not follow the distribution specified in H_0. In particular, the conclusion does *not* state, for example, that Firefox's proportion is significantly greater. Informally, we can compare the observed frequency of 40 with the expected frequency of 30 for the Firefox browser and note that there appears to be evidence of an increase in market share. But this is only informal—and is not part of the hypothesis test. It is a common error in statistical analysis to form conclusions beyond what the hypothesis test is actually testing. ■

The *p*-value for the χ^2 statistic is defined as the area under the χ^2 curve to the right of the test statistic χ^2_{data}, as shown in Figure 11.2. That is,

$$p\text{-value} = P(\chi^2 > \chi^2_{\text{data}})$$

We can use technology to find the exact *p*-value for a particular value of χ^2_{data}. Or, alternatively, the *p*-value may be estimated using the χ^2 table.

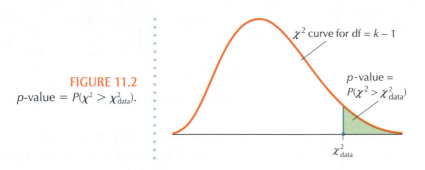

FIGURE 11.2
$p\text{-value} = P(\chi^2 > \chi^2_{\text{data}})$.

χ^2 curve for df = $k - 1$

p-value =
$P(\chi^2 > \chi^2_{\text{data}})$

χ^2_{data}

χ^2 Goodness of Fit Test: *p*-Value Method

Step 1   State the hypotheses and the rejection rule. Check the conditions.
The null hypothesis states that the multinomial random variable follows a particular distribution. The alternative hypothesis states that the random variable does not follow that distribution. Reject H_0 if the *p*-value is less than α.

The following conditions must be met:

    a.   None of the expected frequencies is less than 1.

    b.   At most 20% of the expected frequencies are less than 5.

The expected frequency for the *i*th category is $E_i = n \cdot p_i$ where *n* represents the number of trials and p_i represents the population proportion for the *i*th category.

Step 2   Find the test statistic χ^2_{data}.

$$\chi^2_{\text{data}} = \sum \frac{(O_i - E_i)^2}{E_i}$$

where O_i = observed frequency, and E_i = expected frequency.

Step 3   Find the *p*-value.

$p\text{-value} = P(\chi^2 > \chi^2_{\text{data}})$ (see Figure 11.2)

Step 4   State the conclusion and the interpretation. Compare the *p*-value with α.

Example 11.4   *p*-Value method for the χ^2 goodness of fit test using technology

The Pew Internet and American Life Project released the report *Broadband Adoption at Home*, which updated figures on the market share of cable modem, DSL, and wireless broadband from a 2002 survey (Table 11.5). The 2006 survey (Table 11.6) was based on a random sample of 1000 home broadband users. Test whether the population proportions have changed since 2002, using the *p*-value method, and level of significance $\alpha = 0.05$.

Table 11.5 2002 broadband adoption survey

Cable modem	DSL	Wireless/Other
67%	28%	5%

Table 11.6 2006 broadband adoption survey

Cable modem	DSL	Wireless/Other
410	500	90

Solution

Step 1   **State the hypotheses and the rejection rule. Check the conditions.**

$H_0 : p_{\text{Cable}} = 0.67, p_{\text{DSL}} = 0.28, p_{\text{Wireless/Other}} = 0.05$

H_a : The random variable does not follow the distribution specified in H_0.

Reject H_0 if the p-value is less than 0.05.

First we need to find the expected frequencies. We have $n = 1000$, so the expected frequencies are as shown here.

Expected frequencies for broadband access preference in sample of size $n = 1000$

Category	Expected frequency$_i = E_i = n \cdot p_i$
Cable	$E_{\text{Cable}} = 1000 \cdot 0.67 = 670$
DSL	$E_{\text{DSL}} = 1000 \cdot 0.28 = 280$
Wireless/Other	$E_{\text{Wireless/Other}} = 1000 \cdot 0.05 = 50$

What Results Might We Expect?

Before we do the formal hypothesis test, let's try to figure out what the conclusion might be. Figure 11.3 is a clustered bar graph (see Section 2.1) of the observed and expected frequencies for each of the three categories. If H_0 were true, then, for each category, we would expect the green bars (observed frequencies) and yellow bars (expected frequencies) to have similar heights.

Note that the observed frequency for cable is much lower than the expected frequency, while the observed frequency for DSL is much higher than the expected frequency. These both indicate evidence against the null hypothesis. Thus, we might expect to reject H_0. ◼

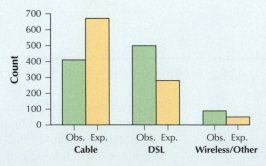

FIGURE 11.3 Graph indicates evidence against H_0.

Next we check whether the requirements for performing the test are met. We verify that (a) none of the expected frequencies is less than 1 and (b) no more than 20% of the expected frequencies are less than 5. We therefore proceed to perform the hypothesis test, using the instructions provided in the Step-by-Step Technology Guide at the end of this section (page 531).

Step 2   **Find the test statistic χ^2_{data}.** The TI-83/84 results in Figure 11.4 tell us that $\chi^2_{\text{data}} = 305.7526652 \approx 305.75$.

Step 3   **Find the *p*-value.** Figure 11.4 also tells us that

$$p\text{-value} = P(\chi^2 > 305.75) \approx \text{``4.0425215E–67''} \approx 0$$

Figure 11.5 illustrates why the *p*-value is so small. There is essentially no area to the right of $\chi^2_{\text{data}} = 305.75$ in the $\chi^2_{\text{df}=2}$ distribution.

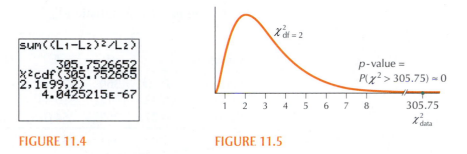

FIGURE 11.4   **FIGURE 11.5**

Step 4   **State the conclusion and the interpretation.** Since the *p*-value is less than α, we reject H_0, which we expected. There is evidence that the proportions of broadband type in 2006 have changed since 2002. As the 2006 Pew report stated, "In a reversal of market share, telephone companies offering digital subscriber line (DSL) services have overtaken cable companies in the broadband world." ■

Example 11.5   Estimated *p*-value method for the χ^2 goodness of fit test

In June 2003, the Pew Internet and American Life Project released their data on the College Students Gaming Survey. Among the questions they asked college students was "Where do you play video/computer/online games the most?" The responses are summarized in the following table. Suppose previous research had indicated that 40% of college students played video/computer/online games the most at their parents' home, 33% at their friend's home, and 27% in their dorm room.

Place	Count
Parents' home	360
Friend's home	314
Dorm room	267

Do the latest findings by Pew indicate that the population proportions have changed? Test, using the estimated *p*-value method, and level of significance $\alpha = 0.10$.

Solution
We note that we have a multinomial random variable, with three categories and df $= k - 1 = 3 - 1 = 2$.

Step 1 **State the hypotheses and the rejection rule. Check the conditions.**

$$H_0 : p_{\text{Parents}} = 0.40, \quad p_{\text{Friends}} = 0.33, \quad p_{\text{Dorm}} = 0.27$$

H_a : The random variable does not follow the distribution specified in H_0.

We reject H_0 if the p-value is less than $\alpha = 0.10$. Let's check the conditions. Here we have $n = 360 + 314 + 267 = 941$. Then, assuming H_0 is correct, we have the following expected frequencies:

$$E_{\text{Parents}} = n \cdot p_{\text{Parents}} = 941 \cdot 0.40 = 376.40$$
$$E_{\text{Friends}} = n \cdot p_{\text{Friends}} = 941 \cdot 0.33 = 310.53$$
$$E_{\text{Dorm}} = n \cdot p_{\text{Dorm}} = 941 \cdot 0.27 = 254.07$$

The conditions are therefore met and we may proceed with the hypothesis test.

Step 2 **Find the test statistic χ^2_{data}.**

Category	p_i	O_i	E_i	$O_i - E_i$	$(O_i - E_i)^2$	$\dfrac{(O_i - E_i)^2}{E_i}$
Parents'	0.40	360	376.40	−16.40	268.96	0.714559
Friend's	0.33	314	310.53	3.47	12.0409	0.038775
Dorm	0.27	267	254.07	12.93	167.1849	0.658027
Total	1.0	941	941	0		$\chi^2_{\text{data}} = 1.411361$

Thus,

$$\chi^2_{\text{data}} = \sum \frac{(O_i - E_i)^2}{E_i} = 1.411361 \approx 1.41$$

Step 3 **Find the estimated p-value.** Our degrees of freedom is $k - 1 = 3 - 1 = 2$. Turning to the χ^2 table, with df = 2, we locate where $\chi^2_{\text{data}} = 1.41$ would be. We see that it would be to the left of 4.605. Since the p-values increase as χ^2_{data} decreases, the estimated p-value for $\chi^2_{\text{data}} = 1.41$ is somewhat greater than 0.10. Figure 11.6 shows that the p-value is actually much greater than 0.10.

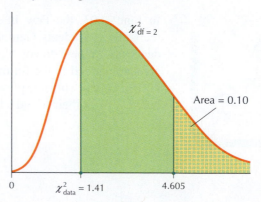

FIGURE 11.6 Estimating the p-value for a χ^2-value.

Step 4 **State the conclusion and the interpretation.** Since our estimated p-value is greater than $\alpha = 0.10$, our conclusion, as expected, is that we do not reject H_0. In other words, there is insufficient evidence that the Pew survey data contains significantly different proportions from those found in earlier research. ■

STEP-BY-STEP TECHNOLOGY GUIDE: The χ^2 Goodness of Fit Test

We illustrate the use of technology, once the observed and expected frequencies are known, for Example 11.4 (page 527).

TI-84

Step 1 Enter observed frequencies in list **L1** and expected frequencies in list **L2**.
Step 2 Press **STAT**, highlight **TESTS**, select **D: χ^2 GOF-Test**, and press **ENTER** (Figure 11.7).

Step 3 Highlight **df**, and enter degrees of freedom **2** (Figure 11.8).
Step 4 Highlight **Calculate** and press **ENTER**. The results are shown in Figure 11.9, including χ^2_{data} and the *p*-value.

Figure 11.7

Figure 11.8

Figure 11.9

TI-83/84

To find χ^2_{data}:

Step 1 Enter observed frequencies in list **L1** and expected frequencies in list **L2**. Press **2nd QUIT**.
Step 2 Press **2nd LIST**, highlight **MATH**, select **5: sum(**, and press **ENTER** (Figure 11.10).

Step 3 Type the following: (L1–L2)²/L2) (see Figure 11.4 in Example 11.4) and press **ENTER**.
Step 4 The TI-83/84 then displays 305.7526652 as χ^2_{data} (see Figure 11.4, page 529).

To find the *p*-value:

Step 1 Select **2nd DISTR**, then χ^2 **cdf(**, and press **ENTER**.
Step 2 To get the *p*-value, that is, the area to the right of 305.7526652, enter **305.7526652, comma, 1E99, comma, 2)**, as shown in Figure 11.4, page 529.

```
NAMES OPS MATH
1:min(
2:max(
3:mean(
4:median(
5:sum(
6:prod(
7↓stdDev(
```

Figure 11.10

EXCEL

To find χ^2_{data}:

Step 1 Enter the observed and expected frequencies in rows 1 and 2 (Figure 11.11).
Step 2 In cell B3, enter: **=(B1-B2)^2/B2** (Figure 11.11).

	B3	▼	f_x =(B1-B2)^2/B2	
	A	B	C	D	E
1	Observed	410	500	90	
2	Expected	670	280	50	
3		100.8955	172.8571	32	305.7527
4	4.04252E-67				

Figure 11.11

Step 3 Copy the contents of cell B3 to cells C3 and D3.
Step 4 Select an empty cell, enter **=SUM(B3:D3)**, and press **ENTER**. Excel then displays the value $\chi^2_{\text{data}} = 305.7527$ (Figure 11.12).

	E3	▼	f_x =SUM(B3:D3)	
	A	B	C	D	E
1	Observed	410	500	90	
2	Expected	670	280	50	
3		100.8955	172.8571	32	305.7527
4	4.04252E-67				

Figure 11.12

To find the *p*-value:

Step 1 Select a cell and enter **=CHITEST(B1:D1,B2:D2)** and press **ENTER**.
Step 2 Excel then provides the *p*-value (Figure 11.13).

	A4	▼	f_x =CHITEST(B1:D1,B2:D2)		
	A	B	C	D	E	F
1	Observed	410	500	90		
2	Expected	670	280	50		
3		100.8955	172.8571	32	305.7527	
4	4.04252E-67					

Figure 11.13

MINITAB

To find χ^2_{data}:

Step 1 Enter the observed frequencies (**O**) into **C1** and the hypothesized proportions into **C2**.

Step 2 Click **Stat** > **Tables** > **Chi-Square Goodness of Fit Test** (one variable).

Step 3 For observed counts, enter **C1**.

Step 4 Click **Specific Proportions** and enter **C2** in box.

Step 5 Click **OK**.

Section 11.1 SUMMARY

1 A random variable is multinomial if (a) each independent trial has k possible outcomes, $k = 2,3,4, \cdots$; (b) the ith outcome (category) occurs with probability p_i, where $i = 1,2, \cdots, k$; and (c) $\sum_{i=1}^{k} p_i = 1$ (Law of Total Probability).

2 A goodness of fit test is a hypothesis test used to ascertain whether a random variable follows a particular distribution. In a goodness of fit test, the hypotheses are

H_0 : The random variable follows a particular distribution.
H_a : The random variable does not follow the distribution specified in H_0.

Compare the observed frequencies (actual data from the field) with the expected frequencies when H_0 is true. If the difference between the observed and expected frequencies is large, reject H_0.

3 The χ^2 goodness of fit test is performed using (a) the critical-value method, (b) the p-value method, or (c) the estimated p-value method.

Section 11.1 EXERCISES

CLARIFYING THE CONCEPTS

1. What are the conditions required for a random variable to be multinomial?

2. Explain in your own words what is meant by a goodness of fit test.

3. Explain the meaning of the term "expected frequency." (*Hint:* Use the idea of the long-run mean in your answer.)

4. State the hypotheses for a χ^2 goodness of fit test.

PRACTICING THE TECHNIQUES

For Exercises 5–8, determine whether the random variable is multinomial.

5. We take a sample of 10 M&M's at random and with replacement and observe the color of each. Let $X =$ color of the M&M's, where the possible values of X are green, blue, yellow, orange, red, and brown.

6. We select 5 students from a group of 25 statistics students at random and without replacement, and we define our random variable to be $X =$ student's class, where the possible values of X are freshman, sophomore, junior, and senior.

7. We choose 10 stocks at random and with replacement, and we define our random variable to be $X =$ the exchange that the stock is traded on. The possible values of X are New York Stock Exchange, NASDAQ, London Stock Exchange, and Shenzhen Stock Exchange.

8. We pick 10 stocks at random and with replacement, and we define our random variable to be $X =$ the amount that the stock price increased or decreased since the last trading day.

For Exercises 9–12, the alternative hypothesis takes the form

H_a : The random variable does not follow the distribution specified in H_0.

    a. Find the expected frequencies.
    b. Determine whether the conditions for performing the χ^2 goodness of fit test are met.

9. $H_0 : p_1 = 0.50, p_2 = 0.25, p_3 = 0.25; n = 100$

10. $H_0 : p_1 = 0.2, p_2 = 0.3, p_3 = 0.4, p_4 = 0.1; n = 10$

11. $H_0 : p_1 = 0.9, p_2 = 0.05, p_3 = 0.04, p_4 = 0.01; n = 100$

12. $H_0 : p_1 = 0.4, p_2 = 0.35, p_3 = 0.10, p_4 = 0.10, p_5 = 0.05; n = 200$

For Exercises 13–16, do the following.

    a. Calculate the expected frequencies and verify that the conditions for performing the χ^2 goodness of fit test are met.
    b. Find χ^2_{crit} for the χ^2 distribution with the given degrees of freedom. State the rejection rule.
    c. Calculate χ^2_{data}.
    d. Compare χ^2_{data} with χ^2_{crit}. State the conclusion and the interpretation.

13. $H_0 : p_1 = 0.4, p_2 = 0.3, p_3 = 0.3; O_1 = 50, O_2 = 25,$
    $O_3 = 25$; level of significance $\alpha = 0.05$
14. $H_0 : p_1 = 1/3, p_2 = 1/3, p_3 = 1/3; O_1 = 40, O_2 = 30,$
    $O_3 = 20$; level of significance $\alpha = 0.01$
15. $H_0 : p_1 = 0.4, p_2 = 0.35, p_3 = 0.10, p_4 = 0.10, p_5 = 0.05;$
    $O_1 = 90, O_2 = 75, O_3 = 15, O_4 = 15, O_5 = 5$; level
    of significance $\alpha = 0.10$
16. $H_0 : p_1 = 0.3, p_2 = 0.2, p_3 = 0.2, p_4 = 0.2, p_5 = 0.1;$
    $O_1 = 63, O_2 = 42, O_3 = 40, O_4 = 38, O_5 = 17$; level
    of significance $\alpha = 0.05$

For Exercises 17–20, do the following.
a. State the rejection rule for the *p*-value method, calculate the expected frequencies, and verify that the conditions for performing the χ^2 goodness of fit test are met.
b. Calculate χ^2_{data}.
c. Estimate the *p*-value.
d. Compare the *p*-value with level of significance α. State the conclusion and the interpretation.

17. $H_0 : p_1 = 0.50, p_2 = 0.50; O_1 = 40, O_2 = 60; \alpha = 0.05$
18. $H_0 : p_1 = 0.50, p_2 = 0.25, p_3 = 0.25; O_1 = 52, O_2 = 23,$
    $O_3 = 25; \alpha = 0.10$
19. $H_0 : p_1 = 0.5, p_2 = 0.25, p_3 = 0.15, p_4 = 0.1;$
    $O_1 = 90, O_2 = 55, O_3 = 40, O_4 = 15; \alpha = 0.10$
20. $H_0 : p_1 = 0.4, p_2 = 0.2, p_3 = 0.2, p_4 = 0.1, p_5 = 0.1;$
    $O_1 = 90, O_2 = 45, O_3 = 40, O_4 = 15, O_5 = 10;$
    $\alpha = 0.05$

APPLYING THE CONCEPTS

21. Adult Education. The National Center for Education Statistics reported on the percentages of adults who enrolled in personal-interest courses in 2005, by the highest education level completed.[2] Of these, 8% had less than a high school diploma, 23% had a high school diploma, 32% had some college, 24% had a bachelor's degree, and 13% had a graduate or professional degree. A survey taken this year of 200 randomly selected adults who enrolled in personal-interest courses showed the following numbers for the highest education level completed. Test whether the distribution of education levels has changed since 2005, using level of significance $\alpha = 0.05$.

Less than high school	High school diploma	Some college	Bachelor's degree	Graduate or professional degree
12	40	62	54	32

22. Mall Restaurants. The International Council of Shopping Centers publishes monthly sales data for malls in North America. In May 2002, the proportions of meals eaten at food establishments in malls were as follows: fast food, 30%; food court, 46%; and restaurants, 24%. A survey taken this year of 100 randomly selected meals eaten at malls showed that 32 were eaten at fast-food places, 49 were eaten at food courts, and the rest were eaten at restaurants. Test

whether the population proportions have changed since 2002, using the critical-value method, and level of significance $\alpha = 0.10$.

23. Spinal Cord Injuries. A study in 2002 found that, of the minority patients who suffered spinal cord injury, 30% had a private health insurance provider, 55.6% used Medicare or Medicaid, and 14.4% had other arrangements.[3] Suppose that a study conducted this year of 1000 randomly selected minority patients with spinal cord injuries found that 350 had a private health insurance provider, 500 used Medicare or Medicaid, and 150 had other arrangements. Test whether the proportions have changed since 2002, using level of significance $\alpha = 0.05$.

24. The College Experience. A 2007 *New York Times* poll of Americans with at least a four-year college degree asked them how they would rate their overall experience as an undergraduate student. The results were 54% excellent, 39% good, 6% only fair, and 1% poor. A survey held this year of 500 randomly selected Americans with at least a four-year college degree found 275 rated their overall experience as an undergraduate student as excellent, 200 as good, 20 as only fair, and 5 as poor. Test whether the proportions have changed since 2007, using level of significance $\alpha = 0.05$.

25. Dormitory Life. The U.S. Census Bureau reported in 2000 that 957,581 males and 1,105,547 females were living in college dormitories. Suppose that a survey taken this year indicates that 1027 males and 1406 females were living in college dormitories. Test whether the population proportions have changed, using level of significance $\alpha = 0.05$.

26. Car Rentals. In 2002, the Web site **www .consumerwebwatch.org** reported that the market shares held by the largest car rental agencies were as follows.

Hertz	Avis	National	Budget	Alamo	Dollar	All others
28%	21.4%	12.1%	10.5%	10.5%	8.5%	9%

This year, a new survey showed that, out of every $100 spent on car rentals, $28 went to Hertz, $22 went to Avis, $12 went to National, $10 went to Budget, $10 went to Alamo, $8 went to Dollar, and $10 went to all others. Test whether the population proportions have changed since 2002, using the estimated *p*-value method and level of significance $\alpha = 0.05$.

27. Census Data. According to the U.S. Census Bureau, the population of the United States in 2002 was 288,369,000. Of these, 232,647,000 were white, 36,746,000 were black, 2,752,000 were American Indian, 11,559,000 were Asian, 484,000 were Hawaiian or other Pacific Islander, and 4,181,000 were two or more races. A survey of 1000 randomly selected Americans taken this year showed 770 white, 130 black, 10 American Indian, 50 Asian, 1 Hawaiian, and 39 of two or more races. Test whether the population proportions have changed since 2002, using the estimated *p*-value method and level of significance $\alpha = 0.10$.

28. Weekly Religious Services. A 2007 *New York Times* poll found that 31% of Americans attend religious services every week, 12% almost every week, 14% once or twice a month, 24% a few times a year, and 19% never. A survey taken this year of 100 randomly selected Americans showed 32 who attend religious services every week, 10 almost every week, 15 once or twice a month, 25 a few times a year, and 18 never. Test whether the population proportions have changed since 2007, using level of significance $\alpha = 0.10$.

29. Community College Advising. In 2007, the Community College Survey of Student Engagement found that 50% of students had met with an adviser by the end of their first four weeks at college, while 41% did not do so and 9% did not recall. A survey this year of 1000 randomly selected community college students found that 550 students had met with an adviser by the end of their first four weeks at college, while 370 did not do so and 80 did not recall. Test whether the population proportions have changed since 2007, using the *p*-value method and level of significance $\alpha = 0.05$.

30. Believing in Angels. Do you believe in angels? A Gallup Poll for May 2004 found that 78% of respondents believed in angels, 12% were not sure or had no opinion, and 10% didn't believe in angels. Suppose that a survey taken this year of 1000 randomly selected people found that 820 people believed in angels, 110 were not sure or had no opinion, and 70 didn't believe in angels. Test whether the population proportions have changed since 2004, using the critical-value method and level of significance $\alpha = 0.05$.

31. U.S. Defense Budget. The Department of Defense reported that the 1985 defense budget included 24% for personnel costs, 28% for operations, and the remainder for other budget items. Former Secretary of Defense Donald Rumsfeld reported that, in 2004, the defense budget included $99 billion for personnel costs, $117 billion for operations, and $163.4 billion for other budget items. Test whether the population proportions changed from 1985 to 2004, using the *p*-value method and level of significance $\alpha = 0.05$.

32. U.S. Defense Budget. Refer to the previous exercise. *What if* the amount spent in 2004 on personnel costs was not $99 billion but some amount less than $99 billion. How would that affect the following, and why? Would the following increase, decrease, stay the same, or is there insufficient information to determine?

  a. χ^2_{data}
  b. *p*-value
  c. Conclusion

11.2 χ^2 Tests for Independence and for Homogeneity of Proportions

Objectives: **By the end of this section, I will be able to…**

1 Explain what a χ^2 test for the independence of two variables is.

2 Perform and interpret a χ^2 test for the independence of two variables using the critical-value method and the *p*-value method.

3 Perform and interpret the results of a test for the homogeneity of proportions.

1 Introduction to the χ^2 Test for Independence

In Section 11.1, we learned that the χ^2 distribution could help us determine a model's goodness of fit to the data. In Section 11.2, we will learn two more hypothesis tests that use the χ^2 distribution. Recall from Chapter 4 that a *contingency table,* also known as a *crosstabulation* or a *two-way table,* is a tabular summary of the relationship between two categorical variables. The categories of one variable label the rows, and the categories of the other variable label the columns. Each cell in the table contains the number of observations that fit the categories of that row and column. The term *contingency table* derives from the fact that the table covers all possible combinations of the values for the two variables, that is, all possible contingencies. Table 11.7 is a contingency table based on the study *How Young People View Their Lives, Futures, and Politics: A Portrait of "Generation Next."*[4] The researchers asked 1500 randomly selected respondents, "How are things in your life?" Subjects were categorized by age and response. The researchers identified those aged 18–25 in 2007 as representing "Generation Next."

Table 11.7 Contingency table showing relative frequencies of variable categories

Response	Age Group			
	Gen Nexter (18–25)	**26+**	**Total**	**Relative frequency**
Very happy	180	330	**510**	$\frac{510}{1500} = 0.34$
Pretty happy	378	435	**813**	$\frac{813}{1500} = 0.542$
Not too happy	42	135	**177**	$\frac{177}{1500} = 0.118$
Total	**600**	**900**	**1500**	
Relative frequency	$\frac{600}{1500} = 0.4$	$\frac{900}{1500} = 0.6$		

We can use contingency tables like Table 11.7 to determine whether two random variables are independent. Recall that two random variables are *independent* if the value of one variable does not affect the probabilities of the values of the other variable. For example, is a "Gen Nexter" (someone aged 18–25 in 2007) less likely to report that he or she is "very happy" and more likely to report that he or she is "pretty happy" than someone older? If so, then age affects the response, and the variables *age group* and *response* are not independent.

To determine whether two categorical variables are independent, using the data in a contingency table, we use a χ^2 test for independence. Just like our χ^2 goodness of fit test from Section 11.1, the χ^2 **test for independence** is based on a comparison of the observed frequencies with the frequencies that are expected if the null hypothesis is assumed true.

χ^2 Test for Independence

To determine whether two categorical variables are independent, using the data from a contingency table, we use a χ^2 **test for independence**. The hypotheses take the form

H_0 : Variable A and Variable B are independent.

H_a : Variable A and Variable B are not independent.

We compare the observed frequencies with the frequencies that we expect if we assume that H_0 is correct. Large differences lead to the rejection of the null hypothesis.

Here, we are testing whether the variables *age group* and *response* are independent. Thus, the hypotheses are

H_0 : *Age group* and *response* are independent.

H_a : *Age group* and *response* are not independent.

H_0 states that a response to the survey question is not affected by age group. H_a says that a response is affected by age group. To calculate the expected frequencies, we begin by recalling the Multiplication Rule for Two Independent Events from Chapter 5:

If A and B are any two independent events, $P(A \cap B) = P(A)\,P(B)$.

To illustrate, let our events be defined as $A = 18$–25 age group, and $B =$ reported "very happy." Then, on the assumption that these events are independent, we have

$$P(\text{Gen Nexter} \cap \text{very happy}) = P(A \cap B) = P(A)P(B) = \frac{600}{1500} \cdot \frac{510}{1500}$$
$$= 0.4 \cdot 0.34 = 0.136$$

Thus, the probability that a randomly chosen young person is both a Gen Nexter and is very happy is 0.136. Then, to find the expected frequency of this cell (Gen Nexters who are very happy), we multiply this probability 0.136 by the total sample size $n = 1500$, using the result from Section 11.1 that the expected frequency is

$$E = \text{expected frequency} = n \cdot p = 1500 \cdot 0.136 = 204$$

In other words, if the random variables *age group* and *response* are independent, then the expected frequency of Gen Nexters who report being very happy is 204. To recapitulate: the expected frequency for the Gen-Nexters-who-are very-happy cell is calculated as follows:

$$\text{expected frequency}_{\text{Gen Nexter/very happy}} = 1500 \cdot \frac{600}{1500} \cdot \frac{510}{1500} = 204$$

But note that two of the 1500s cancel, providing us with the shortcut

$$\text{expected frequency}_{\text{Gen Nexter/very happy}} = \frac{(600)(510)}{1500} = 204$$

Generalizing, this provides us with the following shortcut method for finding expected frequencies. Table 11.8 shows the expected frequencies calculated using the shortcut method.

Expected Frequencies for a χ^2 Test for Independence

The expected frequencies for the cells of a contingency table in a χ^2 test for independence are given by

$$\text{expected frequency} = \frac{(\text{row total})(\text{column total})}{\text{grand total}}$$

Example 11.6 Calculating expected frequencies using the shortcut method

Calculate the expected frequencies from Table 11.7 using the shortcut method.

Solution

Table 11.8 contains the expected frequencies calculated using the shortcut method. ∎

Table 11.8 Expected frequencies using the shortcut method

Response	Age Group		Total
	Gen Nexter (18–25)	**26+**	
Very happy	$\dfrac{(510)(600)}{1500} = 204$	$\dfrac{(510)(900)}{1500} = 306$	**510**
Pretty happy	$\dfrac{(813)(600)}{1500} = 325.2$	$\dfrac{(813)(900)}{1500} = 487.8$	**813**
Not too happy	$\dfrac{(177)(600)}{1500} = 70.8$	$\dfrac{(177)(900)}{1500} = 106.2$	**177**
Total	**600**	**900**	**1500**

The χ^2 test for independence measures the difference between the observed frequencies and the expected frequencies using the following test statistic.

Test Statistic for the χ^2 Test for Independence

Let O_i represent the observed frequency in the ith cell, and E_i represent the expected frequency in the ith cell. Then the **test statistic for the independence of two categorical variables**

$$\chi^2_{\text{data}} = \sum \frac{(O_i - E_i)^2}{E_i}$$

approximately follows a χ^2 (chi-square) distribution with $(r-1)(c-1)$ degrees of freedom, where r is the number of categories in the row variable and c is the number of categories in the column variable, if the following conditions are satisfied:

  a. None of the expected frequencies is less than 1.

  b. At most 20% of the expected frequencies are less than 5.

---

② Performing the χ^2 Test for Independence

The χ^2 test for independence may be performed using either the critical-value method or the p-value method. We provide examples of each.

χ^2 Test for Independence: Critical-Value Method

Step 1 **State the hypotheses and check the conditions.**

$$H_0 : \text{Variable } A \text{ and Variable } B \text{ are independent.}$$
$$H_a : \text{Variable } A \text{ and Variable } B \text{ are not independent.}$$

The following conditions must be met:

  a. None of the expected frequencies is less than 1.

  b. At most 20% of the expected frequencies are less than 5.

The expected frequency for a given cell is

$$\text{expected frequency} = \frac{(\text{row total}) \cdot (\text{column total})}{\text{grand total}}$$

Step 2   **Find the critical value χ^2_{crit} and state the rejection rule.** Reject H_0 if $\chi^2_{data} > \chi^2_{crit}$. Use $(r - 1)(c - 1)$ degrees of freedom, where r is the number of categories in the row variable and c is the number of categories in the column variable. Do not include the row or column totals in your counts.

Step 3   **Find the test statistic χ^2_{data}.**

$$\chi^2_{data} = \sum \frac{(O_i - E_i)^2}{E_i}$$

where O_i = observed frequency and E_i = expected frequency for each cell.

Step 4   **State the conclusion and the interpretation.** Compare χ^2_{data} with χ^2_{crit}.

Example 11.7   Performing the χ^2 test for independence using the critical-value method

Using Table 11.7, test whether *age group* is independent of *response*, using level of significance $\alpha = 0.05$.

Solution

Step 1   **State the hypotheses and check the conditions.**

$$H_0 : \textit{Age group and response are independent.}$$

$$H_a : \textit{Age group and response are not independent.}$$

We note from Table 11.8 that none of the expected frequencies are less than either 1 or 5. Therefore, the conditions are met, and we may proceed with the hypothesis test.

𝒲hat Results Might We Expect?

Note from Figure 11.14 that, for the Gen Nexters, the observed (green) frequency of the response "very happy" is *lower* than the expected (yellow) frequency, while for the 26+ group, the observed frequency of the response "very happy" is *higher* than the expected frequency. A similar situation occurs for the "not too happy" response, while the reverse situation occurs for the "pretty happy" response. In other words, the response proportions of very happy, pretty happy, and not too happy seem to *depend* on the age group. Thus, we might expect to reject H_0, which states that the response does not depend on the age group. ∎

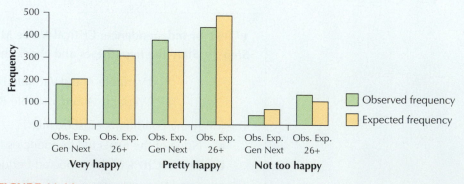

FIGURE 11.14 Graph shows evidence against H_0.

Step 2   **Find the critical value χ^2_{crit} and state the rejection rule.** The row variable, *response*, has three categories, so $r = 3$. The column variable, *age group*, has two categories, so $c = 2$. Thus,

$$\text{degrees of freedom} = (r - 1)(c - 1) = (3 - 1)(2 - 1) = 2$$

With level of significance $\alpha = 0.05$, this gives us $\chi^2_{crit} = 5.991$ from the χ^2 table. The rejection rule is therefore

$$\text{Reject } H_0 \text{ if } \chi^2_{data} > 5.991$$

Step 3   **Find the test statistic χ^2_{data}.** The observed frequencies are found in Table 11.7 and the expected frequencies are found in Table 11.8. Then

$$\chi^2_{data} = \sum \frac{(O_i - E_i)^2}{E_i} = \frac{(180 - 204)^2}{204} + \frac{(330 - 306)^2}{306} + \frac{(378 - 325.2)^2}{325.2}$$

$$+ \frac{(435 - 487.8)^2}{487.8} + \frac{(42 - 70.8)^2}{70.8} + \frac{(135 - 106.2)^2}{106.2}$$

$$\approx 38.5192$$

Step 4   **State the conclusion and the interpretation.** Our χ^2_{data} of 38.5192 is greater than our χ^2_{crit} of 5.991, and so we reject H_0, which is what we expected. The interpretation is: "There is evidence that *age group* is not independent of *response*." In other words, there does appear to be evidence that age group affects the response to the question "How are things in your life?"

χ^2 **Test for Independence: *p*-Value Method**

Step 1   **State the hypotheses and the rejection rule. Check the conditions.**

$$H_0 : \text{Variable } A \text{ and Variable } B \text{ are independent.}$$

$$H_a : \text{Variable } A \text{ and Variable } B \text{ are not independent.}$$

Reject H_0 if the *p*-value is less than α.

The following conditions must be met:

  a.   None of the expected frequencies is less than 1.

  b.   At most 20% of the expected frequencies are less than 5.

The expected frequency for a given cell is

$$\text{expected frequency} = \frac{(\text{row total}) (\text{column total})}{\text{grand total}}$$

Step 2   **Find the test statistic χ^2_{data}.**

$$\chi^2_{data} = \sum \frac{(O_i - E_i)^2}{E_i}$$

where O_i = observed frequency and E_i = expected frequency for each cell.

Step 3   **Find the *p*-value.**

$$\text{*p*-value} = P(\chi^2 > \chi^2_{data})$$

Step 4   **State the conclusion and the interpretation.** Compare the *p*-value with α.

Example 11.8 χ^2 test for independence using the p-value method and technology

Table 11.9 contains the numbers of work-related homicides that took place in the United States in 2002, according to the Bureau of Labor Statistics, categorized by the age group of the victim and the type of homicide. Test whether homicide type and age group of victim are independent, using the TI-83/84, Minitab, the *p*-value method, and level of significance $\alpha = 0.01$.

Table 11.9 Contingency table of age group of victim versus type of homicide

Type of homicide	Age Group of Victim			
	Under 25	**25 to 44**	**Over 44**	**Total**
Shooting	31	258	180	469
Stabbing	5	21	37	63
Total	36	279	217	532

Solution

Step 1 **State the hypotheses and the rejection rule. Check the conditions.**

H_0 : Age group of victim and homicide type are independent.

H_a : Age group of victim and homicide type are not independent.

Reject H_0 if the *p*-value is less than 0.01.

Note that Minitab provides the expected counts (frequencies) below the observed counts. We can then verify that none of the expected frequencies is less than 1. We do have 1 of the 6 expected frequencies (4.26) with a value less than 5. But this represents $1/6 \approx 0.1667$, which is less than 20%, as required. The χ^2 hypothesis test is therefore valid.

Step 2 **Find the test statistic χ^2_{data}.** We use the instructions found in the Step-by-Step Technology Guide at the end of this section. The TI-83/84 results in Figure 11.15 tell us that $\chi^2_{data} = 10.76001797$. The Minitab results in Figure 11.16 round this to "Chi-Sq" $= \chi^2_{data} = 10.760$.

```
Expected counts are printed below observed counts
Chi-Square contributions are printed below expected counts
                    Age 25
          Age < 25   - 44  Age > 44   Total
    1          31     258      180      469
            31.74  245.96   191.30
            0.017   0.589    0.668

    2           5      21       37       63
             4.26   33.04    25.70
            0.127   4.387    4.971

Total          36     279      217      532

Chi-Sq = 10.760, DF = 2, P-Value = 0.005
1 cells with expected counts less than 5.
```

```
χ²-Test
  χ²=10.76001797
  P=.0046077805
  df=2
```

FIGURE 11.15 TI-83/84 χ^2 results.

FIGURE 11.16 Minitab χ^2 results.

Step 3 **Find the *p*-value.** From the TI-83/84 results in Figure 11.15, we have

$$p\text{-value} = P(\chi^2 > \chi^2_{data}) = 0.0046077805$$

The Minitab results in Figure 11.16 round this to *p*-value $= 0.005$.

Step 4   **State the conclusion and the interpretation.** Since *p*-value ≈ 0.0046 < 0.01, we reject H_0. There is evidence that the age group and homicide type are not independent. ∎

3 Test for the Homogeneity of Proportions

Recall the two-sample *Z* test for $p_1 - p_2$ from Chapter 10, where we compared the proportions of two independent populations. When we extend that hypothesis test to *k* independent populations, the result uses a test statistic that follows a χ^2 distribution. Just as the null hypothesis for the two-sample test assumed no difference between the population proportions, the null hypothesis for the *k*-sample test also assumes that all *k* population proportions are equal. The alternative hypothesis states that not all the population proportions are equal. When performing the **test for the homogeneity of proportions,** we use the same steps as for the χ^2 test for independence. The difference between the test for homogeneity of proportions and the test for independence (page 537) has to do with how the data are collected. If a single sample is taken and two variables are measured, then the test for independence is appropriate. If several (*k*) samples are taken and the sample proportion is measured for each sample, then the test for homogeneity of proportions is appropriate.

Example 11.9	Airline on-time performance

The Bureau of Transportation Statistics (**www.bts.gov**) reports on the proportion of airline passenger flights that are on time, for each major airline. The January–April 2007 statistics for the three busiest carriers are shown in Table 11.10. Test whether the population proportions of on-time flights are the same for the three airlines, using the *p*-value method, Minitab, and level of significance $\alpha = 0.05$.

Table 11.10 Observed on-time statistics for three major airlines, January–April 2007

	Southwest Airlines	American Airlines	Skywest Airlines	Total
Number of on-time flights	146,607	68,939	60,298	275,844
Number of flights not on-time	36,697	35,688	32,497	104,882
Total flights	183,304	104,627	92,795	380,726

*W*hat Results Might We Expect?

The observed sample proportions of on-time flights are as follows:

$$p_{\text{Southwest}} = \frac{146,607}{183,304} \approx 0.80 \quad p_{\text{American}} = \frac{68,939}{104,627} \approx 0.66 \quad p_{\text{Skywest}} = \frac{60,298}{92,795} \approx 0.65$$

The 80% on-time proportion of Southwest Airlines does seem to be somewhat higher than the on-time proportions of the other airlines. Thus, we would not be surprised if the hypothesis test found evidence that not all the population proportions were equal. ∎

Solution

The Minitab results are shown here. We use the same steps as for the χ^2 test for independence.

```
Expected counts are printed below observed counts
Chi-Square contributions are printed below expected counts

         Southwest   American    Skywest    Total
     1     146607      68939      60298     275844
         132807.61   75804.46   67231.93
          1433.828    621.792    715.127

     2      36697      35688      32497     104882
          50496.39   28822.54   25563.07
          3771.027   1635.338   1880.815

  Total    183304     104627      92795     380726

Chi-Sq = 10057.927, DF = 2, P-Value = 0.000
```

Step 1   **State the hypotheses and the rejection rule. Check the conditions.**

$$H_0 : p_{\text{Southwest}} = p_{\text{American}} = p_{\text{Skywest}}$$
$$H_a : \text{Not all the proportions in } H_0 \text{ are equal.}$$

Reject H_0 if the p-value is less than 0.05.

None of the expected frequencies are less than either 1 or 5. Therefore, the conditions are met, and we may proceed with the hypothesis test.

Step 2   **Find the test statistic χ^2_{data}.** χ^2_{data} is shown as "Chi-Sq" = 10,057.927. There are $r = 2$ rows and $c = 3$ columns, so the degrees of freedom are $(r - 1)(c - 1) = (2 - 1)(3 - 1) = 2$.

Step 3   **Find the p-value.** Minitab provides the p-value, which is essentially 0.000.

Step 4   **State the conclusion and the interpretation.** The p-value of 0.000 is less than $\alpha = 0.05$. We therefore reject H_0, as expected. There is evidence that not all population proportions of on-time flights are equal.

STEP-BY-STEP TECHNOLOGY GUIDE: Test for Independence or Test for the Homogeneity of Proportions

We demonstrate using Example 11.8 (page 540).

TI-83/84
Entering Matrix Data
Step 1   Press **2nd**, then **MATRIX**.
Step 2   Highlight **EDIT**, and press **ENTER**.
Step 3   Set the dimensions of **MATRIX[A]** (number of rows × number of columns). Table 11.9 has 2 rows and 3 columns, so enter **2**, press **ENTER**, enter **3**, and press **ENTER**.
Step 4   Press the down-arrow key. Enter the first number in the first cell, **31**, and press **ENTER**.
Step 5   Continue entering the data row by row until the matrix is complete (Figure 11.17).

Figure 11.17

χ^2 Test for Independence or Test for Homogeneity of Proportions
Step 1   Enter the data into **Matrix[A]**.
Step 2   Press **STAT**, highlight **TESTS**, select **C: χ^2 Test**, and press **ENTER**.
Step 3   The expected frequencies are automatically generated and put into **MATRIX[B]**. Highlight **Calculate**, and press **ENTER**. The results are shown in Figure 11.15 in Example 11.8.
Step 4   To view the expected frequencies, press **2nd MATRIX**, highlight **EDIT**, choose **2** for **MATRIX[B]**, and press **ENTER**.

EXCEL
χ^2 Test for Independence or Test for Homogeneity of Proportions Using the WHFStat Add-ins
Step 1   Enter the data from Table 11.9, *including row and column totals*, in cells A1 to D3.
Step 2   Load the **WHFStat Add-ins**.

Step 3   Select **Add-Ins > Macros > Tables > Two Way Tables/ Chi Squared Test**.
Step 4   Select cells A1 to D3 as the **Dataset Range**.
Step 5   Select **Chi-squared Test**, and click **OK**.

MINITAB
χ^2 Test for Independence or Test for Homogeneity of Proportions
Step 1   Enter the observed frequencies from Table 11.9 into the Minitab worksheet, as shown here.
Step 2   Click **Stat > Tables > Chi-Square Test**.
Step 3   Choose each of columns C1, C2, and C3 as the **Columns containing the table**. Then click **OK**. The results are

shown in Figure 11.16 in Example 11.8.

C1	C2	C3
Age < 25	Age 25 - 44	Age > 44
31	258	180
5	21	37

Section 11.2 SUMMARY

1 To determine whether two categorical variables are independent, using the data from a contingency table, we use a χ^2 test for independence. The hypotheses take the form

H_0 : Variable A and Variable B are independent.

H_a : Variable A and Variable B are not independent.

2 The χ^2 test for independence is performed using either the critical-value method or the p-value method. The observed frequencies are compared with the expected frequencies on the assumption that H_0 is correct. Large differences lead to the rejection of the null hypothesis.

3 The k-sample test, called the test for the homogeneity of proportions, determines whether all k population proportions are equal. The result uses a test statistic that follows a χ^2 distribution. The null hypothesis for the k-sample test assumes that all k population proportions are equal. The alternative hypothesis states that not all the population proportions are equal. When performing the test for the homogeneity of proportions, the same steps are used as for the χ^2 test for independence.

Section 11.2 EXERCISES

CLARIFYING THE CONCEPTS
1. Explain what a contingency table is.
2. Explain in your own words what is meant by a test for independence.
3. What is the difference between the χ^2 test for homogeneity of proportions and the two-sample Z test for the difference in proportions from Chapter 10?
4. Explain how the expected frequencies are calculated without using the shortcut method.

PRACTICING THE TECHNIQUES
For Exercises 5–10, the observed frequencies are provided in a contingency table of two categorical variables. Find the expected frequencies, on the assumption that the variables are independent.

5.

	A1	A2
B1	10	20
B2	12	18

6.

	C1	C2
D1	50	100
D2	60	90

7.

	E1	E2	E3
F1	30	20	10
F2	35	24	8

8.

	G1	G2
H1	10	8
H2	8	10
H3	9	9

9.

	I1	I2	I3
J1	100	90	105
J2	50	60	55
J3	25	15	20

10.

	K1	K2	K3	K4
L1	40	70	90	100
L2	20	40	60	70
L3	30	65	65	70

For Exercises 11–14, test whether or not the variables are independent.

a. State the hypotheses.
b. Verify that the conditions for performing the χ^2 test for independence are met.
c. Find χ^2_{crit} and state the rejection rule.
d. Calculate χ^2_{data}.
e. Compare χ^2_{data} with χ^2_{crit}. State the conclusion and the interpretation.

11. Exercise 5, level of significance $\alpha = 0.05$.
12. Exercise 7, level of significance $\alpha = 0.10$.
13. Exercise 9, level of significance $\alpha = 0.01$.
14. Exercise 9, level of significance $\alpha = 0.10$.

For Exercises 15–18, test whether or not the proportions of successes are the same for all populations.

a. State the hypotheses.
b. Calculate the expected frequencies and verify that the conditions for performing the χ^2 test for homogeneity of proportions are met.
c. Find χ^2_{crit} and state the rejection rule. Use level of significance $\alpha = 0.05$.
d. Find χ^2_{data}.
e. Compare χ^2_{data} with χ^2_{crit}. State the conclusion and the interpretation.

15.

	Sample 1	Sample 2	Sample 3
Successes	10	20	30
Failures	20	45	62

16.

	Sample 1	Sample 2	Sample 3
Successes	50	50	100
Failures	200	210	425

17.

	Sample 1	Sample 2	Sample 3	Sample 4
Successes	10	15	20	25
Failures	15	24	32	40

18.

	Sample 1	Sample 2	Sample 3	Sample 4
Successes	100	150	200	250
Failures	150	240	320	400

APPLYING THE CONCEPTS

19. Conditioning Mice. A psychologist is conducting research using white mice, brown mice, a classical conditioning stimulus, and an operant conditioning stimulus. The psychologist is interested in whether type of stimulus is independent of the type of mouse. One hundred mice were tested. The following table shows the number of each type of mice that completed their assigned task satisfactorily, given the type of stimulus. Test at level of significance $\alpha = 0.10$ whether type of stimulus and type of mouse are independent.

	Type of Stimulus		
Type of mouse	**Classical**	**Operant**	**Total**
White	20	40	60
Brown	10	30	40
Total	30	70	100

20. Information Technology Jobs. In their report *The Digital Economy 2003*, the Economics and Statistics Administration of the U.S. Department of Commerce tracked the number of information technology–related jobs by year (1999 and 2002) and the level of education and training required (high, moderate, and low). Test whether year and level are independent, using the critical-value method and level of significance $\alpha = 0.05$.

	High	Moderate	Low
1999	3435	1652	1151
2002	3512	1411	1080

21. Cable TV Content Restrictions. A June 2004 *Chicago Tribune* Poll asked, "Should government restrict violence and sexual content that appears on cable TV, or should government not impose restrictions?" The responses were categorized by political affiliation. Test whether the population proportion favoring restriction is the same for all three groups, using level of significance $\alpha = 0.05$.

	Restrict	Not restrict/don't know
Republicans	59	41
Independents	52	48
Democrats	53	47

22. History Course Preferences. A history professor teaches three types of history courses: American history, world history, and European history. The professor would like to assess whether student preference for these courses differs among four ethnic groups. A random sample of 600 students resulted in the following table of preferences. Test using the critical-value method at level of significance $\alpha = 0.05$ whether type of course and ethnic group are independent.

	Type of History Course			
Ethnic group	**American history**	**World history**	**European history**	**Total**
Caucasian	100	40	60	200
African American	100	40	20	160
Asian American	50	60	30	140
Hispanic	50	20	30	100
Total	300	160	140	600

23. Gender Gap on Campus? Much has been written about the gender gap in the electoral process. Is there a gender gap on campus? A random sample of 500 students were selected from one campus, and their gender and political party preference were noted. Test using the *p*-value method at level of significance $\alpha = 0.01$ whether a gender gap exists on this campus (that is, whether or not gender and political party preference are independent).

	Political Party Preference			
Gender	**Democrat**	**Republican**	**Independent**	**Total**
Female	100	50	100	250
Male	50	100	100	250
Total	150	150	200	500

24. Immigrant Origins and Preferences. Does the state where immigrants wish to settle depend on where the immigrant is coming from? The U.S. Department of Homeland Security tracks the continent of origin and the desired state of settlement for immigrants. Some of the 2002 data are shown here, in thousands. Test using the critical-value method whether continent of origin and state of settlement are independent, using level of significance $\alpha = 0.01$.

	California	Florida	New York
Europe	24.0	9.8	23.2
Asia	112.6	9.0	31.3
South America	8.0	16.1	17.7

25. Email, Phone, or in Person? What is the most effective way to handle a task at work: by email, by phone, or in person? Well, you probably say, it depends on the task. The Pew Internet and American Life Project Email at Work Survey surveyed 1000 randomly selected work email users, who chose the following methods as the best for handling certain work tasks. Test whether the proportions who favor email differ between the two tasks, using level of significance $\alpha = 0.05$.

Task	By email	By phone or in person
Edit or review documents	670	330
Arrange meetings or appointments	630	370

26. Using Graphs as Evidence for Your Conclusion. Sick of spam (unsolicited broadcast email)? Do you get more spam at your work, school, or home email address? The Pew Internet and American Life Project Email at Work Survey examined the proportion of spam in email users' work and home email accounts. Using only the information in the clustered bar graph below, would you conclude that the proportion of those who report "a lot of spam" is the same for work email and personal email? Why?

27. Spam, Spam, Spam. Continue your work from the previous exercise. The following contingency table shows the actual percentages in the graph above based on samples of size 100 for each of work email and personal email. Test whether the proportions who report "a lot of spam" are the same for

	None	Some	A lot
Work email	53%	36%	11%
Personal email	22%	48%	30%

work email and personal email, using level of significance $\alpha = 0.01$. Does your conclusion agree with your conjecture in the previous exercise?

28. Gender Differences in Computer/Video/Online Gaming. In June 2003, the Pew Internet and American Life Project released their data on the College Students Gaming Survey. Among the questions they asked 1720 randomly selected college students was "Which one of the following do you play the most: video games, computer games, or online games?" The results are summarized by gender in the following contingency table.

	Video games	Computer games	Internet games
Male	616	221	139
Female	198	372	174

   a. Before you carry out the hypothesis test, *what result might you expect?* Look over the data set carefully to see whether you can detect significant differences between the levels of the variables. Then see whether your hypothesis test bears out your intuition.

   b. Test whether *gender* and *game type* are independent, using level of significance $\alpha = 0.01$.

29. Online Dating. A Pew Internet and American Life Project study reported that the proportion of urban residents who use online dating is 13%, while the proportion for suburban residents is 10% and the proportion for rural residents is 9%.[5] Test using level of significance $\alpha = 0.05$ whether there are differences among the population proportions of residents from the three categories who use online dating. Assume the study sample size was 1000. (*Hint:* The null hypothesis assumes that all proportions are equal.)

Use Minitab or Excel for each of Exercises 30 and 31.

Goals of Middle School Students. Open the **Goals** data set. The subjects are students in grades four, five, and six, from three schools districts in Michigan. The students were asked which of the following was most important to them: good grades, athletic ability, or popularity. Information about the students' age, gender, race, and grade was also gathered, as well as whether their school was in an urban, suburban, or rural setting.[6]

30. Comparing gender and goals.

   a. Looking at the data, do you think that boys and girls at this age differ in what is most important to them: grades, popularity, or sports? In other words, do you think that the variables *gender* and *goals* are dependent or independent?

   b. Perform the χ^2 test for independence, using level of significance $\alpha = 0.05$.

31. Comparing goals and school setting.

   a. Looking at the data, do you think that the setting of the school (urban, suburban, or rural) affects the goals of the students? Or do you think that it has no effect? In other words, do you think that the variables *urb_rur* and *goals* are independent or dependent?

   b. Perform the χ^2 test for independence, using level of significance $\alpha = 0.10$.

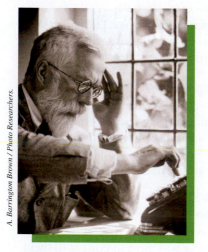

A. Barrington Brown / Photo Researchers.

11.3   Analysis of Variance

Objectives: **By the end of this section, I will be able to...**

1 Describe the characteristics of the *F* distribution

2 Explain that ANOVA works by comparing the variation between samples to the variation within the samples

3 Perform analysis of variance using the *p*-value method.

4 Perform analysis of variance using the critical-value method.

In Sections 11.1 and 11.2, we used the χ^2 distribution to analyze categorical data. Here, in Section 11.3, we need to learn about a new distribution, the *F* distribution, which will help us with the analytic methods we will learn in Sections 11.3 and 11.4.

1 *F* Distribution

The ***F* distribution** was named in honor of the "grandfather of statistics," Sir Ronald A. Fisher. Like the χ^2 distribution, the *F* distribution is right-skewed, never takes negative values, and has an infinite number of different *F* curves (Figure 11.18). The shape of the curve depends on two different degrees of freedom.

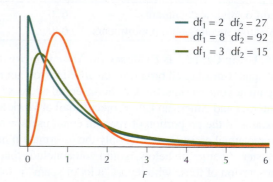

FIGURE 11.18
Shape of the *F* distribution for various degrees of freedom.

Note that the *F* distribution resembles the χ^2 distribution. This is not surprising since the values of the *F* distribution represent ratios of two χ^2 distributions. Moreover, the *F* distribution has two different degrees of freedom, which we shall call df_1 and df_2, derived from the degrees of freedom of the two χ^2 distributions represented in the ratio. Often, df_1 is called the *numerator degrees of freedom*, and df_2 is called the *denominator degrees of freedom*.

Properties of the *F* Curve

1. The total area under the *F* curve equals 1.

2. The value of the *F* random variable is never negative, so the *F* curve starts at 0. However, it extends indefinitely to the right. The curve approaches but never quite meets the horizontal axis.

3. Because of the characteristics described in (2), the *F* curve is right-skewed.

4. There is a different *F* curve for each different pair of degrees of freedom, df_1 and df_2.

Since the *F* distribution is continuous, we can find probabilities associated with values of *F*, and vice versa, just as we did with the normal, *t*, and χ^2 distributions.

Just as for any continuous distribution, probability is represented by the area below the *F* curve above an interval.

How Analysis of Variance (ANOVA) Works

Analysis of variance (ANOVA) is an inferential method for testing whether the means of different populations are equal. Why is it called *analysis of variance* if we use it for testing *means*? The answer lies in the way ANOVA works—by comparing different variance estimates, as we shall see. But first, we learn conceptually how ANOVA works.

Suppose we are interested in determining whether there are significant differences in grade point averages (GPAs) among residents of three dormitories, A, B, and C. Table 11.11 displays three random samples of GPAs of ten residents from each dormitory.

Table 11.11 Sample GPAs from Dorms A, B, and C

A	0.60	3.82	4.00	2.22	1.46	2.91	2.20	1.60	0.89	2.30
B	2.12	2.00	1.03	3.47	3.70	1.72	3.15	3.93	1.26	2.62
C	3.65	1.57	3.36	1.17	2.55	3.12	3.60	4.00	2.85	2.13

The sample mean GPA for Dormitory A is

$$\bar{x}_A = \frac{0.60 + 3.82 + 4.00 + 2.22 + 1.46 + 2.91 + 2.20 + 1.60 + 0.89 + 2.30}{10} = 2.2$$

Similarly, we can find the sample mean GPAs for the other dormitories: $\bar{x}_B = 2.5$ and $\bar{x}_C = 2.8$. We note that the sample means are not equal. The question is, Are the population means equal? Let μ_A, μ_B, and μ_C represent the population mean GPAs for Dormitories A, B, and C, respectively. We are interested in the following hypotheses, where μ_i represents the population mean GPA for dormitory *i*:

Note: In analysis of variance, the null hypothesis *always* states that all the population means are equal and the alternative hypothesis *always* states that not all the population means are equal. Note that H_a is *not* stating that the population means are all different. For H_a to be true, it is sufficient for a single population mean to be different, even though all the other population means may be equal.

$$H_0 : \mu_A = \mu_B = \mu_C \quad \text{versus} \quad H_a : \text{not all the population means are equal}$$

Sufficient differences in the sample means would represent evidence that the population means were not equal. The question is, What represents "sufficiently" different? We need something to compare against, such as the *spread* of each sample. One measure of spread or variability is the *range*:

$$\text{range} = \text{max} - \text{min}$$

We have

$$\text{range (Dorm A)} = 4.00 - 0.60 = 3.40$$
$$\text{range (Dorm B)} = 3.93 - 1.03 = 2.90$$
$$\text{range (Dorm C)} = 4.00 - 1.17 = 2.83$$

These ranges are rather large spreads, and there is a considerable amount of overlap among the different dormitory GPAs, as shown in Figure 11.19.

Figure 11.19 shows the difference among the means for the three dorm GPAs compared with the spread of each dorm's GPAs, as measured by the range. The red triangles represent the sample means, $\bar{x}_A = 2.2$, $\bar{x}_B = 2.5$, and $\bar{x}_C = 2.8$. The spread of the sample means (shown by the red arrows) is much less than the spreads of the individual dorm GPAs (shown by the green arrows). Thus, the sample means $\bar{x}_A = 2.2$, $\bar{x}_B = 2.5$, and $\bar{x}_C = 2.8$ are not sufficiently different when compared against the spread of the GPAs. This graph would therefore not provide evidence to reject the null hypothesis that the population mean GPAs are all equal.

FIGURE 11.19
Comparison dotplot of GPAs for Dorms A, B, and C.

Now we make a similar comparison for the GPAs for Dormitories D, E, and F in Table 11.12.

Table 11.12 Sample GPAs from Dorms D, E, and F

D	2.16	2.23	2.09	2.17	2.25	2.19	2.24	2.28	2.25	2.14
E	2.45	2.34	2.58	2.49	2.60	2.42	2.55	2.62	2.45	2.50
F	2.80	2.75	2.93	2.68	2.88	2.75	2.87	2.81	2.73	2.80

The sample mean GPAs for Dormitories D, E, and F are the *same* as those for Dormitories, A, B, and C, respectively: $\bar{x}_D = 2.2$, $\bar{x}_E = 2.5$, and $\bar{x}_F = 2.8$. Again we are interested in whether the population means are equal.

$$H_0 : \mu_D = \mu_E = \mu_F \quad \text{versus} \quad H_a : \text{not all the population means are equal}$$

Consider the comparison dotplot in Figure 11.20. There now seems to be better evidence for concluding that the three population means are not all equal. There is no overlap among the three samples because the spread *within* each dormitory is much smaller than for Dormitories A, B, and C.

$$\text{range (Dorm D)} = 2.28 - 2.09 = 0.19$$
$$\text{range (Dorm E)} = 2.62 - 2.34 = 0.28$$
$$\text{range (Dorm F)} = 2.93 - 2.68 = 0.25$$

Figure 11.20 shows the difference among the means for the three dorm GPAs compared with the range of each dorm's GPAs. The red triangles represent the sample means, $\bar{x}_D = 2.2$, $\bar{x}_E = 2.5$, and $\bar{x}_F = 2.8$. The spread of the sample means (red arrows) is much greater than the spreads of the individual dorm GPAs (green arrows). Thus, the sample means $\bar{x}_D = 2.2$, $\bar{x}_E = 2.5$, and $\bar{x}_F = 2.8$ are sufficiently different when compared against the range of the GPAs. This graph would, therefore, provide some evidence to reject the null hypothesis that the population mean GPAs are all equal.

FIGURE 11.20
Comparison dotplot of GPAs for Dorms D, E, and F.

Note that we arrived at *opposite conclusions* for the two sets of dormitories, *even though the sample means of the first group are identical to the sample means of the second group.* Here is the key: compared to the large within-sample spread of Dormitories A, B, and C, the difference in sample means did not seem large. However, compared to the small within-sample spread of Dormitories D, E, and F, the difference in sample means did seem large. These are the types of comparisons that the ANOVA method makes.

Instead of using the range as the measure of spread, analysis of variance uses the standard deviation of the individual samples. Recall that samples with larger spread have larger standard deviations, just as they have larger range.

How Does Analysis of Variance Work?

The key to how analysis of variance works is the following comparison. Compare

a. the variability in the sample means—that is, how large the differences are between the sample means (indicated by the lengths of the red arrows in Figures 11.19 and 11.20)—with

b. the variability within each sample—that is, the within-sample spreads (indicated by the lengths of the green arrows in Figures 11.19 and 11.20).

When (a) is much larger than (b), this is evidence that the population means are not all equal and that we should reject the null hypothesis. Thus, our analysis depends on measuring variability. And hence the term *analysis of variance.* ■

Just as for hypothesis-testing procedures from previous chapters, analysis of variance can be performed only if certain conditions are satisfied.

Requirements for Performing Analysis of Variance

1. Each of the k populations is normally distributed.
2. The variances (σ^2) of the populations are all equal.
3. The samples are independently drawn.

Our hypotheses for testing for the equality of the population mean GPA for Dormitories A, B, and C are

$$H_0 : \mu_A = \mu_B = \mu_C \quad \text{versus} \quad H_a : \text{not all the population means are equal}$$

Think about what H_0 is saying, in the light of the requirements specified. If H_0 is true, then all three dormitories would have the same population mean GPA: $\mu_A = \mu_B = \mu_C = \mu$, where we denote the hypothesized common mean as μ. Further, requirement 1 states that each population is normally distributed and requirement 2 states that all the population variances are equal. Let's call this common variance σ^2. Putting all this together, the null hypothesis for performing analysis of variance assumes that the observations from each population come from the same distribution, and this distribution is normal with mean μ and variance σ^2.

Suppose we then take samples of size n from each group. Fact 3 in Chapter 7 states that the sampling distribution of the sample mean $\bar{x}$ for a sample of size n taken from a normal population with mean μ and standard deviation σ (that is, variance σ^2) is also normal, with mean μ and standard deviation $\sigma/\sqrt{n}$ (that is, variance σ^2/n), as shown in Figure 11.21. Since each dormitory's GPA is assumed (under H_0) to come from the same sampling distribution, we would expect the sample means to be fairly close together.

On the other hand, when H_0 is not true, then not all the population means are equal (Figure 11.22). In this case, there is no sampling distribution common to all sample means. We might not expect the sample means to be close together. Note in Figure 11.22 that each distribution nevertheless has the same shape (normal) and spread (i.e., variance) because of the requirements.

FIGURE 11.21 Common sampling distribution when H_0 is true.

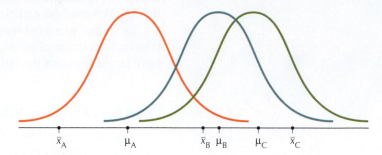

FIGURE 11.22 No common sampling distribution when H_0 is not true.

Note: Normal probability plots were introduced in Chapter 7.

We verify the normality requirement using normal probability plots for each sample. The equal variances requirement is met if the largest sample standard deviation is not larger than twice the smallest sample standard deviation.

> **Procedure for Verifying the Requirements for Analysis of Variance**
>
> **Step 1   Normality.** Check that the data from each group are normally distributed.
>
> **Step 2   Equal Variances.** Compute the sample standard deviation for each group to verify that the largest standard deviation is not larger than twice the smallest standard deviation.
>
> **Step 3   Independence.** Verify that the samples drawn from each group are independently drawn.

Example 11.10   Verify the requirements for performing an analysis of variance

Verify the requirements for performing an analysis of variance using the hypotheses

$$H_0 : \mu_A = \mu_B = \mu_C \quad \text{versus} \quad H_a : \text{not all the population means are equal}$$

where μ_i represents the population mean GPA for Dormitory i, using data from Table 11.11.

Solution

Step 1   Normality. To verify that each of the $k = 3$ populations is normally distributed, we examine normal probability plots of each sample, shown in Figure 11.23. Each plot indicates acceptable normality.

Step 2   Equal Variances. To find the standard deviation for Dorm A, we first find

$$\sum(x - \bar{x})^2 = (0.60 - 2.2)^2 + (3.82 - 2.2)^2 + (4.00 - 2.2)^2 + (2.22 - 2.2)^2$$
$$+ (1.46 - 2.2)^2 + (2.91 - 2.2)^2 + (2.20 - 2.2)^2 + (1.60 - 2.2)^2$$
$$+ (0.89 - 2.2)^2 + (2.30 - 2.2)^2 = 11.5626$$

FIGURE 11.23 Normal probability plots verify normality requirement.

Then

$$s_A = \sqrt{\frac{\sum(x - \bar{x})^2}{n - 1}} = \sqrt{\frac{11.5626}{10 - 1}} \approx 1.133460777$$

Note: We retain many decimal places when calculating s_A, s_B, and s_C because these values are used to calculate other quantities later on.

We similarly find $s_B \approx 1.030857248$ and $s_C \approx 0.9370284$. The largest, $s_A \approx 1.133460777$, is not larger than twice the smallest, $s_C \approx 0.9370284$. Thus, the equal variance requirement is satisfied.

Step 3   **Independence.** Since the students are randomly sampled from each dormitory, with the selection of students in one dormitory not affecting the selection of students sampled from the other dormitories, the independence assumption is also validated. ■

Assuming that H_0 is true, we estimate the common population mean μ using the **overall sample mean, $\bar{\bar{x}}$**:

$$\bar{\bar{x}} = \frac{(n_1\bar{x}_1 + n_2\bar{x}_2 + \cdots + n_k\bar{x}_k)}{n_t}$$

where there are k samples and n_t is the "total sample size" (sum of the k sample sizes). The overall sample mean $\bar{\bar{x}}$ is simply the mean of all the observations from all the samples. For the special case when all the sample sizes are equal, the overall sample mean $\bar{\bar{x}}$ is simply the mean of the k sample means,

$$\bar{\bar{x}} = \frac{(\bar{x}_1 + \bar{x}_2 + \cdots + \bar{x}_k)}{k}$$

Example 11.11   Calculating the overall sample mean $\bar{\bar{x}}$

For the sample GPA data given in Table 11.11 for Dorms A, B, and C, calculate the overall sample mean, $\bar{\bar{x}}$.

Solution
We have $k = 3$ dormitories, with sample mean GPAs $\bar{x}_A = 2.2$, $\bar{x}_B = 2.5$, $\bar{x}_C = 2.8$. Also, $n_A = n_B = n_C = 10$, and $n_t = 10 + 10 + 10 = 30$. Thus,

$$\bar{\bar{x}} = \frac{(10(2.2) + 10(2.5) + 10(2.8))}{30} = 2.5$$

Since all the sample sizes are equal, we can also calculate $\bar{\bar{x}}$ as follows:

$$\bar{\bar{x}} = \frac{(2.2 + 2.5 + 2.8)}{3} = 2.5$$

■

*hat Does This
Number Mean?*

$\overline{\overline{x}} = 2.5$ is the mean GPA for all 30 students from all three samples. We can use $\overline{\overline{x}}$ as our estimate of the common population mean μ assumed in H_0. ■

Recall that analysis of variance works by comparing the variability in the sample means to the variability within each sample. We use the following statistics to measure these variabilities.

The **mean square treatment (MSTR)** measures the variability in the sample means. MSTR is the sample variance of the sample means, weighted by sample size.

$$\text{MSTR} = \frac{\sum n_i(\overline{x}_i - \overline{\overline{x}})^2}{k - 1}$$

where n_i and $\overline{x}_i$ are the sample size and mean of the ith sample, $\overline{\overline{x}}$ is the overall sample mean, and there are k populations.

The **mean square error (MSE)** measures the variability within the samples. MSE is the mean of the sample variances, weighted by sample size.

$$\text{MSE} = \frac{\sum(n_i - 1)s_i^2}{n_t - k}$$

where n_i and s_i^2 are the sample size and variance of the ith sample, n_t is the total sample size, and there are k populations.

The **test statistic for analysis of variance (F_{data})** is

$$F_{\text{data}} = \frac{\text{MSTR}}{\text{MSE}}$$

F_{data} measures the variability among the sample means, compared to the variability within the samples. F_{data} follows an F distribution with $df_1 = k - 1$ and $df_2 = n_t - k$, when the following requirements are met: (1) each of the k populations is normally distributed, (2) the variances of the populations are all equal, and (3) the samples are independently drawn.

The term *mean square* represents a weighted mean of quantities that are squared. Each mean square itself consists of two parts: the *sum of squares* in the numerator and the *degrees of freedom* in the denominator. The numerator for MSTR is called the **sum of squares treatment (SSTR)**, and the numerator for MSE is called the **sum of squares error (SSE)**.

$$\text{MSTR} = \frac{\text{sum of squares treatment}}{df_1} = \frac{\text{SSTR}}{df_1} = \frac{\sum n_i(\overline{x}_i - \overline{\overline{x}})^2}{k - 1}$$

$$\text{MSE} = \frac{\text{sum of squares error}}{df_2} = \frac{\text{SSE}}{df_2} = \frac{\sum(n_i - 1)s_i^2}{n_t - k}$$

Finally, the **total sum of squares (SST)** is found by adding SSTR and SSE:

$$\text{SST} = \text{SSTR} + \text{SSE}$$

Example 11.12 Calculating SSTR, SSE, SST, MSTR, MSE, and F_{data}

Using the summary statistics in Table 11.13 for the sample GPAs for Dorms A, B, and C, calculate the following: (a) SSTR, (b) SSE, (c) SST, (d) MSTR, (e) MSE, (f) F_{data}.

Table 11.13 Summary statistics for sample GPAs for Dorms A, B, and C

	Dorm A	**Dorm B**	**Dorm C**
Mean	$\bar{x}_A = 2.2$	$\bar{x}_B = 2.5$	$\bar{x}_C = 2.8$
Standard deviation	$s_A \approx 1.133460777$	$s_B \approx 1.030857248$	$s_C \approx 0.9370284$
Sample size	$n_1 = 10$	$n_2 = 10$	$n_3 = 10$

Solution

We have $k = 3$ dormitories, and total sample size $n_t = 10 + 10 + 10 = 30$. Thus,

a. $\text{SSTR} = \sum n_i(\bar{x}_i - \bar{\bar{x}})^2 = 10(2.2 - 2.5)^2 + 10(2.5 - 2.5)^2 + 10(2.8 - 2.5)^2$

$\qquad = 10[(-0.3)^2 + (0)^2 + (0.3)^2] = 1.8$

b. $\text{SSE} \approx (10 - 1)(1.133460777)^2 + (10 - 1)(1.030857248)^2 + (10 - 1)(0.9370284)^2$

$\qquad \approx 29.0288$

c. $\text{SST} = \text{SSTR} + \text{SSE} = 1.8 + 29.0288 = 30.8288$

d. $\text{MSTR} = \dfrac{\text{SSTR}}{k - 1} = \dfrac{1.8}{3 - 1} = 0.9$

e. $\text{MSE} = \dfrac{\text{SSE}}{n_t - k} = \dfrac{29.0288}{30 - 3} = 1.0751407407$

f. $F_{data} = \dfrac{\text{MSTR}}{\text{MSE}} = \dfrac{0.9}{1.0751407407} = 0.8370997079 \approx 0.84$

The ANOVA table shown in Table 11.14 is a convenient way to display the various statistics calculated during an analysis of variance. Note that the quantities in the "Mean square" column equal the ratio of the two columns to its left.

Table 11.14 ANOVA table

Source of variation	Sum of squares	Degrees of freedom	Mean square	F-test statistic
Treatment	SSTR	$\text{df}_1 = k - 1$	$\text{MSTR} = \dfrac{\text{SSTR}}{k - 1}$	$F_{data} = \dfrac{\text{MSTR}}{\text{MSE}}$
Error	SSE	$\text{df}_2 = n_t - k$	$\text{MSE} = \dfrac{\text{SSE}}{n_t - k}$	
Total	SST			

3 ANOVA: *p*-Value Method

Now that we know how analysis of variance works, in this and the following section we will learn how to perform analysis of variance.

Analysis of Variance for the Equality of k Population Means: p-Value Method

We have taken random samples from each of k populations and want to test whether the population means of the k populations are all equal.

Required conditions:

1. Each of the k populations is normally distributed.
2. The variances (σ^2) of the populations are all equal.
3. The samples are independently drawn.

Step 1   State the hypotheses, and state the rejection rule.

$$H_0 : \mu_1 = \mu_2 = \cdots = \mu_k \quad \text{versus} \quad H_a : \text{not all the population means are equal}$$

where the μ's represent the population mean from each population. The rejection rule is *Reject H_0 if the p-value is less than α.*

Step 2   Calculate the value of the test statistic, F_{data}.

$$F_{\text{data}} = \frac{\text{MSTR}}{\text{MSE}}$$

where

$$\text{MSTR} = \frac{\sum n_i(\bar{x}_i - \bar{\bar{x}})^2}{k - 1} \quad \text{and} \quad \text{MSE} = \frac{\sum (n_i - 1)s_i^2}{n_t - k}$$

F_{data} follows an F distribution with $df_1 = k - 1$ and $df_2 = n_t - k$ if the required conditions are satisfied, where n_t represents the total sample size.

Step 3   Find the p-value. p-value $= P(F > F_{\text{data}})$, as shown in Figure 11.24. Use technology to find the p-value.

Step 4   State the conclusion and the interpretation. Compare the p-value with α.

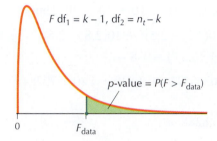

$F\ df_1 = k - 1,\ df_2 = n_t - k$

$p\text{-value} = P(F > F_{\text{data}})$

0 F_{data}

FIGURE 11.24 p-Value for the ANOVA F test.

The *Analysis of Variance* applet allows you to experiment with various values for the sample means and the sample variability in order to see how changes in these values affect F_{data} and the p-value.

Example 11.13   Analysis of variance using the p-value method

Test using the p-value method at $\alpha = 0.05$ whether the population mean GPAs from Example 11.10 differ among the students in dormitories A, B, and C.

What Result Might We Expect?

Recall that the comparison dotplot in Figure 11.19 (page 548) showed a large amount of overlap in the GPAs among the three dormitories. The large ranges illustrate the large within-dormitory spread of the GPAs for these dorms. When compared against this large within-sample variability, the variability in sample means may not seem large. Therefore, we might expect that the null hypothesis of no difference will not be rejected. ■

Solution

We already verified the requirements for performing the analysis of variance in Example 11.10.

Step 1   State the hypotheses, and state the rejection rule. Define μ_i.

$$H_0 : \mu_A = \mu_B = \mu_C \quad \text{versus} \quad H_a : \text{not all the population means are equal}$$

where μ_i represents the population mean GPA of students from dormitory i. The rejection rule is *Reject H_0 if the p-value is less than α.*

Step 2   **Calculate the value of the test statistic, F_{data}.** From Example 11.12, we have MSTR = 0.9, MSE = 1.0751407407, and

$$F_{data} = \frac{\text{MSTR}}{\text{MSE}} = \frac{0.9}{1.0751407407} = 0.8370997079$$

F_{data} follows an F distribution with $df_1 = k - 1 = 3 - 1 = 2$ and $df_2 = n_t - k = 30 - 3 = 27$.

When calculating the p-value for analysis of variance, always retain as many decimal places in the value of F_{data} as you can. This will make the p-value as accurate as possible. Rounding F_{data} too much will make the p-value less accurate.

Step 3   **Find the p-value.** We use the instructions provided in the Step-by-Step Technology Guide at the end of this section (page 559). From Figures 11.25 and 11.26, we have

$$\text{p-value} = P(F > F_{data}) = P(F > 0.8370997079) = 0.4438929572 \approx 0.4439$$

FIGURE 11.25 *p*-Value = $P(F > 0.8370997079)$.

FIGURE 11.26 TI-83/84 *p*-value.

Step 4   **State the conclusion and the interpretation.** Compare the p-value with α. Since the p-value of 0.4439 is not less than $\alpha = 0.05$, we do not reject H_0. There is not enough evidence to conclude that not all population mean GPAs are equal.

Note that this conclusion confirms our earlier intuition from the graphical evidence in Figure 11.19 that the population mean GPAs would not be found significantly different. ■

4 ANOVA: Critical-Value Method

ANOVA can also be performed using the critical-value method, as we have done for other hypothesis tests.

Analysis of Variance for the Equality of *k* Population Means: Critical-Value Method

We have taken random samples from each of *k* populations, and we want to test whether the population means of the *k* populations are all equal. Required conditions:

1. Each of the *k* populations is normally distributed.
2. The variances (σ^2) of the populations are all equal.
3. The samples are independently drawn.

Step 1   **State the hypotheses.**

$$H_0 : \mu_1 = \mu_2 = \cdots = \mu_k \quad \text{versus} \quad H_a : \text{not all the population means are equal}$$

where the μ's represent the population mean from each population.

Step 2   **Find the critical value F_{crit} and state the rejection rule.** The rejection rule is always: Reject H_0 if $F_{\text{data}} > F_{\text{crit}}$.

Step 3   **Calculate the value of the test statistic, F_{data}.**

$$F_{\text{data}} = \frac{\text{MSTR}}{\text{MSE}}$$

where

$$\text{MSTR} = \frac{\sum n_i (\bar{x}_i - \bar{\bar{x}})^2}{k - 1} \quad \text{and} \quad \text{MSE} = \frac{\sum (n_i - 1)s_i^2}{n_t - k}$$

F_{data} follows an F distribution with $\text{df}_1 = k - 1$ and $\text{df}_2 = n_t - k$ if the required conditions are satisfied, where n_t represents the total sample size.

Step 4   **State the conclusion and the interpretation.** Compare F_{data} with F_{crit}.

Example 11.14   Analysis of variance using the critical-value method

A study compared the size of ants from different colonies.[7] Researchers measured the masses (in milligrams) of random samples of ants from three different colonies. The summary statistics are shown in Table 11.15. Test using the critical-value method at $\alpha = 0.05$ whether the population mean sizes differ among the three ant colonies.

Table 11.15   Summary statistics for ant sizes from Colonies 1, 2, and 3

	Colony 1	Colony 2	Colony 3
Mean	$\bar{x}_1 = 89.66$ mg	$\bar{x}_2 = 99.99$ mg	$\bar{x}_3 = 92.16$ mg
Standard deviation	$s_1 = 29.00$ mg	$s_2 = 26.96$ mg	$s_3 = 31.40$ mg
Sample size	$n_1 = 130$	$n_2 = 111$	$n_3 = 120$

Solution

First we verify whether the requirements are satisfied.

- The normal probability plots indicate acceptable normality (Figure 11.27).

FIGURE 11.27 Normal probability plots.

- The largest sample standard deviation, $s_3 = 31.40$, is not larger than twice the smallest, $s_2 = 26.96$. Thus, the equal variance requirement is satisfied.
- Since the ants are randomly sampled from each colony, the independence assumption is also validated. We therefore proceed with the ANOVA.

Step 1 $H_0 : \mu_1 = \mu_2 = \mu_3$ versus H_a : not all the population means are equal where μ_i represents a population mean mass of ants from each colony.

Step 2 We have $k = 3$ colonies, and total sample size $n_t = 130 + 111 + 120 = 361$. Thus, $df_1 = 3 - 1 = 2$ and $df_2 = 361 - 3 = 358$. Referring to the F tables in the Appendix (pages T-13–T-16), there are no entries for $df_2 = 358$, so we take the closest value, $df_2 = 200$. For $\alpha = 0.05$, $F_{crit} = 3.04$. We reject H_0 if $F_{data} > 3.04$.

Step 3 **Calculate F_{data}.** Minitab yields the ANOVA table provided here.

```
Source    DF        SS      MS     F      P
Colony     2      6798    3399  3.98  0.020
Error    358    305802     854
Total    360    312600
```

We see that $df_1 = 2$ and $df_2 = 358$, as we found earlier. The row labeled "Colony" represents the "treatment" row in the ANOVA table. Also, Minitab switches the "DF" and "SS" columns. Notice the relationships among the statistics, as shown in the ANOVA table below.

Source	Sum of squares	Degrees of freedom	Mean square	F
Treatment	$SSTR = 6798$	$df_1 = 2$	$MSTR = \dfrac{SSTR}{df_1} = \dfrac{6798}{2} = 3399$	$F_{data} = \dfrac{MSTR}{MSE}$ $= \dfrac{3399}{854} = 3.98$
Error	$SSE = 305{,}802$	$df_2 = 358$	$MSE = \dfrac{SSE}{n_t - k} = \dfrac{305{,}802}{358} = 854$	
Total	$SST = SSTR + SSE$ $= 312{,}600$			

Step 4 **State the conclusion and the interpretation.** Compare F_{data} with F_{crit}. In Step 2, we said that we would reject H_0 if $F_{data} > 3.04$. Since $F_{data} = 3.98 > 3.04$, we reject H_0. There is evidence that not all the population mean sizes are equal among the three ant colonies. ■

Developing Your Statistical Sense

Do Not Draw the Wrong Conclusion

Note that the following interpretation is incorrect: "There is evidence that all the population mean sizes are different." Our conclusion is simply that the population means are not all equal, which does *not* imply that they are all different. For example, two of the three colony mean sizes might still be equal, and if the third mean size is different, we would still reject H_0.

Also, we cannot determine which colony has the larger population mean size, even though the sample mean of Colony 2 ($\bar{x}_2 = 99.99$ mg) is greater than the sample mean of Colony 1 ($\bar{x}_1 = 89.66$ mg). Since the original hypothesis stated that all population means are equal, the ANOVA results alone do not allow us to conclude formally that the population mean for Colony 2 is greater than that of Colony 1. To test these types of hypotheses, we would need to apply a type of analysis known as multiple comparisons.[8] ■

Photo Illustration by
Chris Jackson/Getty Images

CASE STUDY

Professors on Facebook

A recent study investigated whether the amount of information a professor posts about himself or herself (that is, *self-disclosure*) on the online social network Facebook is related to student motivation. A professor constructed three different Facebook sites, one offering low self-disclosure, one offering medium self-disclosure, and one offering high self-disclosure. For example, the low-disclosure site offered information only about her position at the university. The medium-disclosure site also showed the professor's favorite movies, books, and quotes. On the high-disclosure site, fictitious comments from "friends" were posted on "the Wall," highlighting social gatherings.

Study participants (students not enrolled in the professor's courses) were then randomly assigned to access and browse one of the three Facebook sites, develop an impression of the professor, and complete the research questionnaire. Student motivation was measured using a set of 16 items, and the sum of the 16 items was calculated to form the total motivation score. The items measured student interest, involvement, stimulation, level of excitement, and whether the student was inspired or challenged. Use technology to test at $\alpha = 0.05$ whether the population mean motivation scores are equal for the three types of Facebook pages, low, medium, and high self-disclosure.

Solution

First we verify whether the requirements are satisfied.

- The normal probability plots in Figure 11.28 indicate acceptable normality.

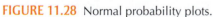

FIGURE 11.28 Normal probability plots.

- The standard deviations are shown in the Minitab output in Figure 11.29. The largest, $s_{\text{medium}} = 14.70$, is not larger than twice the smallest, $s_{\text{low}} = 11.88$. Thus, the equal variance requirement is satisfied.

FIGURE 11.29
Minitab output for Facebook ANOVA.

```
Source    DF      SS     MS      F      P
Group      2    2712   1356   8.05  0.001
Error    127   21386    168
Total    129   24098


S = 12.98   R-Sq = 11.25%   R-Sq(adj) = 9.86%



                              Individual 95% CIs For Mean Based on
                              Pooled StDev
Level    N    Mean   StDev   ------+---------+---------+---------+--
High    43   81.09   12.12                       (-------*------)
Medium  44   79.36   14.70                   (-------*------)
Low     43   70.63   11.88   (-------*------)
                              ------+---------+---------+---------+--
                                 70.0      75.0      80.0      85.0


Pooled StDev = 12.98
```

- Since the student participants were randomly selected for each level of self-disclosure, the independence assumption is also validated.

We therefore proceed with the ANOVA.

$$H_0 : \mu_{\text{high}} = \mu_{\text{medium}} = \mu_{\text{low}} \quad \text{versus} \quad H_a : \text{not all the population means are equal}$$

where μ_i represents a population mean motivation score for each self-disclosure level. Reject H_0 if the p-value is less than $\alpha = 0.05$.

From Figure 11.29, we get $F_{\text{data}} = 8.05$, with an associated p-value of approximately 0.001. Since this p-value is less than $\alpha = 0.05$, we reject H_0. There is evidence that not all population mean motivation scores are equal across all levels of self-disclosure. Informally, we may observe that the sample mean motivation score was lowest for the Facebook Web site with low self-disclosure.

STEP-BY-STEP TECHNOLOGY GUIDE: Analysis of Variance

TI-83/84

Performing ANOVA with Technology
(Example 11.13, page 554, is used to illustrate the procedure.)
Step 1   Enter the Dormitory A data in **L1**, the Dormitory B data in **L2**, and the Dormitory C data in **L3**.
Step 2   Press **STAT**, highlight **TESTS**, select "ANOVA(" , and press **ENTER**.
Step 3   On the home screen, enter "**L1, L2, L3**)" and press **ENTER** (Figure 11.30).

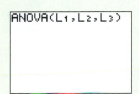

Finding the p-Value for a Given F_{data}.
(Example 11.13, page 554, is used to illustrate the procedure.)
p-value $= P(F > F_{\text{data}}) = P(F > 0.8370997079)$, where $df_1 = 2$ and $df_2 = 27$.
Step 1   Press **2nd > DISTR**.
Step 2   Select **Fcdf(** and press **ENTER**.
Step 3   Enter "**0.8370997079, 1E99, 2, 27)**" and press **ENTER**. The results are shown in Figure 11.26 (page 555).

Figure 11.30

EXCEL

Performing ANOVA with Technology
(Example 11.13, page 554, is used to illustrate the procedure.)
Step 1   Enter the Dormitory A data in column A, the Dormitory B data in column B, and the Dormitory C data in column C.
Step 2   Click **Data > Data Analysis > Anova: Single Factor**, and click **OK**.
Step 3   Select the input range of the data by clicking and dragging over the data in columns A, B, and C. Then click **OK**.

Finding the p-Value for a Given F_{data}.
(Example 11.13, page 554, is used to illustrate the procedure.)
p value $= P(F > F_{\text{data}}) = P(F > 0.8370997079)$, where $df_1 = 2$ and $df_2 = 27$.
Step 1   Select cell **A1**. Click the **Insert Function** icon f_x.
Step 2   For **Search for a Function**, type **FDIST** and click **OK**.
Step 3   For **X**, enter **0.8370997079**, for **Deg_freedom 1**, enter **2**, and for **Deg_freedom 2**, enter **27**. Then click **OK**. The cell now contains the p-value: 0.4438929572.

MINITAB

Performing ANOVA with Technology
(Example 11.13, page 554, is used to illustrate the procedure.)
   Minitab accepts data in two different forms for performing ANOVA, *stacked* or *unstacked*. *Unstacked* refers to the data

of each group being in a separate column. *Stacked* merges each group's data together in a single column, with the group numbers in a different column.

ANOVA (Stacked)

Step 1   Enter the **GPA** data for all three groups in C1 and the values for the categorical variable **Dorm** in C2.
Step 2   Click on **Stat > ANOVA > One-Way**.

Step 3   Choose the quantitative variable **GPA** as your **response** and the categorical variable **Dorm** as your **factor**. Then click **OK**.

ANOVA (Unstacked)

Step 1   Enter the Dormitory B data in **C1**, the Dormitory B data in **C2**, and the Dormitory C data in **C3**.
Step 2   Click **Stat > ANOVA > One-Way (Unstacked)**.

Step 3   For **Responses (in separate columns)**, select columns C1–C3 and click **OK**.

Finding the *p*-Value for a Given F_{data}

(Example 11.13, page 554, is used to illustrate the procedure.)
p-value = $P(F > F_{data}) = P(F > 0.8370997079)$, where $df_1 = 2$ and $df_2 = 27$.
Step 1   Click **Calc > Probability Distributions > F**.
Step 2   Select **Cumulative Probability**, enter **2** for **Numerator degrees of freedom** and **27** for **Denominator degrees of freedom**.
Step 3   Select **Input Constant**, enter **0.8370997079**, and click **OK**.

Step 4   Minitab then displays the cumulative probability $P(F < 0.8370997079) = 0.5561070428$. This cumulative probability represents the area to the left of 0.8370997079 (the unshaded area in Figure 11.25, page 555). Since the entire area under the curve equals 1, to get the *p*-value we need to subtract $P(F < 0.8370997079) = 0.5561070428$ from 1:

$$p\text{-value} = P(F > 0.8370997079) = 1 - P(F < 0.8370997079)$$
$$= 1 - 0.5561070428 = 0.4438929572$$

Section 11.3 SUMMARY

1   The F distribution is right-skewed, never takes negative values, and has an infinite number of different F curves, the shape of a curve depending on two different degrees of freedom, df_1 and df_2.

2   Analysis of variance (ANOVA) is an inferential method for testing whether the means of different populations are equal. The null hypothesis always states that all the population means are equal, and the alternative hypothesis always states that not all the population means are equal. ANOVA works by comparing (a) the variability in the sample means and (b) the variability within each sample.

If (a) is large *compared with* (b), this is evidence that the true means are not all equal and we should reject the null hypothesis.

3   ANOVA may be performed using the critical-value method or the *p*-value method. In the *p*-value method, technology is found to calculate the exact *p*-value and compare it to α.

4   In the critical-value method, F_{data} is compared with F_{crit}. If F_{data} is larger than F_{crit}, the null hypothesis is rejected.

Section 11.3 EXERCISES

CLARIFYING THE CONCEPTS

1.   Does the overall sample mean always equal the mean of the sample means? Explain.
2.   What does MSTR measure? What does MSE measure?
3.   In your own words, explain how ANOVA works.
4.   What are the required conditions for performing an analysis of variance?
5.   A comparison dotplot of the SAT scores of three sororities shows no overlap at all between the groups. Does this represent evidence for or against the null hypothesis that all population means are equal?

6.   True or false: If we reject the null hypothesis in an analysis of variance, then there is evidence that all the population mean sizes are different. If the statement is false, explain why it is false.

For Exercises 7–12, **(a)** find the value of the F critical value F_{crit} for the given α, df_1, df_2, and **(b)** draw the F curve, showing the critical value on the number line and the area to the right of it equaling α.
7. $\alpha = 0.05$, $df_1 = 1$, $df_2 = 7$
8. $\alpha = 0.10$, $df_1 = 1$, $df_2 = 7$
9. $\alpha = 0.01$, $df_1 = 2$, $df_2 = 15$

10. $\alpha = 0.05$, $df_1 = 2$, $df_2 = 15$

11. $\alpha = 0.05$, $df_1 = 3$, $df_2 = 16$

12. $\alpha = 0.10$, $df_1 = 3$, $df_2 = 16$

PRACTICING THE TECHNIQUES

For Exercises 13–16, assume that the ANOVA assumptions are verified. Test whether the population means differ, using the critical-value method and $\alpha = 0.05$.

a. State the hypotheses.

b. Find F_{crit} and the rejection rule.

c. Find F_{data}.

d. Compare F_{data} with F_{crit}. State the conclusion and the interpretation.

13.

Sample A	Sample B	Sample C
$\bar{x}_A = 10$	$\bar{x}_B = 12$	$\bar{x}_C = 8$
$s_A = 1$	$s_B = 1$	$s_C = 1$
$n_A = 5$	$n_B = 5$	$n_C = 5$

14.

Sample A	Sample B	Sample C	Sample D
$\bar{x}_A = 10$	$\bar{x}_B = 12$	$\bar{x}_C = 8$	$\bar{x}_D = 14$
$s_A = 1$	$s_B = 1$	$s_C = 1$	$s_D = 1$
$n_A = 5$	$n_B = 5$	$n_C = 5$	$n_D = 5$

15.

Sample A	Sample B	Sample C	Sample D
$\bar{x}_A = 50$	$\bar{x}_B = 75$	$\bar{x}_C = 100$	$\bar{x}_D = 125$
$s_A = 5$	$s_B = 4$	$s_C = 6$	$s_D = 5$
$n_A = 100$	$n_B = 150$	$n_C = 200$	$n_D = 250$

16.

Sample A	Sample B	Sample C	Sample D
$\bar{x}_A = 0$	$\bar{x}_B = 10$	$\bar{x}_C = 20$	$\bar{x}_D = 10$
$s_A = 1.5$	$s_B = 2.25$	$s_C = 1.75$	$s_D = 2.0$
$n_A = 50$	$n_B = 100$	$n_C = 50$	$n_D = 100$

17. Part of an ANOVA table for an analysis of variance involving seven groups for a study follows. Each sample contained 100 data values.

Source of variation	Sum of squares	Degrees of freedom	Mean square	F
Treatment	120	___	___	___
Error	315	___	___	
Total	___			

a. Find all six missing values in the table and fill in the blanks.

b. Perform the appropriate hypothesis test using the critical-value method and $\alpha = 0.05$.

18. Part of an ANOVA table for an analysis of variance involving three groups follows. Each sample contained 6 data values.

Source of variation	Sum of squares	Degrees of freedom	Mean square	F
Treatment	___	___	___	___
Error	90	___	___	
Total	150			

a. Find all six missing values in the table.

b. Perform the appropriate hypothesis test using the critical value method and $\alpha = 0.01$.

19. Part of an ANOVA table follows.

Source of variation	Sum of squares	Degrees of freedom	Mean square	F
Treatment	___	4	10	1.0
Error	___	___	___	
Total	440			

a. Find all four missing values in the table and fill in the blanks.

b. Perform the appropriate hypothesis test using the critical-value method and $\alpha = 0.10$.

20. Part of an ANOVA table follows.

Source of variation	Sum of squares	Degrees of freedom	Mean square	F
Treatment	___	2	___	2.0
Error	480	___	24	
Total	___			

a. Find all four missing values in the table and fill in the blanks.

b. Perform the appropriate hypothesis test using the critical-value method and $\alpha = 0.05$.

APPLYING THE CONCEPTS

For Exercises 21–23, assume the ANOVA assumptions are valid. Perform the appropriate analysis of variance using $\alpha = 0.05$.

21. Education and Religious Background. The General Social Survey collected data on the number of years of education and the religious preference of the respondents. The summary statistics are shown here.

	N	Mean	Std. Deviation
PROTESTANT	1660	13.10	2.87
CATHOLIC	683	13.51	2.74
JEWISH	68	15.37	2.80
NONE	339	13.52	3.22
OTHER	141	14.46	3.18

22. ANOVA can be applied to two populations. Researchers are interested in whether the mean heart rates of women and men differ. The following table summarizes the data.

	Females	Males
n	65	65
$\bar{x}$	98.384	98.104
s	0.743	0.699

23. The Full Moon and Emergency Room Visits. Is there a difference in emergency room visits before, during, and after a full moon? A study looked at the admission rate (number of patients per day) to the emergency room of a Virginia mental health clinic over a series of 12 full moons.[9] The data are provided in the table. Is there evidence of a difference in emergency room visits before, during, and after the full moon.?

Before	During	After
6.4	5	5.8
7.1	13	9.2
6.5	14	7.9
8.6	12	7.7
8.1	6	11.0
10.4	9	12.9
11.5	13	13.5
13.8	16	13.1
15.4	25	15.8
15.7	14	13.3
11.7	14	12.8
15.8	20	14.5

Breakfast Cereals. Use this information for Exercises 24–26. Does the nutritional rating of breakfast cereals vary by manufacturer? The data set **Breakfast Cereals 2** contains information on the nutritional rating and manufacturer for 68 breakfast cereals.[10] Assume that the cereals were selected randomly from the manufacturers so that the selection of a cereal from one manufacturer did not affect the selection of cereals from other manufacturers.

24. The largest standard deviation is, $s_{Quaker\ Oats} = 10.603$, while the smallest is $s_{Nabisco} = 5.509$. State whether the equal variance requirement is satisfied.

25. Explain why the independence assumption is validated.

26. The Minitab output in Figure 11.31 contains the ANOVA results. F_{data} and the p-value are in boldface. Use the p-value method and $\alpha = 0.05$ to test whether all population mean nutritional ratings are equal across all manufacturers. Provide all steps.

```
Source         DF      SS      MS      F       P
Manufacturer    5   5717.7  1143.5  15.71   0.000
Error          62   4511.7    72.8
Total          67  10229.4

S = 8.531   R-Sq = 55.89%   R-Sq(adj) = 52.34%
```

Figure 11.31 Minitab output for nutritional ratings ANOVA.

27. Online, Hybrid, and Traditional Classrooms. A researcher randomly selected six students for each of three different treatment groups. The first group of students took elementary statistics online. The second group of students took the same course in a traditional class. The third group

of students took a hybrid course which met once each week in person and also had an online component. The table shows the grade results. Researchers are interested in whether significant differences exist among the mean grades for the three groups. Assume that the assumptions are satisfied. Perform the appropriate ANOVA using the critical-value method at $\alpha = 0.01$.

Online	Traditional	Hybrid
70	75	95
75	75	60
60	95	90
90	60	75
85	60	85
50	80	75
$\bar{x} = 71.6667$	$\bar{x} = 74.1667$	$\bar{x} = 80$
$s = 15.0555$	$s = 13.1972$	$s = 12.6491$

28. Store Sales. The district sales manager would like to determine whether there are significant differences in the mean sales among the four franchise stores in her district. Sales (in thousands of dollars) over 5 days at each of the four stores are summarized in the following table. Assume that the assumptions are satisfied. Perform the appropriate analysis of variance using the $\alpha = 0.05$.

Store A sales	Store B sales	Store C sales	Store D sales
10	20	3	30
15	20	7	25
10	25	5	30
20	15	10	35
20	20	4	30
$\bar{x} = 15$	$\bar{x} = 20$	$\bar{x} = 5.8$	$\bar{x} = 30$
$s = 5$	$s = 3.5355$	$s = 2.7749$	$s = 3.5355$

? **29.** Refer to Exercise 28. *What if* the data are all wrong and all the stores actually have a sample mean sales of $30,000. Try to answer the following questions without touching your calculator.
  a. Find the value of SSTR, MSTR, and F_{data}.
  b. What would be the p-value of the ANOVA hypothesis test?
  c. What would be the conclusion?

30. Gas Mileage for European, Japanese, and American Cars. The following table shows the vehicle mileage (in miles per gallon) for random samples of automobiles manufactured in Europe, Japan, and the United States. Test using $\alpha = 0.01$ whether population mean gas mileage differs among automobiles from the three regions. Assume that the assumptions are satisfied.

Mpg	Sample 1: Europe	Sample 2: Japan	Sample 3: USA
Sample mean	$\bar{x}_1 = 27.603$	$\bar{x}_2 = 30.451$	$\bar{x}_3 = 20.033$
Sample standard deviation	$s_1 = 6.58$	$s_2 = 6.09$	$s_3 = 6.440$
Sample size	$n_1 = 68$	$n_2 = 79$	$n_3 = 245$

31. Do Taller People Have Lower Voices? The heights of the members of the New York Choral Society were recorded.[11] Choral singers are roughly divided into four groups, from highest-pitched voices to lowest: sopranos, altos (usually female), tenors, and basses (usually male). Assume that the assumptions are satisfied. The summary statistics are shown here. Test at $\alpha = 0.01$ whether the population mean heights of the groups differ.

```
Variable    n     Mean    StDev
Sopranos   36   64.250   1.873
Altos      35   64.886   2.795
Tenors     20   69.150   3.216
Basses     39   70.718   2.361
```

Use technology to solve Exercises 32 and 33. Perform the appropriate hypothesis test using the *p*-value method.

Open the **Crash** data set, which contains information about the severity of injuries sustained by crash dummies when the National Transportation Safety Board crashed automobiles into a wall at 35 miles per hour.

32. Head Injuries and Vehicle Size. The variable *head_inj* contains a measure of the severity of the head injury

sustained by the dummies. The variable *size2* categorizes the type of vehicle, such as light, medium, heavy, pick-up truck, mpv (suv), and so on. The values of the variable *size2* are as follows: 1 = compact car, 2 = light car, 3 = medium car, 4 = heavy car, 5 = mini-compact car, 6 = van, 7 = pickup truck, and 8 = mpv (suv). Would you expect the population mean severity of head injuries suffered by the dummies to be the same across all the size categories? Use Excel or Minitab to perform the analysis of variance, using $\alpha = 0.05$. Comment on the results. *Note:* Omit the observations where the *head_inj* value is missing or null.

33. Head Injuries and Vehicle Year. The variable *year* indicates the model year of the vehicle (87–91). Would you expect the population mean severity of head injuries suffered by the dummies to be the same across all model years? Use Excel or Minitab to perform the analysis of variance, using $\alpha = 0.05$. Comment on the results.

Use the *Analysis of Variance* applet for Exercises 34 and 35.

34. Move the group means so that they are about the same by clicking and dragging the black dots so that they are about even horizontally.

   a. What happens to the value of F (F_{data})?

   b. Explain why this happens, using the concept of between-sample variability and the statistics SSTR, MSTR, and F_{data}.

35. Click **Reset.** Increase the **Within-Sample Variability.**

   a. What happens to the value of F (F_{data})?

   b. Explain why this change happens, using the concept of within-sample variability and the statistics SSE, MSE, and F_{data}.

11.4 Inference in Regression

Objectives: **By the end of this section, I will be able to…**

1 Explain and apply the regression model.

2 Construct and interpret the confidence interval for the slope of the regression line.

3 Perform and interpret the *t* test for the relationship between the predictor variable and the response variable.

In Chapter 4 we learned descriptive methods for regression analysis. In this section we learn about a hypothesis test and a confidence interval for the slope of the regression line.

1 Regression Model

Recall Example 4.12 (page 178), where we examined time and score data from a sample of ten students. The data are shown here in Table 11.16.

Table 11.16 Results of short-term memory test

Student	Time to memorize (in minutes) (x)	Short-term memory score (y)
1	1	9
2	1	10
3	2	11
4	3	12
5	3	13
6	4	14
7	5	19
8	6	17
9	7	21
10	8	24

Based on the sample, we obtained the regression line $\hat{y} = 7 + 2x$. Suppose we examine a different sample of student times and scores. It is likely that this second sample will differ from the first, giving us a different regression line and different values for b_0 and b_1. In fact, for every different sample, b_0 and b_1 take different values since b_0 and b_1 are sample statistics. However, every sample comes from a population. Since we do not have data on all students, we are not able to find the characteristics of the population regression equation for time and score for all students. The y intercept β_0 and slope β_1 of this "true" population regression equation are unknown population parameters, just as μ and p are parameters in other contexts. The values of β_0 and β_1 are unknown, so we need to perform inference to learn about them.

The **regression model** may be used to approximate the relationship between the predictor variable x and the response variable y for the *entire population* of (x, y) pairs.

Regression Model

The **population regression equation** is defined as

$$y = \beta_0 + \beta_1 x + \varepsilon$$

where β_0 is the y intercept of the population regression line, β_1 is the slope of the population regression line, and ε is the error term.

Students 1 and 2 in Table 11.16 both studied for the same $x = 1$ minute, but Student 1 had a score of $y = 9$ while Student 2 had a score of $y = 10$. Thus, any linear approximation of the true relationship between x and y will introduce a certain amount of error. This is why the error term ε is needed.

We can estimate the population (true) regression equation $y = \beta_0 + \beta_1 x + \varepsilon$ by generating a regression equation from a sample. What if the value of β_1 was 0? Then the population regression equation would be

$$y = \beta_0 + (0)x + \varepsilon$$

that is,

$$y = \beta_0 + \varepsilon$$

Note that there is no longer any x term in the population regression equation if $\beta_1 = 0$. In other words, *if β_1 equals 0, then there is no relationship between x and y.* If β_1 takes on any other value than 0, then there does exist a linear relationship between x and y. This idea forms the basis for the inference we perform in this section.

Regression Model Assumptions

The regression model operates under a set of four assumptions that must be valid in order to perform the inference in this section.

> **Regression Model Assumptions**
>
> 1. **Zero mean assumption.** The error term ε is a random variable, with a mean of 0. That is, the expected value of the random variable ε is 0: $E(\varepsilon) = 0$.
> 2. **Constant variance assumption.** The variance of ε, which is denoted as σ^2, is the same regardless of the value of x.
> 3. **Independence assumption.** The values of ε are independent of each other.
> 4. **Normality assumption.** The error term ε is a normal random variable.

To summarize, for each value of x, the values of y come from a normally distributed population with a mean on the population regression line $E(y) = \beta_0 + \beta_1 x$ and constant standard deviation σ^2. Figure 11.32 illustrates how y is distributed for each value of x. Note that each normal curve has the same shape, indicating constant variance for each x.

FIGURE 11.32
Illustrating the regression assumptions.

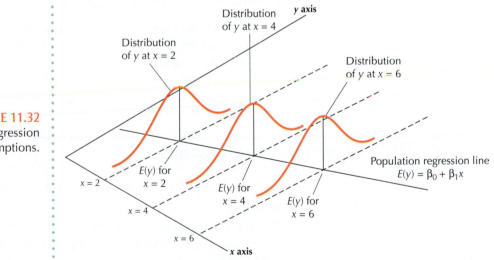

Verifying the Regression Assumptions

To check the regression model assumptions, we construct two graphs:
1. Scatterplot of the residuals (prediction errors $y - \hat{y}$) against the fitted values (**fitted values** refers to the predicted values, $\hat{y}$)

2. Normal probability plot of the residuals

Figure 11.33 shows four types of patterns that might be observed in the residuals versus fitted values plots.

- Plot (a) is a "healthy" plot, displaying no noticeable patterns.

- In plot (b) we see a curve, which indicates a violation of the independence assumption. Independence implies that knowing the value of a particular y does not help to predict the value of a different y. However, a curve suggests that knowing the value of a previous y helps in knowing the value of the next y.

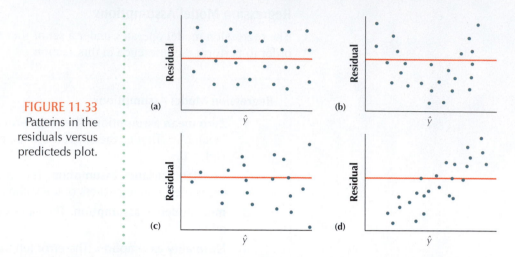

FIGURE 11.33
Patterns in the residuals versus predicteds plot.

- Plot (c) shows a "funnel" pattern, which contradicts the constant variance assumption. The residuals on the left are close together vertically (small variability), while the residuals on the right are far apart vertically (large variability).

- In plot (d) we see an increasing pattern, that violates the zero-mean assumption. The residuals on the left are all below the midline, so $E(y) < \beta_0 + \beta_1 x$, while the residuals on the right are all above the midline, so $E(y) > \beta_0 + \beta_1 x$.

With small data sets, it is difficult to ascertain whether or not patterns really exist. Be wary of seeing patterns where none exist. If one or more regression assumptions is violated, we should not proceed with inferential methods such as hypothesis tests or confidence intervals. However, even if one or more regression assumptions is violated, we can still report and interpret the descriptive regression statistics that we learned in Sections 4.3 and 4.4.

Example 11.15   Verifying the regression assumptions

Verify the regression assumptions for the data in Table 11.16.

Solution
Figure 11.34 is a scatterplot of the residuals versus fitted values (predicted values), and Figure 11.35 contains a normal probability plot of the residuals for the data in Table 11.16.

FIGURE 11.34   Residuals versus fitted values plot.

FIGURE 11.35   Normal probability plot of the residuals.

Figure 11.34 contains no strong evidence of the unhealthy patterns shown in Figure 11.33. Thus, the independence assumption, the constant variance assumption, and the zero-mean assumption are verified. Also, Figure 11.35 indicates no evidence of departures from normality in the residuals. Therefore, we conclude that the regression assumptions are verified. ■

Confidence Interval for Slope β_1

The slope b_1 of the regression equation is a statistic that has a sampling distribution.

Sampling Distribution of b_1

When the regression assumptions are satisfied, then the quantity

$$t = \frac{b_1 - \beta_1}{s/\sum(x - \bar{x})^2}$$

has approximately a t distribution, with $n - 2$ degrees of freedom, where n is the sample size, b_1 is the estimate of the slope β_1 of the population regression equation, and s is the standard error of the estimate.

Recall that in Chapter 8 we constructed a confidence interval estimate for a population parameter, consisting of an interval of numbers that contain the parameter with a certain confidence level. We can find a similar **confidence interval for β_1**.

Confidence Interval for β_1

When the regression assumptions are met, a $100(1 - \alpha)\%$ confidence interval for β_1 is given by

$$b_1 \pm t_{\alpha/2} \cdot \frac{s}{\sqrt{\sum(x - \bar{x})^2}}$$

where b_1 is the estimate of the true slope β_1, s is the standard error of the estimate, and $t_{\alpha/2}$ has $n - 2$ degrees of freedom.

Margin of Error E

The margin of error for a $100(1 - \alpha)\%$ confidence interval for β_1 is given by

$$E = t_{\alpha/2} \cdot \frac{s}{\sqrt{\sum(x - \bar{x})^2}}$$

Thus, the confidence interval for β_1 takes the form $b_1 \pm E$.

Example 11.16   Confidence interval for β_1

Find a 95% confidence interval for the true slope β_1 of the relationship between time and score.

Solution

The regression assumptions were verified in Example 11.15. From Table 11.16 we can calculate $\bar{x} = \sum x/n = 40/10 = 4$. Table 11.17 shows that

$$\sum(x - \bar{x})^2 = 54$$

In Example 4.14 (page 180), we found that $s = \sqrt{\text{SSE}/(n-2)} = \sqrt{12/8} \approx 1.2247$. Hence,

$$\frac{s}{\sqrt{\sum(x-\bar{x})^2}} = \frac{\sqrt{\frac{12}{8}}}{\sqrt{54}} = \frac{1}{6} \approx 0.1667$$

Table 11.17 Calculation of $\sum(x-\bar{x})^2$

Student	x	$(x-\bar{x})^2$
1	1	$(1-4)^2 = 9$
2	1	$(1-4)^2 = 9$
3	2	$(2-4)^2 = 4$
4	3	$(3-4)^2 = 1$
5	3	$(3-4)^2 = 1$
6	4	$(4-4)^2 = 0$
7	5	$(5-4)^2 = 1$
8	6	$(6-4)^2 = 4$
9	7	$(7-4)^2 = 9$
10	8	$(8-4)^2 = 16$
		$\sum(x-\bar{x})^2 = 54$

We have $b_1 = 2$ from Example 4.12 (page 178), and from the t table, we find that $t_{\alpha/2}$ for $n - 2 = 10 - 2 = 8$ degrees of freedom is $t_{\alpha/2} = 2.306$. Hence, our margin of error E is

$$E = t_{\alpha/2} \cdot \frac{s}{\sqrt{\sum(x-\bar{x})^2}} = (2.306)\left(\frac{1}{6}\right) \approx 0.3843$$

The 95% confidence interval for β_1 is then given by

$$b_1 \pm E = 2 \pm 0.3843 = (1.6157, 2.3843)$$

What Do These Numbers Mean?

The margin of error $E = 0.3843$ means that, when we repeatedly take samples from this population, 95% of the time the sample estimate b_1 will be within $E = 0.3843$ of the unknown value of the true slope β_1. In other words, we are 95% confident that the interval (1.6157, 2.3843) captures the true slope β_1 of the relationship between time and score. ■

3 Hypothesis Tests for Slope β_1

We next construct the hypothesis test to determine whether or not β_1 equals 0. The hypotheses are

$H_0 : \beta_1 = 0$ There is no linear relationship between x and y.

$H_a : \beta_1 \neq 0$ There is a linear relationship between x and y.

The hypothesis test is based on the same approximate t distribution as the t interval for β_1. Assuming H_0 ($\beta_1 = 0$), the test statistic is

$$t_{\text{data}} = \frac{b_1 - \beta_1}{\dfrac{s}{\sqrt{\sum(x-\bar{x})^2}}} = \frac{b_1 - 0}{\dfrac{s}{\sqrt{\sum(x-\bar{x})^2}}} = \frac{b_1}{\dfrac{s}{\sqrt{\sum(x-\bar{x})^2}}}$$

If the **test statistic** t_{data} is extreme (far from 0), then it represents evidence against the null hypothesis.

Example 11.17   Calculating the test statistic t_{data}

For the data from Table 11.16, calculate the value of the test statistic t_{data}.

Solution

From Example 4.12, we have $b_1 = 2$, and from Example 11.16, we have $\dfrac{s}{\sqrt{\sum(x-\bar{x})^2}}$ $= 1/6$. Therefore,

$$t_{\text{data}} = \frac{b_1}{\dfrac{s}{\sqrt{\sum(x-\bar{x})^2}}} = \frac{2}{1/6} = 12$$

We illustrate performing the t test for the slope β_1 using the *p*-value method and the critical value method.

Hypothesis Test for Slope β_1 : *p*-Value Method

Step 1   **State the hypotheses, and state the rejection rule.**

$H_0 : \beta_1 = 0$ There is no linear relationship between x and y.

$H_a : \beta_1 \neq 0$ There is a linear relationship between x and y.

The rejection rule is *Reject H_0 if the p-value is less than α*.

Step 2   **Calculate the value of the test statistic t_{data}.**

$$t_{\text{data}} = \frac{b_1}{\dfrac{s}{\sqrt{\sum(x-\bar{x})^2}}}$$

which follows an approximate t distribution with $n-2$ degrees of freedom.

Step 3   **Find the *p*-value.** If you have access to technology, use it. Otherwise, estimate the p-value using the following formula.

$$p\text{-value} = P(t > |t_{\text{data}}|) + P(t < -|t_{\text{data}}|)$$
$$= 2 \cdot P(t > |t_{\text{data}}|)$$

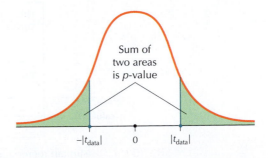

Step 4   **State the conclusion and the interpretation.** Compare the *p*-value with α.

Example 11.18 Hypothesis test for the slope β_1 using the *p*-value method and technology

Test using the *p*-value method whether a linear relationship exists between time and score, using $\alpha = 0.01$.

Solution

We use the instructions provided in the Step-by-Step Technology Guide at the end of this section. The regression assumptions were verified in Example 11.15.

Step 1 $H_0 : \beta_1 = 0$ There is no linear relationship between time and score.
$H_a : \beta_1 \neq 0$ There is a linear relationship between time and score.

Reject H_0 if the p-value is less than 0.01.

Step 2 From Example 11.17, we have $t_{\text{data}} = 12$.

Step 3 Find the *p*-value. The regression results (including the *p*-value) for the TI-83/84, Excel, and Minitab are shown in Figures 11.36, 11.37, and 11.38. (Differing results are due to rounding.)

Regression equation $\hat{y} = b_0 + b_1 x$
(TI-83/84 expresses as $y = a + bx$)

$t_{\text{data}} = 12$

p-value of 2.1438667E-6 = 0.0000021439

Degrees of freedom, $n - 2 = 8$

$a = b_0 = 7$

$b = b_1 = 2$

Standard error of the estimate $s \approx 1.2247$

Coefficient of determination $r^2 \approx 0.9474$

Correlation coefficient $r \approx 0.9733$

FIGURE 11.36 TI-83/84 regression results.

Regression Statistics	
Multiple R	0.9733
R Square	0.9474
Adjusted R Square	0.9408
Standard Error	1.2247
Observations	10

Correlation coefficient $r = 0.9733$
Coefficient of determination $r^2 = 0.9474$
Standard error of the estimate $s = 1.2247$
Sample size $n = 10$

	Coefficients	Standard Error	t Stat	P-value
Intercept	7	0.7710	9.0791	0.00001738
X Variable - Time	2	0.1667	12.0000	0.00000214

$b_0 = 7$
$b_1 = 2$
$t_{\text{data}} = 12$
p-value = 0.00000214

FIGURE 11.37 Excel regression results.

```
The regression equation is
Score = 7.00 + 2.00 Time

Predictor    Coef    SE Coef        T       P
Constant   7.0000    0.7710     9.08   0.000
Time       2.0000    0.1667    12.00   0.000

S = 1.22474    R-Sq = 94.7%    R-Sq(adj) = 94.1%
```

$b_0 = 7$
$b_1 = 2$
$t_{\text{data}} = 12$
p-value ≈ 0

FIGURE 11.38 Minitab regression results.

Step 4 Since the *p*-value of about 0.000 is less than $\alpha = 0.01$, we reject H_0. There is evidence for a linear relationship between time and score. ∎

Hypothesis Test for Slope β_1: Critical-Value Method

Step 1   State the hypotheses.

$$H_0 : \beta_1 = 0 \quad \text{There is no linear relationship between } x \text{ and } y.$$
$$H_a : \beta_1 \neq 0 \quad \text{There is a linear relationship between } x \text{ and } y.$$

Step 2   Find the t critical value t_{crit} and the rejection rule. To find t_{crit}, use the t-distribution table (Table D in the Appendix) for a two-tailed test and degrees of freedom $n - 2$. The rejection rule is *Reject H_0 if $t_{data} > t_{crit}$ or $t_{data} < -t_{crit}$.*

Step 3   Calculate the value of the test statistic t_{data}.

$$t_{data} = \frac{b_1}{\dfrac{s}{\sqrt{\sum (x - \bar{x})^2}}}$$

which follows an approximate t distribution with $n - 2$ degrees of freedom.

Step 4   State the conclusion and the interpretation. Compare t_{data} with t_{crit}.

Example 11.19   Hypothesis test for the slope β_1 using the critical value method

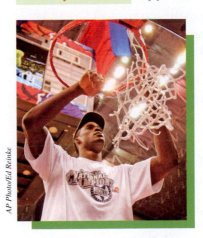

AP Photo/Ed Reinke

The following table shows the winning percentage (x) and power rating (y) for ten teams in the 2004 NCAA Men's Basketball Tournament, according to **www.teamrankings .com**. Is there a linear relationship between winning percentage and power rating? Test using the critical value method using $\alpha = 0.05$.

School	Win % (x)	Rating (y)
Duke	0.838	96.020
St. Joseph's	0.938	95.493
Connecticut	0.846	95.478
Oklahoma State	0.882	95.320
Pittsburgh	0.857	94.541
Georgia Tech	0.737	93.091
Stanford	0.938	92.862
Kentucky	0.844	92.692
Gonzaga	0.903	92.609
Mississippi State	0.867	91.912

What Results Might We Expect?

The scatterplot in Figure 11.39 indicates no evidence for a linear relationship between winning percentage and power rating. As winning percentage increases (as we move left to right), the power rating tends to remain unchanged. We might expect to *not reject* the null hypothesis that there is no linear relationship. ∎

FIGURE 11.39 Scatterplot of power rating versus winning percentage.

FIGURE 11.40 Plot of residuals versus fitted values.

FIGURE 11.41 Normality plot of residuals.

There is no strong evidence in Figures 11.40 and 11.41 that the regression assumptions have been violated. We thus proceed with the hypothesis test.

Step 1 $H_0 : \beta_1 = 0$ There is no linear relationship between winning percentage and power rating.

$H_a : \beta_1 \neq 0$ There is a linear relationship between winning percentage and power rating.

Step 2   We find t_{crit} in the t table for a two-tailed test with $\alpha = 0.05$ and df $= n - 2 = 10 - 2 = 8$: $t_{crit} = 2.306$.   Reject H_0 if $t_{data} > 2.306$ or $t_{data} < -2.306$.

Step 3   **Find t_{data}.** We first calculate $\bar{x} = \sum x / n = 8.65/10 = 0.865$ and $\sum (x - \bar{x})^2 = 0.030374$. The calculations are provided in the accompanying table. In Exercise 17 of Section 4.4 (page 186), we found that $b_1 = 1.587$ and $s = 1.60388$. Thus,

$$\frac{s}{\sqrt{\sum (x - \bar{x})^2}} = \frac{1.60388}{\sqrt{0.030374}} \approx 9.2028$$

and

$$t_{data} = \frac{b_1}{\dfrac{s}{\sqrt{\sum (x - \bar{x})^2}}} = \frac{1.587}{9.2028} \approx 0.1724$$

Calculating $\sum (x - \bar{x})^2$		
School	x	$(x - \bar{x})^2$
Duke	0.838	$(0.838 - 0.865)^2 = 0.000729$
St. Joseph's	0.938	$(0.938 - 0.865)^2 = 0.005329$
Connecticut	0.846	$(0.846 - 0.865)^2 = 0.000361$
Oklahoma State	0.882	$(0.882 - 0.865)^2 = 0.000289$
Pittsburgh	0.857	$(0.857 - 0.865)^2 = 0.000064$
Georgia Tech	0.737	$(0.737 - 0.865)^2 = 0.016384$
Stanford	0.938	$(0.938 - 0.865)^2 = 0.005329$
Kentucky	0.844	$(0.844 - 0.865)^2 = 0.000441$
Gonzaga	0.903	$(0.903 - 0.865)^2 = 0.001444$
Mississippi State	0.867	$(0.867 - 0.865)^2 = 0.000004$
	$\sum x = 8.65$	$\sum (x - \bar{x})^2 = 0.030374$

Step 4   Since $t_{data} = 0.1724$ is not greater than 2.306 and is not less than -2.306, we therefore do not reject H_0. There is insufficient evidence for a linear relationship between winning percentage and power rating.   ■

STEP-BY-STEP TECHNOLOGY GUIDE: Regression Analysis

Data from Example 11.18 (page 570) are used to illustrate the steps.

TI-83/84

Step 1 Enter the **X (Time)** data in **L1** and the **Y (Score)** data in **L2**.
Step 2 Press **STAT**, highlight **CALC**, and press **8** (not 4!) to choose **LinReg(a+bx)**. On the home screen, the following command appears: **LinReg(a+bx)**.
Step 3 Press **ENTER**. The output shows **y = a +bx, a=7, b=2**. The TI-83/84 denotes the y intercept b_0 as **a** and the slope b_1 as **b**. Thus the TI-83/84 is telling you that the estimated regression equation is $\hat{y} = 7 + 2x$.
Step 4 Now Press **STAT** again and press the right arrow key until **TESTS** is highlighted.

Step 5 Press the **down arrow** key until **E** is highlighted (for **LinRegTTest**).
Step 6 Press **ENTER**. The LinRegTTest menu appears.
Step 7 For **Xlist**, enter **L1** (or whichever list you entered the X data in).
Step 8 For **Ylist**, enter **L2** (or whichever list you entered the Y data in).
Step 9 For **Freq**, enter **1**, and for β & ρ highlight "$\neq 0$".
Step 10 Move the cursor over **Calculate**, make sure all your entries are correct, and press **ENTER**. The results are as shown in Figure 11.36 (page 570).

EXCEL

Step 1 Enter the "Time" variable in column **A** and the "Score" variable in column **B**.
Step 2 Click on **Data > Data Analysis > Regression** and click **OK**.
Step 3 For **Input Y Range**, select cells **B1 – B10**. For **Input X Range**, select cells **A1 – A10**.

Step 4 If you would like to verify the regression assumptions, then select **Residual Plots** and **Normal Probability Plots**.
Step 5 Click **OK**. The results are as shown in Figure 11.37 (page 570).

MINITAB

Step 1 Enter the "Time" variable in **C1** and the "Score" variable in **C2**.
Step 2 Click on **Stat > Regression > Regression**.
Step 3 Select "Score" as your **Response Variable** and "Time" as your **Predictor Variable**.

Step 4 If you would like to verify the regression assumptions, click the button labeled **Graphs** and select **Four in One**.
Step 5 Click **OK** twice. The results are as shown in Figure 11.38 (page 570).

Section 11.4 SUMMARY

1 This section examines inferential methods for regression analysis. The regression model, or the (population) regression equation, is $y = \beta_0 + \beta_1 x + \varepsilon$, where β_0 is the y intercept of the population regression line, β_1 is the slope of the population regression line, and ε is the error term.

2 We can construct confidence intervals for the true value of the population regression slope β_1 since it is unknown.

3 A hypothesis test may be performed to determine whether a linear relationship exists between x and y.

Section 11.4 EXERCISES

CLARIFYING THE CONCEPTS

1. What is the difference between the regression equation (calculated using the sample) and the population regression equation?
2. What are the four regression model assumptions?
3. How do we go about verifying the regression model assumptions?
4. What is the difference between b_0 and b_1 on the one hand and β_0 and β_1 on the other hand?

PRACTICING THE TECHNIQUES

For Exercises 5–10, follow these steps.
    a. Construct a scatterplot of the residuals versus the predicted values.
    b. Use technology to construct a normal probability plot of the residuals.
    c. Verify that the regression assumptions are valid.

5.

x	y
1	15
2	20
3	20
4	25
5	25

6.

x	y
0	10
5	20
10	45
15	50
20	75

7.

x	y
−5	0
−4	8
−3	8
−2	16
−1	16

8.

x	y
−3	−5
−1	−15
1	−20
3	−25
5	−30

9.

x	y
10	100
20	95
30	85
40	85
50	80

10.

x	y
0	11
20	11
40	16
60	21
80	26

For Exercises 11–14, follow these steps. Assume that the regression model assumptions are valid.

a. Find $t_{\alpha/2}$ for a 95% confidence interval for β_1.
b. Construct a 95% confidence interval for β_1.

11. Data in Exercise 5, where $b_1 = 2.5$
12. Data in Exercise 6, where $b_1 = 3.2$
13. Data in Exercise 7, where $b_1 = 4.0$
14. Data in Exercise 8, where $b_1 = -3.0$

For Exercises 15–17, follow these steps. Assume that the regression model assumptions are valid.

a. Calculate t_{data}.
b. Find p-value $= 2 \cdot P(t > |t_{data}|)$.
c. Perform the hypothesis test for the linear relationship between x and y using the p-value method and $\alpha = 0.05$.

15. Data in Exercise 5, where $b_1 = 2.5$
16. Data in Exercise 6, where $b_1 = 3.2$
17. Data in Exercise 7, where $b_1 = 4.0$

For Exercises 18–20, follow these steps. Assume that the regression model assumptions are valid.

a. Find t_{crit} for a two-tailed test with $\alpha = 0.05$ and df $= n - 2$.
b. Calculate t_{data}.
c. Perform the hypothesis test for the linear relationship between x and y, using the critical-value method, and $\alpha = 0.05$.

18. Data in Exercise 8, where $b_1 = -3.0$
19. Data in Exercise 9, where $b_1 = -0.5$
20. Data in Exercise 10, where $b_1 = 0.2$

APPLYING THE CONCEPTS

For Exercises 21–26, follow steps **(a)–(c)**. Assume that the regression model assumptions are valid.

a. Find $t_{\alpha/2}$ for a 95% confidence interval for β_1.
b. Construct a 95% confidence interval for β_1.
c. Interpret the interval.

21. Volume and Weight. The table contains the volume (x, in cubic meters) and weight (y, in kilograms) of five randomly chosen packages shipped to a local college.

Volume (x)	Weight (y)
4	10
8	16
12	25
16	30
20	35

22. Family Size and Pets. Shown in the table are the number of family members (x) in a random sample taken from a suburban neighborhood and the number of pets (y) belonging to each family.

Family size (x)	Pets (y)
2	1
3	2
4	2
5	3
6	3

23. SAT Math and Verbal Scores. The College Board reported the SAT I Verbal scores (x) and SAT I Math scores (y) for the five states with the best participation rates:

State	Verbal (x)	Math (y)
New York	497	510
Connecticut	515	515
Massachusetts	518	523
New Jersey	501	514
New Hampshire	522	521

24. World Temperatures. Listed in the table are the low (x) and high (y) temperatures for a particular day in 2006, measured in degrees Fahrenheit, for a random sample of cities worldwide.

City	Low (x)	High (y)
Kolkata, India	57	77
London, England	36	45
Montreal, Quebec	7	21
Rome, Italy	39	55
San Juan, Puerto Rico	70	83
Shanghai, China	34	45

25. NCAA Power Ratings. The table shows the team's winning percentage (x) and power rating (y) for the 2004 NCAA Basketball Tournament, according to **www.teamrankings.com**.

School	Win% (x)	Rating (y)
Duke	0.838	96.020
St. Joseph's	0.938	95.493
Connecticut	0.846	95.478
Oklahoma State	0.882	95.320
Pittsburgh	0.857	94.541
Georgia Tech	0.737	93.091
Stanford	0.938	92.862
Kentucky	0.844	92.692
Gonzaga	0.903	92.609
Mississippi State	0.867	91.912

26. Stock Prices. Would you expect there to be a relationship between the price (x) of a stock and its change (y) in price on a particular day? The table provides stock price and stock price change for August 2, 2004, for a random sample of ten stocks listed on the New York Stock Exchange.

Stock	Price (x)	Change (y)
Nortel Networks	3.86	+0.04
Qwest Communications	3.41	−0.56
Tyco International	32.02	+0.78
Lucent Technologies	3.04	−0.03
Vishay Intertechnology	13.96	−1.96
Tenet Healthcare	10.36	−0.82
Select Medical Group	14.74	+1.62
Cox Communications	33.19	+0.03
Verizon Communications	39.15	+0.46
General Electric	33.05	−0.21

SAT Verbal and Math Scores. Use this information for Exercises 27–31. The table shows the SAT I scores for the five states with the best participation rates, as reported by the College Board in 2006. We are interested in whether a linear relationship exists between the SAT I Verbal score (x) and the SAT I Math score (y).

State	SAT I Verbal	SAT I Math
New York	497	510
Connecticut	515	515
Massachusetts	518	523
New Jersey	501	514
New Hampshire	522	521

27. Consider Figures 11.42 and 11.43. Is there strong evidence that the regression assumptions are violated?

Figure 11.42 Plot of residuals versus fitted values.

Figure 11.43 Normality plot of residuals.

28. Write out the hypotheses for testing whether or not a linear relationship exists between the verbal score and the math score.

29. Find the t critical value t_{crit} and the rejection rule. Use level of significance $\alpha = 0.05$.

30. Calculate the value of the test statistic t_{data}. Note that $\bar{x} = 510.6$ and $\sum(x - \bar{x})^2 = 481.2$. Let $b_1 = 0.4264$ and $s = 2.92665$.

31. Compare t_{data} with t_{crit}. State the conclusion and the interpretation.

For Exercises 32–34, follow steps **(a)–(e)**.
  a. Verify the regression model assumptions. (*Hint:* You can use either Excel or Minitab; see the Step-by-Step Technology Guide on page 573.)
  b. Based on the confidence interval constructed in the indicated exercise, would you expect the hypothesis test to reject the null hypothesis that $\beta_1 = 0$?
  c. Calculate t_{data}.
  d. Find p-value $= 2 \cdot P(t > |t_{data}|)$.
  e. Perform the hypothesis test for the linear relationship between x and y using the p-value method and $\alpha = 0.05$.

32. Data from Exercise 21; $b_1 = 1.6$
33. Data from Exercise 23; $b_1 = 0.4264$
34. Data from Exercise 25; $b_1 = 1.587$

For Exercises 35–37, follow steps **(a)–(e)**.
  a. Verify the regression model assumptions.
  b. Based on the confidence interval constructed in the indicated exercise, would you expect the hypothesis test to reject the null hypothesis that $\beta_1 = 0$?
  c. Find t_{crit} for a two-tailed test with $\alpha = 0.05$ and df $= n - 2$.
  d. Calculate t_{data}.
  e. Perform the hypothesis test for the linear relationship between x and y, using the critical value method and $\alpha = 0.05$.

35. Data from Exercise 22; $b_1 = 0.5$
36. Data from Exercise 24; $b_1 = 1.04658$
37. Data from Exercise 26; $b_1 = 0.01944$

38. Batting Average and Runs Scored. The accompanying table (see page 576) shows the top ten hitters in Major League Baseball for 2007. We are interested in estimating the number of runs scored (y) using the player's batting average (x).

Player	Team	Batting average (x)	Runs scored (y)
M. Ordonez	Detroit Tigers	.363	117
I. Suzuki	Seattle Mariners	.351	111
P. Polanco	Detroit Tigers	.341	105
M. Holliday	Colorado Rockies	.340	120
J. Posada	New York Yankees	.338	91
C. Jones	Atlanta Braves	.337	108
D. Ortiz	Boston Red Sox	.332	116
H. Ramirez	Florida Marlins	.332	125
E. Renteria	Atlanta Braves	.332	87
C. Utley	Philadelphia Phillies	.332	104

a. What type of pattern do you see in the residuals versus predicted values plot?
b. Which regression assumption is violated?
c. Should we construct a confidence interval or perform a hypothesis test for the slope of the regression line?
d. Is it still appropriate to report the descriptive statistics we learned in Chapter 4?

For Exercises 39–41, you can use Excel, Minitab, or some other statistical computing software to solve the following problems.
   a. Verify the regression model assumptions.
   b. Construct and interpret a 95% confidence interval for β_1.
   c. Based on the confidence interval constructed in (b), would you expect the hypothesis test to reject the null hypothesis that $\beta_1 = 0$?
   d. Test at $\alpha = 0.05$ whether a linear relationship exists between x and y.

39. Open the **Darts** data set, which we used for the Chapter 3 Case Study. Use the Dow Jones Industrial Average (x) to estimate the pros' performance (y).
40. Open the **Nutrition** data set. Estimate the number of calories per gram (y) using the amount of fat per gram (x).
41. Open the **Pulse and Temp** data set. Estimate body temperature (y) using heart rate (x).

Try This in Class! In-class activities to enhance your understanding of statistics

(*Note to instructor:* The small sample sizes involved in the following activities may not meet the assumptions for the χ^2 test. You can alleviate this by taking a larger sample or by collapsing the categories with the smaller expected frequencies.)

A χ^2 GOODNESS OF FIT TEST FOR THE PROPORTIONS OF FRESHMEN, SOPHOMORES, JUNIORS, AND SENIORS AT YOUR COLLEGE

1. What do you think are the proportions of freshmen, sophomores, juniors, and seniors at your college? Write out the hypotheses for the test to determine whether these proportions are correct.
2. Use a sample of 25 students to test this hypothesis. What are the expected frequencies for each category? Do you have any data yet to evaluate your claim in (1)?

3. Each student should generate a sample of 25 students from the class, using either (a) technology, (b) the *Random Sample* applet, or (c) the method shown in the Chapter 1 Try This in Class!

4. For each student in your sample, ask which class he or she belongs to. Then add up the counts for each class. Which type of frequencies are these?
5. You will use a level of significance of $\alpha = 0.10$ for this test. What is the critical value χ^2_{crit} for this test? Will it be the same for each student?
6. Carry out the χ^2 goodness of fit test using your sample data. Will the value of χ^2_{data} be the same for each student?
7. Record your name and your value for the test statistic χ^2_{data} calculated in (6) on a Post-It Note®.
8. Place your Post-It Note on the number line that the instructor prepared on the blackboard, noting the location of the cutoff value χ^2_{crit}. We thus have a collection of values for the statistic χ^2_{data}, representing an approximation of the sampling distribution of χ^2_{data}. Can you make any comment on the shape of the distribution of χ^2_{data} values?
9. What proportion of students rejected the null hypothesis?
10. Reach a class consensus as to why most students either did or did not reject the null hypothesis.

CHAPTER 11 FORMULAS AND VOCABULARY

SECTION 11.1
- χ^2 **Goodness of fit test** (p. 523)
- **Multinomial random variable** (p. 521)
- **Test statistic for the goodness of fit test** (p. 524).

$$\chi^2_{data} = \sum \frac{(O_i - E_i)^2}{E_i}$$

where O_i represents the observed frequency for category i, and E_i represents the expected frequency for category i.

SECTION 11.2
- χ^2 **test for independence** (p. 535)
- **Test for the homogeneity of proportions** (p. 541)
- **Test statistic for both the test for independence and the test for the homogeneity of proportions** (p. 537).

$$\chi^2_{data} = \sum \frac{(O_i - E_i)^2}{E_i}$$

SECTION 11.3
- **Analysis of variance (ANOVA)** (p. 547)
- F_{data} (p. 552).

$$F_{data} = \frac{MSTR}{MSE}$$

- **F distribution** (p. 546)
- **Hypotheses for analysis of variance** (p. 554)
- **Mean square error (MSE)** (p. 552).

$$MSE = \frac{\sum (n_i - 1)s_i^2}{n_t - k}$$

- **Mean square treatment (MSTR)** (p. 552).

$$MSTR = \frac{\sum n_i (\bar{x}_i - \bar{\bar{x}})^2}{k - 1}$$

- **Overall sample mean,** $\bar{\bar{x}}$ (p. 551).

$$\bar{\bar{x}} = \frac{(n_1\bar{x}_1 + n_2\bar{x}_2 + \cdots + n_k\bar{x}k)}{n_t}$$

- ***p*-Value** (p. 554). p-value $= P(F > F_{data})$.

- **Sum of squares error (SSE)** (p. 552).

$$SSE = \sum (n_i - 1)s_i^2$$

- **Sum of squares treatment (SSTR)** (p. 552).

$$SSTR = \sum n_i(\bar{x}_i - \bar{\bar{x}})^2$$

- **Total sum of squares (SST)** (p. 552).

$$SST = SSTR + SSE$$

SECTION 11.4
- **Confidence interval for the true slope β_1 of the regression line** (p. 567)
- **Fitted values** (p. 565)
- **Regression model, or population regression equation** (p. 564)
- **Test statistic t_{data}** (p. 569)

CHAPTER 11 REVIEW EXERCISES

SECTION 11.1
For Exercises 1–5, perform the χ^2 goodness of fit test.

1. Truck-Hauled Trade. In 2002, according to the U.S. Census Bureau, 32% of North American international truck-hauled trade (in dollars) went from the United States to Canada, 22% went from the United States to Mexico, 31% went from Canada to the United States, and 15% went from Mexico to the United States. Suppose that a survey this year showed that $25 billion went from the United States to Canada, $15 billion went from the United States to Mexico, $20 billion went from Canada to the United States, and $10 billion went from Mexico to the United States. Test whether the population proportions of truck-hauled trade have changed since 2002, using the p-value method and level of significance $\alpha = 0.05$.

2. Death Row Inmates. According to the report *Death Row, USA,* published by the National Association for the Advancement of Colored People, 42% of the people on death row in 2004 were black, 46% were white, 10% were Hispanic, and 2% were other. Suppose that a survey of randomly selected death row inmates this year found 1500 blacks, 1600 whites, 375 Hispanics, and 100 inmates otherwise classified. Test whether the population proportions have changed since 2004, using the critical-value method and level of significance $\alpha = 0.10$.

3. Alcohol Abuse and Dependence in College. According to a 2002 report, 25% of college students had abused alcohol in the last 12 months, while a further 6% (not counted in the 25%) were alcohol dependent.[12] Suppose that a survey of 1000 randomly selected college students taken this year finds 275

who had abused alcohol in the last 12 months and a further 50 (not counted in the 275) who are alcohol dependent. Test whether the population proportions have changed since 2002, using the p-value method and level of significance $\alpha = 0.10$.

4. Truly Random Lottery Drawing? Have you ever wondered whether lottery drawings are truly random? Shown here is the relative frequency distribution of the third digit in the Maryland lottery's Pick 3 game (218 drawings from September 1989 to April 1990). In a Pick 3 game, you choose a three-digit number between 000 and 999, and if your number comes up, you win the cash prize. Notice that 1 appears as the third digit least of all the digits, and quite a bit less often than some of the other digits. Does the relative scarcity of 1s indicate that the system is flawed? We would of course expect each digit to show up 10% of the time. Test whether the population proportions of digits are all 0.10, using level of significance $\alpha = 0.05$.

Digit	Count	Percent
0	26	11.93
1	12	5.50
2	26	11.93
3	18	8.26
4	23	10.55
5	19	8.72
6	18	8.26
7	27	12.39
8	30	13.76
9	19	8.72
$N =$	218	

5. Alternative Medicine Use. A study conducted in 2006 examined the prevalence of alternative medicine usage by age group among persons with diabetes.[13] In the study, 5.7% of the subjects were aged 18–34 years, 20.7% were aged 35–49 years, 38.8% were aged 50–64 years, and 34.8% were age 65 or older. Suppose that a study conducted this year found that, of the 1000 randomly selected respondents with diabetes, 70 were 18–34 years old, 220 were 35–49 years old, 440 were 50–64 years old, and 270 were over age 65. Test using level of significance $\alpha = 0.05$ whether the proportions have changed since 2006.

SECTION 11.2

6. Grades and the SAT. In their 2003 "Profile of College-Bound Seniors," the College Board provided the following data on high school grade point average and gender for the students taking the SAT exam. We are interested in testing whether the proportion of females is the same across the six grade categories.

Gender	A+	A	A–	B	C	D–F
Female	60	62	59	53	43	43
Male	40	38	41	47	57	57

High School Grade Point Average

a. Before you perform any calculations, *what result might we expect*? Examine the table carefully to see whether you can identify any differences in the proportions of females.

b. Test whether the proportion of females is the same across the six grade categories, using level of significance $\alpha = 0.05$.

7. Homicide Victims by Race. The U.S. Bureau of Justice Statistics reported the following numbers of homicide victims, by gender and race, for the state of California in the year 2000. Test whether gender and race are independent, using level of significance $\alpha = 0.10$.

Gender	White	Black	Other
Male	1089	516	64
Female	297	85	23

8. September 11 and Pearl Harbor. The terrorist attacks on New York City and Washington, D.C., on September 11, 2001, were often compared to the Japanese attack on Pearl Harbor on December 7, 1941. In an NBC News Terrorism Poll, the following question was asked: Would you say that Tuesday's attacks are more serious than, equal to, or not as serious as the Japanese attack on Pearl Harbor? This poll was conducted on September 12, 2001, and the results are given in the following table (see the second column). Were there systematic differences in the way men and women responded to this question? In other words, are the variables *poll response* and *gender* independent? Perform

the χ^2 test for independence between *poll response* and *gender*, using the *p*-value method and level of significance $\alpha = 0.01$.

	Male	Female	Total
More serious	200	212	412
Equal	70	84	154
Not as serious	23	6	29
Not sure	11	12	23
Total	304	314	618

Gender

9. Radio-Listening Trends. The Arbitron Corporation tracks radio-listening trends among demographic groups. In a 2003 survey, 240 teens aged 12–17 listened to pop contemporary hit radio stations while 170 teens listened to alternative radio stations. Also, 250 young adults aged 18–24 listened to pop contemporary hit radio stations while 260 young adults listened to alternative radio stations. Test whether age and radio station type are independent, using level of significance $\alpha = 0.05$.

10. Happiness in Marriage. The General Social Survey tracks trends in American society. The accompanying crosstabulation shows the responses to a question that asked people to characterize their feelings about being married. Test whether happiness in marriage is independent of gender, using level of significance $\alpha = 0.05$.

Respondents' gender	Very happy	Pretty happy	Not too happy	Total
Male	242	115	9	366
Female	257	149	17	423
Total	499	264	26	789

Happiness in Marriage

SECTION 11.3

For Exercises 11 and 12, assume that the ANOVA assumptions are met. Perform the appropriate analysis of variance using the critical method and $\alpha = 0.05$. Note that you will have to find the summary statistics yourself, so make use of a computer or calculator, if available. For each exercise, provide the following:

a. Hypotheses
b. *F* critical value
c. Rejection rule
d. Test statistic
e. Conclusion
f. Interpretation

11. Diversity Awareness Test Scores. A corporate multicultural affairs officer would like to assess whether there are differences in the diversity awareness test scores for new employee training at three corporate locations, Branches A, B, and C. She randomly selects eight new employees from each of the three different branches. The following table summarizes the test score results. Test whether significant differences exist in the mean diversity awareness test scores for the three branches.

Branch A	Branch B	Branch C
50	80	85
90	90	95
60	95	90
90	60	75
85	60	85
50	80	75
60	70	65
70	75	80

12. Differences in Medical Treatments. A psychologist is interested in investigating whether differences in mean client improvement exist for three medical treatments. Seven clients undergoing each medical treatment were asked to rate their level of satisfaction on a scale of 0 to 100. The data are summarized in the following table.

Medical treatment 1	Medical treatment 2	Medical treatment 3
75	75	100
100	100	100
0	25	50
50	75	90
50	50	75
40	75	75
25	60	90

For Exercises 13 and 14, assume that the ANOVA assumptions are met. Perform the appropriate analysis of variance using $\alpha = 0.05$ and the p-value method. For each ANOVA, provide the following measures.
  a. Hypotheses and the rejection rule
  b. F_{data}
  c. p-Value
  d. Conclusion and interpretation

13. Customer Satisfaction. The district sales manager of a local chain store would like to determine whether there are significant differences in the mean customer satisfaction among the four franchise stores in her district. Customer satisfaction data were gathered over 7 days at each of the four stores. The resulting data are summarized in the following table.

Store A	Store B	Store C	Store D
50	60	25	75
40	45	30	60
60	70	50	80
60	70	30	90
50	60	40	70
45	65	25	85
55	70	45	95
$\bar{x}_A = 51.43$	$\bar{x}_B = 62.86$	$\bar{x}_C = 35.00$	$\bar{x}_D = 79.29$
$s_A = 7.48$	$s_B = 9.06$	$s_C = 10.00$	$s_D = 12.05$

14. The Price of Apples. A market researcher is interested in whether differences exist in mean price per pound for apples at local supermarket chains. Random samples of apples from each chain were taken, and their prices per pound (in dollars) are summarized in Table 11.18. Does it pay to comparison shop?

SECTION 11.4
For Exercises 15–18, construct and interpret a 95% confidence interval for β_1.
15. Education and Earnings. The U.S. Census Bureau reports the mean annual earnings of American citizens according to the number of years of education. Here is an excerpt from the 2003 findings. We are interested in the relationship between earnings (y, in thousands of dollars) and years of education (x).

Education (x)	Annual earnings (y)
8	18.6
10	18.9
12	27.3
13	29.7
14	34.2
16	51.2
18	60.4

TABLE 11.18 Apple prices

Supermarket A	Supermarket B	Supermarket C	Supermarket D	Supermarket E
1.50	2.00	2.60	3.00	3.30
1.90	2.10	2.40	2.90	3.90
1.50	2.00	2.70	3.50	3.30
1.00	2.30	1.80	2.60	3.50
1.50	3.00	2.50	3.00	3.20
2.40		2.90		

16. High School GPA and College GPA. The college admissions office would like to determine if there is a relationship between the high school grade point average and the first-year college grade point average of first-year college students, using the data in the following table.

Student	High school GPA (x)	First-year college GPA (y)
1	2.4	2.6
2	2.5	1.9
3	2.9	2.7
4	2.7	2.5
5	3.0	2.4
6	3.5	2.9
7	3.0	2.7
8	3.6	3.1
9	3.4	3.0
10	3.9	3.3

17. NBA Points Leaders. The table lists the 2003–2004 National Basketball Association leaders in points per game, as reported by **nba.com**. Would you expect there to be a linear relationship between field goals (x) and free throws (y)?

Player	Field goals (x)	Free throws (y)
1 Tracy McGrady	653	398
2 Predrag Stojakovic	665	394
3 Kevin Garnett	804	368
4 Kobe Bryant	516	454
5 Paul Pierce	602	517
6 Baron Davis	554	237
7 Vince Carter	608	336
8 Tim Duncan	592	352
9 Dirk Nowitzki	605	371
10 Michael Redd	633	383

18. Used Cars: Price versus Age. Do you think you can predict the price of a used car based on how old it is? The accompanying table shows the price (in thousands of dollars) and the age (in years) of 10 previously owned vehicles of the same make and model.

Car	Age (x)	Price (y)
1	1	18.0
2	2	16.0
3	3	15.5
4	4	13.5
5	4	14.5
6	5	10.5
7	5	12.0
8	6	9.5
9	7	8.5
10	8	7.0

For Exercises 19 and 20, follow steps **(a)**–**(e)**.
   a. Verify the regression model assumptions.
   b. Based on the confidence intervals constructed in Exercises 15 and 16, would you expect the hypothesis test to reject the null hypothesis that $\beta_1 = 0$?
   c. Calculate t_{data}.
   d. Find p-value $= 2 \cdot P(t > |t_{data}|)$.
   e. Perform the hypothesis test for the linear relationship between x and y using the p-value method and $\alpha = 0.05$.

19. Data in Exercise 15
20. Data in Exercise 16

For Exercises 21 and 22, follow steps **(a)**–**(d)**.
   a. Based on the confidence intervals constructed in Exercises 17 and 18, would you expect the hypothesis test to reject the null hypothesis that $\beta_1 = 0$?
   b. Find t_{crit} for a two-tailed test with $\alpha = 0.05$ and df $= n - 2$.
   c. Calculate t_{data}.
   d. Perform the hypothesis test for the linear relationship between x and y using the critical-value method and $\alpha = 0.05$.

21. Data in Exercise 17
22. Data in Exercise 18

CHAPTER 11 *Quiz*

TRUE OR FALSE

1. True or false: The F curve is symmetric.
2. True or false: In a χ^2 test for independence, the degrees of freedom equals $k - 1$.
3. True or false: If we reject the null hypothesis in an ANOVA, we conclude that there is evidence that all the population means are different.

FILL IN THE BLANK

4. In the test for the homogeneity of proportions, the null hypothesis states that all k population proportions are _____.
5. In ANOVA the _____ _____ _____ [three words] measures the variability in the sample means.
6. In ANOVA the _____ _____ _____ [three words] measures the variability within the samples.

SHORT ANSWER

7. In ANOVA what do we use for an estimate of the overall population mean?

8. In the test for the homogeneity of proportions, which hypothesis states that not all population proportions are equal?

9. How does one calculate the degrees of freedom for the χ^2 test for independence?

CALCULATIONS AND INTERPRETATIONS

10. Pregnancy and HIV Testing. A study examined the proportions of pregnant women in the United States who have had an HIV test in the past 12 months.[14] The proportions for the Northeast, Midwest, South, and West were 56.8%, 49.3%, 58.5%, and 50.2%. Test whether the population proportions of pregnant women who have had an HIV test in the past 12 months are the same across all four regions, using level of significance $\alpha = 0.01$. Assume that each sample size equals 1000.

11. Illicit Drug Use Among Young People. Monitoring the Future (**www.monitoringthefuture.org**), at the University of Michigan, is an "an ongoing study of the behaviors, attitudes, and values of American secondary school students, college students, and young adults." They reported the lifetime prevalence of the use of any illicit drug among 8th-graders, 10th-graders, and 12th-graders, as shown in the table. Test using level of significance $\alpha = 0.01$ for differences among the proportions of children in those grades who have ever used an illicit drug.

	8th-graders	10th-graders	12th-graders
Have used an illicit drug	3,655	6,527	7,461
Have never used an illicit drug	13,345	9,873	7,139

12. Beef Cattle and Farm Size. The National Agricultural Statistics Service publishes data on farm products in the United States.[15] The accompanying table shows the number of beef cattle on smaller-scale operations (farms having fewer than 50 head) for three states. Test whether the proportions of cattle on smaller farms are the same across all three states, using level of significance $\alpha = 0.05$.

	Texas	Oklahoma	Pennsylvania
Beef cattle on smaller scale operations	103,000	3,600	11,400
Beef cattle on operations that are not smaller scale	28,000	44,400	600

For Exercises 13–16, perform the appropriate analysis of variance using $\alpha = 0.05$.

13. Gas Mileage and Number of Cylinders. When it comes to getting good gas mileage, does the number of cylinders in your engine make a difference? The following table provides the summary statistics regarding miles per gallon for 4-cylinder, 6-cylinder, and 8-cylinder cars.

	4 cylinders	6 cylinders	8 cylinders
n	199	83	103
$\bar{x}$	29.3	20.0	15.0
s	5.7	3.8	2.9

14. Hours Worked and Marital Status. The General Social Survey tracks demographic trends. Here we are interested in whether the mean number of hours worked differs by marital status. The summary statistics are shown here.

	N	Mean	Std. Deviation
MARRIED	964	42.76	14.08
WIDOWED	72	40.13	14.28
DIVORCED	342	43.69	13.93
SEPARATED	79	41.66	15.71
NEVER MARRIED	478	41.03	14.03

15. Calories in Breakfast Cereals. A dietary researcher is interested in whether differences exist in the mean number of calories in breakfast cereals made by different manufacturers. The summary statistics for the samples from three manufacturers appear in the following table.

	Kellogg's	Quaker	Ralston Purina
n	23	8	8
$\bar{x}$	109	95	115
s	22	29	23

16. GPA Differences by Class. A researcher is interested in whether differences exist in the mean grade point averages of freshmen, sophomores, juniors, seniors, and graduate students at a local university. Random samples of students from each category are taken, and their grade point averages are summarized in Table 11.19, on the following page.

For Exercises 17 and 18, construct and interpret a 95% confidence interval for the slope β_1 of the regression line.

17. Men's Heights and Weights. The university medical unit is collecting data on the heights and weights of the male students on campus. A random sample of ten male students showed the following heights (in inches) and weights (in pounds).

Student	Height (x)	Weight (y)
1	66	150
2	68	145
3	69	160
4	70	165
5	70	165
6	71	180
7	72	175
8	72	180
9	73	195
10	75	210

TABLE 11.19 Grade point averages

Freshmen	Sophomores	Juniors	Seniors	Graduate students
1.0	2.2	2.5	2.9	3.5
1.5	1.5	2.5	3.0	4.0
1.5	2.0	2.7	3.5	3.3
1.0	2.3	1.8	2.6	3.5
1.5	2.0	2.5	3.0	3.2
2.5		3.0		
$\bar{x} = 1.5$	$\bar{x} = 2.0$	$\bar{x} = 2.5$	$\bar{x} = 3.0$	$\bar{x} = 3.5$
$s = 0.5477$	$s = 0.3082$	$s = 0.3950$	$s = 0.3240$	$s = 0.3082$

18. Ratio Accounting Grades. An accounting professor is trying to predict the performance of her students in the second semester of the introductory accounting course by their performance in the first semester. The first-semester grade and second-semester grade were recorded for a random sample of 12 students taking the two-semester course at a local college. The results are shown in the table.

Student	First-semester grade (x)	Second-semester grade (y)
1	75	65
2	80	90
3	50	75
4	70	60
5	90	80
6	75	80
7	50	60
8	95	90
9	80	75
10	50	40
11	60	55
12	75	70

19. For the data in Exercise 17, perform the hypothesis test for the linear relationship between x and y using $\alpha = 0.05$

20. For the data in Exercise 18, perform the hypothesis test for the linear relationship between x and y using $\alpha = 0.05$

CHAPTER 1

Section 1.1

1. Answers will vary.
3. Answers will vary.
5. Answers will vary.
7. Note the large differences in the comparative heights of the rectangles that measure responses of sadness, anger and disbelief.
9. Yes; the point corresponding to Palm Beach County votes lies significantly above the rest of the data and outside the otherwise linear trend.
11. About 3500 votes.

Section 1.2

1. Answers will vary.
3. Elements
5. Categorical variable
7. A population is the collection of all elements (persons, items, or data) of interest in a particular study. A sample is a subset of the population from which the information is collected.
9. The value of a parameter is constant but usually unknown. The value of a statistic may vary from sample to sample but is usually known.
11. (a) Quantitative (b) Interval
13. (a) Quantitative (b) Ratio
15. (a) Quantitative (b) Ratio
17. (a) Qualitative (b) Ordinal
19. (a) Qualitative (b) Nominal
21. (a) Qualitative (b) Nominal
23. (a) Quantitative (b) Ratio
25. Population: all home sales in Tarrant County, Texas; sample: 100 home sales selected.
27. Population: all 4-H clubs in Utah; sample: ten selected 4-H clubs.
29. Population: all students at Portland Community College; sample: 50 selected Portland Community College students.
31. Descriptive statistics; the variable describes a sample.
33. Statistical inference; the sample was used to draw a conclusion about the entire population.
35. Descriptive statistics; the variable describes a sample.
37. Descriptive statistics; the variable describes a sample.
39. Sample (out of a total of 11 species)
41. (a) Elements are the companies. (b) Employees and Industry. (c) Employees (d) Industry (e) It has 1,755 employees and is a health services industry.
43. (a) Elements are the types of crime. (b) 2005 Total, per 100,000 People, National per 100,000 People, and Compared to National Average. (c) 2005 Total, per 100,000 People, and National per 100,000 People. (d) Compared to National Average (e) In 2005, there were 55 automobile thefts in Stillwater, Oklahoma, at a rate of 134.1 per 100,000 people. This was better than the national rate of 526.5 thefts per 100,000 people.
45. (a) Elements are the hospitals. (b) Beds, City, and Zip. (c) Beds (d) City and Zip (e) It has 72 beds and is in the city of Kosciusko with a zip code of 39090.
47. Births, Average Maternal Age; quantitative.

49. (a) Elements are the types of music that *all* radio listeners listened to in the winter of 2004; we can only list Mainstream AC, Hot AC, Mod AC, and Soft AC—other elements are not given. (b) Percentage of listeners for each type; quantitative. (c) The proportion is (% Mainstream AC ÷ % AC) *100% = (7.8/14)100% = 55.71%.
51. They compared the average lifetime of a sample of their own light bulb to the reported average lifetimes of other current models of light bulbs.
53. (a) Years. (b) Deaths; quantitative (c) No; years were not selected randomly.

Section 1.3

1. Answers will vary.
3. Answers will vary; could have chosen a random sample of houses and apartments and surveying the people door to door, for instance.
5. Answers will vary.
7. Stratified sampling
9. Systematic sampling
11. Random sampling
13. Leading question
15. Observational study; explanatory variable: whether or not the company gives large bonuses to their CEO's; response variable: stock price.
17. Answers will vary; samples are different because each was randomly selected.
19. Target population: all high schools in New England; potential population: all high schools in greater Boston. Potential population is not a random sample of all high schools in New England; drop out rate for target population may differ from potential population.
21. Desired response type is open to interpretation: preference or yes/no.
23. No; the conclusion leaves out the details of the survey question
25. (a) No, different random samples may contain different lowest stock prices. (b) Answers will vary. (c) No; see answer (a). (d) Answers will vary.
27. Predictor variable: patient diet, Mediterranean or Western; response variable: risk for a second heart attack.
29. (a) Type of bracelet worn, placebo or ionized (b) Wearing the ionized bracelet. (c) Measure of pain.
31. Answers will vary.
33. (a) Answers will vary. (b) No. (c) Answers will vary. (d) We cannot tell in advance.

Chapter 1 Review

1. (a) The ten elements are the countries. (b) Continent, Climate, Water Use (Per Capita Gallons per Day), and Main Use. (c) It is on the continent of North America, it has a temperate climate, its water use is 1268 gallons per capita per day, and its main use of water is industry. (d) Iraq (e) "Per person."
3. United States and Canada. (a) No; the United States uses 1565 gallons per capita per day, Canada uses 1268 gallons per capita per day. (b) No; the United States has more farming and industry than Canada so it uses more water.

5. (a) Turn on all one million light bulbs, leave them all on until they burn out, compute the average of all bulb lifetimes. **(b)** Use the average lifetime of a sample to estimate the average lifetime of the population.
7. (a) All statistics students **(b)** The students in the statistics class who were selected for the sample **(c)** Left-handed or not; qualitative **(d)** No; not likely to be very far away from the population proportion since enrollment in a specific statistics class is not dependent on being left-handed or not.
9. (a) Observational study **(b)** It would be impractical to randomly reassign people to a statistics class after classes have started.
11. (a) Randomization **(b)** Without randomization, results may be biased.
13. No; there may be other factors that determine a child's cognitive skills.

Chapter 1 Quiz

1. False
2. False
3. collecting
4. observation
5. sample
6. Observational study
7. Experimental study
8. Predictor variable: drug given to an elderly patient with Alzheimer's, new or placebo; response variable: whether or not the patient's Alzheimer's symptoms are reduced.
9. Predictor variable: form of pesticide used, new or traditional; response variable: the amount of insect damage to crops.
10. Experimental study
11. Treatment: new pesticide; control: traditional pesticide.
12. Different people have different interpretations of the words often, occasionally, sometimes, and seldom.

CHAPTER 2

Section 2.1

1. We use graphical and tabular form to summarize data in order to organize it in a format where we can better assess the information. If we just report the raw data, it may be extremely difficult to extract the information contained in the data.
3. True.

5.

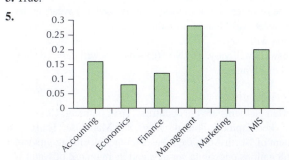

7. (a) and **(b)**

Climate	Frequency	Relative frequency
Arid	5	0.50
Temperate	4	0.40
Tropical	1	0.10

(c)

(d)

(e)

(f)

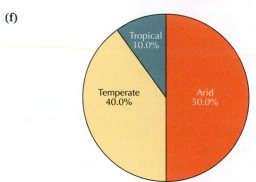

9. (a)

End-use sector	Relative frequency
Residential	0.2049
Commercial	0.1746
Industrial	0.2931
Transportation	0.3274

(b)

(c)

(d)

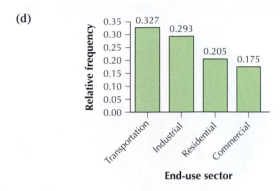

(e) *Note:* In the chart, the relative frequencies are expressed as percentages.

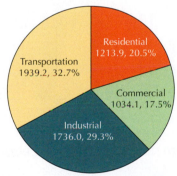

11. The sample may be a representative sample of all U.S. teenagers if one assumes that teenagers of all walks of life visit this particular Web site and participate in the survey. If only teenagers of a particular social class visit this Web site, then although the sample is "large" it will not be a representative sample. Bias may be introduced if only teenagers of a particular social standing, region, ethnicity, and so on visit this Web site.

13. A pie chart cannot be constructed for the data since the percentages for the different classes do not add up to 100%. Respondents could select more than one category.

15.

Response	Frequency
Yes	1860
No	2700
I've never thought about it.	1440

17. *Note:* Relative frequencies are expressed as percentages.

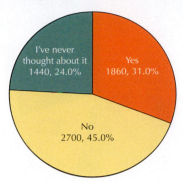

19. Virgo occurs the most and Aries occurs the least.

Sign	Relative frequency	Frequency
Pisces	0.081	118
Aries	0.075	110
Taurus	0.083	121
Gemini	0.087	127
Cancer	0.081	118
Leo	0.079	116
Virgo	0.091	133
Libra	0.077	113
Scorpio	0.082	120
Sagittarius	0.088	129
Capricorn	0.087	127
Aquarius	0.090	132

21. Yes.

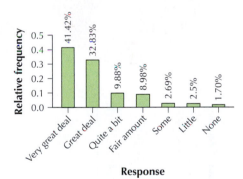

23.

Investor	Frequency	Relative frequency
Intel Capital Corp.	41	0.299
New Enterprise Associates	22	0.161
Sequoia Capital	19	0.139
Venrock Associates	19	0.139
Canaan Partners	18	0.131
Draper Fisher Jurvetson	18	0.131

25.

27.

29.

31.

33.

35.

37. *Note:* The relative frequencies are expressed as percents in the graph.

39. (a) Girls: 0.525; boys: 0.475

(b) Grades: 51.67%; popular: 29.50%; sports: 18.83%

41. There are 51 observations and 8 variables.

43. Each relative frequency in the relative frequency distribution would be the same, each slice in the pie chart would be the same size, and each rectangle in the bar graph would be the same height.

45. Answers will vary.

Section 2.2

1. Both: frequency distribution, relative frequency distribution; quantitative data only: histograms, frequency polygons, stem-and-leaf displays, dotplot.

3. Between 5 and 20

5. Answers will vary.

7. Answers will vary.

9. (a) 4 **(b)** 1 and 6 **(c)** 15 times **(d)** 15% of the times

11. (a) 46 **(b)** 33 (not including a frequency of 0) **(c)** highest: 49: lowest: 33. **(d)** left-skewed

13. (a) Divide the frequency values by the total frequency—classes not affected **(b)** change the scale along the relative frequency (vertical) axis by multiplying the relative frequency values by the total frequency—shape of distribution not affected **(c)** 19

15. (a) 0 **(b)** 0 **(c)** $25 to $27.5 has the largest relative frequency, 4/19 = 0.2105. **(d)** 3 **(e)** 0

17. Data set: 23 24 25 26 27 28 28 29 30 31 31 32 32 32 33 35 36 37 39 40

19. Histogram with five classes

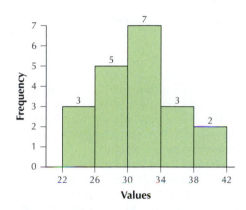

21. (a) 15 **(b)** 37.5 **(c)** 52.5 **(d)** 67.5 to 82.5 **(e)** 22.5 to 37.5

23. (a) 2000 **(b)** 1000 **(c)** 1000 to 3000 **(d)** 17,000 to 19,000.

25. (a) and **(b)**

Continent	Frequency	Relative frequency
Africa	1	0.1
Asia	5	0.5
Europe	1	0.1
North America	2	0.2
South America	1	0.1

(c) Cannot construct because the variable is qualitative. **(d)** Cannot construct because the variable is qualitative. **(e)** We cannot construct because the variable is qualitative.

27.

29.

31. Highest: Class 750 to 799 and Class 800 to 849. Lowest: Class 550 to 599, Class 600 to 649, 650 to 699, and Class 850 to 899 (disregarding a frequency of zero)

33. Fairly symmetrical

35. $n \approx 690$.

37.

39. (a) The horizontal scale would shift to the left. The vertical scale and size of the rectangles will be the same. **(b)** The horizontal scale would shift to the left. The vertical scale and the number of dots in each column will remain the same. **(c)** The numbers in the stem-and-leaf display will all be less by the same amount. This may change the number of numbers in each row, depending on the increments of the leaves.

41. The histogram is right-skewed.

43. The histogram is right-skewed.

45. (a) 9 **(b)** 0.30

47. (a) Decreases **(b)** Increases

49. (a) 3 **(b)** 57. There are 57 observations in the data set.

51. Answers will vary.

Section 2.3

1. A frequency distribution gives the frequency counts for each class (grouped or ungrouped). A cumulative frequency distribution gives the number of values which are less than or equal to the upper limit of a given class for grouped data or it gives the number of values which are less than or equal to a given number for ungrouped data.

3. Ogive.

5. Time series data.

7. (a), (b), and **(c)**

Value	Frequency	Cumulative frequency	Relative frequency	Cumulative relative frequency
1	13	13	0.13	0.13
2	20	33	0.20	0.33
3	15	48	0.15	0.48
4	24	72	0.24	0.72
5	15	87	0.15	0.87
6	13	100	0.13	1.00

9. (a)

(b)

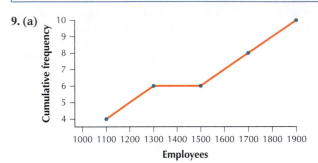

11. (a) 0.8 **(b)** 2.39 **(c)** 1.99
13. (a) Approximately 950 **(b)** Approximately 50 **(c)** Between 1996 and 1997 **(d)** Encouraging news
15. (a) 45% **(b)** 0%

Exports	Frequency	Cumulative frequency	Relative frequency	Cumulative relative frequency
0.0–1.49	4	4	0.20	0.20
1.5–2.99	9	13	0.45	0.65
3.0–4.49	6	19	0.30	0.95
4.5–5.99	0	19	0.00	0.95
6.0–7.49	0	19	0.00	0.95
7.5–8.99	0	19	0.00	0.95
9.0–10.49	0	19	0.00	0.95
10.5–11.99	1	20	0.05	1.00

17. (a)

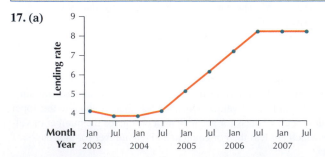

(b) Upward trend
19. The only change would be that the entire time series graph would shift up 3 units. The horizontal scale would stay the same and the shape of the graph would stay the same.

21. (a)

(b) The trade deficit has been increasing.

Section 2.4
1. Answers will vary.
3. Figure 2.28 visually reinforces the magnitude of the differences.
5. Table 2.21 gives the actual number of cars stolen.
7. (a) Biased distortion or embellishment; omitting the zero on the relevant scales; inaccuracy in relative lengths of bars in a bar chart. **(b)** A Pareto chart or pie chart can be used.
9. (a) The percentage (or number) of doctors who are devoted to family practice is decreasing **(b)** inaccuracy in relative lengths of bars in the bar chart **(c)** One may use a bar chart.
11. (a) Manipulating the scale **(b)** We can use a bar chart.

13. (a)

(b)

(c)

15. Answers will vary.

Chapter 2 Review

1. No, because the variable is categorical.

3. The relative frequencies are expressed as percentages.

(a)

(b)

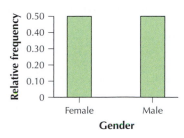

5. No, because we only have the frequencies for the categories inexpensive, moderately priced, and expensive and not the original data.

7. No, we would need the total number of observations.

9. Around a value of 62.

11. (b) tending to be right-skewed

13. Shape would not be affected. The scale would move 0.5 units to the right

15. (a)

(b)

17. (a) Pareto chart or bar graph **(b)** It gives the impression that the NFL category is significantly larger than the rest because of the width of the bar and the fact that other graphics are on that bar. **(c)** Omitting the zero on the *y* scale; inaccuracy in relative lengths of bars; biased distortion or embellishments **(d)** We can use a Pareto chart.

Chapter 2 Quiz

1. False

2. True

3. sample size.

4. frequency distribution

5. Symmetric

6. Right skewed

7–10.

Vowels	Frequency	Cumulative frequency	Relative frequency	Cumulative relative frequency
a	73	73	0.1931	0.1931
e	130	203	0.3439	0.5370
i	74	277	0.1958	0.7328
o	74	351	0.1958	0.9286
u	27	378	0.0714	1.0000

11.

12.

13.

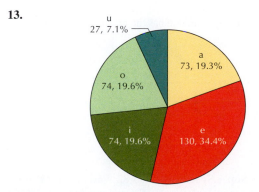

15. Can't construct because variable is qualitative.

CHAPTER 3

Section 3.1

1. A value that locates the center of the data set.
3. Because the mean depends in part on the sum of all data values, an outlier will skew the mean (pull it in one direction or another). Since the median simply depends on position in an ordered list, it is not sensitive to outliers.
5. (a) 22.5 **(b)** 20 **(c)** 20
7. (a) 16.6 **(b)** 17 **(c)** 5 and 15
9. (a) 7 **(b)** 7.5 **(c)** 4 and 9
11. (a) 507.38 **(b)** 503 **(c)** 501
13. (a) English; no. **(b)** No, the data are qualitative. **(c)** Economics does not occur with the highest frequency.
15. (a) \$119.84 **(b)** \$119.84 = \$23.968 $\times$ 5 **(c)** If a set of values is multiplied by a positive constant then the mean for this new data set will equal the mean of the original data set times the constant.
17. (a) 2 **(b)** Mean = 3.5; Median = 3.
19. (a) Mean < Median < Mode **(b)** Mode < Median < Mean **(c)** Mean = Median = Mode.
21. The mean is larger than the median, which implies the distribution is positively skewed.
23. The mean, median and mode will all be halved. Each statistic will be affected equally.
25. (a) 73 **(b)** 76 **(c)** Females
27. (a) Decrease **(b)** Unchanged **(c)** Unchanged
29. \$2,400,000; 11.0462%.
31. Answers will vary.
33. Answers will vary.
35. (a) Answers will vary. **(b)** The mean increases. **(c)** The median remains the same.

Section 3.2

1. Deviation for a data value gives the distance the value is from the mean.
3. Less variability will lead to more precise estimates, higher confidence in conclusions, and greater discerning power.
5. 82.5
7. 15
9. 6.45
11. Variance = 254.84 wins squared; standard deviation = 15.96 wins. The standard deviation is more easily understood because it has the same units as the variable *total*.
13. 19 mpg
15. 6.2 years
17. The Super Speedways variable has the greater variability.

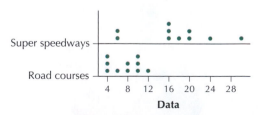

Data

19. Variance for Super Speedway = 55.82; variance for Road Courses = 8; standard deviation for Super Speedway = 7.47; standard deviation for Road Courses = 2.828. Super Speedway. Yes, the results agree with what we saw in the dotplots.

21. Zooplankton: 1.992; phytoplankton: 2.819. **(a)** Phytoplankton **(b)** Phytoplankton because this variable has the larger standard deviation; yes. **(c)** Phytoplankton because it has the larger standard deviation.
23. Colony A = 480.04; Colony B = 694.49. **(a)** Colony B **(b)** Colony B; yes **(c)** Colony B, because it has the larger variance.
25. Range for SAT I Verbal = 25; range for SAT I Math = 24. Variance for SAT I Verbal = 90.55; variance for SAT I Math = 84.84. Standard deviation for SAT I Verbal = 9.52; standard deviation for SAT I Math = 9.21. These values support the judgment in the previous exercise since they are similar in magnitude.
27. (a) 6 inches **(b)** 2 inches
29. (a) Increase the range, variance and standard deviation. **(b)** New range = 10. New variance = 15.2. New standard deviation = 3.9. Yes, the judgment in **(a)** was supported.
31. (a) They are the same. **(b)** 8.11 **(c)** Slight difference of 0.11. **(d)** 5.875 **(e)** They are the same. **(f)** Female; yes **(g)** Females
33. Answers will vary.
35. Answers will vary.

Section 3.3

1. These formulas will provide only estimates because we will not know the exact data values.
3. Population mean
5. 69
7. 3.2
9.

Class limits	Midpoints
0 – 1.99	1
2 – 3.99	3
4 – 5.99	5
6 – 7.99	7
8 – 9.99	9

11. 14.2857
13. Estimated standard deviation = 6.226998; estimated variance = 38.7755.
15. (a)

Age	Frequency	Midpoints
0–4.99	63,422	2.5
5–17.99	240,629	11.5
18–64.99	540,949	41.5

(b) Estimated mean = 30.0298 years **(c)** Estimated standard deviation = 15.455909 years; estimated variance = 238.8851 years squared

17. (a)

Dollar value	Housing units (frequency)	Midpoints
\$0–\$49,999	5,430	25,000
\$50,000–\$99,000	90,605	75,000
\$100,000–\$149,999	90,620	125,000
\$150,000–\$199,999	54,295	175,000
\$200,000–\$299,000	34,835	250,000
\$300,000–\$499,999	15,770	400,000
\$500,000–\$999,999	5,595	750,000

(b) Estimated mean = \$158,079.25 **(c)** Estimated standard deviation = \$61,682.53; estimated variance = 384,734,862 (dollars squared)
19. 82.8%
21. \$70,940

Section 3.4

1. Answers will vary.

3. Answers will vary.

5. It is possible for the 1st percentile to equal the 99th percentile if all of the data values are the same.

7. (a) 20 **(b)** 24 **(c)** 18

9. (a) 0.21 **(b)** –0.78 **(c)** −0.64

11. No; z score is 1.92 and $-2 < 1.92 < 2$.

13. (a) 102.5 **(b)** 130

15. (a) −1.65 **(b)** 0.93

17. (a) 4,200,000 **(b)** 7,100,000

19. (a) 1.11 **(b)** 0.6033

21. (a)

(b)

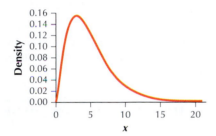

(c) 503

23. (a) 33.55 **(b)** 34.35 **(c)** 34.55

25. None.

Section 3.5

1. When $k \le 1$ or when the data set is not continuous

3. True. *Note:* Does not pertain to all situations.

5. (a) At least 75% **(b)** At least 88.9%

7. (a) Between 80 and 120 **(b)** 120 **(c)** 80

9. (a) At least 75% **(b)** At least 88.9% **(c)** Can't find since $k = 1$.

11. (a) About 95% **(b)** About 99.7% **(c)** About 68%

13. (a) (i) 102,000,000 (ii) He is very truthful. **(b)** The day corresponding to the 25th percentile. **(c)** At least 75%

15. (a) Cannot be done since $k = 1$. **(b)** At least 75% **(c)** At most 25%

17. Answers in Exercise 16 are more precise; the distribution is bell-shaped.

19. Lower-performing school z-score 0.8, so 4.0 GPA is not considered unusual; higher-performing school z-score 2.8, so a GPA of 4.0 is considered moderately unusual.

Section 3.6

1. We need statistics that can summarize a data set while being less sensitive to outliers; mean and the standard deviation might not be very representative of center and spread of the data because of sensitivity to outliers.

3. False

5. (a) 68 **(b)** 76 **(c)** 85.5

7. Min = 51, Q1 = 68, median = 76, Q3 = 85.5, max = 98

9.

Final exam scores

11. WMU: IQR = 0; NCU: IQR = 3. For WMU, the spread of the middle 50% of the data is 0, and for NCU, the spread of the middle 50% of the data is 3.

13. Min = 8.33, Q1 = 13.69, median = 23.375, Q3 = 37.79, max = 55.46

15. Q1 − 1.5 * IQR = −22.46 and Q3 + 1.5 * IQR = 73.94. There are no values outside this interval, so there are no outliers.

17. Min = −4.09, Q1 = −0.14, median = −0.015, Q3 = 0.08, max = 0.1

19. Q1 − 1.5 * IQR = −0.47 and Q3 + 1.5 * IQR = 0.41. The change −4.09 is an outlier.

21. Min = 2,000,000; Q1 = 2,800,000; median = 4,200,000; Q3 = 7,100,000; max = 14,700,000

23. Q1 − 1.5 * IQR = −3.65 and Q3 + 1.5 * IQR = 13.55. Usage of 14,700,000 is the only outlier.

25. Mean: 5,073,000; standard deviation: 3,359,300

27. (a) 961 **(b)** 25

29. Homemade dark fruitcake (1 cake)

Chapter 3 Review

1. 3.1227

3. 2.55

5. The mode, since the value with the largest frequency is unaffected by the deletion of values less than 60 calories.

7. (a) Increase because the contribution of a value of 84 to the computation is greater than the contribution of a value of 72. **(b)** 7.27; yes.

9. (a) Range = 12; standard deviation = 4.9 **(b)** Multiplying each value in a data set by a constant will increase the value of the original range or standard deviation by a factor of the constant. **(c)** (i) Increase (ii) Yes

11. Estimated mean = 41.2055. Estimated standard deviation = 10.7061

13. $936

15. 2.3575

17. −0.7548

19. Journal of Pure and Applied Algebra

21. About 95%

23. Q1 = 25, Q2 = 34.5, Q3 = 48

25. (a) Min = 8, Q1 = 25, median = 34.5, Q3 = 48, max = 60

(b)

Ragweed pollen index

(c) Close to symmetric, slightly right-skewed. (d) The mean should be close to the median or a little above the median. (e) Mean = 36.30, standard deviation = 16.51. The value of the mean is slightly above the median of 34.50.
27. Unchanged; robustness

Chapter 3 Quiz

1. False
2. False
3. False
4. outlier
5. center
6. mean
7. robust measures
8. mode
9. Zero
10. Class midpoint
11. (a) Mean = 87,453 (b) Median = 98,008
12. (a) Range = 86,910 (b) Standard deviation = 33,857
13. Estimated mean = 61.6527; estimated standard deviation = 18.4518.
14. (a) 1.5 (b) −1 (c) 1 (d) −1.5 (e) 0
15. (a) 60 (b) between 34% and 81.5% (c) No, furthermore we must assume that one of the values of k is less than 1. (d) Between 2.5% and 16%.
16. (a) 501.5 (b) 512 (c) 518
17. IQR = 16.5
18. Min = 499, Q1 = 501.5, median = 512, Q3 = 518, max = 523.
19. Q1 − 1.5 * IQR = 476.75 and Q3 + 1.5 * IQR = 542.75. All the SAT scores lie between 476.75 and 542.75 so there are no outliers.
20.

SAT scores

CHAPTER 4

Section 4.1

1. Answers will vary.
3. The headings of the columns and the rows of a crosstabulation are categories, not numbers. We could recode the variables in order to use crosstabulation by dividing the numbers up into classes, as we did for grouped frequency distributions. Then we could code each class as a category.
5.

	Beverages	Food	Household/ personal	Tobacco	All
Glass	2	0	0	0	2
Metal	0	0	2	0	2
Paper	0	0	1	3	4
Plastic	1	2	1	0	4
All	3	2	4	3	12

7. (a)

(b) Negative relationship. (c) As x increases, y decreases. Smaller values of x are associated with larger values of y; larger values of x are associated with smaller values of y.

9. (a)

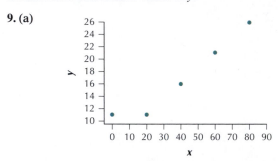

(b) Positive relationship. (c) As x increases, y increases. Smaller values of x are associated with smaller values of y; larger values of x are associated with larger values of y.

11. (a)

	Freshman	Junior	Senior	Sophomore	All
Art	0	1	0	1	2
Business	0	1	0	2	3
Communication	1	0	0	2	3
Economics	0	0	2	0	2
English	4	0	0	1	5
History	0	3	0	0	3
Philosophy	0	0	1	0	1
Political Science	0	0	1	0	1
Psychology	0	2	0	3	5
All	5	7	4	9	25

No apparent relationship; no patterns

(b)

Tally for Discrete Variables: Major	
Major	**Count**
Art	2
Business	3
Communication	3
Economics	2
English	5
History	3
Philosophy	1
Political Science	1
Psychology	5
N =	25

Tally for Discrete Variables: Class	
Class	**Count**
Freshman	5
Junior	7
Senior	4
Sophomore	9
N =	25

13.

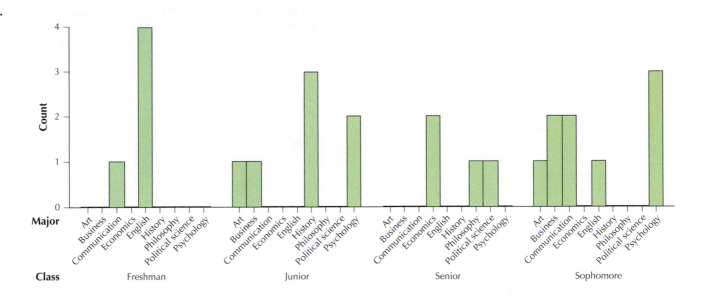

15. Missing values are in red.

"How much do you enjoy shopping?"	Gender		
	Male	**Female**	**Total**
A lot	388	950	1338
Some	528	673	1255
Only a little	662	497	1159
Not at all	497	220	717
Don't know/refused	20	25	45
Total	2149	2365	4514

17. (a) Women (b) Women (c) Men (d) Men

19. (a) .7100 (b) .5363 (c) .4288 (d) .3068

21. (a)

(b) The relationship between overall birth rate and teenage birth rate appears to be positive.

23. There appears to be no relationship between batting average and age.

25. (a)

(b) There appears to be no relationship between body weight and brain weight. (c) Yes (d) Humans

27. Missing values are in red.

Visited Memorial Web Site			
Student Status	**Yes**	**No**	**Total**
Full time	45	171	216
Part time	53	189	242
No	371	1629	2000
Total	469	1989	2458

29. People who weren't students; full-time students

31. (a)

(b) There is a negative relationship between the variables.

Section 4.2

1. Scatterplot

3. Between −1 and 1, inclusive

5. i

7. ii

9. iv

11. (a) (6,8), (7,7), (8,7), (9,8), (10,8), (11,8), (12,7), (13,8), (14,8), (15,7) (b) Minitab: Correlations: x, y; Pearson correlation of x and $y = 0.000$. TI-84: $r = 0$

13. (a) (1,11), (2,9), (3,7), (4,7), (5,6), (6,6), (7,4), (8,3), (9,3), (10,1) (b) Minitab: Correlations: x, y; Pearson correlation of x and $y = -0.978$. TI-84: $r = -0.9781316853$

15. Positively correlated

17. SAT I Verbal scores and SAT I Math scores are positively correlated. As the SAT I Verbal score increases, the SAT I Math score increases. Yes.

19. Minitab: Correlations: Body Weight (kg), Brain Weight (g); Pearson correlation of body weight (kg) and brain weight (g) = 0.099. TI-84: $r = 0.0992032294$

21. Positive

23. The number of calories in a serving of breakfast cereal and the amount of sugar in a serving of breakfast cereal are positively correlated. As the amount of sugar in a serving of breakfast cereal increases the number of calories in a serving of the breakfast cereal tends to increase.

25. A player's age and a player's batting average are not correlated. As a player's age increases the player's batting average tends to remain unchanged.

27. (a) $\bar{x} = 9.16666667$; $\bar{y} = 5.5$; $s_x = 9.432214303$; $s_y = 4.679743583$; $r = -0.7453498716$. Minitab: Correlations: Hip-Hop CDs owned, Country CDs owned; Pearson correlation of Hip-Hop CDs owned and Country CDs owned = -0.745. TI-84: $r = -0.7453498716$. (b) Yes (c) The variables number of Hip-Hop CDs owned and number of Country CDs owned are negatively correlated. As the number of Country CDs owned increases, the number of Hip-Hop CDs owned decreases.

29. (a) Yes (b) The variables birth weight and time born are not correlated. As the time born increases, the birth weight tends to remain the same.

31. Answers will vary.

33. (a) Positively correlated (b) Negatively correlated (c) Not correlated

Section 4.3

1. To approximate the relationship between two numerical variables using the regression line and the regression equation.

3. We can find the predicted value of y by plugging a given value of x into the regression equation and simplifying.

5. Extrapolation is the process of making predictions based on x-values that are beyond the range of the x-values in our data set.

7. (a) $b_1 = 0.0337643678$ (b) $b_0 = 18.24784483$ (c) $\hat{y} = b_0 + b_1 x = 8.24784483 + 0.0337643678 x$. The estimated value of the number of Super Speedways won is 18.24784483 plus 0.0337643678 times the number of Short Tracks won.

9. (a) $\cong 19$ (b) $x = 50$ is not in the range of the x-values of the data set. (c) $\cong 19$

11. (a) The slope $b_1 = 0.4264339152$ means that the estimated SAT I Math Score increases by 0.4264339152 points for every increase of one point in the SAT I Verbal Score. (b) The y-intercept $b_0 = 298.8628429$ means that the estimated SAT I Math Score is 298.8628429 when the SAT I Verbal Score is 0. (c) The SAT I Verbal Score can't be 0, so this situation will never happen.

13. (a) $\hat{y} = 193.5 + 1.595x$ (b) The estimated brain weight of a mammal is 193.5 grams plus 1.595 times the body weight.

15. (a) 273.25 grams (b) 353 grams (c) The value $x = 200$ kg is outside the range of the x-values of the data set, which is between 27.66 kg and 100 kg.

17. Curved

19. (a) 10.5. In a state with 0 households, 10.5% of the households are headed by women. Since all states have households, the value $x = 0$ would not occur. (b) This estimate would be considered extrapolation since the value of $x = 0$ is outside the range of x-values in the data set. (c) 0.000000282. For each increase of one household, the percentage of households headed by women increases by 0.000000282. (d) (Percentage of households headed by women) = $10.5 + 0.000000282$ (Total number of households). The estimated percentage of households headed by women equals 10.5 plus 0.000000282 times the total number of households. (e) Positive since the slope is positive.

21. (a) 12.474% (b) $x = 100,000$ is not in the range of the x-values in the data set.

23. (a) 1.032. The estimated percent change in the stock prices selected by the Darts increases by 1.032% for each unit percent increase in the Dow Jones Industrial Average. (b) Darts = $-2.49 + 1.032$ DJIA. The estimated percent change in the stock prices selected by the Darts is equal to -2.49 plus 1.032 times the percent change in the Dow Jones Industrial Average. (c) Positive since the slope of the regression line is positive.

25. (a) 20.214% (b) -12.81% (c) $x = -22$ is not in the range of the x-values of the data set.

27, 29. Answers will very.

Section 4.4

1. The prediction error, or residual $(y - \hat{y})$, measures how far the predicted $\hat{y}$ value is from the actual value of y observed in the data set. We want our prediction errors small.

3. (a)

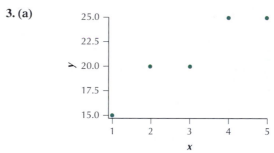

(b) $b_0 = 13.5$, $b_1 = 2.5$, $\hat{y} = 13.5 + 2.5 x$

5. (a)

(b) $b_0 = 21.6$, $b_1 = 4$, $\hat{y} = 21.6 + 4 x$

7. (a)

x	y	Predicted value $\hat{y} = 13.5 + 2.5\,x$	Residual $(y - \hat{y})$	(Residual)^2 $(y - \hat{y})^2$
1	15	16	-1	1
2	20	18.5	1.5	2.25
3	20	21	-1	1
4	25	23.5	1.5	2.25
5	25	26	-1	1

(b) SSE = 7.5. The quantity SSE measures the sum of the squared distances between the predicted value $\hat{y}$ and the actual value y. **(c)** $s = 1.5811$. The typical error in prediction is 1.5811.

9. (a)

x	y	Predicted value $\hat{y} = 21.6 + 4\,x$	Residual $(y - \hat{y})$	(Residual)^2 $(y - \hat{y})^2$
-5	0	1.6	-1.6	2.56
-4	8	5.6	2.4	5.76
-3	8	9.6	-1.6	2.56
-2	16	13.6	2.4	5.76
-1	16	17.6	-1.6	2.56

(b) SSE = 19.2. The quantity SSE measures the sum of the squared distances between the predicted value $\hat{y}$ and the actual value y. **(c)** $s = 2.5298$. The typical error in prediction is 2.5298.

11. (a)

y	$y - \bar{y}$	$(y - \bar{y})^2$
1	-6	36
2	-1	1
3	-1	1
4	4	16
5	4	16

(b) $SST = 70$. This quantity measures the variability in y. **(c)** $SSR = 62.5$. This quantity measures the amount of improvement in the accuracy of our estimates when using the regression equation compared with relying only on the y-values and ignoring the x information. **(d)** $r^2 = 0.8928571429$. Thus 0.8928571429 of the variability in y is accounted for by the linear relationship between y and x. **(e)** $r = 0.9449111825$. Thus y and x are strongly positively correlated.

13. (a)

y	$y - \bar{y}$	$(y - \bar{y})^2$
0	-9.6	92.16
8	-1.6	2.56
8	-1.6	2.56
16	6.4	40.96
16	6.4	40.96

(b) $SST = 179.2$. This quantity measures the variability in y. **(c)** $SSR = 160$. This quantity measures the amount of improvement in the accuracy of our estimates when using the regression equation compared with relying only on the y-values and ignoring the x information. **(d)** $r^2 = 0.8928571429$. Thus 0.8928571429 of the variability in y is accounted for by the linear relationship between y and x. **(e)** $r = 0.9449111825$. Thus y and x are strongly positively correlated.

15. (a)

(b) $\hat{y} = 4 + 1.6\,x$ **(c)** $b_0 = 4$ means that the estimated weight of a package that has a volume of 0 cubic meters is 4 kg. $b_1 = 1.6$ means that for every additional cubic meter of volume the weight of the package is increased by 1.6 kg. **(d)** $s = 1.3166$ kg. If we know the volume (x) of the package, then our estimate of the weight of the package will typically differ from the actual weight of the package by 1.3166 kg.

17. (a)

(b) $\hat{y} = 92.629 + 1.587\,x$ **(c)** $b_0 = 92.629$ means that the estimated rating for a team that has an 0% win percentage is 92.629. $b_1 = 1.587$ means that for every additional percentage point of wins, the rating increases by 1.587 percentage points. **(d)** $s = 1.60388$. If we know the percentage of wins, then our estimate of the rating will typically differ from the actual rating by 1.60388.

19. (a) $r^2 = 0.987$. **(b)** $r = 0.9935$.

21. (a) $r^2 = 0.004$. **(b)** $r = 0.0632$.

23. (a)

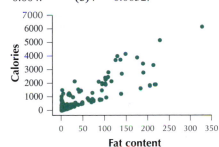

(b) $\hat{y} = 94.400 + 14.0842\,x$ **(c)** $s = 277.345$. If we know the fat content of a serving of a food (x), then our estimate of the number of calories per serving of the food will differ from the actual number of calories per serving by 277.345 grams. **(d)** $r^2 = 0.739$, meaning that 73.9% of the variability in the number of calories is accounted for by the fat content. **(e)** $r = 0.8587$. The calories and the fat content are positively correlated.

25. Answers will vary.

27. Answers will vary.

29. (a) Answers will vary. **(b)** That your adjusted line matches the least squares regression line **(c)** Yes

Chapter 4 Review

1. 0.6612

3. 0.0246

5. Answers will vary. May have clustered bar graph by happiness of marriage or clustered bar graph by sex.

7. (a) Positive. **(b)** As midterm exam score increases, overall grade tends to increase.

9. Midterm exam scores and overall grades are positively correlated. Low (high) midterm exam scores are associated with low (high) overall grades.

11. Positive

13. (a) 216 **(b)** 192 **(c)** Since all the points lie on a straight line, this *x* variable is extremely useful in predicting this *y* variable.
15. $b_0 \cong 0$
17. SSE = 598.9.
19. SST = 2006.3. SSR = 1407.3.
21. $r = 0.8373$. Midterm exam scores and overall grades are positively correlated.

Chapter 4 Quiz

1. False
2. False
3. estimate
4. unit
5. extrapolation
6. negative
7. Missing values are in red.

	Exciting	Routine	Dull	Total
Protestant	264	326	33	623
Catholic	107	128	6	241
Jewish	12	8	0	20
None	38	29	2	69
Other	9	5	0	14
Total	430	496	41	967

8.

	Exciting	Routine	Dull
Protestant	61.40%	65.73%	80.49%
Catholic	24.88%	25.81%	14.63%
Jewish	2.79%	1.61%	0%
None	8.84%	5.85%	4.88%
Other	2.09%	1.01%	0%
Total	100.00%	100.01%	100.00%

9. Protestant
10.

11. Positive
12. (a)

(b) No apparent relationship. **(c)** Near zero; near zero.
13. (a) $\bar{x} = 4$; $\bar{y} = 0$; $s_x = 0.894427191$; $s_y = 8.694826048$;

$$r = \frac{\sum(x - \bar{x})(y - \bar{y})}{(n-1)s_x s_y} = \frac{0}{(6-1)(0.894427191)(8.694826048)} = 0$$

(b) Yes **(c)** There is no apparent relationship between the net price change for equities and the net price change for guaranteed income investments. As the net price change for guaranteed income investments increases, the net price change for equities remains the same.
14. (a) *y* intercept $b_1 = 97.1$ **(b)** A breakfast cereal with 0 grams of fat will have 97.1 calories per serving. Yes, because there are other ingredients in cereal besides fat and they have calories. **(c)** No, since $x = 0$ is an *x*-value in the data set. **(d)** The number of calories in a serving of breakfast cereal increases by 9.65 calories for every increase of 1 gram of fat. **(e)** Calories = 97.1 + 9.65 fat. The estimated number of calories in a serving of breakfast cereal equals 97.1 plus 9.65 times the number of grams of fat. **(f)** Positive, since the slope of the regression line is positive.
15. (a) Cereal A has 2 (9.65) = 19.3 more calories per serving than Cereal B has. **(b)** Cereal C has 3 (9.65) = 28.95 fewer calories per serving than Cereal D has.
16. SSR = 3.7726, SSE = 0.5899, SST = 4.3625.
17. $s = 0.2429$. The typical difference between the predicted 2004 percentage and the actual 2004 percentage is 0.2429 percentage points.
18. $r^2 = 0.865$, meaning that 86.5% of the variability in 2004 percentage is accounted for by the 2002 percentage.
19. $r = 0.9300$. The 2002 percentage and the 2004 percentage are positively correlated.
20. (a) $-0.1980, 0.3020$. **(b)** $0.1456, -0.0544$. **(c)** 0.0511.

CHAPTER 5

Section 5.1

1. Answers will vary.
3. When we have prior knowledge about the relative frequency of an outcome
5. Use the relative frequency method. First find out how many students are at your college and find out how many of them like hip-hop music. Then calculate the relative frequency of students who like hip-hop music.
7. No, probability for females is greater than 1.
9. No, sum of probabilities is greater than 1.
11. No, sum of probabilities is less than 1.
13. 1/2
15. 1/3
17. 0
19. 1/4
21. 1/2
23. {HHH, HHT, HTH, THH, TTH, THT, HTT, TTT}
25. Choose a branch for the first toss, then choose a branch for the second toss, then choose a branch for the third toss. This will give one of the possible outcomes. If we follow all possible choices of branches for each toss, we get all the outcomes in the sample space.
27. outcome; event; event; event; event; event
29. (a) a sum of 7 **(b)** 1/6
31. (a) 0.33 **(b)** 0.67 **(c)** Relative frequency method
33. (a) 0.025 **(b)** Relative frequency method **(c)** The events "cutting class" and "not cutting class" are not equally likely.

35. (a)

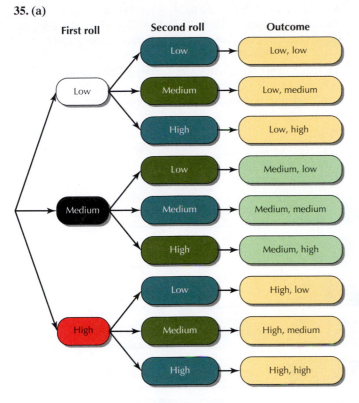

First roll	Second roll	Outcome
Low	Low	Low, low
	Medium	Low, medium
	High	Low, high
Medium	Low	Medium, low
	Medium	Medium, medium
	High	Medium, high
High	Low	High, low
	Medium	High, medium
	High	High, high

(b) {Low and Low, Low and Medium, Low and High, Medium and Low, Medium and Medium, Medium and High, High and Low, High and Medium, High and High} **(c)** 1/9; Classical

37. (a)

	Frequency	Relative frequency
Girls	18	18/44 = 0.4091
Boys	26	26/44 = 0.5909
Total	44	44/44 = 1.0000

(b)

Outcome	Probability
Girl	18/44 = 0.4091
Boy	26/44 = 0.5909

(c) Both P (Girl) = 18/44 = 0.4091 and P (Boy) = 26/44 = 0.5909 are between 0 and 1. P (Girl) + P (Boy) = 18/44 + 26/44 = 0.4091 + 0.5909 = 44/44 = 1.0000.
39. Answers will vary.
41. (a) Half of the most popular letters are vowels, but only 5/26 of all of the letters are vowels. **(b)** T, N, R, S, D
43. (a) 5/18 **(b)** 13/18 **(c)** $1.39
45. (a)

Type of music	Probability
Hip-hop/rap	0.27
Pop	0.23
Rock/punk	0.17
Alternative	0.07
Christian/gospel	0.06
R&B	0.06
Country	0.05
Techno/house	0.04
Jazz	0.01
Other	0.04

(b) Yes. **(c)** Answers will vary. **(d)** As the sample size increases the relative frequencies approach the probabilities.
47. (a)–(d) Answers will vary.

Section 5.2

1. Two events are mutually exclusive if they have no outcomes in common.
3. Yes
5. Answers will vary; $1 - P$ (will rain); they are complementary events.
7. 25/36   **9.** 0   **11.** 25/36   **13.** 4/13
15. 1/2   **17.** 51/52   **19.** 1/2   **21.** 9/52
23. 3/52   **25.** 7/8   **27.** 5/8   **29.** 3/8
31. 3/4   **33.** 1/8   **35. (a)** 1/3 **(b)** 2/3
37. (a) 3/5 **(b)** 3/4 **(c)** 2/5
39. (a) 1/9 **(b)** 1/18 **(c)** 1/6
41. (a) 1/2 **(b)** 7/10 **(c)** 9/10
43. (a) 1/2 **(b)** 7/13 **(c)** 11/26 **(d)** 0 **(e)** 3/4
45. (a) 1/4 **(b)** 9/12 = 3/4; 1–1/4 = 3/4
47. (a) 0.85 **(b)** 0.50 **(c)** 0.65 **(d)** 0
49. (a) 0.57 **(b)** 0.55
51. (a) 0.2727 **(b)** 0.7273
53. Yes

Section 5.3

1. (a) Yes. **(b)** The probability of winning the football game depends on whether or not the star quarterback can play in the game.
3. For $P(A \mid B)$, we assume that the event B has occurred, and now need to find the probability of event A, given event B. On the other hand, for $P(A \cap B)$, we do not assume that event B has occurred, and instead need to determine the probability that both events occurred.
5. $A \cap B$ is a subset of A, $P(A \cap B) \le P(A)$, and $A \cap B$ is a subset of B, $P(A \cap B) \le P(B)$. Therefore, $P(A \cap B)$ cannot equal 0.30 if $P(A) = 0.25$ and $P(B) = 0.25$.
7. 0.24   **9.** 0.4   **11.** 0.1   **13.** 0.2
15. 0.1667   **17.** 1/8   **19.** 1/32   **21.** 0.2
23. 0.5   **25.** 0.8   **27.** 0.7
29. Yes. $P(C \text{ and } D) = 0.21 = (0.7)(0.3) = P(C) P(D)$, $P(C \mid D) = 0.7 = P(C)$, and $P(D \mid C) = 0.3 = P(D)$.
31. $P(E \text{ and } F)$ **33.** 1/6   **35.** 1/2
37. They are independent.
39. (a) 1/6 **(b)** 1/3 **(c)** P (rolling a 6) = 1/6; P (rolling a 6 | odd number) = 0
41. (a) 1/3 **(b)** 1/10 **(c)** 1/6 **(d)** 1/15
43. No; $P(C) = 1/2 \ne 5/9 = P(C \mid F)$, $P(C) = 1/2 \ne 5/12 = P(C \mid M)$, $P(F \text{ and } C) = 1/3 \ne 3/10 = P(F) P(C)$, and $P(M \text{ and } C) = 1/6 \ne 1/5 = P(M) P(C)$.
45. (a) 16/49 **(b)** 2/7
47. (a) 0.000361 **(b)** 0.962361 **(c)** 0.9085 **(d)** 0.0559
49. (a) 0.055 **(b)** 0.3025 **(c)** 0.55 **(d)** 0.55 **(e)** 0.003025 **(f)** It is not possible since we are given only the percent of female students that are business majors, not the percent of all students that are business majors.

Section 5.4

1. Tree diagram
3. In a permutation, order is important. In a combination, order is not important.
5. Answers will vary.
7. 720   **9.** 1   **11.** 1
13. 210   **15.** 6720   **17.** 100

19. 100!

21. 35   **23.** 165   **25.** 11   **27.** 1

29. $_7C_3 = \dfrac{7!}{3! \cdot 4!} = \dfrac{7!}{4! \cdot 3!} = {_7C_4}$

31. {Amy, Bob, Chris}, {Amy, Chris, Bob}, {Bob, Amy, Chris}, {Bob, Chris, Amy}, {Chris, Amy, Bob}, {Chris, Bob, Amy}, {Amy, Bob, Danielle}, {Amy, Danielle, Bob}, {Bob, Amy, Danielle}, {Bob, Danielle, Amy}, {Danielle, Amy, Bob}, {Danielle, Bob, Amy}, {Amy, Chris, Danielle}, {Amy, Danielle, Chris}, {Chris, Amy, Danielle}, {Chris, Danielle, Amy}, {Danielle, Amy, Chris}, {Danielle, Chris, Amy}, {Bob, Chris, Danielle}, {Bob, Danielle, Chris}, {Chris, Bob, Danielle}, {Chris, Danielle, Bob}, {Danielle, Bob, Chris}, {Danielle, Chris, Bob}. $_4P_3 = 24$

33. {Amy, Bob, Chris}, {Amy, Chris, Bob}, {Chris, Amy, Bob}, {Chris, Bob, Amy}, {Bob, Amy, Chris}, and {Bob, Chris, Amy} are all different permutations but the same combination.

35. $r!$

37. (a)

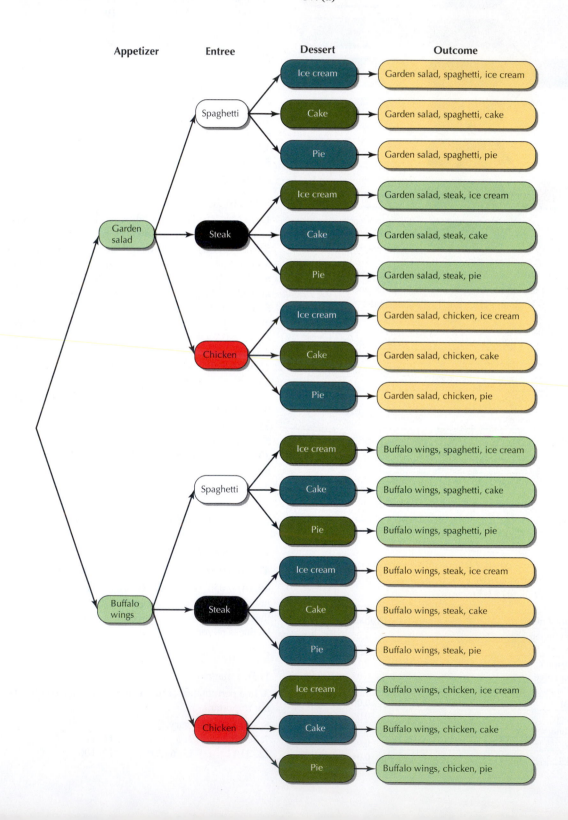

(b) 18

39. 362,880

41. 120

43. 720

45. 20

47. 300

49. 20

51. 184,756

53. 6720

55. (a) Combination; order is unimportant. (b) 260 (c) 0.0640

57. (a) 0.0265 (b) 0.0013 (c) 0.2652 (d) 0.4419

Chapter 5 Review

1. 3/8

3. 0

5. 1/2

7. (a) 0.213 (b) 0.656 (c) 0

9. (a) Yes, since no student in the sample repeated twelfth grade.
(b) 0.0290 (c) 0.0150

11. (a) 0.9269 (b) 0 (c) 1

13. (a) 0.3729 (b) 0.6271

15. (a) 1/2 (b) 1/3

17. (a) 5/18 (b) 5/12

19. 34,650

21. (a) Combination; order is not important. (b) 3 (c) 0.0196.

Chapter 5 Quiz

1. False

2. True

3. 0, 1

4. or, and

5. 0.5

6. 1

7. With replacement

8. Intersection of A and B.

9. (a) 1/36 (b) 1/36 (c) 5/18 (d) 1/6

10. (a) 1/9 (b) 8/9 (c) 5/18 (d) 1/18 (e) 1/3

11. 0.2

12. 0.2125

13. (a) 1/4 (b) 3/13 (c) 1/13 (d) 1/2 (e) 1/52 (f) 1/26

14. (a) 0.3830 (b) 0.1654 (c) 0.1075 (d) 0.4409

15. (a) 0.5361 (b) 0.4639 (c) 0.0330

16. (a) 0.0215 (b) 0.0114

17. (a) 0.0402 (b) 0.0246

18. No, P (Not too happily married) $= 0.0330 \neq 0.0402 = P$ (Not too happily married | Female) and P (Not too happily married) $= 0.0330 \neq 0.0246 = P$ (Not too happily married | Male)

19. 4

20. (a) Permutation; the order in which the numbers are selected is important. (b) 6840 (c) 1/6840

CHAPTER 6

Section 6.1

1. Answers will vary.

3. Discrete: takes finite or a countable number of values that can be graphed as separate points on the number line; continuous takes infinitely many values that form an interval on the number line.

5. Discrete

7. Continuous

9. Discrete

11. {0, 1, 2, 3, 4, 5, 6, 7, 8, 9, 10, 11, 12, 13, 14, 15}

13. {0, 1, 2, 3, 4}

15. No, the probabilities don't add up to 1.

17. No, $P(X = 1)$ is negative.

19. (a) 0.30 (b) 0.06 (c) 0

21. (a) 2/3 (b) Gaining $10,000 (c) 1/3

23. Mean: $\mu = 3$; $Var(X) = 0.5$; $SD(X) = 0.7071$; yes

25. Mean: $\mu = 3.1$; $Var(X) = 1.29$; $SD(X) = 1.1358$; yes

27. (a)

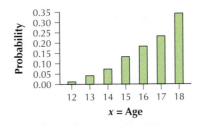

(b) About 16 years (c) 16.48 years (d) The "typical" age of a young person who has tried tobacco products in the last month is 16.48 years.

29. (a)

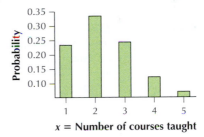

(b) 2 classes; 5 classes (c) $\mu = E(X) = 2.458$ classes (d) The "typical" number of courses taught by faculty at all degree-granting institutions of higher learning in the United States is 2.458.

31. (a) $SD(X) = 1.1689$ classes
(b) Z-score is 2.17; moderately unusual.

33. (a)

(b) 1 vehicle; 0 vehicles (c) $\mu = 1.547$ vehicle (d) The "typical" number of vehicles owned by residents of Florida is 1.547.

35.

$X =$ sum of dice	2	3	4	5	6	7	8	9	10	11	12
$P(X)$	1/36	1/18	1/12	1/9	5/36	1/6	5/36	1/9	1/12	1/18	1/36

37. 7. The estimate is equal to the actual value.

39.

x = Sum of dice

41. (a) No. The mean is 2 but the most likely value is 0.

X	0	2	8
P (X)	0.6	0.2	0.2

(b) Symmetric, one mode

Section 6.2

1. Not binomial; the events "Person A comes to party" and "Person B comes to party" may not be independent.

3. Not binomial; more than two different ethnicities.

5. Not binomial; more than two possible total number of spots.

7. Binomial; $n = 3$, $X =$ number of even numbers, $p = 1/2$, $1 - p = 1/2$.

9. Binomial; $n = 2$, $X =$ number of games won, $p = 0.25$, $1 - p = 0.75$

11. $1/8 = 0.125$

13. $3/8 = 0.375$

15. 0.2461

17. 0.1216

19. $\mu = 10$, $\sigma^2 = 5$, $\sigma = 2.2361$. The expected number of heads in 20 tosses of a fair coin is 10.

21. 0.0741

23. 0.1563

25. $\mu = 2.5$, $\sigma^2 = 1.25$, $\sigma = 1.1180$. The expected number of cars obeying the speed limit in a random sample of 5 cars on the Interstate is 2.5.

27. 0.2668

29. 0.0170

31. $\mu = 3$, $\sigma^2 = 2.55$, $\sigma = 1.5969$. The expected number of people who will support an Independent candidate for president in the next election in a random sample of 20 people is 3.

33. (a) 0.2581 **(b)** 0.6489 **(c)** 3 sophomores

35. (a) $\mu = 3$; $\sigma^2 = 2.25$; $\sigma = 1.5$. The expected number of sophomores in a random sample of 12 statistics students is 3. **(b)** Z-score is –2; moderately unusual.

37. (a) 0.1171 **(b)** 0.2447 **(c)** 8 women

39. (a) 0.2496 **(b)** 9 women **(c)** $\mu = 8.7$ women. The expected number of women who reported that "Music I Like" is the biggest factor in deciding which radio station to listen to in a random sample of 10 women is 8.7.

41. (a) $\sigma^2 = 1.131$; $\sigma = 1.0635$. **(b)** Z-score is −6.30; unusual.

43. (a) 0.0795 **(b)** 0.2414 **(c)** 61 women **(d)** $\mu = 61$ women. The expected number of women with access to the Internet in a random sample of 100 women is 61.

45. (a) $\sigma^2 = 23.79$; $\sigma = 4.8775$ **(b)** Z-score is −2.46; moderately unusual.

47. (a) 0.3337 **(b)** 12 women **(c)** $\mu = 12$ women. The expected number of women who are affected by a depressive disorder in a random sample of 100 women is 12.

49. (a) $\sigma^2 = 10.56$; $\sigma = 3.2496$ **(b)** Z-score is −0.62; not unusual.

51. (a) 0.0311 **(b)** 68 music downloaders **(c)** $\mu = 68$ music downloaders. The expected number of music downloaders who say that they are using paid services in a sample of 400 music downloaders is 68.

53. (a) $\sigma^2 = 56.44$; $\sigma = 7.5127$ **(b)** Z-score is 2.93; moderately unusual.

Section 6.3

1. The probability that X equals some particular value is zero.

3. Area under the normal distribution curve above an interval.

5. 0

7. 0.5

9. Greater than 0.5. Since $X = 4285$ is greater than the mean of 3285 and the area to the left of $\mu = 3285$ is 0.5, the area to the left of $X = 4285$ is greater than the area to the left of $X = 3285$.

11. About 0.997.

13. A has mean 10; B has mean 25. The peak of a normal curve is at the mean; from the graphs we see that the mean of A is less than the mean of B.

15. $\mu = 0$, $\sigma = 1$

17. $\mu = -10$, $\sigma = 10$

19. $\mu = 100$, $\sigma = 12.5$

21. $\mu = 0$, $\sigma = 2$

23. (a)

x = Wind speed (in mph)

(b) About 0.68

25. (a)

x = Number of viewers (in millions)

(b) About 0.025

27. (a)

x = Temperature (in degrees Fahrenheit)

(b) About 0.95

29. About 0.135

31. About 0.025

33. There is no minimum value of X for a normal distribution, but birth weights can't be less than 0.

Section 6.4

1. $\mu = 0$

3. True

5. (a) Greater than 0.5 **(b)** 0.9750 **(c)** 0.9750 is greater than 0.5

7. (a) Less than 0.5 **(b)** 0.4821 **(c)** 0.4821 is less than 0.5

9. (a) Less than 0.5 **(b)** 0.0179 **(c)** 0.0179 is less than 0.5

11. (a) Greater than 0.5 **(b)** 0.8020 **(c)** 0.8020 is greater than 0.5

13. (a) Less than 0.5 **(b)** 0.1832 **(c)** 0.1832 is less than 0.5

15. (a)

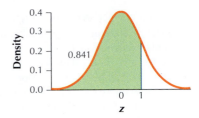

(b) Greater than 0.5 **(c)** 0.8413 **(d)** 0.8413 is greater than 0.5

17. (a)

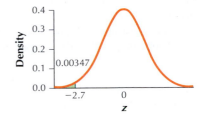

(b) Less than 0.5 **(c)** 0.0035 **(d)** 0.0035 is less than 0.5

19. (a)

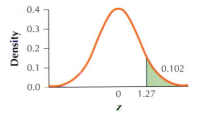

(b) Less than 0.5 **(c)** 0.1020 **(d)** 0.1020 is less than 0.5

21. (a)

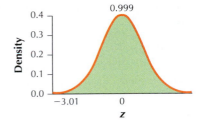

(b) Greater than 0.5 **(c)** 0.9987 **(d)** 0.9987 is greater than 0.5

23. (a)

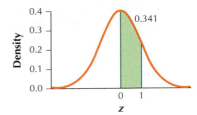

(b) Less than 0.5 **(c)** 0.3413 **(d)** 0.3413 is less than 0.5

25. (a)

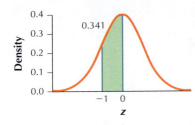

(b) Less than 0.5 **(c)** 0.3413 **(d)** 0.3413 is less than 0.5

27. (a)

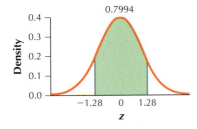

(b) Greater than 0.5 **(c)** 0.7994 **(d)** 0.7994 is greater than 0.5

29. (a)

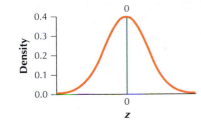

(b) 0 **(c)** 0 **(d)** the estimate equals the actual value

31. (a)

(b) 1 **(c)** 1 **(d)** the estimate equals the actual value

33. (a)

(b) Less than 0.5 **(c)** 0.0150 **(d)** 0.0150 is less than 0.5

35. (a)

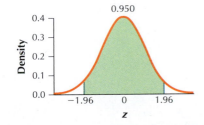

(b) Greater than 0.5 **(c)** 0.9500 **(d)** 0.9500 is greater than 0.5

37. (a)

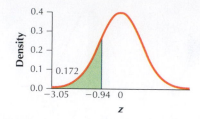

(b) Less than 0.5 **(c)** 0.1725 **(d)** 0.1725 is less than 0.5

39. (a)

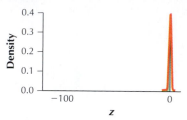

(b) About 0.5 **(c)** 0.5000 **(d)** The estimate equals the actual value.

41. Less than 0; $Z = -0.43$; -0.43 is less than 0
43. Less than 0; $Z = -0.45$; -0.45 is less than 0
45. Less than 0; $Z = -0.87$; -0.87 is less than 0
47. Less than 0; $Z = -2.03$; -2.03 is less than 0
49. $Z = 0$
51. $Z = 2.58$
53. The area between $Z = -2$ and $Z = 2$ is 0.9544. By the Empirical Rule, the area between $Z = -2$ and $Z = 2$ is about 0.95.
55. (a) 0.0668 **(b)** 0.9332 **(c)** 0.8664
57. $Z = -2.58$ and $Z = 2.58$.
59. -0.67; 0; 0.67

Section 6.5

1. To standardize things means to make them all the same, uniform, or equivalent. To standardize a normal random variable X, we transform X into the standard normal random variable Z using the formula $Z = \dfrac{X - \mu}{\sigma}$. We do this so that we can use the standard normal table to find the probabilities.
3. You would get the wrong probability.
5. 0.5
7. 0.8413
9. 0.0062
11. 0.9332
13. 0.8400
15. 0.0049
17. $X = 86.45$
19. $X = 82.8$
21. $X = 50.4$ and $X = 89.6$.
23. Appropriate
25. Not appropriate
27. Not appropriate
29. 0.1272
31. 0.4364
33. 0.4364
35. 0.3616
37. 0.0992
39. 0.6772
41. 0.0853
43. 0.0992
45. (a) 0.1359 **(b)** 0.0228
47. (a) 0.1423 (TI: 0.1431) **(b)** 0.1423 (TI: 0.1431) **(c)** 26.67% (TI: 26.47%)

49. (a) $X = 27.6$ mph **(b)** Z-score is -2.27; moderately unusual
51. (a) 0.3821 **(b)** 1.39% **(c)** 0.1151
53. (a) $X = 130.55$ **(b)** The Z-score for $X = 118$ is -2.9. Since $-3 < -2.9 \leq -2$, a score of 118 on the test is considered moderately unusual.
55. (a) 0.0139 **(b)** 18.41% **(c)** 0.5249
57. (a) 4.87 days; the mean equals the median. **(b)** The Z-score for $X = 8$ days is 3.23. Since $3.23 \geq 3$, a hospital stay of 8 days is unusual.
59. (a) 0.0228 **(b)** 21.19% **(c)** 0.3056 (TI: 0.3057)
61. (a) $X = 1.78$ **(b)** The Z-score for $X = 4.0$ is 2.8. Since $2 \leq 2.8 < 3$, a GPA of 4.0 is moderately unusual.
63. (a) 0.7506 **(b)** 0.2963 **(c)** 0.7734 **(d)** \$24,880 (TI: \$24,873.79) **(e)** \$26,640 and \$33,360 (TI: \$26,633.52 to \$33,366.48)
65. (a) -1.04 calories per gram **(b)** No, food cannot have negative calories per gram. **(c)** Yes **(d)** The distribution of the number of calories per gram is right-skewed, so it is not normal.
67. (a) 0.0026 **(b)** 0.2313 **(c)** 80.40 and 119.60
69. (a) 0.7157 **(b)** 0.0548 **(c)** 0.0003 **(d)** 0.0951
71. (a) 0.1350 **(b)** 0.5675 **(c)** 0.4325
73. (a) No **(b)** The normal distribution is not a good approximation to the binomial distribution ($n \cdot p = 2 < 10$), so not appropriate.

Chapter 6 Review

1. (a) 0.95 **(b)** 2 games **(c)** 0.60
3. (a) $\mu = \$1500$, $Var(X) = 60,250,000$, $SD(X) = \$7762.09$
(b)

$x = $ Profit

(c) This value for the mean makes sense as the point where the distribution balances.
5. (a) 0.1887 **(b)** 0.8113 **(c)** 0.7880
7. (a) 0.2123 **(b)** 0.1792 **(c)** 0.5025
9. $\mu = 5.7$, Var $(X) = 2.451$, $SD(X) = 1.5656$. The expected number of females in a random sample of 10 students from the College of DuPage is 5.7.
11. 0
13. 0.5
15. Greater than 0.5. Since the area to the left of $\mu = 106$ mm is 0.5 and $X = 110$ mm is greater than the mean $\mu = 106$ mm, the area to the left of $X = 110$ mm is greater than the area to the left of the mean $\mu = 106$.
17. About 0.95
19. (a)

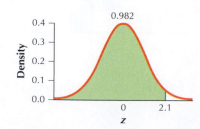

(b) Greater than 0.5 **(c)** 0.9821 **(d)** 0.9821 is greater than 0.5.

21. (a)

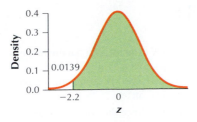

(b) Greater than 0.5 **(c)** 0.9861 **(d)** 0.9861 is greater than 0.5.

23. (a)

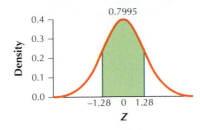

(b) Greater than 0.5 **(c)** 0.7995 **(d)** 0.7995 is greater than 0.5.
25. (a) 0.0736 **(b)** 0.3821 **(c)** 0.3085 **(d)** 0.6915 **(e)** 0.6179
(f) 0.3557
27. (a) 0.1192 **(b)** 0.0003 **(c)** 0.4404 **(d)** 0.5596 **(e)** 0.4404
(f) 0.6157
29. (a) 0.1587 **(b)** 0.2018 **(c)** $X = 17$ **(d)** Z-score is -3; unusal.
The producer will have to file a report.

Chapter 6 Quiz

1. True
2. False
3. False
4. 0.5
5. 0
6. 0
7. discrete
8. binomial
9. $\mu = 0, \sigma = 1$
10. (a)

X = Amount won	0	5
P(X)	13/18	5/18

(b) \$1.39 **(c)** \$1.39
11. (a) 0.0962 **(b)** 19 CEOs **(c)** $\mu = 19$ CEOs, $Var(X) = 15.39$, $SD(X)$ = 3.9230. The expected number of CEOs who drive luxury cars in a random sample of 100 CEOs is 19. **(d)** Z-score is 5.35; unusual
12. (a) 0.1003 **(b)** 33.22% **(c)** \$4329.50 **(d)** Z-score is -2.05; moderately unusual
13. (a) 0.0250 **(b)** 0.0082 **(c)** For cats, $\mu = 314$ households, $Var(X) = 215.404$, $SD(X) = 14.6766$. The expected number of American households that have a cat in a random sample of 1000 households is 314. For dogs, $\mu = 343$ households, $Var(X) =$ 225.351, $SD(X) = 15.0117$. The expected number of American households that have a dog in a random sample of 1000 households is 343. **(d)** Z-score is -1.64; not unusual
14. (a) 0.4207 **(b)** 0.2358 **(c)** $X = 50,800$ visitors **(d)** Z-score is -2.3; moderately unusual

CHAPTER 7

Section 7.1

1. *Sampling error* is the distance between the point estimate and its target parameter. It measures how far the point estimate misses the actual target parameter. In the real world, the only time we know

the actual value of the sampling error is when we know the true value of the target parameter. Usually, the true value of the target parameter is unknown.
3. For a given sample size n, it is normal with mean μ and standard deviation $\sigma_{\bar{x}} = \sigma/\sqrt{n}$.
5. 5
7. 5
9. 0.05
11. $\mu_{\bar{x}} = 100, \sigma_{\bar{x}} = 4$
13. $\mu_{\bar{x}} = 0, \sigma_{\bar{x}} = 3.3333$
15. $\mu_{\bar{x}} = -10, \sigma_{\bar{x}} = 0.5$
17. Normal with mean $\mu_{\bar{x}} = \mu = 10$ and standard deviation $\sigma_{\bar{x}} = 1$
19. 0.1587
21. 11.96
23. 0.95
25. (a) $\bar{x} = 3021$ pounds **(b)** Probably not; different samples may have different average weights. **(c)** Not likely; since samples are subsets of the population, they are not perfect representations of the population.
27. (a) 2930.34 pounds **(b)** 90.66 pounds
29. (a) 0.1587 **(b)** 0.9974 **(c)** 0.1587
31. (a) 404.20 **(b)** 325.80 **(c)** The middle 95% of sample mean carbon dioxide concentrations of samples of size $n = 25$.
33. (a) $_5C_2 = 10$ samples of size $n = 2$ **(b)** $\mu = 64.12$. No; we don't usually know all values from the population. **(c)** $\sigma = 2.9991$.
35. (a–b) About 64.1
37. (a) 3.5 **(b)** 0.4270
39. (a) $\mu_{\bar{x}} = 515$ **(b)** $\sigma_{\bar{x}} = 29$ **(c)** Since the sampling distribution of math SAT scores is normal with mean $\mu = 515$ and standard deviation $\sigma = 116$, the sampling distribution of the sample mean is normal with a mean of $\mu_{\bar{x}} = 515$ and a standard deviation of $\sigma_{\bar{x}} = 29$, by Fact 3.
41. (a–b) Increase **(c)** Decrease
43. The sampling error decreases as the sample size increases.
45. The sampling error decreases as the sample size increases.

Section 7.2

1. Symmetric; answers will vary.
3. The variability of the sampling distribution becomes smaller as the sample size becomes larger.
5. Case 1
7. Case 1
9. Case 2
11. (a) 516 **(b)** 38.6667 **(c)** Normal
13. (a) \$60,000 **(b)** \$2500 **(c)** Normal
15. (a) 80 **(b)** 1 **(c)** Approximately normal
17. Not possible since the exchange rate for US \$100 in euros for 2007 is not normally distributed and the sample size of $n = 9$ is small ($n < 30$). The sampling distribution of $\bar{x}$ is unknown by Case 3.
19. 0.0918
21. 0.9544
23. 579.61
25. \$56,800
27. Not possible since the pollen count distribution for Los Angeles in September is not normally distributed and the sample size of $n = 16$ is small ($n < 30$). Case 3 applies and the sampling distribution of $\bar{x}$ is unknown.
29. No, 0.0918 (TI: 0.0912)
31. (a) $P(\bar{x} < 100) = 0$. **(b)** 1
33. For $x = 30$
35. 0
37. (a) Unchanged **(b)–(d)** Decrease

Section 7.3

1. Example 7.15 shows that a sample of size $n = 30$ may not be enough for the sampling distribution of $\hat{p}$ to be approximately normal.

3. (a) 0.5 (b) 0.05 (c) Approximately normal

5. (a) 0.01 (b) 0.0099 (c) Unknown

7. 10

9. 50

11. 500

13. 0.1587

15. Not possible; sampling distribution of $\hat{p}$ is unknown.

17. Not possible; sampling distribution of $\hat{p}$ is unknown.

19. 0.564

21. 0.962

23. 0.052

25. (a) $\mu_{\hat{p}} = 0.25$ (b) $\sigma_{\hat{p}} = 0.0433$ (c) Approximately normal

27. (a) 0.1251 (b) 0.321

29. (a) $np = (100)(0.61) = 61 \geq 5$ and $n(1 - p) = 100(1 - 0.61) = 39 \geq 5$. (b) $\mu_{\hat{p}} = 0.61$ and $\sigma_{\hat{p}} = 0.0488$. (c) 0.4207

31. (a) $n = 13$ (b) $np = (13)(0.40) = 5.2 \geq 5$ and $n(1 - p) = (13)(1 - 0.40) = 7.8 \geq 5$. (c) The sampling distribution of $\hat{p}$ is approximately normal with mean $\mu_{\hat{p}} = 0.40$ and standard deviation $\sigma_{\hat{p}} = 0.1359$. The Central Limit Theorem for Proportions allows us to say this. (d) $\mu_{\hat{p}} = 0.40$ and $\sigma_{\hat{p}} = 0.1095$. (e) 0.4286 (TI: 0.4276)

33. (a) An increase in the sample size n will result in a decrease of $\sigma_{\hat{p}}$. (b) Any value greater than 20

Chapter 7 Review

1. 0

3. 0.1

5. $\mu_{\bar{x}} = 50$ and $\sigma_{\bar{x}} = 10$.

7. 0.1056

9. 0.7888

11. 0.8185

13. 100

15. We have $np = (40)(0.1) = 4 < 5$, so the sampling distribution of $\hat{p}$ is unknown. Thus $P(\hat{p} < 0.12)$ can't be found.

17. 0.0153

19. (a) We have $np = (120)(0.11) = 13.2 \geq 5$ and $n(1 - p) = (120)(1 - 0.11) = 106.8 \geq 5$, so the sampling distribution of $\hat{p}$ is approximately normal. (b) $\mu_{\hat{p}} = 0.11$ and $\sigma_{\hat{p}} = 0.0286$. (c) 0.1762

Chapter 7 Quiz

1. True

2. False

3. Sampling error

4. Approximately normal

5. No

6. $np \geq 5$ and $n(1 - p) \geq 5$

7. (a) 0.1587 (b) 0.9500 (c) 0.1056

8. (a) 45.15 grams (b) 34.85 grams (c) 34.85 grams and 45.15 grams

9. (a) 0.0228 (b) 0.0228 (c) 0.9544

10. (a) 68.77 inches (b) 67.23 inches (c) 67.23 and 68.77 inches

11. (a) We have $np = (100)(0.12) = 12 \geq 5$ and $n(1 - p) = (100)(1 - 0.12) = 88 \geq 5$, so the sampling distribution of $\hat{p}$ is approximately normal. (b) $\mu_{\hat{p}} = 0.12$ and $\sigma_{\hat{p}} = 0.0325$. (c) 0.7324

12. (a) 0.0563 (b) 0.1837 (c) 0.0563 and 0.1837.

13. (a) We have $np = (400)(0.17) = 68 \geq 5$ and $n(1 - p) = (400)(1 - 0.17) = 332 \geq 5$, so the sampling distribution of $\hat{p}$ is approximately normal. (b) $\mu_{\hat{p}} = 0.17$ and $\sigma_{\hat{p}} = 0.0188$. (c) 0

14. (a) 0.1391 (b) 0.2009 (c) 0.1391 and 0.2009.

CHAPTER 8

Section 8.1

1. A *confidence interval estimate* of a parameter consists of an interval of numbers generated by a point estimate, together with an associated *confidence level* specifying the probability that the interval contains the parameter.

3. (a) $Z_{\alpha/2}$ increases. (b) Since the confidence level is $(1 - \alpha) \times 100\%$, as the confidence level increases, $1 - \alpha$ increases. Thus α and $\alpha/2$ will decrease. Since $\alpha/2$ is the area underneath the standard normal curve to the right of $Z_{\alpha/2}$, a decrease in $\alpha/2$ will result in an increase in $Z_{\alpha/2}$.

5. Increases, Decreases

7. (a) No (b) Yes (c) Yes

9. $Z_{\alpha/2} = 2.576$

11. $Z_{\alpha/2} = 2.576$

13. $Z_{\alpha/2} = 1.645$

15. $Z_{\alpha/2} = 1.645$

17. (a) 0.5 (b) $Z_{\alpha/2} = 1.96$ (c) $E = 0.98$. We can estimate μ to within 0.98 with 95% confidence. (d) (9.02, 10.98). We are 95% confident that the true mean μ lies between 9.02 and 10.98.

19. (a) 0.5 (b) $Z_{\alpha/2} = 1.96$ (c) $E = 0.98$. We can estimate μ to within 0.98 with 95% confidence. (d) (34.02, 35.98). We are 95% confident that the true mean μ lies between 34.02 and 35.98.

21. (a) 2 (b) $Z_{\alpha/2} = 1.96$ (c) $E = 3.92$. We can estimate μ to within 3.92 with 95% confidence. (d) (96.08, 103.92). We are 95% confident that the true mean μ lies between 96.08 and 103.92.

23. (a) (9.342, 10.658). We are 90% confident that the true mean μ lies between 9.342 and 10.658. (b) (9.216, 10.784). We are 95% confident that the true mean μ lies between 9.216 and 10.784. (c) (8.9696, 11.0304) (TI: (8.9697, 11.03)). We are 99% confident that the true mean μ lies between 8.9696 and 11.0304 (TI: 8.9697 and 11.03). (d) The confidence interval for a given sample size becomes wider as the confidence level increases.

25. 1

27. 5

29. 5

31. 11

33. (a) 69 gallons (b) 3.65 gallons (c) $Z_{\alpha/2} = 1.96$ (d) 7.16 gallons. We can estimate μ to within 7.16 gallons with 95% confidence. (e) (61.84, 76.16). We are 95% confident that μ lies between 61.84 gallons and 76.16 gallons.

35. (a) 107 seconds (b) 19.5 seconds (c) $Z_{\alpha/2} = 1.96$ (d) $E = 38.22$ seconds. We can estimate μ, the mean length of time that boys remain engaged with a science exhibit at a museum, to within 38.22 seconds with 95% confidence. (e) (68.78, 145.22). We are 95% confident that the true mean length of time that boys remain engaged with a science exhibit at a museum μ lies between 68.78 seconds and 145.22 seconds.

37. (a) 3 (b) 62

39. (a) 59 (b) 9604 days, approximately 26.31 years

41. (a) 7 bits (b) 2 bits (c) $Z_{\alpha/2} = 1.96$ (d) 1 bit (e) $\bar{x} \pm E = 7 \pm 2 = (5, 9)$. We are 95% confident that μ lies between 5 bits and 9 bits.

43. (a) \$3325.60. We can estimate μ to within \$3325.60 with 99% confidence. (b) (46674.40, 53325.60). We are 99% confident that μ lies between \$46,674.40 and 53,325.60.

45. (a) $Z_{\alpha/2} = 1.96$ (b) 5.91 (c) 3.1948. We can estimate the true mean quality of life score within 3.1948 with 95% confidence. (d) 9.78

47. (a) 97 (b) 385

49. Answers will vary.

51. Answers will vary.

Section 8.2

1. In most real-world problems, the population standard deviation σ is unknown, so we can't use the Z interval.

3. The t curve approaches closer and closer to the Z curve.

5. Conservative: $t_{\alpha/2} = 2.009$. Interpolation: $t_{\alpha/2} = 2.0054$.

7. Conservative: $t_{\alpha/2} = 1.684$. Interpolation: $t_{\alpha/2} = 1.680$.

9. (a) $t_{\alpha/2} = 3.182$ **(b)** $E = 9.546$. **(c)** $(40.454, 59.546)$

11. (a) $t_{\alpha/2} = 1.860$ **(b)** $E = 3.720$. **(c)** $(-23.720, -16.280)$

13. (a) $t_{\alpha/2} = 2.660$ **(b)** $E = 2.660$ **(c)** $(32.340, 37.660)$

15. (a) $t_{\alpha/2} = 1.664$ **(b)** $E = 1.1093$ **(c)** $(-21.1093, -18.8907)$

17. Since the distribution of the population is unknown, Case 1 does not apply. Since the sample size of $n = 25$ is small ($n < 30$), Case 2 does not apply. Thus we cannot construct the indicated confidence interval.

19. $(9.3387, 10.6613)$

21. Since the distribution of the population is unknown, Case 1 does not apply. Since the sample size of $n = 16$ is small ($n < 30$), Case 2 does not apply. Thus we cannot construct the indicated confidence interval.

23. $(47.97, 52.03)$

25. (a) $t_{\alpha/2} = 1.994$ **(b)** $E = \$6.91$. We can estimate μ, the true mean revenue collected from all parking meters, within $\$6.91$ with 95% confidence. **(c)** $(113.09, 126.91)$. We are 95% confident that the true mean revenue collected from all parking meters μ lies between $\$113.09$ and $\$126.91$.

27. (a) See the graph. All the data points lie between the curved lines. In fact all the points lie close to the center line. Thus the distribution appears to be normal. **(b)** $t_{\alpha/2} = 2.015$ **(c)** $E = 3.276$ miles per gallon. We can estimate μ, the true mean city gas mileage for hybrid cars, within 3.276 miles per gallon with 90% confidence. **(d)** $(27.057, 33.609)$. We are 90% confident that the true mean city gas mileage for hybrid cars μ lies between 27.057 miles per gallon and 33.609 miles per gallon.

29. (a)

(b) Yes, the points do not appear to lie in a straight line. **(c)** Since the data do not appear to be normal, Case 1 does not apply. Since the sample size of $n = 12$ is small ($n < 30$), Case 2 does not apply. Thus a t interval cannot be used.

31. (a) $t_{\alpha/2} = 2.045$ **(b)** $E = 1.1003$ miles. We can estimate μ, the true mean commuting distance, within 1.1003 miles with 95% confidence. **(c)** $(8.3330, 11.0336)$. We are 95% confident that the true mean commuting distance lies between 8.3330 miles and 11.0336 miles.

33. (a) t interval: $(29.6671, 54.3329)$. Z interval $= (32.2248, 51.7752)$. We are 99% confident that the population mean weekly number of HIV-related deaths in all large American cities lies between 29.6671 and 54.3329. **(b)** Lower the confidence level or increase the sample size.

35. (a) The histogram indicates that the data are left-skewed and therefore not normally distributed. The normal probability plot indicates that the data points all lie between the curved bounds but do not all lie along a straight line, so the normality assumption may not be valid. **(b)** $n = 10, \bar{x} = -0.306, s = 0.3435$ **(c)** $t_{\alpha/2} = 2.262$ **(d)** $(-0.5517, -0.0603)$ **(e)** $E = 0.2457$. We can estimate μ, the true mean net change of all NYSE stocks, within 0.2457 with 95% confidence.

37. (a) Except for one outlier on the left, the data appear to be normally distributed. **(b)** $(28.9902, 37.7098)$. We are 95% confident that the true mean rate of minority mortgage lending refusals for all banks lies between 28.9902% and 37.7098%.

39. (a) Since the margin of error is $E = t_{\alpha/2}(s/\sqrt{n})$, a decrease in the confidence level will result in a decrease in $t_{\alpha/2}$, which will result in a decrease in the margin of error. **(b)** Since the width of the confidence interval is $2E$, a decrease in the confidence level will result in a decrease in E, which will result in a decrease in the width of the confidence interval.

Section 8.3

1. No, unless there is some reason to suspect that the value of p has changed.

3. (a) $(0.304, 0.696)$

(b)

5. (a) $(0.8939, 1.0061)$

(b)

7. (a) $(0.280, 0.520)$

(b)

9. (a) $(0.1923, 0.4077)$

(b)

11. (a) $Z_{\alpha/2} = 1.96$ **(b)** Yes **(c)** 0.0535 **(d)** $(0.0465, 0.1535)$

13. (a) $Z_{\alpha/2} = 1.645$ **(b)** Yes **(c)** 0.2056 **(d)** $(0.2944, 0.7056)$

15. (a) $Z_{\alpha/2} = 1.96$ **(b)** No **(c)–(d)** Not applicable.

17. (a) $Z_{\alpha/2} = 1.645$ **(b)** Yes **(c)** 0.0979 **(d)** $(0.0521, 0.2479)$

19. (a) 0.3099 **(b)** 0.098 **(c)** 0.0310 **(d)** 0.0098

21. (a) Since the margin of error is $E = Z_{\alpha/2} \cdot \sqrt{\hat{p}(1 - \hat{p})/n}$, an increase in the sample size while $\hat{p}$ remains constant results in a decrease in the margin of error. **(b)** Since the width of the confidence interval is $2E$, an increase in the sample size while $\hat{p}$ remains constant results in a decrease in the width of the confidence interval.

23. 897.

25. 897.

27. 43.

29. Decreases.

31. 1068.

33. 4269.

35. 68,296.

37. Increases by about a factor of 4.

39. (a) 1.96 **(b)** We have $n\hat{p} = 12(0.38) = 4.56 < 5$, so we cannot use the Z interval for p. **(c)** From **(b)**, it is not appropriate to calculate the margin of error. **(d)** We have $n\hat{p} = 12(0.38) = 4.56 < 5$, so we cannot use the Z interval for p.

41. (a) $Z_{\alpha/2} = 1.96$ **(b)** Yes **(c)** $E = 0.0304$. We can estimate p, the true proportion of NASCAR racing attendees who own a pickup truck, within 0.0304 with 95% confidence. **(d)** (0.3696, 0.4304). We are 95% confident that the true proportion of NASCAR racing attendees who own a pickup truck lies between 0.3696 and 0.4304.
43. (a) Yes. **(b)** $E = 0.0353$. We can estimate p, the true proportion of all white twelfth-graders who are using Ecstasy, within 0.0353 with 99% confidence. **(c)** (0.0037, 0.0743). We are 99% confident that the true proportion of all white twelfth-graders who are using Ecstasy lies between 0.0037 and 0.0743.
45. (a) 271 **(b)** 6766.
47. 1770.
49. (a) 0.0318. We can estimate p within 0.0318 with 99% confidence. **(b)** (0.1838, 0.2474)
51. (0.5565, 0.7435)
53. (a)–(c) Increases
55. (a) Decrease **(b)** Decrease **(c)** Decrease
57. (0.7198, 0.8694)

Section 8.4

1. False. The χ^2 curve is *not* symmetric. It is right-skewed.
3. True
5. $\chi^2_{1-\alpha/2} = \chi^2_{0.95} = 13.848$ and $\chi^2_{\alpha/2} = \chi^2_{0.05} = 36.415$.
7. $\chi^2_{1-\alpha/2} = \chi^2_{0.995} = 9.886$ and $\chi^2_{\alpha/2} = \chi^2_{0.005} = 45.559$.
9. $\chi^2_{1-\alpha/2} = \chi^2_{0.975} = 5.629$ and $\chi^2_{\alpha/2} = \chi^2_{0.025} = 26.119$.
11. For a given sample size, $\chi^2_{1-\alpha/2}$ decreases and $\chi^2_{\alpha/2}$ increases as the confidence level increases.
13. Lower bound = 6.59, upper bound = 17.33
15. Lower bound = 5.27, upper bound = 24.28
17. Lower bound = 2.47, upper bound = 4.40
19. As the confidence level increases but the sample size stays the same, the lower bound for the confidence interval for σ^2 decreases and the upper bound for the confidence interval for σ^2 increases.
21. Lower bound = 4.73, upper bound = 33.33
23. Lower bound = 5.78, upper bound = 21.33
25. Lower bound = 2.32, upper bound = 4.99
27. As the sample size increases but the confidence level stays the same, the lower bound of a confidence interval for σ^2 increases and the upper bound of a confidence interval for σ^2 decreases.
29. (a) $\chi^2_{1-\alpha/2} = \chi^2_{0.975} = 0.484$ and $\chi^2_{\alpha/2} = \chi^2_{0.025} = 11.143$. **(b)** Lower bound = 154.82, upper bound = 3564.46. We are 95% confident that the population variance σ^2 lies between 154.82 prisoners squared and 3564.46 prisoners squared. **(c)** Lower bound = 12.44, upper bound = 59.70. We are 95% confident that the population standard deviation σ lies between 12.44 prisoners and 59.70 prisoners.
31. (a) Prisoners squared **(b)** No **(c)** Prisoners **(d)** Yes
33. (a) $\chi^2_{1-\alpha/2} = \chi^2_{0.975} = 8.907$ and $\chi^2_{\alpha/2} = \chi^2_{0.025} = 32.852$. **(b)** Lower bound = 0.95, upper bound = 3.49. We are 95% confident that the population variance σ^2 lies between 0.95 centimeters squared and 3.49 centimeters squared. **(c)** Lower bound = 0.97, upper bound = 1.87. We are 95% confident that the population standard deviation σ lies between 0.97 centimeters and 1.87 centimeters. **(d)** The interpretation in (c) is easier to understand since it is in the same units as our data.
35. (a) $\chi^2_{1-\alpha/2} = \chi^2_{0.995} = 1.735$ and $\chi^2_{\alpha/2} = \chi^2_{0.005} = 23.589$. **(b)** Lower bound = 1.85, upper bound = 6.81. We are 99% confident that the population standard deviation σ lies between 1.85 trillion BTUs and 6.81 trillion BTUs.

Chapter 8 Review

1. (a) $\sigma/\sqrt{n} = 2$ **(b)** $Z_{\alpha/2} = 1.96$ **(c)** $E = 3.92$. We can estimate μ to within 3.92 with 95% confidence. **(d)** (46.08, 53.92). We are 95% confident that the true mean μ lies between 46.08 and 53.92.
3. (a) 7 points **(b)** 0.2981 points **(c)** 1.645 **(d)** 0.4904 points. We can estimate μ to within 0.4904 points with 90% confidence. **(e)** (6.5096, 7.4904). We are 90% confident that the true mean increase in IQ points for all children after listening to a Mozart piano sonata for about 10 minutes μ lies between 6.5096 points and 7.4904 points.
5. 97.
7. 4.
9. Since the population is not normal, Case 1 does not apply. Since the sample size of $n = 25$ is small ($n < 30$), Case 2 does not apply. Thus we cannot use the t interval.
11. (21.169, 22.831)
13. (a) $t_{\alpha/2} = 1.667$ **(b)** $E = \$5.77$. We can estimate μ, the true mean amount of money collected by all parking meters, within \$5.77 with 90% confidence. **(c)** (114.23, 125.77). We are 90% confident that the true mean amount of money collected by all parking meters lies between \$114.23 and \$125.77.
15. (a) 1.96 **(b)** We have $n\hat{p} = 500(0.99) = 495 \geq 5$ and $n(1 - \hat{p}) = 500(1 - 0.99) = 5 \geq 5$. Thus we can use the Z interval for p. **(c)** 0.0087. We can estimate p within $E = 0.0087$ with 95% confidence. **(d)** (0.9813, 0.9987) (TI: (0.98128, 0.99872)). Thus we are 95% confident that the true proportion lies between 0.9813 and 0.9987 (TI: 0.98128 and 0.99872).
17. (a) The margin of error is $E = 0.0255$. We can estimate p, the true proportion of all erasures that resulted in a correct answer, within $E = 0.0255$ with 99% confidence. **(b)** (0.8645, 0.9155) (TI: (0.86451, 0.91549)) **(c)** Since 0.70 does not lie in our confidence interval and none of the values below 0.70 lie in our confidence interval, we are 99% confident that no number less than or equal to 0.70 is a plausible value for p. All numbers less than or equal to 0.70 lie below our confidence interval. This is an indication that tampering may have occurred.
19. 43.
21. 271.
23. 6766.
25. Lower bound = 65.786, upper bound = 170.159
27. Lower bound = 8.111, upper bound = 13.044

Chapter 8 Quiz

1. False
2. True
3. 4
4. less
5. α is a probability.
6. Either the population is normal or the sample size is large ($n \geq 30$).
7. (a) $E = \$705$. We can estimate μ, the mean cost of a college education, to within \$705 with 90% confidence. **(b)** (29,795, 31,205). We are 90% confident that the true mean cost of a college education lies between \$29,795 and \$31,205.
8. (a) $E = 49.35$ pounds (TI: using t: 49.32, using Z: 49.35). We can estimate μ, the mean femur load number in a frontal crash for the passenger in a 2005 Ford Equinox SUV, within 49.35 pounds with 90% confidence. **(b)** (953.65, 1052.35) (TI: t interval: (952.68, 1052.3), Z interval: (952.65, 1052.3)). We are 90% confident that the true mean femur load number in a frontal crash for the passenger in a 2005 Ford Equinox SUV lies between 953.65 pounds and 1052.35 pounds.

9. (55.15, 58.85) (TI: (55.151, 58.849)). We are 90% confident that the true mean day count of wolves in Isle Royale lies between 55.15 wolves and 58.85 wolves.

10. (9.30, 9.70) (TI: (9.3002, 9.6998)). We are 95% confident that the true mean fog index for all the books of this horror author lies between 9.30 and 9.70.

11. (a) $E = 0.0142$. We can estimate p, the true proportion of all Americans who attended a religious service in response to the attacks on the World Trade Center and the Pentagon, within 0.0142 with 95% confidence. **(b)** (0.2513, 0.2797). We are 95% confident that the true proportion of all Americans who attended a religious service in response to the attacks on the World Trade Center and the Pentagon lies between 0.2513 and 0.2797.

12. (a) $E = 0.0386$. We can estimate p, the true proportion of all Québecois who favor independence for the Province of Quebec, within 0.0386 with 99% confidence. **(b)** (0.3014, 0.3786)

13. (a) (1.863, 4.944). We are 95% confident that the population standard deviation of Amanda's delivery times lies between 1.8623 minutes and 4.941 minutes. **(b)** (6.083, 16.147). We are 95% confident that the population standard deviation of Bethany's delivery times lies between 6.083 minutes and 16.147 minutes.

14. 363

15. 554

16. 752

CHAPTER 9

Section 9.1

1. The null hypothesis is assumed to be true unless the sample evidence indicates that the alternative hypothesis is true instead. It represents what has been tentatively assumed about the value of the parameter. It is the status quo hypothesis.

3.

Form	Null hypothesis	Alternative hypothesis
1	$H_0 : \mu = \mu_0$ vs.	$H_a : \mu > \mu_0$
2	$H_0 : \mu = \mu_0$ vs.	$H_a : \mu < \mu_0$
3	$H_0 : \mu = \mu_0$ vs.	$H_a : \mu \neq \mu_0$

5. $H_0 : \mu = 10$ vs. $H_a : \mu > 10$

7. $H_0 : \mu = 0$ vs. $H_a : \mu \neq 0$

9. $H_0 : \mu = 36$ vs. $H_a : \mu \neq 36$

11. (a) $H_0 : \mu = 43.9$ vs. $H_a : \mu > 43.9$ **(b)** A Type II error was made.

13. (a) $H_0 : \mu = \mu_0$ vs. $H_a : \mu > \mu_0$ **(b)** A Type I error was made.

15. (a) $H_0 : \mu = 1{,}602{,}000{,}000$ vs. $H_a : \mu > 1{,}602{,}000{,}000$ **(b)** Conclude that the mean is greater than 1.602 billion when the population mean is actually greater than 1.602 billion, and conclude that the mean is equal to 1.602 billion when the population is actually equal to 1.602 billion.

(c) Concluding that the mean is greater than 1.602 billion when the population mean is actually equal to 1.602 billion

(d) Concluding that the mean is equal to 1.602 billion when the population mean is actually greater than 1.602 billion

17. (a) $H_0 : \mu = 3.24$ vs. $H_a : \mu > 3.24$ **(b)** Conclude that the mean is greater than $3.24 when it actually is greater than $3.24, and conclude that the mean is equal to $3.24 when it actually is equal to $3.24. **(c)** Concluding that the mean is greater than $3.24 when it actually is equal to $3.24 **(d)** Concluding that the mean is equal to $3.24 when it actually is greater than $3.24

19. (a) $H_0 : \mu = 673$ vs. $H_a : \mu < 673$ **(b)** Conclude that the mean is less than 673 when it actually is less than 673, and conclude that the mean is equal to 673 when it actually is equal to 673. **(c)** Concluding that the mean is less than 673 when it actually is equal to

673 **(d)** Concluding that the mean is equal to 673 when it actually is less than 673.

Section 9.2

1. Answers will vary.

3. The p-value is small when it is less than α.

5. (a) Since the p-value $< \alpha$, we reject the null hypothesis. **(b)** There is evidence that the population mean is less than μ_0.

7. (a) We do not reject the null hypothesis. **(b)** There is insufficient evidence that the population mean is less than μ_0.

9. (a)

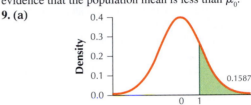

(b) p-value $= 0.1587$

11. (a)

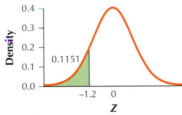

(b) p-value $= 0.0228$

13. The p-value decreases as Z_{data} increases.

15. (a) $H_0 : \mu = 3.0$ vs. $H_a : \mu < 3.0$. Reject H_0 if the p-value < 0.10.

(b) -1.2 **(c)** 0.1151

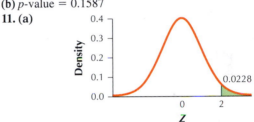

(d) Since the p-value $\geq \alpha$, do not reject H_0. There is insufficient evidence that the population mean is less than 3.0.

17. (a) $H_0 : \mu = 0$ vs. $H_a : \mu > 0$. Reject H_0 if the p-value < 0.05.

(b) 2 **(c)** 0.0228

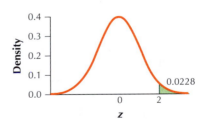

(d) Since the p-value $< \alpha$, reject H_0. There is evidence that the population mean is greater than 0.

19. No evidence

21. Slight evidence

23. Solid evidence

25. (a) $H_0 : \mu = 43.9$ vs. $H_a : \mu > 43.9$. Reject H_0 if the p-value < 0.10. **(b)** 19.61 **(c)** ≈ 0 **(d)** Since the p-value $< \alpha$, reject H_0. There is evidence that the population mean referral rate is greater than 43.9 per 1000 children.

27. (a) $H_0 : \mu = 700$ vs. $H_a : \mu < 700$. Reject H_0 if the p-value $<$ 0.10. **(b)** -20 **(c)** ≈ 0 **(d)** Since the p-value $< \alpha$, reject H_0. There is evidence that the population mean number of meals prepared and eaten at home is less than 700.

29. (a) $H_0 : \mu = 1.4261$ vs. $H_a : \mu < 1.4261$. Reject H_0 if the p-value < 0.05. **(b)** -17.32 **(c)** ≈ 0 **(d)** Since the p-value $< \alpha$, reject H_0. There is evidence that the population mean annual ring growth in the tree's later years is less than 1.4261 mm per year.

31. (a) Yes, the normal probability plot indicates acceptable normality. **(b)** $H_0 : \mu = 210$ vs. $H_a : \mu < 210$. Reject H_0 if p-value < 0.01. $Z_{data} = -1.69$. p-value $= 0.0455$. Since p-value $\geq \alpha$, do not reject H_0. There is insufficient evidence that the population mean sodium content per serving of breakfast cereal is less than 210 grams.

33. There is solid evidence against the null hypothesis.

35. (a) p-value $= 0.0571$ **(b)** Since the p-value ≥ 0.05, we would not reject H_0.

37. As sample size increases, the p-value decreases.

Section 9.3

1. Threshold for significance.

3. Threshold for significance.

5. Probability.

7. (a) -1.28 **(b)** Reject H_0 if $Z_{data} < -1.28$

(c)

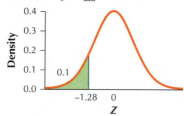

9. (a) 1.96 **(b)** Reject H_0 if $Z_{data} < -1.96$ or $Z_{data} > 1.96$.

(c)

11. (a) 1.645 **(b)** Reject H_0 if $Z_{data} > 1.645$

(c)

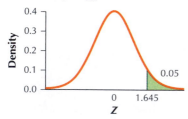

(d) $Z_{data} > 1.645$, so we reject H_0. **(e)** There is evidence that the population mean is greater than μ_0.

13. (a) -1.28 **(b)** Reject H_0 if $Z_{data} < -1.28$

(c)

(d) Since $Z_{data} < -1.28$, we reject H_0. **(e)** There is evidence that the population mean is less than μ_0.

15. (a) $H_0 : \mu = 100$ vs. $H_a : \mu > 100$ **(b)** $Z_{crit} = 1.645$. Reject H_0 if $Z_{data} > 1.645$. **(c)** $Z_{data} = 2$

(d) Since $Z_{data} > 1.645$, we reject H_0. There is evidence that the population mean is greater than 100.

17. (a) Since $\mu_0 = -3$ does not lie in the confidence interval, we reject H_0. **(b)** Since $\mu_0 = -2$ lies in the confidence interval we do not reject H_0. **(c)** Since $\mu_0 = 0$ lies in the confidence interval, we do not reject H_0. **(d)** Since $\mu_0 = 5$ lies in the confidence interval, we do not reject H_0. **(e)** Since $\mu_0 = 7$ does not lie in the confidence interval, we reject H_0.

19. (a) $H_0 : \mu = 3.24$ vs. $H_a : \mu > 3.24$ **(b)** $Z_{crit} = 1.645$. Reject H_0 if $Z_{data} > 1.645$ **(c)** $Z_{data} = 6.4$

(d) Since $Z_{data} > 1.645$ we reject H_0. There is evidence that the population mean price of a gallon of milk this year is greater than \$3.24 per gallon.

21. (a) $H_0 : \mu = 50{,}000$ vs. $H_a : \mu > 50{,}000$ **(b)** $Z_{crit} = 2.33$. Reject H_0 if $Z_{data} > 2.33$. **(c)** $Z_{data} = 15.75$

(d) Since $Z_{data} > 2.33$, we reject H_0. There is evidence that the population mean salary of assistant professors in science is greater than \$50,000.

23. (a) (23,160, 24,840) **(b)** (i) Since $\mu_0 = 24{,}000$ lies in the confidence interval, we do not reject H_0. (ii) Since $\mu_0 = 23{,}000$ does not lie in the confidence interval, we reject H_0. (iii) Since $\mu_0 = 23{,}200$ lies in the confidence interval, we do not reject H_0. (iv) Since $\mu_0 = 25{,}000$ does not lie in the confidence interval, we reject H_0.

25. (a)–(c) Unchanged **(d)** Value of Z_{data} greater than 1.67 **(e)** Unchanged

27. (a) To not reject the null hypothesis that the population mean credit score is at least 673, when this hypothesis is false. **(b)** 0.1480 **(c)** 0.4821 **(d)** 0.8512, 0.4179

Section 9.4

1. The population standard deviation σ is known.

3. Case 1: The population is normal. Case 2: The sample size is large ($n \geq 30$).

5. $t_{crit} = -1.363$

7. $t_{crit} = -2.718$

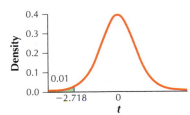

9. $t_{data} = -3$

11. $t_{data} = -2.67$

13. (a) $H_0 : \mu = 0$ vs. $H_a : \mu \neq 0$. Reject H_0 if the p-value < 0.05.
(b) $t_{data} = 6$ **(c)** p-value $= 0.0003233933213$ **(d)** Since the p-value < 0.05, we reject H_0. There is evidence that the population mean is different than 0.

15. (a) $H_0 : \mu = 28$ vs. $H_a : \mu < 28$. Reject H_0 if the p-value < 0.05. **(b)** $t_{data} = -1$ **(c)** p-value $= 0.1598742373$. **(d)** Since the p-value ≥ 0.05, we do not reject H_0. There is insufficient evidence that the population mean is less than 28.

17. (a) $H_0 : \mu = 100$ vs. $H_a : \mu > 100$ **(b)** $t_{crit} = 2.492$. Reject H_0 if $t_{data} > 2.492$.

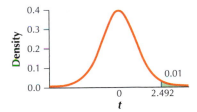

(c) 2 **(d)** Since $t_{data} \leq 2.492$, we do not reject H_0. There is insufficient evidence that the population mean is greater than 100.

19. (a) $H_0 : \mu = 9$ vs. $H_a : \mu \neq 9$ **(b)** $t_{crit} = 1.690$. Reject H_0 if $t_{data} < -1.690$ or $t_{data} > 1.690$.

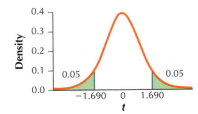

(c) 2 **(d)** Since $t_{data} > 1.690$, we reject H_0. There is evidence that the population mean is different than 9.

21. Since neither Case 1 nor Case 2 applies, we cannot proceed with the t test.

23. (a) $H_0 : \mu = 60$ vs. $H_a : \mu > 60$ **(b)** $t_{crit} = 1.771$. Reject H_0 if $t_{data} > 1.771$ **(c)** $t_{data} = 2.50$ **(d)** Since $t_{data} > 1.771$ we reject H_0. There is evidence that the population mean response time is greater than 60 milliseconds.

25. (a) Yes. Since the prices per bottle for Obsession by Calvin Klein are normally distributed, Case 1 applies. **(b)** No, a sample mean that is greater than the hypothesized mean is not enough to reject H_0. **(c)** 3.87 **(d)** $H_0 : \mu = 46.42$ vs. $H_a : \mu > 46.62$. Reject H_0 if the p-value < 0.01. $t_{data} = 3.87$. p-value < 0.005. Since the p-value < 0.01, we reject H_0. There is evidence that the population mean price per bottle of Obsession is greater than \$46.42.

27. $H_0 : \mu = 30$ vs. $H_a : \mu > 30$. Reject H_0 if p-value < 0.01. $t_{data} = 3.54$. p-value $= 0.0031570524$. Since p-value < 0.01, we reject H_0. There is evidence that the population mean gas mileage is greater than 30 mpg.

29. (a) Yes. Since all the points lie within the curved lines, the normal probability plot indicates acceptable normality. Case 1 applies. **(b)** $n = 10$, $\bar{x} = \$2538.92$, $s = \$404.75$ **(c)** $t_{data} = 2.09$. **(d)** It says, "Test of mu $= 2272$ vs. not $= 2272$" across the top of the printout. The researchers wanted to test the hypotheses "$H_0 : \mu = 2272$ vs. $H_a : \mu > 2272$."

31. (a) Case 2 applies, so it is appropriate to apply the t test. **(b)** 0.3192 years **(c)** 1.57. The sample mean age at first marriage among females $\bar{x} = 26.5$ years is 1.57 standard errors above the hypothesized mean age at first marriage among females $\mu_0 = 26$ years. **(d)** There is insufficient evidence that the population mean age at first marriage is not equal to 26 years.

33. (a) There is insufficient evidence that the population mean marriage age has changed, but there is evidence that the population mean marriage age has increased. **(b)** Two-tailed test

Section 9.5

1. $\hat{p}$ is the sample proportion and p is the population proportion.
3. Between 0 and 1 inclusive: $0 \leq p_0 \leq 1$
5. 0
7. -4.47
9. (a) We have $np_0 = 100(0.4) = 40 \geq 5$ and $n(1 - p_0) = 100(1 - 0.4) = 60 \geq 5$, so we can use the Z test for proportions. **(b)** $H_0 : p = 0.4$ vs. $H_a : p > 0.4$. Reject H_0 if the p-value < 0.05. **(c)** 0.82 **(d)** p-value $= 0.2061$ **(e)** Since the p-value ≥ 0.05, we do not reject H_0. There is insufficient evidence that the population proportion is greater than 0.4.

11. (a) We have $np_0 = 900(0.5) = 450 \geq 5$ and $n(1 - p_0) = 900(1 - 0.5) = 450 \geq 5$, so we may use the Z test for proportions. **(b)** $H_0 : p = 0.5$ vs. $H_a : p \neq 0.5$. Reject H_0 if the p-value < 0.05. **(c)** 1.67 **(d)** p-value $= 0.095$ **(e)** Since the p-value ≥ 0.05, we do not reject H_0. There is insufficient evidence that the population proportion is not equal to 0.5.

13. (a) We have $np_0 = 225(0.5) = 112.5 \geq 5$ and $n(1 - p_0) = 225(1 - 0.5) = 112.5 \geq 5$, so we can use the Z test for proportions. **(b)** $H_0 : p = 0.5$ vs. $H_a : p < 0.5$ **(c)** $Z_{crit} = -1.645$. Reject H_0 if $Z_{data} < -1.645$. **(d)** -1.67 **(e)** Since $Z_{data} < -1.645$, we reject H_0. There is evidence that the population proportion is less than 0.5.

15. (a) We have $np_0 = 400(0.6) = 240 \geq 5$ and $n(1 - p_0) = 400(1 - 0.6) = 160 \geq 5$, so we can use the Z test for proportions. **(b)** $H_0 : p = 0.6$ vs. $H_a : p > 0.6$ **(c)** $Z_{crit} = 1.645$. Reject H_0 if $Z_{data} > 1.645$. **(d)** 2.04 **(e)** Since $Z_{data} > 1.645$, we reject H_0. There is evidence that the population proportion is greater than 0.6.

17. $np_0 = 1000(0.368) = 368 \geq 5$ and $n(1 - p_0) = 1000(1 - 0.368) = 632 \geq 5$, so we may use the Z test for proportions. $H_0 : p = 0.368$ vs. $H_a : p > 0.368$. Reject H_0 if p-value < 0.05. $Z_{data} = 0.79$. p-value $= 0.2148$. Since p-value ≥ 0.05, we do not reject H_0. There is insufficient evidence that the population proportion of births to unmarried women is greater than 0.368.

19. $np_0 = 900\,(0.048) = 43.2 \geq 5$ and $n(1 - p_0) = 900\,(1 - 0.048) = 856.8 \geq 5$, so we may use the Z test for proportions. $H_0 : p = 0.048$ vs. $H_a : p > 0.048$. Reject H_0 if p-value < 0.01. $Z_{data} = 1.68$. p-value $= 0.0465$. Since p-value ≥ 0.01, we do not reject H_0. There is insufficient evidence that the population proportion of persons 12 or older that have used a prescription pain reliever non-medically is greater than 0.048.

21. $H_0 : p = 0.123$ vs. $H_a : p > 0.123$. Reject H_0 if p-value < 0.05. $Z_{data} = -0.43$. p-value $= 0.3336$. Since p-value ≥ 0.05, we do not reject H_0. There is insufficient evidence that the population proportion of California residents of Asian ethnicity is greater than 0.123.

23. (a) We have $n\,p_0 = 100\,(0.456) = 45.6 \geq 5$ and $n\,(1 - p_0) = 100\,(1 - 0.456) = 54.4 \geq 5$, so we can use the Z test for proportions. (b) 0.0498. It is a measure of the variability of the sampling distribution of $\hat{p}$ if H_0 is true. (c) 0.92; $\hat{p} = 0.41$ is not extreme. (d) $H_0 : p = 0.456$ vs. $H_a : p \neq 0.456$. Reject H_0 if p-value < 0.05. $Z_{data} = 0.92$. p-value $= 0.3576$ (TI: 0.3557). Since p-value ≥ 0.05, we do not reject H_0. There is insufficient evidence that the population proportion of eighth-graders who have used alcohol is not equal to 0.456.

25. (a) We have $n\,p_0 = 100\,(0.11) = 11 \geq 5$ and $n\,(1 - p_0) = 100\,(1 - 0.11) = 89 \geq 5$, so we can use the Z test for proportions. (b) 0.0313. It is a measure of the variability of the sampling distribution of $\hat{p}$ if H_0 is true. (c) 1.60; $\hat{p} = 0.06$ is not extreme. (d) Near the center

27. (a) Type II; answers will vary. (b) Since we did not reject H_0, our hypothesis test does not support this headline.

29. (a) Unchanged (b)–(c) Decrease (d) Unchanged (e) Depends on new value of $\hat{p}$.

Section 9.6

1. Answers will vary.
3. No, σ will never be less than 0.
5. $H_0 : \sigma = 10$ vs. $H_a : \sigma > 10$
7. $H_0 : \sigma = 3$ vs. $H_a : \sigma \neq 3$
9. (a) $\chi^2_{data} = 60$
(b)

(c) p-value $= 7.121750863 \times 10^{-6}$ (d) Since the p-value < 0.05, we reject H_0. There is evidence that the population standard deviation is greater than 1.
11. (a) $\chi^2_{data} = 10.417$

(b)

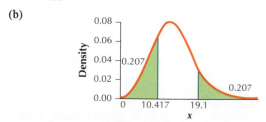

(c) p-value $= 0.4145552434$ (d) Since the p-value ≥ 0.05, we do not reject H_0. There is insufficient evidence that the population standard deviation is different than 3.

13. (a) $\chi^2_{data} = 6.125$
(b)

(c) p-value $= 0.4747679539$ (d) Since the p-value ≥ 0.05, we do not reject H_0. There is insufficient evidence that the population standard deviation is less than 20.
15. (a) $\chi^2_{\alpha} = \chi^2_{0.05} = 31.410$ (b) Reject H_0 if $\chi^2_{data} > 31.41$ (c) Since $\chi^2_{data} > 31.410$, we reject H_0. There is evidence that the population standard deviation is greater than 1.
17. (a) $\chi^2_{\alpha/2} = \chi^2_{0.025} = 27.488$ and $\chi^2_{1 - \alpha/2} = \chi^2_{0.975} = 6.262$ (b) Reject H_0 if $\chi^2_{data} < 6.262$ or $\chi^2_{data} > 27.488$. (c) Since χ^2_{data} is not less than 6.262 and χ^2_{data} is not greater than 27.488, we do not reject H_0. There is insufficient evidence that the population standard deviation is different than 3.
19. (a) $\chi^2_{1 - \alpha} = \chi^2_{0.90} = 2.833$ (b) Reject H_0 if $\chi^2_{data} < 2.833$. (c) Since $\chi^2_{data} \geq 2.833$, we do not reject H_0. There is insufficient evidence that the population standard deviation is less than 20.
21. (a) $H_0 : \sigma = 36.5$ vs. $H_a : \sigma > 36.5$. Reject H_0 if the p-value < 0.01. (b) 982.75 (c) p-value ≈ 0 (d) Since the p-value < 0.01, we reject H_0. (e) There is evidence that the population standard deviation of DDT level in the breast milk of Hispanic women in the Yakima valley is greater than 36.5 parts per billion.
23. (a) $H_0 : \sigma = 30{,}000$ vs. $H_a : \sigma \neq 30{,}000$. Reject H_0 if the p-value < 0.05. (b) 0.00001497113333 (c) p-value ≈ 0 (d) Since the p-value < 0.05, we reject H_0. (e) There is evidence that the population standard deviation of union membership differs from 30,000.
25. *Test using the p-value method:* $H_0 : \sigma = 50$ vs. $H_a : \sigma > 50$. Reject H_0 if p-value < 0.05. $\chi^2_{data} = 104$. p-value $= 0.3721497012$. Since p-value ≥ 0.05, we do not reject H_0. There is insufficient evidence that the population standard deviation of test scores for boys is greater than 50 points. *Test using the critical value method:* $H_0 : \sigma = 50$ vs. $H_a : \sigma > 50$. $\chi^2_{\alpha} = \chi^2_{0.05} = 124.342$. Reject H_0 if $\chi^2_{data} > 124.342$. $\chi^2_{data} = 104$. Since $\chi^2_{data} \leq 124.342$, we do not reject H_0. There is insufficient evidence that the population standard deviation of test scores for boys is greater than 50 points.

Chapter 9 Review

1. (a) $H_0 : \mu = 2.58$ vs. $H_a : \mu \neq 2.58$ (b) Concluding that the population mean has changed from 2.58 when it actually is different from 2.58, and concluding that the population mean has not changed from 2.58 when it actually is not different from 2.58. (c) Concluding that the population mean has changed from 2.58 when it hasn't (d) Concluding that the population mean has not changed from 2.58 when it actually has changed from 2.58.
3. -1
5. -10
7. (a) $H_0 : \mu = -10$ vs. $H_a : \mu < -10$. Reject H_0 if the p-value < 0.01. (b) -5 (c) p-value $= 2.87105 \times 10^{-7}$

(d) Since the *p*-value < 0.01, reject H_0. There is evidence that the population mean is less than -10.

9. (a) 2.58 **(b)** Reject H_0 if $Z_{data} < -2.58$ or $Z_{data} > 2.58$.
(c)

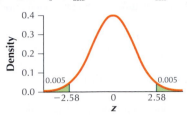

(d) Since Z_{data} is not less than -2.58 and Z_{data} is not greater than 2.58, we do not reject H_0. There is insufficient evidence that the population mean is different than μ_0.

11. (a) 1.645 **(b)** Reject H_0 if $Z_{data} > 1.645$.
(c)

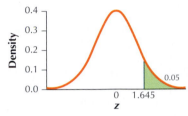

(d) Since $Z_{data} \leq 1.645$, we do not reject H_0. There is insufficient evidence that the population mean is greater than μ_0.

13. (a) $H_0 : \mu = 52,200$ vs. $H_a : \mu > 52,200$ **(b)** $Z_{crit} = 1.28$. Reject H_0 if $Z_{data} > 1.28$. **(c)** $Z_{data} = 9.33$

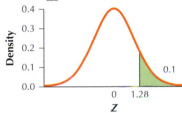

(d) Since $Z_{data} > 1.28$, we reject H_0. There is evidence that the population mean salary of college graduates is greater than \$52,200.

15. $H_0 : \mu = 45$ vs. $H_a : \mu \neq 45$. Reject H_0 if *p*-value < 0.10. $t_{data} = 0$. *p*-value > 0.20. Since *p*-value ≥ 0.10, we do not reject H_0. There is insufficient evidence that the population mean differs from 45.

17. (a) We have $n p_0 = 100(0.25) = 25 \geq 5$ and $n(1 - p_0) = 100(1 - 0.25) = 75 \geq 5$. **(b)** $H_0 : p = 0.25$ vs. $H_a : p < 0.25$. Reject H_0 if the *p*-value < 0.05. **(c)** 0 **(d)** 0.5 **(e)** Since the *p*-value ≥ 0.05, we do not reject H_0. There is insufficient evidence that the population proportion is less than 0.25.

19. (a) We have $n p_0 = 900(0.2) = 180 \geq 5$ and $n(1 - p_0) = 900(1 - 0.2) = 720 \geq 5$. **(b)** $H_0 : p = 0.2$ vs. $H_a : p < 0.2$ **(c)** 1.645; reject H_0 if $Z_{data} < -1.645$. **(d)** -1.67 **(e)** Since $Z_{data} < -1.645$, we reject H_0. There is evidence that the population proportion is less than 0.2.

21. (a) $H_0 : \sigma = 35$ vs. $H_a : \sigma < 35$. Reject H_0 if the *p*-value $< \alpha$. **(b)** 6.857 **(c)** *p*-value $= 0.5560805474$

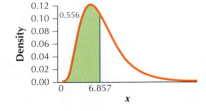

(d) Since the *p*-value ≥ 0.05, we do not reject H_0. There is insufficient evidence that the population standard deviation is less than 35.

23. (a) $H_0 : \sigma = 6$ vs. $H_a : \sigma > 6$ **(b)** $\chi^2 = \chi^2_{0.05} = 30.144$. Reject H_0 if $\chi^2_{data} > 30.144$ **(c)** $\chi^2_{data} = 42.75$

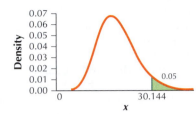

(d) Since $\chi^2_{data} > 30.144$, we reject H_0. There is evidence that the population standard deviation is greater than 6.

25. $H_0 : \sigma = 50$ vs. $H_a : \sigma \neq 50$. Reject H_0 if the *p*-value < 0.01. $\chi^2_{data} = 0.690$. *p*-value $= 0.094887$. Since the *p*-value ≥ 0.01, we do not reject H_0. There is insufficient evidence that the population standard deviation differs from 50.

Chapter 9 Quiz

1. False
2. True
3. True
4. I
5. small
6. α
7. $n p_0 \geq 5$ and $n(1 - p_0) \geq 5$
8. A small *p*-value indicates that there is strong evidence against the null hypothesis. A large *p*-value indicates that there is no evidence against the null hypothesis.
9. No
10. $H_0 : \mu = 47$ vs. $H_a : \mu \neq 47$. Reject H_0 if the *p*-value < 0.10. $Z_{data} = 5.83$, *p*-value $= 5.560346071 \times 10^{-9}$. Since the *p*-value < 0.10, we reject H_0. There is evidence that the population mean prison sentence length for felons convicted on drug offenses is different than 47 months.
11. $H_0 : \mu = 1.14$ vs. $H_a : \mu < 1.14$. Z_{crit} -1.645. Z_{data} -1.68. Since $Z_{data} < -1.645$, we reject H_0. There is evidence that the population mean fee charged by banks when withdrawing funds from an ATM machine not owned by your bank is less than \$1.14.
12. The conclusion remains the same since the *p*-value method and the critical-value method are equivalent.
13. $H_0 : \mu = 32$ vs. $H_a : \mu \neq 32$. $Z_{crit} = 1.690$. $Z_{data} = 1.80$. Since $t_{data} > 1.690$, we reject H_0. There is evidence that the population mean years of potential life lost in alcohol-related fatal automobile accidents is different than 32 years.
14. (a) $0.05 < $ *p*-value ≤ 0.10 **(b)** There is mild evidence against the null hypothesis that the population mean years of potential life lost in alcohol-related fatal automobile accidents is 32 years.
15. $H_0 : p = 0.127$ vs. $H_a : p > 0.127$. $Z_{data} = 0.93$, *p*-value $= 0.1762$. Since the *p*-value ≥ 0.05, we do not reject H_0. There is insufficient evidence that the population proportion of preterm births is greater than 0.127.
16. $H_0 : \sigma = 0.25$ vs. $H_a : \sigma < 0.25$. $\chi^2_{1-\alpha} = \chi^2_{0.90} = 4.168$. Reject H_0 if $\chi^2_{data} < 4.168$. $\chi^2_{data} = 16.992$. Since $\chi^2_{data} \geq 4.168$, we do not reject H_0. There is insufficient evidence that the population standard deviation of net price change is less than 25 cents.

CHAPTER 10

Section 10.1

1. When the subjects selected for the first sample do not determine the subjects in the second sample

3. Matched pairs or paired samples

5. Since both samples of games were based on the same players, this is an example of dependent sampling.

7. Since the same students are taking both tests, this is an example of dependent sampling.

9. (a) $\bar{x}_d = 1.1$, $s_d = 0.9618$ **(b)** $(-0.0940, 2.294)$

11. (a) $\bar{x}_d = -7.1429$, $s_d = 2.6726$ **(b)** $(-9.6127, -4.6711)$

13. p-value $= 0.0315$. Since the p-value < 0.05, we reject H_0. There is evidence that the population mean of the differences is greater than 0.

15. p-value $= 0.143$. Since the p-value ≥ 0.05, we do not reject H_0. There is insufficient evidence that the population mean of the differences is greater than 0.

17. Since $t_{data} \leq -1.943$, we reject H_0. There is evidence that the population mean of the differences is different than 0.

19. (a) $\bar{x}_d = 753.75$, $s_d = 986.1658$ **(b)** $(-815.460, 2322.960)$

21. There is insufficient evidence that 2007 models are on average more expensive.

23. (a) $\bar{x}_d = -28,825$, $s_d = 16,806.2241$ **(b)** $(-48,600.5766, -9,049.4234)$

25. Since $\mu_d = 0$ does not lie in the confidence interval, we reject H_0. There is evidence that the population mean difference between the 1st quarter 2007 median price and the 1st quarter 2008 median price differs from 0.

27. (a) $(1.9618, 23.5048)$ **(b)** Since 0 is not in the interval, there is evidence that the 2003 science test scores for fourth-graders are higher than the 1995 science test scores for fourth-graders.

29. *Critical value method:* $H_0 : \mu_d = 0$ vs. $H_a : \mu_d < 0$. $t_{crit} = 3.365$. Reject H_0 if $t_{data} < -3.365$. $t_{data} = -4.843$. Since $t_{data} < -3.365$, we reject H_0. There is evidence that the population mean number of goals scored per game has increased.

Section 10.2

1. Case 1: The two populations are normally distributed. Case 2: The sample sizes are large (at least 30).

3. (a) Since both sample sizes are large ($n_1 \geq 30$ and $n_2 \geq 30$), it is appropriate to construct a 95% confidence interval for $\mu_1 - \mu_2$. **(b)** 2 **(c)** 0.9569 (TI: 0.9402). **(d)** $(1.0431, 2.9569)$ (TI: $(1.0598, 2.9402)$). We are 95% confident that the interval captures the difference in population means.

5. (a) Since both sample sizes are large ($n_1 \geq 30$ and $n_2 \geq 30$), it is appropriate to construct a 95% confidence interval for $\mu_1 - \mu_2$. **(b)** -5 **(c)** 2.011 (TI: 1.99). **(d)** $(-7.011, -2.989)$ (TI: $(-6.99, -3.01)$). We are 95% confident that the interval captures the difference in population means.

7. (a) Since both sample sizes are large ($n_1 \geq 30$ and $n_2 \geq 30$), it is appropriate to construct a 95% confidence interval for $\mu_1 - \mu_2$. **(b)** -1 **(c)** 0.811 (TI: 0.7985). **(d)** $(-1.811, -0.189)$ (TI: $(-1.799, 0.2015)$). We are 95% confident that the interval captures the difference in population means.

9. (a) $H_0 : \mu_1 = \mu_2$ vs. $H_a : \mu_1 \neq \mu_2$ **(b)** $t_{crit} = 1.690$. Reject H_0 if $t_{data} < -1.690$ or $t_{data} > 1.690$. **(c)** $t_{data} = 4.243$. **(d)** Since $t_{data} > 1.690$, we reject H_0. There is evidence that the population mean for Population 1 is different than the population mean for Population 2.

11. (a) $H_0 : \mu_1 = \mu_2$ vs. $H_a : \mu_1 \neq \mu_2$ **(b)** $t_{crit} = 2.756$. Reject H_0 if $t_{data} < -2.756$ or $t_{data} > 2.756$. **(c)** $t_{data} = -5.085$. **(d)** Since $t_{data} < -2.756$, we reject H_0. There is evidence that the population mean for Population 1 is different than the population mean for Population 2.

13. (a) $H_0 : \mu_1 = \mu_2$ vs. $H_a : \mu_1 > \mu_2$ **(b)** $t_{crit} = 2.423$. Reject H_0 if $t_{data} > 2.423$. **(c)** $t_{data} = -2.492$. **(d)** Since $t_{data} \leq 2.423$, we do not reject H_0. There is insufficient evidence that the population mean for Population 1 is greater than the population mean for Population 2.

15. Since neither population is normal, Case 1 does not apply. Since both sample sizes are small ($n_1 < 30$ and $n_2 < 30$), Case 2 does not apply. Therefore, it is not appropriate to perform two-sample t inference.

17. (a) Since both populations are normally distributed Case 1 applies. Thus it is appropriate to apply two-sample inference. **(b)** -57 **(c)** 59.3489 (TI: 56.1131) **(d)** $(-116.3489, 2.3489)$ (TI: $(-113.1, -0.8738)$). We are 95% confident that the interval captures the difference between the population mean math scores of students from the United States and Hong Kong.

19. (a) 1.6 **(b)** 2.647 (TI: 2.5848) **(c)** $(-1.047, 4.247)$ (TI: $(-0.9848, 4.1548)$). We are 99% confident that the interval captures the difference in the population mean number of children per teacher in the towns of Cupertino, California, and Santa Rosa, California.

21. $(-13.9365, 33.9365)$ (TI: $(-13.57, 33.567)$). We are 95% confident that the interval captures the difference between the population mean verbal SAT score of males and the population mean verbal SAT score of females.

23. Since 0 is in the 95% confidence interval $(-13.9365, 33.9365)$, we would not reject H_0.

25. (a) $(11,442.85, 15,479.15)$ (TI: $(11,458, 15,464)$). We are 95% confident that the interval captures the difference between the population mean starting salary for college graduates majoring in information systems and the population mean starting salary for college graduates majoring in psychology. **(b)** Wider; the higher the confidence level, the wider the confidence interval

27. (a) $(-791.006, -206.042)$ (TI: $(-766, -231)$). We are 90% confident that the interval captures the difference in the population mean daily calorie intake of children in the inner city and the suburbs. **(b)** $H_0 : \mu_1 = \mu_2$ vs. $H_a : \mu_1 < \mu_2$. Reject H_0 if p-value < 0.10. $t_{data} = -3.434$. p-value $= 0.0092775929$ (TI: 0.0039880443). Since the p-value < 0.10, we reject H_0. There is evidence that the population mean daily calorie intake of inner city children is less than that of children from the suburbs.

29. (a) $(-3.161, 0.561)$ (TI: $(-3.136, 0.53598)$). We are 99% confident that the interval captures the difference of the population mean number of hours worked in 2003 and 1994. **(b)** $H_0 : \mu_1 = \mu_2$ vs. $H_a : \mu_1 < \mu_2$. Reject H_0 if p-value < 0.01. $t_{data} = -1.838$. p-value $= 0.0342678723$ (TI: 0.0336138699). Since the p-value ≥ 0.01, we do not reject H_0. There is insufficient evidence that the population mean number of hours worked in 2003 is less than the population mean number of hours worked in 1994.

31. (a) This will result in an increase in $\bar{x}_1 - \bar{x}_2$, but $\bar{x}_1 - \bar{x}_2$ will still be negative. **(b)** This will result in an increase in t_{data}, but t_{data} will still be negative. **(c)** Increase in p-value **(d)** Unchanged

33. (a) $\bar{x}_1 = 31,363$, $s_1 = 135,270$ **(b)** $\bar{x}_2 = 18,305$, $s_2 = 260,938$ **(c)** $H_0 : \mu_1 = \mu_2$ vs. $H_a : \mu_1 \neq \mu_2$. Reject H_0 if the p-value < 0.05. $t_{data} = 1.259$; p-value $= 0.2084026891$. Since the p-value ≥ 0.05, we do not reject H_0. There is insufficient evidence that the population mean number of residents in towns and cities in California is different than the population mean number of residents in towns and cities in New York.

Section 10.3

1. $\hat{p}_1$ and $\hat{p}_2$

3. Z_{data} measures the standardized distance between sample proportions. Extreme values of Z_{data} indicate evidence against the null hypothesis.

5. (a) $x_1 = 80 \geq 5$, $n_1 - x_1 = 20 \geq 5$, $x_2 = 30 \geq 5$, and $n_2 - x_2 = 10 \geq 5$, so it is appropriate. **(b)** 0.05 **(c)** 0.1554. The point estimate $\hat{p}_1 - \hat{p}_2$ will lie within $E = 0.1554$ of the difference in population proportions $p_1 - p_2$ 95% of the time. **(d)** $(-0.1054, 0.2054)$. We are 95% confident that the difference in population proportions lies between -0.1054 and 0.2054.

7. (a) $x_1 = 60 \geq 5$, $n_1 - x_1 = 140 \geq 5$, $x_2 = 40 \geq 5$, and $n_2 - x_2 = 210 \geq 5$, so it is appropriate. **(b)** 0.14 **(c)** 0.078. The point estimate $\hat{p}_1 - \hat{p}_2$ will lie within $E = 0.078$ of the difference in population proportions $p_1 - p_2$ 95% of the time. **(d)** $(0.062, 0.218)$. We are 95% confident that the difference in population proportions lies between 0.062 and 0.218.

9. (a) $x_1 = 490 \geq 5$, $n_1 - x_1 = 510 \geq 5$, $x_2 = 620 \geq 5$, and $n_2 - x_2 = 380 \geq 5$, so it is appropriate. **(b)** -0.13 **(c)** 0.0431 (TI: 0.0432). The point estimate $\hat{p}_1 - \hat{p}_2$ will lie within $E = 0.0431$ of the difference in population proportions $p_1 - p_2$ 95% of the time. **(d)** $(-0.1731, -0.0869)$ (TI: $(-0.1732, -0.0808)$). We are 95% confident that the difference in population proportions lies between -0.1731 and -0.0869.

11. (a) $H_0 : p_1 = p_2$ vs. $H_a : p_1 \neq p_2$; $Z_{crit} = 1.645$. Reject H_0 if $Z_{data} < -1.645$ or $Z_{data} > 1.645$. **(b)** 0.7857 **(c)** 0.65 **(d)** Since $Z_{data} \geq -1.645$ and $Z_{data} \leq 1.645$, we do not reject H_0. There is insufficient evidence that the population proportion from Population 1 is different than the population proportion from Population 2.

13. (a) $H_0 : p_1 = p_2$ vs. $H_a : p_1 > p_2$; $Z_{crit} = 1.645$. Reject H_0 if $Z_{data} > 1.645$. **(b)** 0.2222 **(c)** 3.55 **(d)** Since $Z_{data} > 1.645$, we reject H_0. There is evidence that the population proportion from Population 1 is greater than the population proportion from Population 2.

15. (a) $H_0 : p_1 = p_2$ vs. $H_a : p_1 > p_2$. Reject H_0 if the p-value < 0.10. **(b)** 0.5625 **(c)** 3.56 **(d)** 0.000185467351 **(e)** Since the p-value < 0.10, we reject H_0. There is evidence that the population proportion from Population 1 is greater than the population proportion from Population 2.

17. (a) $H_0 : p_1 = p_2$ vs. $H_a : p_1 \neq p_2$. Reject H_0 if the p-value < 0.05. **(b)** 0.7982 **(c)** -1.28 **(d)** 0.2005452669 **(e)** Since the p-value ≥ 0.05, we do not reject H_0. There is insufficient evidence that the population proportion from Population 1 is different than the population proportion from Population 2.

19. (a) $x_1 = 360.38 \geq 5$, $n_1 - x_1 = 126.62 \geq 5$, $x_2 = 404.21 \geq 5$, and $n_2 - x_2 = 82.79 \geq 5$, so it is appropriate. **(b)** -0.09 **(c)** 0.0514. The point estimate of the difference in the population proportion of teenage boys who post their photo on their online profile and the population proportion of teenage girls who do so will lie within $E = 0.0514$ of the difference in population proportions $p_1 - p_2$ 95% of the time. **(d)** $(-0.1414, -0.0386)$. We are 95% confident that the difference in the population proportion of teenage boys who post their photo on their online profile and the population proportion of teenage girls who do so lies between -0.1414 and -0.0386.

21. (a) $H_0 : p_1 = p_2$ vs. $H_a : p_1 \neq p_2$ where $p_1 =$ the population proportion of teenage boys who posted their photo online and $p_2 =$ the population proportion of teenage girls who posted their

photo online. **(b)** Zero does not fall within the confidence interval. **(c)** Since 0 does not fall within the confidence interval, we reject H_0. There is evidence that the population proportion of teenage boys who post their photo on their online profile differs from the population proportion of teenage girls who do so.

23. $H_0 : p_1 = p_2$ vs. $H_a : p_1 \neq p_2$. Reject H_0 if p-value < 0.01. $Z_{data} = 2.65$. p-value $= 0.008$. Since p-value < 0.01, we reject H_0. There is evidence of a generation gap in the perception of same-sex marriage.

25. (a) $(-0.4827, -0.1099)$ **(b)** $H_0 : p_1 = p_2$ vs. $H_a : p_1 \neq p_2$. Since 0 does not fall in the confidence interval, we reject H_0. There is evidence that there is a difference in the population proportions of women who had fetal cells.

27. $H_0 : p_1 = p_2$ vs. $H_a : p_1 \neq p_2$. Reject H_0 if the p-value < 0.05. $\hat{p}_{pooled} = 0.7705$. $Z_{data} = 0.21$. p-value $= 0.8336$. Since p-value ≥ 0.05, we do not reject H_0. There is insufficient evidence that the proportion of the people who wore the ionized bracelets who reported improvement in their maximum pain index is different than the proportion of the people who wore the placebo bracelets who reported improvement in their maximum pain index.

29. (a) $H_0 : p_1 = p_2$ vs. $H_a : p_1 \neq p_2$, where $p_1 =$ the population proportion of 18- to 24-year-old males who listen to the radio each week and $p_2 =$ the population proportion of males age 65 or older who listen to the radio each week. **(b)** No **(c)** Since 0 does not fall within the confidence interval, we reject H_0. There is evidence that the population proportion of 18–24-year-old males who listen to the radio each week is different than the population proportion of males age 65 or older who listen to the radio each week.

Chapter 10 Review

1. (a) $\bar{x}_d = -2.6875$, $s_d = 1.6146$ **(b)** $(-4.0376, -1.3374)$

3. $H_0 : \mu_d = 0$ vs. $H_a : \mu_d < 0$. Reject H_0 if p-value < 0.05. $t_{data} = -4.708$. p-value $= 0.0010939869$. Since the p-value < 0.05, we reject H_0. There is evidence that the population mean of the differences is less than 0.

5. 0.1

7. $(0.094, 0.106)$. We are 95% confident that the interval captures the difference in population means.

9. (a) $(0.0096, 0.2304)$ **(b)** $H_0 : p_1 = p_2$ vs. $H_a : p_1 \neq p_2$. Reject H_0 if p-value < 0.10. $\hat{p}_{pooled} = 0.0637$, $Z_{data} = 2.48$. p-value $= 0.013138259$. Since the p-value < 0.10, we reject H_0. There is evidence that the population proportion packet loss from Asian Web sites is different than the population proportion packet loss from North American Web sites.

11. (a) $(0.1610, 0.2110)$ **(b)** $H_0 : p_1 = p_2$ vs. $H_a : p_1 \neq p_2$. Reject H_0 if p-value < 0.01. $\hat{p}_{pooled} = 0.4586$, $Z_{data} = 18.85$. p-value ≈ 0. Since p-value < 0.01, we reject H_0. There is evidence that the population proportion of suspects who were denied bail for violent crimes is different than the population proportion of suspects who were denied bail for property-related crimes.

Chapter 10 Quiz

1. True
2. True
3. False
4. normal; large (greater than or equal to 30)
5. margin of error
6. $\bar{x}_d$
7. $\mu_1 - \mu_2$
8. $\hat{p}_{pooled}$

9. No difference

10. (a) (6.6680, 21.3320) **(b)** Since 0 does not lie in the confidence interval we reject H_0. There is evidence that the population mean difference in the number of cigarettes smoked before and after attending Butt-Enders is different than 0.

11. (a) $(-23,304.69, -6,695.31)$ (TI: $(-23,171, -6829)$) p-value $= 0.000499$. We are 95% confident that the interval captures the difference of the population mean income of Suburb A and the population mean income of Suburb B.
(b) $H_0 : \mu_1 = \mu_2$ vs. $H_a : \mu_1 \neq \mu_2$. $t_{crit} = 2.030$. Reject H_0 if $t_{data} < -2.030$ or $t_{data} > 2.030$. $t_{data} = -3.667$. Since $t_{data} < -2.030$, we reject H_0. There is evidence that the population mean income of Suburb A is different than the population mean income of Suburb B.

12. (2.2406, 17.7594). We are 95% confident that the interval captures the difference of the population mean number of bottles processed by the updated machine and the population mean number of bottles processed by the non-updated machine.

13. (a) Since $t_{data} > 1.662$, we reject H_0. There is evidence that the population mean number of bottles processed by the updated machine is greater than the population mean number of bottles processed by the non-updated machine. **(b)** Since confidence intervals can be used only to perform two-tailed tests and the hypothesis test in (a) is a one-tailed test, the confidence interval in Exercise 12(a) cannot be used to perform the hypothesis test in (a).

14. (a) $H_0 : \mu_1 = \mu_2$ vs. $H_a : \mu_1 \neq \mu_2$. $t_{crit} = 1.662$. Reject H_0 if $t_{data} < -1.662$ or $t_{data} > 1.662$. $t_{data} = -6.129$. Since $t_{data} < -1.662$, we reject H_0. There is evidence that the population mean income of people 18 to 24 years old who never married is different than the population mean income of people 18 to 24 years old who are married. **(b)** No, the conclusion of the two-tailed hypothesis test for $\alpha = 0.10$ is "Reject H_0."

15. $H_0 : p_1 = p_2; H_a : p_1 < p_2$. Reject H_0 if the p-value < 0.05. $Z_{data} = 0.28$; p-value $= 0.3897$ (TI: 0.3912). Since the p-value ≥ 0.05, we do not reject H_0. There is insufficient evidence that the population proportion of 18- to 20-year-olds who used an illicit drug decreased from 2004 to 2005.

16. $H_0 : p_1 = p_2$ vs. $H_a : p_1 > p_2$. $Z_{crit} = 1.645$. Reject H_0 if $Z_{data} > 1.645$. $\hat{p}_1 - \hat{p}_2 = 0.051$, $\hat{p}_{pooled} = 0.7155$, $Z_{data} = 2.53$. Since $Z_{data} > 1.645$, we reject H_0. There is evidence that the population proportion of people aged 35 to 44 who were married was lower in 2000 than in 1990.

CHAPTER 11

Section 11.1

1. (1) Each independent trial of the experiment has k possible outcomes, $k = 2, 3, \ldots$ (2) The ith outcome (category) occurs with probability p_i, where $i = 1, 2, \ldots, k$ (3) $\sum_{i=1}^{k} p_i = 1$.

3. It is the long-run mean of that random variable after an arbitrarily large number of trials.

5. Multinomial

7. Multinomial

9. (a) $E_1 = 50, E_2 = 25, E_3 = 25$ **(b)** Conditions are met.

11. (a) $E_1 = 90, E_2 = 5, E_3 = 4, E_4 = 1$ **(b)** Conditions are not met.

13. (a) $E_1 = 40, E_2 = 30, E_3 = 30$; conditions are met. **(b)** $\chi^2_{crit} = \chi^2_{0.05} = 5.991$. Reject H_0 if $\chi^2_{data} > 5.991$. **(c)** 4.167 **(d)** Since $\chi^2_{data} \leq 5.991$, we do not reject H_0. There is insufficient evidence that the random variable does not follow the distribution specified in H_0.

15. (a) $E_1 = 80, E_2 = 70, E_3 = 20, E_4 = 20, E_5 = 10$; conditions are met. **(b)** $\chi^2_{crit} = \chi^2_{0.10} = 7.779$. Reject H_0 if $\chi^2_{data} > 7.779$. **(c)** 6.607 **(d)** Since $\chi^2_{data} \leq 7.779$, we do not reject H_0. There is insufficient evidence that the random variable does not follow the distribution specified in H_0.

17. (a) Reject H_0 if the p-value < 0.05. $E_1 = 50, E_2 = 50$; conditions are met. **(b)** 4 **(c)** $0.025 < p$-value < 0.05 **(d)** Since the p-value < 0.05, we reject H_0. There is evidence that the random variable does not follow the distribution specified in H_0.

19. (a) Reject H_0 if the p-value < 0.10. $E_1 = 100, E_2 = 50, E_3 = 30$, $E_4 = 20$; conditions are met. **(b)** 6.083 **(c)** p-value > 0.10 **(d)** Since the p-value ≥ 0.10, we do not reject H_0. There is insufficient evidence that the random variable does not follow the distribution specified in H_0.

21. Since $\chi^2_{data} \leq 9.488$, we do not reject H_0. There is insufficient evidence that the distribution of education levels has changed since 2005.

23. $H_0 : p_{phip} = 0.30, p_{mm} = 0.556, p_{other} = 0.144$. H_a : The random variable does not follow the distribution specified in H_0. $E_{phip} = 300, E_{mm} = 556, E_{other} = 144$. Since none of the expected frequencies is less than 1 and none of the expected frequencies is less than 5, the conditions for performing the χ^2 goodness of fit test are met. $\chi^2_{crit} = \chi^2_{0.05} = 5.991$. Reject H_0 if $\chi^2_{data} > 5.991$. $\chi^2_{data} = 14.224$. Since $\chi^2_{data} > 5.991$, we reject H_0. There is evidence that the population proportions of minority patients who suffered spinal cord injuries, who had a private health insurance provider, Medicare, Medicaid, or other arrangements, have changed.

25. $H_0 : p_{males} = 0.4641403733, p_{females} = 0.5358596267$. H_a : The random variable does not follow the distribution specified in H_0. $E_{males} = 1129.25, E_{females} = 1303.75$. Since none of the expected frequencies is less than 1 and none of the expected frequencies is less than 5, the conditions for performing the χ^2 goodness of fit test are met. $\chi^2_{crit} = \chi^2_{0.05} = 3.841$. Reject H_0 if $\chi^2_{data} > 3.841$. $\chi^2_{data} = 17.278$. Since $\chi^2_{data} > 3.841$, we reject H_0. There is evidence that the population proportions of males and females living in college dormitories have changed.

27. $H_0 : p_{White} = 0.8068, p_{Black} = 0.1274, p_{Indian} = 0.0095, p_{Asian} = 0.0401, p_{Hawaiian} = 0.0017, p_{Twomore} = 0.0145$. H_a : The random variable does not follow the distribution specified in H_0. $E_{White} = 806.8, E_{Black} = 127.4, E_{Indian} = 9.5, E_{Asian} = 40.1, E_{Hawaiian} = 1.7, E_{Twomore} = 14.5$. Since none of the expected frequencies is less than 1 and only one of the six expected frequencies is less than 5, the conditions for performing the χ^2 goodness of fit test are met. Reject H_0 if p-value < 0.10. $\chi^2_{data} = 45.887$. p-value < 0.005. Since p-value < 0.10, we reject H_0. There is evidence that the population proportions have changed since 2002.

29. $H_0 : p_{sawadv} = 0.50, p_{notseeadv} = 0.41, p_{notrecall} = 0.09$. H_a : The random variable does not follow the distribution specified in H_0. $E_{sawadv} = 500, E_{notseeadv} = 410, E_{notrecall} = 90$. Since none of the expected frequencies is less than 1 and none of the expected frequencies is less than 5, the conditions for performing the χ^2 goodness of fit test are met. Reject H_0 if p-value < 0.05. $\chi^2_{data} = 10.014$. p-value $= 0.0066924517$. Since p-value < 0.05, we reject H_0. There is evidence that the population proportions have changed since 2007.

31. $H_0 : p_{personnel} = 0.24, p_{operations} = 0.28, p_{other} = 0.48$. H_a : The random variable does not follow the distribution specified in H_0. $E_{personnel} = 91.056, E_{operations} = 106.232, E_{other} = 182.112$. Since none of the expected frequencies is less than 1 and none of the expected frequencies is less than 5, the conditions for performing the χ^2 goodness of fit test are met. $\chi^2_{crit} = \chi^2_{0.05} = 5.991$. Reject H_0 if $\chi^2_{data} > 5.991$. $\chi^2_{data} = 3.7072$. Since $\chi^2_{data} \leq 5.991$, we do not reject

H_0. There is insufficient evidence that the population proportions of budget items have changed from 1985 to 2004.

Section 11.2

1. Tabular summary of the relationship between two categorical variables

3. The two-sample Z test for the difference in proportions from Chapter 10 is for comparing proportions of two independent populations, and the χ^2 test for homogeneity of proportions is for comparing proportions of k independent populations.

5.

	A1	A2	Total
B1	11	19	30
B2	11	19	30
Total	22	38	60

7.

	E1	E2	E3	Total
F1	30.71	20.79	8.50	60
F2	34.29	23.21	9.50	67
Total	65	44	18	127

9.

	I1	I2	I3	Total
J1	99.2788	93.6058	102.1154	295
J2	55.5288	52.3558	57.1154	165
J3	20.1923	19.0385	20.7692	60
Total	175	165	180	520

11. (a) H_0: Variable A and Variable B are independent. H_a: Variable A and Variable B are not independent.

(b)

	A1	A2	Total
B1	11	19	30
B2	11	19	30
Total	22	38	60

Since none of the expected frequencies is less than 1 and none of the expected frequencies is less than 5, the conditions for performing the χ^2 test for independence are met. **(c)** 3.841. Reject H_0 if $\chi^2_{data} > 3.841$. **(d)** 0.2871 **(e)** Since $\chi^2_{data} \leq 3.841$, we do not reject H_0. There is insufficient evidence that variable A and variable B are not independent.

13. (a) H_0: Variable I and Variable J are independent. H_a: Variable I and Variable J are not independent.

(b)

	I1	I2	I3	Total
J1	99.2788	93.6058	102.1154	295
J2	55.5288	52.3558	57.1154	165
J3	20.1923	19.0385	20.7692	60
Total	175	165	180	520

Since none of the expected frequencies is less than 1 and none of the expected frequencies is less than 5, the conditions for performing the χ^2 test for independence are met. **(c)** 13.277. Reject H_0 if $\chi^2_{data} > 13.277$. **(d)** 4.000 **(e)** Since $\chi^2_{data} \leq 13.277$, we do not reject H_0. There is insufficient evidence that variable I and variable J are not independent.

15. (a) H_0: $p_1 = p_2 = p_3$. H_a: Not all the proportions in H_0 are equal.

(b)

	Sample 1	Sample 2	Sample 3	Total
Successes	9.63	20.86	29.52	60.01
Failures	20.37	44.14	62.48	126.99
Total	30	65	92	187

Since none of the expected frequencies is less than 1 and none of the expected frequencies is less than 5, the conditions for perform-

ing the χ^2 test for homogeneity of proportions are met. **(c)** 5.991. Reject H_0 if $\chi^2_{data} > 5.991$. **(d)** 0.0847 **(e)** Since $\chi^2_{data} \leq 5.991$, we do not reject H_0. There is insufficient evidence that not all the proportions in H_0 are equal.

17. (a) H_0: $p_1 = p_2 = p_3 = p_4$. H_a: Not all the proportions in H_0 are equal.

(b)

	Sample 1	Sample 2	Sample 3	Sample 4	Total
Successes	9.67	15.08	20.11	25.14	70
Failures	15.33	23.92	31.89	39.86	111
Total	25	39	52	65	181

Since none of the expected frequencies is less than 1 and none of the expected frequencies is less than 5, the conditions for performing the χ^2 test for homogeneity of proportions are met. **(c)** 7.815. Reject H_0 if $\chi^2_{data} > 7.815$. **(d)** 0.0215 **(e)** Since $\chi^2_{data} \leq 7.815$, we do not reject H_0. There is insufficient evidence that not all the proportions in H_0 are equal.

19. H_0: *Type of stimulus* and *type of mouse* are independent. H_a: *Type of stimulus* and *type of mouse* are not independent. Reject H_0 if *p*-value < 0.05. Since none of the expected frequencies is less than 1 and none of the expected frequencies is less than 5, the conditions for performing the χ^2 test for independence are met. $\chi^2_{data} = 0.7937$. *p*-value > 0.10. Since *p*-value ≥ 0.10, we do not reject H_0. There is insufficient evidence that *type of stimulus* and *type of mouse* are not independent.

21. H_0: $p_{Republicans} = p_{Independents} = p_{Democrats}$. H_a: Not all the proportions in H_0 are equal. Since none of the expected frequencies is less than 1 and none of the expected frequencies is less than 5, the conditions for performing the χ^2 test for homogeneity of proportions are met. $\chi^2_{crit} = \chi^2_{0.05} = 5.991$. Reject H_0 if $\chi^2_{data} > 5.991$. $\chi^2_{data} = 1.1568$. Since $\chi^2_{data} \leq 5.991$, we do not reject H_0. There is insufficient evidence that the population proportion favoring restriction is not the same for all three groups.

23. H_0: *Gender* and *political party preference* are independent. H_a: *Gender* and *political party preference* are not independent. Reject H_0 if *p*-value < 0.01. Since none of the expected frequencies is less than 1 and none of the expected frequencies is less than 5, the conditions for performing the χ^2 test for independence are met. $\chi^2_{data} = 33.333$. *p*-value < 0.005. Since *p*-value < 0.01, we reject H_0. There is evidence that *gender* and *political party preference* are not independent.

25. H_0: $p_{Edit} = p_{Arrange}$. H_a: Not all the proportions in H_0 are equal. Reject H_0 if *p*-value < 0.05. Since none of the expected frequencies is less than 1 and none of the expected frequencies is less than 5, the conditions for performing the χ^2 test for homogeneity of proportions are met. $\chi^2_{data} = 3.5165$. $0.05 < $ *p*-value < 0.10. Since *p*-value ≥ 0.05, we do not reject H_0. There is insufficient evidence that the proportions who favor email differ between the two tasks.

27. H_0: $p_{Work} = p_{Personal}$. H_a: Not all the proportions in H_0 are equal. Reject H_0 if *p*-value < 0.01. Since none of the expected frequencies is less than 1 and none of the expected frequencies is less than 5, the conditions for performing the χ^2 test for homogeneity of proportions are met. $\chi^2_{data} = 11.0753$. *p*-value < 0.005. Since *p*-value < 0.01, we reject H_0. There is evidence that the population proportions who report "a lot of spam" are not the same for work email and personal email. Yes.

29. H_0: $p_{Urban} = p_{Suburban} = p_{Rural}$. H_a: Not all the proportions in H_0 are equal. Reject H_0 if *p*-value < 0.05. Since none of the expected frequencies is less than 1 and none of the expected frequencies is less than 5, the conditions for performing the χ^2 test for homogeneity of proportions are met. $\chi^2_{data} = 9.095$. $0.01 < $ *p*-value < 0.025.

Since p-value < 0.05, we reject H_0. There is evidence that the population proportions of residents from the three categories who use online dating are not all the same.

31. (a) Dependent **(b)** Since the p-value 0.001, p-value < 0.10. Thus we reject H_0. There is evidence that *urb_rural* and *goals* are not independent.

Section 11.3

1. No. If the sample sizes are not all the same, then we need to calculate the overall sample mean by calculating the weighted mean of the sample means where the weights are the sample sizes.

3. Answers will vary.

5. Against, because this indicates the variability among the sample means is large compared to the within-sample variability.

7. (a) $F_{crit} = 5.59$
(b)

9. (a) $F_{crit} = 6.36$
(b)

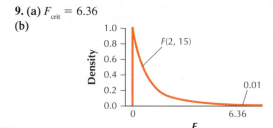

11. (a) $F_{crit} = 3.24$
(b)

13. (a) $H_0 : \mu_1 = \mu_2 = \mu_3$. H_a : Not all the population means are equal. **(b)** $F_{crit} = 3.89$. Reject H_0 if $F_{data} > 3.89$. **(c)** $F_{data} = 20$. **(d)** Since $F_{data} = 20$ is greater than $F_{crit} = 3.89$, the conclusion is reject H_0. There is evidence that not all the population means are equal.

15. (a) $H_0 : \mu_A = \mu_B = \mu_C = \mu_D$. H_a : Not all the population means are equal. **(b)** $F_{crit} = 2.61$. Reject H_0 if $F_{data} > 2.61$. **(c)** 1.225966055 **(d)** Since $F_{data} \leq 2.61$, we do not reject H_0. There is insufficient evidence that not all the population means are equal.

17. (a) Missing values are in red.

Source of variation	Sum of squares	Degrees of freedom	Mean square	F-test statistic
Treatment	SSTR = 120	$df_1 = 6$	MSTR = 20	$F_{data} = 44$
Error	SSE = 315	$df_2 = 693$	MSE = 0.4545454545	
Total	SST = 435			

(b) $H_0 : \mu_1 = \mu_2 = \mu_3 = \mu_4 = \mu_5 = \mu_6 = \mu_7$. H_a : Not all the population means are equal. Reject H_0 if the p-value < 0.05. $F_{data} = 44$;

p-value < 0.001. Since the p-value < 0.05, we reject H_0. There is evidence that not all the population means are equal.

19. (a) Missing values are in red.

Source of variation	Sum of squares	Degrees of freedom	Mean square	F-test statistic
Treatment	SSTR = 40	$df_1 = 4$	MSTR = 10	$F_{data} = 1.0$
Error	SSE = 400	$df_2 = 40$	MSE = 10	
Total	SST = 440			

(b) $H_0 : \mu_1 = \mu_2 = \mu_3 = \mu_4 = \mu_5$. H_a : Not all the population means are equal. $F_{crit} = 2.06$. Reject H_0 if $F_{data} > 2.06$. $F_{data} = 1.0$. Since $F_{data} \leq 2.06$, we do not reject H_0. There is insufficient evidence that not all the population means are equal.

21. $H_0 : \mu_{Protestant} = \mu_{Catholic} = \mu_{Jewish} = \mu_{none} = \mu_{other}$. H_a : Not all the population means are equal. $\mu_{Protestant}$ = the population mean number of years of education of people whose religious preference is Protestant. $\mu_{Catholic}$ = the population mean number of years of education of people whose religious preference is Catholic. μ_{Jewish} = the population mean number of years of education of people whose religious preference is Jewish. μ_{none} = the population mean number of years of education of people whose religious preference is None. μ_{other} = the population mean number of years of education of people whose religious preference is Other. Reject H_0 if the p-value < 0.05. $F_{data} = 17.31488654$ p-value < 0.001. Since the p-value < 0.05, we reject H_0. There is evidence that not all the population mean numbers of years of education are equal.

23. $H_0 : \mu_{before} = \mu_{during} = \mu_{after}$. H_a : Not all the population means are equal. μ_{before} = the population mean number of emergency room visits before a full moon. μ_{during} = the population mean number of emergency room visits during a full moon. μ_{after} = the population mean number of emergency room visits after a full moon. $F_{crit} = 3.39$. Reject H_0 if $F_{data} > 3.39$. $F_{data} = 1.174110786$. Since $F_{data} \leq 3.39$, we do not reject H_0. There is insufficient evidence that not all the population mean numbers of emergency room visits are equal.

25. The cereals were selected randomly from the manufacturers so that the selection of a cereal from one manufacturer did not affect the selection of cereals from other manufacturers.

27. $H_0 : \mu_{online} = \mu_{traditional} = \mu_{hybrid}$. H_a : Not all the population means are equal. $F_{crit} = 6.36$. Reject H_0 if $F_{data} > 6.36$. $F_{data} = 0.586924$. Since $F_{data} \leq 6.36$, we do not reject H_0. There is insufficient evidence that not all the population mean numbers of emergency room visits are equal.

29. (a) $SSTR = 0$; $MSTR = 0$; $F_{data} = 0$. **(b)** p-value $= 1$. **(c)** Since the p-value ≥ 0.05, we would not reject H_0.

31. Since the p-value < 0.01, we reject H_0. There is evidence that not all the population mean heights are equal.

33. (a) $H_0 : \mu_{87} = \mu_{88} = \mu_{89} = \mu_{90} = \mu_{91}$. H_a : Not all the population means are equal. **(b)** Reject H_0 if the p-value < 0.05. **(c)** $F_{data} = 1.34$ **(d)** p-value $= 0.255$ **(e)** Since the p-value ≥ 0.05, we do not reject H_0. **(f)** There is insufficient evidence that not all the population means are equal.
(g)

Source	df	SS	MS	F	P
Year	4	1155151	288788	1.34	0.255
Error	333	71728290	215400		
Total	337	72883441			

35. (a) Decreases **(b)** Since the within-sample variability is increased, SSE and MSE are increased. Since the sample means remain the same, SSTR and MSTR remain unchanged. Thus $F_{data} = \dfrac{MSTR}{MSE}$ decreases.

Section 11.4

1. The regression equation is calculated from a sample and is valid only for values of x in the range of the sample data. The population regression equation may be used to approximate the relationship between the predictor variable x and the response variable y for the entire population of (x, y) pairs.

3. We construct a scatterplot of the residuals against the fitted values and a normal probability plot of the residuals. We must make sure that the scatterplot contains no strong evidence of any unhealthy patterns and that the normal probability plot indicates no evidence of departures from normality in residuals.

5. (a–b) See Student Solutions Manual. (c) The scatterplot of the residuals contains an unhealthy pattern, so the regression assumptions are not verified.

7. (a–b) See Student Solutions Manual. (c) The scatterplot of the residuals contains an unhealthy pattern, so the regression assumptions are not verified.

9. (a–b) See Student Solutions Manual. (c) The scatterplot of the residuals contains an unhealthy pattern, so the regression assumptions are not verified.

11. (a) $t_{\alpha/2} = 3.182$ (b) $(0.909, 4.091)$

13. (a) $t_{\alpha/2} = 3.182$ (b) $(1.4544, 6.5456)$

15. (a) $t_{data} = 5$ (b) p-value $= 0.0153924381$ (c) $H_0 : \beta_1 = 0$: There is no relationship between x and y. $H_a : \beta_1 \neq 0$: There is a linear relationship between x and y. Reject H_0 if the p-value < 0.05. Since the p-value < 0.05, we reject H_0. There is evidence for a linear relationship between x and y.

17. (a) $t_{data} = 5$ (b) p-value $= 0.0153924381$ (c) $H_0 : \beta_1 = 0$: There is no relationship between x and y. $H_a : \beta_1 \neq 0$: There is a linear relationship between x and y. Reject H_0 if the p-value < 0.05. Since the p-value < 0.05, we reject H_0. There is evidence for a linear relationship between x and y.

19. (a) $t_{crit} = 3.182$ (b) $t_{data} = -6.1237$ (c) $H_0 : \beta_1 = 0$: There is no relationship between x and y. $H_a : \beta_1 \neq 0$: There is a linear relationship between x and y. Reject H_0 if $t_{data} < -3.182$ or $t_{data} > 3.182$. Since $t_{data} < -3.182$, we reject H_0. There is evidence for a linear relationship between x and y.

21. (a) $t_{\alpha/2} = 3.182$ (b) $(1.2688, 1.9312)$ (c) We are 95% confident that the interval $(1.2688, 1.9312)$ captures the population slope β_1 of the relationship between volume and weight.

23. (a) $t_{\alpha/2} = 3.182$ (b) $(0.0019, 0.8509)$ (c) We are 95% confident that the interval $(0.0019, 0.8509)$ captures the population slope β_1 of the relationship between SAT I Verbal and SAT I Math.

25. (a) $t_{\alpha/2} = 2.306$ (b) $(-19.6355, 22.8087)$ (c) We are 95% confident that the interval $(-19.6355, 22.8087)$ captures the population slope β_1 of the relationship between Win% and Rating.

27. No, the regression assumptions are not violated.

29. $t_{crit} = 3.182$. Reject H_0 if $t_{data} < -3.182$ or if $t_{data} > 3.182$.

31. Since $t_{data} > 3.182$, we reject H_0. There is evidence of a linear relationship between *SAT I Verbal* and *SAT I Math*.

33. (a) The scatterplot of the residuals contains no strong evidence of unhealthy patterns and the normal probability plot indicates no evidence of departures from normality in the residuals. Therefore we conclude that the regression assumptions are verified. (b) Since 0 does not lie in the confidence interval $(0.0019, 0.8509)$, we would expect to reject the null hypothesis that $\beta_1 = 0$. (c) $t_{data} = 3.20$ (TI: 3.196) (d) p-value $= 0.049$ (TI: 0.0495) (e) Since the p-value < 0.05, we reject H_0. There is evidence for a linear relationship between *SAT I Verbal score* (x) and *SAT I Math score* (y).

35. (a) The scatterplot of the residuals contains an unhealthy pattern, so the regression assumptions are not verified. (b) Yes (c) $t_{crit} = 3.182$ (d) $t_{data} = 5$ (e) $H_0 : \beta_1 = 0$: There is no relationship between Family Size (x) and Pets (y). $H_a : \beta_1 \neq 0$: There is a linear relationship between Family Size (x) and Pets (y). Reject H_0 if $t_{data} < -3.182$ or $t_{data} > 3.182$. Since $t_{data} > 3.182$, we reject H_0. There is evidence for a linear relationship between Family Size (x) and Pets (y).

37. (a) The scatterplot of the residuals contains no strong evidence of unhealthy patterns and the normal probability plot indicates no evidence of departures from normality in the residuals. Therefore we conclude that the regression assumptions are verified. (b) Since 0 lies in the confidence interval $(-0.0332, 0.0721)$, we would expect to not reject the null hypothesis that $\beta_1 = 0$. (c) $t_{crit} = 2.306$ (d) $t_{data} = 0.85$ (TI: 0.8512) (e) Since t_{data} is not less than -2.306 and t_{data} is not greater than 2.306, we do not reject H_0. There is insufficient evidence for a linear relationship between *Price* (x) and *Change* (y).

39. (a) The scatterplot of the residuals contains no strong evidence of unhealthy patterns and the normal probability plot indicates no evidence of departures from normality in the residuals. Therefore we conclude that the regression assumptions are verified. (b) $(1.0203, 1.9577)$. We are 95% confident that the interval $(1.0203, 1.9577)$ captures the population slope β_1 of the relationship between *Dow Jones Industrial Average* (x) and *pros' performance* (y). (c) Since 0 does not lie in the confidence interval, we would expect to reject the null hypothesis that $\beta_1 = 0$. (d) $H_0 : \beta_1 = 0$. There is no relationship between *Dow Jones Industrial Average* (x) and *pros' performance* (y). $H_a : \beta_1 \neq 0$. There is a linear relationship between *Dow Jones Industrial Average* (x) and *pros' performance* (y). Reject H_0 if p-value < 0.05. $t_{data} = 6.31$. p-value ≈ 0. Since the p-value < 0.05, we reject H_0. There is evidence for a linear relationship between *Dow Jones Industrial Average* (x) and *pros' performance* (y).

41. (a) The scatterplot of the residuals contains no strong evidence of unhealthy patterns and the normal probability plot indicates no evidence of departures from normality in the residuals. Therefore we conclude that the regression assumptions are verified. (b) $(0.0087, 0.0439)$. We are 95% confident that the interval $(0.0087, 0.0439)$ captures the population slope β_1 of the relationship between *heart rate* and *body temperature*. (c) Since 0 does not lie in the confidence interval, we would expect to reject the null hypothesis that $\beta_1 = 0$. (d) $H_0 : \beta_1 = 0$. There is no relationship between *heart rate* (x) and *body temperature* (y). $H_a : \beta_1 \neq 0$. There is a linear relationship between *heart rate* (x) and *body temperatue* (y). Reject H_0 if p-value < 0.05. $t_{data} = 2.97$. p-value $= 0.004$. Since the p-value < 0.05, we reject H_0. There is evidence for a linear relationship between *heart rate* (x) and *body temperature* (y).

Chapter 11 Review

1. $H_0 : p_{USCan} = 0.32, p_{USMex} = 0.22, p_{CanUS} = 0.31, p_{MexUS} = 0.15$. H_a: The random variable does not follow the distribution specified in H_0. $E_{USCan} = \$22.4$ billion, $E_{USMex} = \$15.4$ billion, $E_{CanUS} = \$21.7$ billion, $P_{MexUS} = \$10.5$ billion. Since none of the expected frequencies is less than 1 and none of the expected frequencies is less than 5, the conditions for performing the χ^2 goodness of fit test are met. Reject H_0 if p-value < 0.05. $\chi^2_{data} = 0.469$. $0.90 < p$-value < 0.95. Since p-value ≥ 0.05, we do not reject H_0. There is insufficient evidence that the population proportions of truck-hauled trade have changed since 2002.

3. $H_0 : p_{abusedalcohol} = 0.25, p_{alcoholdependent} = 0.06, p_{other} = 0.69$. H_a: The random variable does not follow the distribution specified in

H_0. $E_{abusedalcohol} = 250$, $E_{alcoholdependent} = 60$, $E_{other} = 690$. Since none of the expected frequencies is less than 1 and none of the expected frequencies is less than 5, the conditions for performing the χ^2 goodness of fit test are met. Reject H_0 if p-value < 0.10. $\chi^2_{data} = 4.493$. p-value $= 0.1057687682$. Since p-value ≥ 0.10, we do not reject H_0. There is insufficient evidence that the population proportions have changed since 2002.

5. $H_0: p_{18-34} = 0.057$, $p_{35-49} = 0.207$, $p_{50-64} = 0.388$, $p_{over65} = 0.348$. H_a: The random variable does not follow the distribution specified in H_0. $E_{18-34} = 57$, $E_{35-49} = 207$, $E_{50-64} = 388$, $E_{over65} = 348$. Since none of the expected frequencies is less than 1 and none of the expected frequencies is less than 5, the conditions for performing the χ^2 goodness of fit test are met. $\chi^2_{crit} = \chi^2_{0.05} = 7.815$. Reject H_0 if $\chi^2_{data} > 7.815$. $\chi^2_{data} = 28.233$. Since $\chi^2_{data} > 7.815$, we reject H_0. There is evidence that the proportions have changed since 2006.

7. H_0: *Gender* and *race* are independent. H_a: *Gender* and *race* are not independent. Since none of the expected frequencies is less than 1 and none of the expected frequencies is less than 5, the conditions for performing the χ^2 test for independence are met. $\chi^2_{crit} = \chi^2_{0.05} = 4.605$. Reject H_0 if $\chi^2_{data} > 4.605$. $\chi^2_{data} = 16.918$. Since $\chi^2_{data} > 4.605$, we reject H_0. There is evidence that *gender* and *race* are not independent.

9. H_0: Age and radio station type are independent. H_a: Age and radio station type are not independent. Reject H_0 if p-value < 0.05. Since none of the expected frequencies is less than 1 and none of the expected frequencies is less than 5, the conditions for performing the χ^2 test for independence are met. $\chi^2_{data} = 8.269$. p-value < 0.005. Since p-value < 0.05, we reject H_0. There is evidence that age and radio station type are not independent.

11. (a) $H_0: \mu_A = \mu_B = \mu_C$. H_a: Not all the population means are equal. (b) $F_{crit} = 3.49$ (c) Reject H_0 if $F_{data} > 3.49$. (d) $F_{data} = 1.57$ (e) Since $F_{data} \leq 3.49$, we do not reject H_0. (f) There is insufficient evidence that not all the population means are equal.

13. (a) $H_0: \mu_A = \mu_B = \mu_C = \mu_D$. H_a: Not all the population means are equal. Reject H_0 if the p-value < 0.05. (b) $F_{data} = 25.47$ (c) p-value ≈ 0 (d) Since the p-value < 0.05, we reject H_0. There is evidence that not all the population means are equal.

15. (2.9453, 5.9919). We are 95% confident that the interval (2.9453, 5.9919) captures the population slope β_1 of the relationship between Education and Annual Earnings.

17. (-0.7770, 0.7676). We are 95% confident that the interval (-0.7770, 0.7676) captures the population slope β_1 of the relationship between Field Goals and Free Throws.

19. (a) The scatterplot of the residuals contains an unhealthy pattern, so the regression assumptions are not verified. (b) Yes (c) $t_{data} = 7.54$ (TI: 7.542) (d) p-value $= 0.001$ (TI: 0.0006) (e) $H_0: \beta_1 = 0$: There is no relationship between Education (x) and Annual Earnings (y). $H_a: \beta_1 \neq 0$: There is a linear relationship between Education (x) and Annual Earnings (y). Reject H_0 if p-value < 0.05. Since the p-value < 0.05, we reject H_0. There is evidence for a linear relationship between Education (x) and Annual Earnings (y).

21. (a) Since 0 is in the confidence interval, we would expect to not reject the null hypothesis that $\beta_1 = 0$. (b) $t_{crit} = 2.306$ (c) $t_{data} = -0.014$ (d) $H_0: \beta_1 = 0$: There is no relationship between Field Goals (x) and Free Throws (y). $H_a: \beta_1 \neq 0$: There is a linear relationship between Field Goals (x) and Free Throws (y). Reject H_0 if $t_{data} < -2.306$ or $t_{data} > 2.306$. Since t_{data} is not less than -2.306 and t_{data} is not greater than 2.306, we do not reject H_0. There is insufficient evidence for a linear relationship between Field Goals (x) and Free Throws (y).

Chapter 11 Quiz

1. False
2. False
3. False
4. equal
5. mean square treatment
6. mean square area error
7. $\bar{\bar{x}}$
8. H_a, the alternative hypothesis
9. Degrees of freedom $= (r - 1)(c - 1)$, where $r =$ the number of categories in the row variable and $c =$ the number of categories in the column variable.
10. $H_0: p_{Northeast} = p_{Midwest} = p_{South} = p_{West}$. H_a: Not all the proportions in H_0 are equal. Reject H_0 if p-value < 0.01. Since none of the expected frequencies is less than 1 and none of the expected frequencies is less than 5, the conditions for performing the χ^2 test for homogeneity of proportions are met. $\chi^2_{data} = 11.97$. p-value < 0.005. Since p-value < 0.01, we reject H_0. There is evidence that the population proportions of pregnant women who have had an HIV test in the past 12 months are not the same across all four regions.
Here is minitab's calculation of $\chi^2_{data} = 11$. This value is 25.846.

	Northeast	Midwest	South	West	Total
1	568	493	585	502	2148
	537.00	537.00	537.00	537.00	
	1.790	3.605	4.291	2.281	
2	432	507	415	498	1852
	463.00	463.00	463.00	463.00	
	2.076	4.181	4.976	2.646	
Total	1000	1000	1000	1000	4000

Chi-Sq = 25.846,   DF = 3,   P-Value = 0.000

11. $H_0: p_{8thgraders} = p_{10thgraders} = p_{12thgraders}$. H_a: Not all the proportions in H_0 are equal. Reject H_0 if p-value < 0.01. Since none of the expected frequencies is less than 1 and none of the expected frequencies is less than 5, the conditions for performing the χ^2 test for homogeneity of proportions are met. $\chi^2_{data} = 3060.14226$. p-value < 0.005. Since p-value < 0.01, we reject H_0. There is evidence that the proportions of children in those grades that have ever used an illicit drug are not all the same.

12. $H_0: p_{Texas} = p_{Oklahoma} = p_{Pennsylvania}$. H_a: Not all the proportions in H_0 are equal. Reject H_0 if p-value < 0.05. Since none of the expected frequencies is less than 1 and none of the expected frequencies is less than 5, the conditions for performing the χ^2 test for homogeneity of proportions are met. $\chi^2_{data} = 81,246.70827$. p-value < 0.005. Since p-value < 0.01, we reject H_0. There is evidence that the population proportions of cattle on smaller farms are not the same across all three states.

13. $H_0: \mu_4 = \mu_6 = \mu_8$. H_a: Not all the population means are equal. $\mu_4 =$ population mean miles per gallon for 4-cylinder cars. $\mu_6 =$ population mean miles per gallon for 6-cylinder cars. $\mu_8 =$ population mean miles per gallon for 8-cylinder cars. Reject H_0 if the p-value < 0.05. $F_{data} = 341.4931564$. p-value < 0.001. Since the p-value < 0.05, we reject H_0. There is evidence that not all the population mean gas mileages are equal.

14. $H_0: \mu_{married} = \mu_{widowed} = \mu_{divorced} = \mu_{separated} = \mu_{nevermarried}$. H_a: Not all the population means are equal. $\mu_{married} =$ population mean number of hours worked by people who are married people.

μ_{widowed} = population mean number of hours worked by people who are widowed. μ_{divorced} = population mean number of hours worked by people who are divorced. $\mu_{\text{separated}}$ = population mean number of hours worked by people who are separated. $\mu_{\text{nevermarried}}$ = population mean number of hours worked by people who have never been married. Reject H_0 if the p-value < 0.05. $F_{\text{data}} = 2.510227871$. $0.025 < p$-value < 0.05. Since the p-value < 0.05, we reject H_0. There is evidence that not all the population mean numbers of hours worked are equal.

15. $H_0 : \mu_{\text{Kelloggs}} = \mu_{\text{Quaker}} = \mu_{\text{RalstonPurina}}$. H_a : Not all the population means are equal. μ_{Kelloggs} = population mean number of calories per serving in breakfast cereals made by Kellogg's. μ_{Quaker} = population mean number of calories per serving in breakfast cereals made by Quaker. $\mu_{\text{RalstonPurina}}$ = population mean number of calories per serving in breakfast cereals made by Ralston Purina. $F_{\text{crit}} = 3.39$. Reject H_0 if $F_{\text{data}} > 3.39$. $F_{\text{data}} = 1.557344523$. Since $F_{\text{data}} \leq 3.39$, we do not reject H_0. There is insufficient evidence that not all the population mean numbers of calories per serving are equal.

16. $H_0 : \mu_{\text{freshmen}} = \mu_{\text{sophomores}} = \mu_{\text{juniors}} = \mu_{\text{seniors}} = \mu_{\text{graduate students}}$. H_a : Not all the population means are equal. μ_{freshmen} = population mean GPA of freshmen at this university. $\mu_{\text{sophomores}}$ = population mean GPA of sophomores at this university. μ_{juniors} = population mean GPA of juniors at this university. μ_{seniors} = population mean GPA of seniors at this university. $\mu_{\text{graduate students}}$ = population mean GPA of graduate students at this university. $F_{\text{crit}} = 2.87$. Reject H_0 if $F_{\text{data}} > 2.87$. $F_{\text{data}} = 21.40182689$. Since $F_{\text{data}} > 2.87$, we reject H_0. There is evidence that not all the population mean GPAs are equal.

17. $(5.3021, 9.2675)$. We are 95% confident that the interval $(5.3021, 9.2675)$ captures the population slope β_1 of the relationship between Weight and Height.

18. $(0.2421, 1.1663)$. We are 95% confident that the interval $(0.2421, 1.1663)$ captures the population slope β_1 of the relationship between First-Semester Grade and Second-Semester Grade.

19. $H_0 : \beta_1 = 0$: There is no relationship between Height (x) and Weight (y). $H_a : \beta_1 \neq 0$: There is a linear relationship between Height (x) and Weight (y). Reject H_0 if the p-value < 0.05. $t_{\text{data}} = 8.47$, p-value ≈ 0. Since the p-value < 0.05, we reject H_0. There is evidence for a linear relationship between Height (x) and Weight (y).

20. $H_0 : \beta_1 = 0$: There is no relationship between First-Semester Grade (x) and Second-Semester Grade (y). $H_a : \beta_1 \neq 0$: There is a linear relationship between First-Semester Grade (x) and Second-Semester Grade (y). Reject H_0 if the p-value < 0.05. $t_{\text{data}} = 3.40$, p-value $= 0.007$. Since the p-value < 0.05, we reject H_0. There is evidence for a linear relationship between First-Semester Grade (x) and Second-Semester Grade (y).

Tables Appendix

Table A Random numbers

10480	15011	01536	02011	81647	91646	67179	14194	62590	36207	20969	99570	91291	90700
22368	46573	25595	85393	30995	89198	27982	53402	93965	34095	52666	19174	39615	99505
24130	48360	22527	97265	76393	64809	15179	24830	49340	32081	30680	19655	63348	58629
42167	93093	06243	61680	07856	16376	39440	53537	71341	57004	00849	74917	97758	16379
37570	39975	81837	16656	06121	91782	60468	81305	49684	60672	14110	06927	01263	54613
77921	06907	11008	42751	27756	53498	18602	70659	90655	15053	21916	81825	44394	42880
99562	72905	56420	69994	98872	31016	71194	18738	44013	48840	63213	21069	10634	12952
96301	91977	05463	07972	18876	20922	94595	56869	69014	60045	18425	84903	42508	32307
89579	14342	63661	10281	17453	18103	57740	84378	25331	12566	58678	44947	05584	56941
85475	36857	43342	53988	53060	59533	38867	62300	08158	17983	16439	11458	18593	64952
28918	69578	88231	33276	70997	79936	56865	05859	90106	31595	01547	85590	91610	78188
63553	40961	48235	03427	49626	69445	18663	72695	52180	20847	12234	90511	33703	90322
09429	93969	52636	92737	88974	33488	36320	17617	30015	08272	84115	27156	30613	74952
10365	61129	87529	85689	48237	52267	67689	93394	01511	26358	85104	20285	29975	89868
07119	97336	71048	08178	77233	13916	47564	81056	97735	85977	29372	74461	28551	90707
51085	12765	51821	51259	77452	16308	60756	92144	49442	53900	70960	63990	75601	40719
02368	21382	52404	60268	89368	19885	55322	44819	01188	65255	64835	44919	05944	55157
01011	54092	33362	94904	31273	04146	18594	29852	71585	85030	51132	01915	92747	64951
52162	53916	46369	58586	23216	14513	83149	98736	23495	64350	94738	17752	35156	35749
07056	97628	33787	09998	42698	06691	76988	13602	51851	46104	88916	19509	25625	58104
48663	91245	85828	14346	09172	30168	90229	04734	59193	22178	30421	61666	99904	32812
54164	58492	22421	74103	47070	25306	76468	26384	58151	06646	21524	15227	96909	44592
32639	32363	05597	24200	13363	38005	94342	28728	35806	06912	17012	64161	18296	22851
29334	27001	87637	87308	58731	00256	45834	15398	46557	41135	10367	07684	36188	18510
02488	33062	28834	07351	19731	92420	60952	61280	50001	67658	32586	86679	50720	94953
81525	72295	04839	96423	24878	82651	66566	14778	76797	14780	13300	87074	79666	95725
29676	20591	68086	26432	46901	20849	89768	81536	86645	12659	92259	57102	80428	25280
00742	57392	39064	66432	84673	40027	32832	61362	98947	96067	64760	64584	96096	98253
05366	04213	25669	26422	44407	44048	37937	63904	45766	66134	75470	66520	34693	90449
91921	26418	64117	94305	26766	25940	39972	22209	71500	64568	91402	42416	07844	69618
00582	04711	87917	77341	42206	35126	74087	99547	81817	42607	43808	76655	62028	76630
00725	69884	62797	56170	86324	88072	76222	36086	84637	93161	76038	65855	77919	88006
69011	65797	95876	55293	18988	27354	26575	08625	40801	59920	29841	80150	12777	48501
25976	57948	29888	88604	67917	48708	18912	82271	65424	69774	33611	54262	85963	03547
09763	83473	73577	12908	30883	18317	28290	35797	05998	41688	34952	37888	38917	88050
91567	42595	27958	30134	04024	86385	29880	99730	55536	84855	29080	09250	79656	73211
17955	56349	90999	49127	20044	59931	06115	20542	18059	02008	73708	83517	36103	42791
46503	18584	18845	49618	02304	51038	20655	58727	28168	15475	56942	53389	20562	87338
92157	89634	94824	78171	84610	82834	09922	25417	44137	48413	25555	21246	35509	20468
14577	62765	35605	81263	39667	47358	56873	56307	61607	49518	89656	20103	77490	18062
98427	07523	33362	64270	01638	92477	66969	98420	04880	45585	46565	04102	46880	45709
34914	63976	88720	82765	34476	17032	87589	40836	32427	70002	70663	88863	77775	69348
70060	28277	39475	46473	23219	53416	94970	25832	69975	94884	19661	72828	00102	66794
53976	54914	06990	67245	68350	82948	11398	42878	80287	88267	47363	46634	06541	97809
76072	29515	40980	07391	58745	25774	22987	80059	39911	96189	41151	14222	60697	59583
90725	52210	83974	29992	65831	38857	50490	83765	55657	14361	31720	57375	56228	41546
64364	67412	33339	31926	14883	24413	59744	92351	97473	89286	35931	04110	23726	51900
08962	00358	31662	25388	61642	34072	81249	35648	56891	69352	48373	45578	78547	81788
95012	68379	93526	70765	10593	04542	76463	54328	02349	17247	28865	14777	62730	92277
15664	10493	20492	38391	91132	21999	59516	81652	27195	48223	46751	22923	32261	85653

Reprinted with permission from W. H. Beyer, *Handbook of Tables for Probability and Statistics*, 2nd ed. Copyright CRC Press, Boca Raton, Fla., 1986.

Table B Binomial distribution

n	X	\multicolumn{9}{c}{p}
		0.10	0.15	0.20	0.25	0.30	0.35	0.40	0.45	0.50
2	0	0.8100	0.7225	0.6400	0.5625	0.4900	0.4225	0.3600	0.3025	0.2500
	1	0.1800	0.2550	0.3200	0.3750	0.4200	0.4550	0.4800	0.4950	0.5000
	2	0.0100	0.0225	0.0400	0.0625	0.0900	0.1225	0.1600	0.2025	0.2500
3	0	0.7290	0.6141	0.5120	0.4219	0.3430	0.2746	0.2160	0.1664	0.1250
	1	0.2430	0.3251	0.3840	0.4219	0.4410	0.4436	0.4320	0.4084	0.3750
	2	0.0270	0.0574	0.0960	0.1406	0.1890	0.2389	0.2880	0.3341	0.3750
	3	0.0010	0.0034	0.0080	0.0156	0.0270	0.0429	0.0640	0.0911	0.1250
4	0	0.6561	0.5220	0.4096	0.3164	0.2401	0.1785	0.1296	0.0915	0.0625
	1	0.2916	0.3685	0.4096	0.4219	0.4116	0.3845	0.3456	0.2995	0.2500
	2	0.0486	0.0975	0.1536	0.2109	0.2646	0.3105	0.3456	0.3675	0.3750
	3	0.0036	0.0115	0.0256	0.0469	0.0756	0.1115	0.1536	0.2005	0.2500
	4	0.0001	0.0005	0.0016	0.0039	0.0081	0.0150	0.0256	0.0410	0.0625
5	0	0.5905	0.4437	0.3277	0.2373	0.1681	0.1160	0.0778	0.0503	0.0312
	1	0.3280	0.3915	0.4096	0.3955	0.3602	0.3124	0.2592	0.2059	0.1562
	2	0.0729	0.1382	0.2048	0.2637	0.3087	0.3364	0.3456	0.3369	0.3125
	3	0.0081	0.0244	0.0512	0.0879	0.1323	0.1811	0.2304	0.2757	0.3125
	4	0.0004	0.0022	0.0064	0.0146	0.0284	0.0488	0.0768	0.1128	0.1562
	5		0.0001	0.0003	0.0010	0.0024	0.0053	0.0102	0.0185	0.0312
6	0	0.5314	0.3771	0.2621	0.1780	0.1176	0.0754	0.0467	0.0277	0.0156
	1	0.3543	0.3993	0.3932	0.3560	0.3025	0.2437	0.1866	0.1359	0.0938
	2	0.0984	0.1762	0.2458	0.2966	0.3241	0.3280	0.3110	0.2780	0.2344
	3	0.0146	0.0415	0.0819	0.1318	0.1852	0.2355	0.2765	0.3032	0.3125
	4	0.0012	0.0055	0.0154	0.0330	0.0595	0.0951	0.1382	0.1861	0.2344
	5	0.0001	0.0004	0.0015	0.0044	0.0102	0.0205	0.0369	0.0609	0.0938
	6			0.0001	0.0002	0.0007	0.0018	0.0041	0.0083	0.0156
7	0	0.4783	0.3206	0.2097	0.1335	0.0824	0.0490	0.0280	0.0152	0.0078
	1	0.3720	0.3960	0.3670	0.3115	0.2471	0.1848	0.1306	0.0872	0.0547
	2	0.1240	0.2097	0.2753	0.3115	0.3177	0.2985	0.2613	0.2140	0.1641
	3	0.0230	0.0617	0.1147	0.1730	0.2269	0.2679	0.2903	0.2918	0.2734
	4	0.0026	0.0109	0.0287	0.0577	0.0972	0.1442	0.1935	0.2388	0.2734
	5	0.0002	0.0012	0.0043	0.0115	0.0250	0.0466	0.0774	0.1172	0.1641
	6		0.0001	0.0004	0.0013	0.0036	0.0084	0.0172	0.0320	0.0547
	7				0.0001	0.0002	0.0006	0.0016	0.0037	0.0078
8	0	0.4305	0.2725	0.1678	0.1001	0.0576	0.0319	0.0168	0.0084	0.0039
	1	0.3826	0.3847	0.3355	0.2670	0.1977	0.1373	0.0896	0.0548	0.0312
	2	0.1488	0.2376	0.2936	0.3115	0.2965	0.2587	0.2090	0.1569	0.1094
	3	0.0331	0.0839	0.1468	0.2076	0.2541	0.2786	0.2787	0.2568	0.2188
	4	0.0046	0.0185	0.0459	0.0865	0.1361	0.1875	0.2322	0.2627	0.2734
	5	0.0004	0.0026	0.0092	0.0231	0.0467	0.0808	0.1239	0.1719	0.2188
	6		0.0002	0.0011	0.0038	0.0100	0.0217	0.0413	0.0703	0.1094
	7			0.0001	0.0004	0.0012	0.0033	0.0079	0.0164	0.0313
	8					0.0001	0.0002	0.0007	0.0017	0.0039

Note: Blank entries indicate a binomial probability of less than 0.00005.

(Continued)

Table B Binomial distribution (*continued*)

n	X	p 0.10	0.15	0.20	0.25	0.30	0.35	0.40	0.45	0.50
9	0	0.3874	0.2316	0.1342	0.0751	0.0404	0.0207	0.0101	0.0046	0.0020
	1	0.3874	0.3679	0.3020	0.2253	0.1556	0.1004	0.0605	0.0339	0.0176
	2	0.1722	0.2597	0.3020	0.3003	0.2668	0.2162	0.1612	0.1110	0.0703
	3	0.0446	0.1069	0.1762	0.2336	0.2668	0.2716	0.2508	0.2119	0.1641
	4	0.0074	0.0283	0.0661	0.1168	0.1715	0.2194	0.2508	0.2600	0.2461
	5	0.0008	0.0050	0.0165	0.0389	0.0735	0.1181	0.1672	0.2128	0.2461
	6	0.0001	0.0006	0.0028	0.0087	0.0210	0.0424	0.0743	0.1160	0.1641
	7			0.0003	0.0012	0.0039	0.0098	0.0212	0.0407	0.0703
	8				0.0001	0.0004	0.0013	0.0035	0.0083	0.0176
	9						0.0001	0.0003	0.0008	0.0020
10	0	0.3487	0.1969	0.1074	0.0563	0.0282	0.0135	0.0060	0.0025	0.0010
	1	0.3874	0.3474	0.2684	0.1877	0.1211	0.0725	0.0403	0.0207	0.0098
	2	0.1937	0.2759	0.3020	0.2816	0.2335	0.1757	0.1209	0.0763	0.0439
	3	0.0574	0.1298	0.2013	0.2503	0.2668	0.2522	0.2150	0.1665	0.1172
	4	0.0112	0.0401	0.0881	0.1460	0.2001	0.2377	0.2508	0.2384	0.2051
	5	0.0015	0.0085	0.0264	0.0584	0.1029	0.1536	0.2007	0.2340	0.2461
	6	0.0001	0.0012	0.0055	0.0162	0.0368	0.0689	0.1115	0.1596	0.2051
	7		0.0001	0.0008	0.0031	0.0090	0.0212	0.0425	0.0746	0.1172
	8			0.0001	0.0004	0.0014	0.0043	0.0106	0.0229	0.0439
	9					0.0001	0.0005	0.0016	0.0042	0.0098
	10							0.0001	0.0003	0.0010
12	0	0.2824	0.1422	0.0687	0.0317	0.0138	0.0057	0.0022	0.0008	0.0002
	1	0.3766	0.3012	0.2062	0.1267	0.0712	0.0368	0.0174	0.0075	0.0029
	2	0.2301	0.2924	0.2835	0.2323	0.1678	0.1088	0.0639	0.0339	0.0161
	3	0.0853	0.1720	0.2362	0.2581	0.2397	0.1954	0.1419	0.0923	0.0537
	4	0.0213	0.0683	0.1329	0.1936	0.2311	0.2367	0.2128	0.1700	0.1208
	5	0.0038	0.0193	0.0532	0.1032	0.1585	0.2039	0.2270	0.2225	0.1934
	6	0.0005	0.0040	0.0155	0.0401	0.0792	0.1281	0.1766	0.2124	0.2256
	7		0.0006	0.0033	0.0115	0.0291	0.0591	0.1009	0.1489	0.1934
	8		0.0001	0.0005	0.0024	0.0078	0.0199	0.0420	0.0762	0.1208
	9			0.0001	0.0004	0.0015	0.0048	0.0125	0.0277	0.0537
	10					0.0002	0.0008	0.0025	0.0068	0.0161
	11						0.0001	0.0003	0.0010	0.0029
	12								0.0001	0.0002
15	0	0.2059	0.0874	0.0352	0.0134	0.0047	0.0016	0.0005	0.0001	
	1	0.3432	0.2312	0.1319	0.0668	0.0305	0.0126	0.0047	0.0016	0.0005
	2	0.2669	0.2856	0.2309	0.1559	0.0916	0.0476	0.0219	0.0090	0.0032
	3	0.1285	0.2184	0.2501	0.2252	0.1700	0.1110	0.0634	0.0318	0.0139
	4	0.0428	0.1156	0.1876	0.2252	0.2186	0.1792	0.1268	0.0780	0.0417
	5	0.0105	0.0449	0.1032	0.1651	0.2061	0.2123	0.1859	0.1404	0.0916
	6	0.0019	0.0132	0.0430	0.0917	0.1472	0.1906	0.2066	0.1914	0.1527
	7	0.0003	0.0030	0.0138	0.0393	0.0811	0.1319	0.1771	0.2013	0.1964
	8		0.0005	0.0035	0.0131	0.0348	0.0710	0.1181	0.1647	0.1964
	9		0.0001	0.0007	0.0034	0.0016	0.0298	0.0612	0.1048	0.1527
	10			0.0001	0.0007	0.0030	0.0096	0.0245	0.0515	0.0916
	11				0.0001	0.0006	0.0024	0.0074	0.0191	0.0417
	12					0.0001	0.0004	0.0016	0.0052	0.0139
	13						0.0001	0.0003	0.0010	0.0032
	14								0.0001	0.0005
	15									

Note: Blank entries indicate a binomial probability of less than 0.00005.

Table B Binomial distribution (*continued*)

n	X	0.10	0.15	0.20	0.25	0.30	0.35	0.40	0.45	0.50
18	0	0.1501	0.0536	0.0180	0.0056	0.0016	0.0004	0.0001		
	1	0.3002	0.1704	0.0811	0.0338	0.0126	0.0042	0.0012	0.0003	0.0001
	2	0.2835	0.2556	0.1723	0.0958	0.0458	0.0190	0.0069	0.0022	0.0006
	3	0.1680	0.2406	0.2297	0.1704	0.1046	0.0547	0.0246	0.0095	0.0031
	4	0.0700	0.1592	0.2153	0.2130	0.1681	0.1104	0.0614	0.0291	0.0117
	5	0.0218	0.0787	0.1507	0.1988	0.2017	0.1664	0.1146	0.0666	0.0327
	6	0.0052	0.0301	0.0816	0.1436	0.1873	0.1941	0.1655	0.1181	0.0708
	7	0.0010	0.0091	0.0350	0.0820	0.1376	0.1792	0.1892	0.1657	0.1214
	8	0.0002	0.0022	0.0120	0.0376	0.0811	0.1327	0.1734	0.1864	0.1669
	9		0.0004	0.0033	0.0139	0.0386	0.0794	0.1284	0.1694	0.1855
	10		0.0001	0.0008	0.0042	0.0149	0.0385	0.0771	0.1248	0.1669
	11			0.0001	0.0010	0.0046	0.0151	0.0374	0.0742	0.1214
	12				0.0002	0.0012	0.0047	0.0145	0.0354	0.0708
	13					0.0002	0.0012	0.0045	0.0134	0.0327
	14						0.0002	0.0011	0.0039	0.0117
	15							0.0002	0.0009	0.0031
	16								0.0001	0.0006
	17									0.0001
	18									
20	0	0.1216	0.0388	0.0115	0.0032	0.0008	0.0002			
	1	0.2702	0.1368	0.0576	0.0211	0.0068	0.0020	0.0005	0.0001	
	2	0.2852	0.2293	0.1369	0.0669	0.0278	0.0100	0.0031	0.0008	0.0002
	3	0.1901	0.2428	0.2054	0.1339	0.0716	0.0323	0.0123	0.0040	0.0011
	4	0.0898	0.1821	0.2182	0.1897	0.1304	0.0738	0.0350	0.0139	0.0046
	5	0.0319	0.1028	0.1746	0.2023	0.1789	0.1272	0.0746	0.0365	0.0148
	6	0.0089	0.0454	0.1091	0.1686	0.1916	0.1712	0.1244	0.0746	0.0370
	7	0.0020	0.0160	0.0545	0.1124	0.1643	0.1844	0.1659	0.1221	0.0739
	8	0.0004	0.0046	0.0222	0.0609	0.1144	0.1614	0.1797	0.1623	0.1201
	9	0.0001	0.0011	0.0074	0.0271	0.0654	0.1158	0.1597	0.1771	0.1602
	10		0.0002	0.0020	0.0099	0.0308	0.0686	0.1171	0.1593	0.1762
	11			0.0005	0.0030	0.0120	0.0336	0.0710	0.1185	0.1602
	12			0.0001	0.0008	0.0039	0.0136	0.0355	0.0727	0.1201
	13				0.0002	0.0010	0.0045	0.0146	0.0366	0.0739
	14					0.0002	0.0012	0.0049	0.0150	0.0370
	15						0.0003	0.0013	0.0049	0.0148
	16							0.0003	0.0013	0.0046
	17								0.0002	0.0011
	18									0.0002
	19									
	20									

Note: Blank entries indicate a binomial probability of less than 0.00005.

(Continued)

Table B Binomial distribution (*continued*)

n	X	0.55	0.60	0.65	0.70	0.75	0.80	0.85	0.90	0.95
2	0	0.2025	0.1600	0.1225	0.0900	0.0625	0.0400	0.0225	0.0100	0.0025
	1	0.4950	0.4800	0.4550	0.4200	0.3750	0.3200	0.2550	0.1800	0.0950
	2	0.3025	0.3600	0.4225	0.4900	0.5625	0.6400	0.7225	0.8100	0.9025
3	0	0.0911	0.0640	0.0429	0.0270	0.0156	0.0080	0.0034	0.0010	0.0001
	1	0.3341	0.2880	0.2389	0.1890	0.1406	0.0960	0.0574	0.0270	0.0071
	2	0.4084	0.4320	0.4436	0.4410	0.4219	0.3840	0.3251	0.2430	0.1354
	3	0.1664	0.2160	0.2746	0.3430	0.4219	0.5120	0.6141	0.7290	0.8574
4	0	0.0410	0.0256	0.0150	0.0081	0.0039	0.0016	0.0005	0.0001	
	1	0.2005	0.1536	0.1115	0.0756	0.0469	0.0256	0.0115	0.0036	0.0005
	2	0.3675	0.3456	0.3105	0.2646	0.2109	0.1536	0.0975	0.0486	0.0135
	3	0.2995	0.3456	0.3845	0.4116	0.4219	0.4096	0.3685	0.2916	0.1715
	4	0.0915	0.1296	0.1785	0.2401	0.3164	0.4096	0.5220	0.6561	0.8145
5	0	0.0185	0.0102	0.0053	0.0024	0.0010	0.0003	0.0001		
	1	0.1128	0.0768	0.0488	0.0284	0.0146	0.0064	0.0022	0.0005	
	2	0.2757	0.2304	0.1811	0.1323	0.0879	0.0512	0.0244	0.0081	0.0011
	3	0.3369	0.3456	0.3364	0.3087	0.2637	0.2048	0.1382	0.0729	0.0214
	4	0.2059	0.2592	0.3124	0.3601	0.3955	0.4096	0.3915	0.3281	0.2036
	5	0.0503	0.0778	0.1160	0.1681	0.2373	0.3277	0.4437	0.5905	0.7738
6	0	0.0083	0.0041	0.0018	0.0007	0.0002	0.0001			
	1	0.0609	0.0369	0.0205	0.0102	0.0044	0.0015	0.0004	0.0001	
	2	0.1861	0.1382	0.0951	0.0595	0.0330	0.0154	0.0055	0.0012	0.0001
	3	0.3032	0.2765	0.2355	0.1852	0.1318	0.0819	0.0415	0.0146	0.0021
	4	0.2780	0.3110	0.3280	0.3241	0.2966	0.2458	0.1762	0.0984	0.0305
	5	0.1359	0.1866	0.2437	0.3025	0.3560	0.3932	0.3993	0.3543	0.2321
	6	0.0277	0.0467	0.0754	0.1176	0.1780	0.2621	0.3771	0.5314	0.7351
7	0	0.0037	0.0016	0.0006	0.0002	0.0001				
	1	0.0320	0.0172	0.0084	0.0036	0.0013	0.0004	0.0001		
	2	0.1172	0.0774	0.0466	0.0250	0.0115	0.0043	0.0012	0.0002	
	3	0.2388	0.1935	0.1442	0.0972	0.0577	0.0287	0.0109	0.0026	0.0002
	4	0.2918	0.2903	0.2679	0.2269	0.1730	0.1147	0.0617	0.0230	0.0036
	5	0.2140	0.2613	0.2985	0.3177	0.3115	0.2753	0.2097	0.1240	0.0406
	6	0.0872	0.1306	0.1848	0.2471	0.3115	0.3670	0.3960	0.3720	0.2573
	7	0.0152	0.0280	0.0490	0.0824	0.1335	0.2097	0.3206	0.4783	0.6983
8	0	0.0017	0.0007	0.0002	0.0001					
	1	0.0164	0.0079	0.0033	0.0012	0.0004	0.0001			
	2	0.0703	0.0413	0.0217	0.0100	0.0038	0.0011	0.0002		
	3	0.1719	0.1239	0.0808	0.0467	0.0231	0.0092	0.0026	0.0004	
	4	0.2627	0.2322	0.1875	0.1361	0.0865	0.0459	0.0185	0.0046	0.0004
	5	0.2568	0.2787	0.2786	0.2541	0.2076	0.1468	0.0839	0.0331	0.0054
	6	0.1569	0.2090	0.2587	0.2965	0.3115	0.2936	0.2376	0.1488	0.0515
	7	0.0548	0.0896	0.1373	0.1977	0.2670	0.3355	0.3847	0.3826	0.2793
	8	0.0084	0.0168	0.0319	0.0576	0.1001	0.1678	0.2725	0.4305	0.6634

Note: Blank entries indicate a binomial probability of less than 0.00005.

Table B Binomial distribution (*continued*)

					p				
n	X	0.55	0.60	0.65	0.70	0.75	0.80	0.85	0.90	0.95
9	0	0.0008	0.0003	0.0001						
	1	0.0083	0.0035	0.0013	0.0004	0.0001				
	2	0.0407	0.0212	0.0098	0.0039	0.0012	0.0003			
	3	0.1160	0.0743	0.0424	0.0210	0.0087	0.0028	0.0006	0.0001	
	4	0.2128	0.1672	0.1181	0.0735	0.0389	0.0165	0.0050	0.0008	
	5	0.2600	0.2508	0.2194	0.1715	0.1168	0.0661	0.0283	0.0074	0.0006
	6	0.2119	0.2508	0.2716	0.2668	0.2336	0.1762	0.1069	0.0446	0.0077
	7	0.1110	0.1612	0.2162	0.2668	0.3003	0.3020	0.2597	0.1722	0.0629
	8	0.0339	0.0605	0.1004	0.1556	0.2253	0.3020	0.3679	0.3874	0.2985
	9	0.0046	0.0101	0.0207	0.0404	0.0751	0.1342	0.2316	0.3874	0.6302
10	0	0.0003	0.0001							
	1	0.0042	0.0016	0.0005	0.0001					
	2	0.0229	0.0106	0.0043	0.0014	0.0004	0.0001			
	3	0.0746	0.0425	0.0212	0.0090	0.0031	0.0008	0.0001		
	4	0.1596	0.1115	0.0689	0.0368	0.0162	0.0055	0.0012	0.0001	
	5	0.2340	0.2007	0.1536	0.1029	0.0584	0.0264	0.0085	0.0015	0.0001
	6	0.2384	0.2508	0.2377	0.2001	0.1460	0.0881	0.0401	0.0112	0.0010
	7	0.1665	0.2150	0.2522	0.2668	0.2503	0.2013	0.1298	0.0574	0.0105
	8	0.0763	0.1209	0.1757	0.2335	0.2816	0.3020	0.2759	0.1937	0.0746
	9	0.0207	0.0403	0.0725	0.1211	0.1877	0.2684	0.3474	0.3874	0.3151
	10	0.0025	0.0060	0.0135	0.0282	0.0563	0.1074	0.1969	0.3487	0.5987
12	0	0.0001								
	1	0.0010	0.0003	0.0001						
	2	0.0068	0.0025	0.0008	0.0002					
	3	0.0277	0.0125	0.0048	0.0015	0.0004	0.0001			
	4	0.0762	0.0420	0.0199	0.0078	0.0024	0.0005	0.0001		
	5	0.1489	0.1009	0.0591	0.0291	0.0115	0.0033	0.0006		
	6	0.2124	0.1766	0.1281	0.0792	0.0401	0.0155	0.0040	0.0005	
	7	0.2225	0.2270	0.2039	0.1585	0.1032	0.0532	0.0193	0.0038	0.0002
	8	0.1700	0.2128	0.2367	0.2311	0.1936	0.1329	0.0683	0.0213	0.0021
	9	0.0923	0.1419	0.1954	0.2397	0.2581	0.2362	0.1720	0.0852	0.0173
	10	0.0339	0.0639	0.1088	0.1678	0.2323	0.2835	0.2924	0.2301	0.0988
	11	0.0075	0.0174	0.0368	0.0712	0.1267	0.2062	0.3012	0.3766	0.3413
	12	0.0008	0.0022	0.0057	0.0138	0.0317	0.0687	0.1422	0.2824	0.5404
15	0									
	1	0.0001								
	2	0.0010	0.0003	0.0001						
	3	0.0052	0.0016	0.0004	0.0001					
	4	0.0191	0.0074	0.0024	0.0006	0.0001				
	5	0.0515	0.0245	0.0096	0.0030	0.0007	0.0001			
	6	0.1048	0.0612	0.0298	0.0116	0.0034	0.0007	0.0001		
	7	0.1647	0.1181	0.0710	0.0348	0.0131	0.0035	0.0005		
	8	0.2013	0.1771	0.1319	0.0811	0.0393	0.0138	0.0030	0.0003	
	9	0.1914	0.2066	0.1906	0.1472	0.0917	0.0430	0.0132	0.0019	
	10	0.1404	0.1859	0.2123	0.2061	0.1651	0.1032	0.0449	0.0105	0.0006
	11	0.0780	0.1268	0.1792	0.2186	0.2252	0.1876	0.1156	0.0428	0.0049

Note: Blank entries indicate a binomial probability of less than 0.00005.

(Continued)

Table B Binomial distribution (*continued*)

						p				
n	*X*	0.55	0.60	0.65	0.70	0.75	0.80	0.85	0.90	0.95
	12	0.0318	0.0634	0.1110	0.1700	0.2252	0.2501	0.2184	0.1285	0.0307
	13	0.0090	0.0219	0.0476	0.0916	0.1559	0.2309	0.2856	0.2669	0.1348
	14	0.0016	0.0047	0.0126	0.0305	0.0668	0.1319	0.2312	0.3432	0.3658
	15	0.0001	0.0005	0.0016	0.0047	0.0134	0.0352	0.0874	0.2059	0.4633
18	0									
	1									
	2	0.0001								
	3	0.0009	0.0002							
	4	0.0039	0.0011	0.0002						
	5	0.0134	0.0045	0.0012	0.0002					
	6	0.0354	0.0145	0.0047	0.0012	0.0002				
	7	0.0742	0.0374	0.0151	0.0046	0.0010	0.0001			
	8	0.1248	0.0771	0.0385	0.0149	0.0042	0.0008	0.0001		
	9	0.1694	0.1284	0.0794	0.0386	0.0139	0.0033	0.0004		
	10	0.1864	0.1734	0.1327	0.0811	0.0376	0.0120	0.0022	0.0002	
	11	0.1657	0.1892	0.1792	0.1376	0.0820	0.0350	0.0091	0.0010	
	12	0.1181	0.1655	0.1941	0.1873	0.1436	0.0816	0.0301	0.0052	0.0002
	13	0.0666	0.1146	0.1664	0.2017	0.1988	0.1507	0.0787	0.0218	0.0014
	14	0.0291	0.0614	0.1104	0.1681	0.2130	0.2153	0.1592	0.0700	0.0093
	15	0.0095	0.0246	0.0547	0.1046	0.1704	0.2297	0.2406	0.1680	0.0473
	16	0.0022	0.0069	0.0190	0.0458	0.0958	0.1723	0.2556	0.2835	0.1683
	17	0.0003	0.0012	0.0042	0.0126	0.0338	0.0811	0.1704	0.3002	0.3763
	18		0.0001	0.0004	0.0016	0.0056	0.0180	0.0536	0.1501	0.3972
20	0									
	1									
	2									
	3	0.0002								
	4	0.0013	0.0003							
	5	0.0049	0.0013	0.0003						
	6	0.0150	0.0049	0.0012	0.0002					
	7	0.0366	0.0146	0.0045	0.0010	0.0002				
	8	0.0727	0.0355	0.0136	0.0039	0.0008	0.0001			
	9	0.1185	0.0710	0.0336	0.0120	0.0030	0.0005			
	10	0.1593	0.1171	0.0686	0.0308	0.0099	0.0020	0.0002		
	11	0.1771	0.1597	0.1158	0.0654	0.0271	0.0074	0.0011	0.0001	
	12	0.1623	0.1797	0.1614	0.1144	0.0609	0.0222	0.0046	0.0004	
	13	0.1221	0.1659	0.1844	0.1643	0.1124	0.0545	0.0160	0.0020	
	14	0.0746	0.1244	0.1712	0.1916	0.1686	0.1091	0.0454	0.0089	0.0003
	15	0.0365	0.0746	0.1272	0.1789	0.2023	0.1746	0.1028	0.0319	0.0022
	16	0.0139	0.0350	0.0738	0.1304	0.1897	0.2182	0.1821	0.0898	0.0133
	17	0.0040	0.0123	0.0323	0.0716	0.1339	0.2054	0.2428	0.1901	0.0596
	18	0.0008	0.0031	0.0100	0.0278	0.0669	0.1369	0.2293	0.2852	0.1887
	19	0.0001	0.0005	0.0020	0.0068	0.0211	0.0576	0.1368	0.2702	0.3774
	20			0.0002	0.0008	0.0032	0.0115	0.0388	0.1216	0.3585

Note: Blank entries indicate a binomial probability of less than 0.00005.

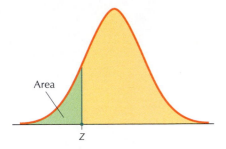

Area

Z

Table C Standard normal distribution

Z	0.00	0.01	0.02	0.03	0.04	0.05	0.06	0.07	0.08	0.09
−3.4	0.0003	0.0003	0.0003	0.0003	0.0003	0.0003	0.0003	0.0003	0.0003	0.0002
−3.3	0.0005	0.0005	0.0005	0.0004	0.0004	0.0004	0.0004	0.0004	0.0004	0.0003
−3.2	0.0007	0.0007	0.0006	0.0006	0.0006	0.0006	0.0006	0.0005	0.0005	0.0005
−3.1	0.0010	0.0009	0.0009	0.0009	0.0008	0.0008	0.0008	0.0008	0.0007	0.0007
−3.0	0.0013	0.0013	0.0013	0.0012	0.0012	0.0011	0.0011	0.0011	0.0010	0.0010
−2.9	0.0019	0.0018	0.0018	0.0017	0.0016	0.0016	0.0015	0.0015	0.0014	0.0014
−2.8	0.0026	0.0025	0.0024	0.0023	0.0023	0.0022	0.0021	0.0021	0.0020	0.0019
−2.7	0.0035	0.0034	0.0033	0.0032	0.0031	0.0030	0.0029	0.0028	0.0027	0.0026
−2.6	0.0047	0.0045	0.0044	0.0043	0.0041	0.0040	0.0039	0.0038	0.0037	0.0036
−2.5	0.0062	0.0060	0.0059	0.0057	0.0055	0.0054	0.0052	0.0051	0.0049	0.0048
−2.4	0.0082	0.0080	0.0078	0.0075	0.0073	0.0071	0.0069	0.0068	0.0066	0.0064
−2.3	0.0107	0.0104	0.0102	0.0099	0.0096	0.0094	0.0091	0.0089	0.0087	0.0084
−2.2	0.0139	0.0136	0.0132	0.0129	0.0125	0.0122	0.0119	0.0116	0.0113	0.0110
−2.1	0.0179	0.0174	0.0170	0.0166	0.0162	0.0158	0.0154	0.0150	0.0146	0.0143
−2.0	0.0228	0.0222	0.0217	0.0212	0.0207	0.0202	0.0197	0.0192	0.0188	0.0183
−1.9	0.0287	0.0281	0.0274	0.0268	0.0262	0.0256	0.0250	0.0244	0.0239	0.0233
−1.8	0.0359	0.0351	0.0344	0.0336	0.0329	0.0322	0.0314	0.0307	0.0301	0.0294
−1.7	0.0446	0.0436	0.0427	0.0418	0.0409	0.0401	0.0392	0.0384	0.0375	0.0367
−1.6	0.0548	0.0537	0.0526	0.0516	0.0505	0.0495	0.0485	0.0475	0.0465	0.0455
−1.5	0.0668	0.0655	0.0643	0.0630	0.0618	0.0606	0.0594	0.0582	0.0571	0.0559
−1.4	0.0808	0.0793	0.0778	0.0764	0.0749	0.0735	0.0721	0.0708	0.0694	0.0681
−1.3	0.0968	0.0951	0.0934	0.0918	0.0901	0.0885	0.0869	0.0853	0.0838	0.0823
−1.2	0.1151	0.1131	0.1112	0.1093	0.1075	0.1056	0.1038	0.1020	0.1003	0.0985
−1.1	0.1357	0.1335	0.1314	0.1292	0.1271	0.1251	0.1230	0.1210	0.1190	0.1170
−1.0	0.1587	0.1562	0.1539	0.1515	0.1492	0.1469	0.1446	0.1423	0.1401	0.1379
−0.9	0.1841	0.1814	0.1788	0.1762	0.1736	0.1711	0.1685	0.1660	0.1635	0.1611
−0.8	0.2119	0.2090	0.2061	0.2033	0.2005	0.1977	0.1949	0.1922	0.1894	0.1867
−0.7	0.2420	0.2389	0.2358	0.2327	0.2296	0.2266	0.2236	0.2206	0.2177	0.2148
−0.6	0.2743	0.2709	0.2676	0.2643	0.2611	0.2578	0.2546	0.2514	0.2483	0.2451
−0.5	0.3085	0.3050	0.3015	0.2981	0.2946	0.2912	0.2877	0.2843	0.2810	0.2776
−0.4	0.3446	0.3409	0.3372	0.3336	0.3300	0.3264	0.3228	0.3192	0.3156	0.3121
−0.3	0.3821	0.3783	0.3745	0.3707	0.3669	0.3632	0.3594	0.3557	0.3520	0.3483
−0.2	0.4207	0.4168	0.4129	0.4090	0.4052	0.4013	0.3974	0.3936	0.3897	0.3859
−0.1	0.4602	0.4562	0.4522	0.4483	0.4443	0.4404	0.4364	0.4325	0.4286	0.4247
−0.0	0.5000	0.4960	0.4920	0.4880	0.4840	0.4801	0.4761	0.4721	0.4681	0.4641

(Continued)

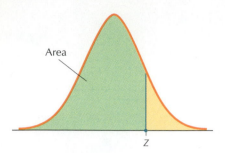

Area

Z

Table C Standard normal distribution (*continued*)

Z	0.00	0.01	0.02	0.03	0.04	0.05	0.06	0.07	0.08	0.09
0.0	0.5000	0.5040	0.5080	0.5120	0.5160	0.5199	0.5239	0.5279	0.5319	0.5359
0.1	0.5398	0.5438	0.5478	0.5517	0.5557	0.5596	0.5636	0.5675	0.5714	0.5753
0.2	0.5793	0.5832	0.5871	0.5910	0.5948	0.5987	0.6026	0.6064	0.6103	0.6141
0.3	0.6179	0.6217	0.6255	0.6293	0.6331	0.6368	0.6406	0.6443	0.6480	0.6517
0.4	0.6554	0.6591	0.6628	0.6664	0.6700	0.6736	0.6772	0.6808	0.6844	0.6879
0.5	0.6915	0.6950	0.6985	0.7019	0.7054	0.7088	0.7123	0.7157	0.7190	0.7224
0.6	0.7257	0.7291	0.7324	0.7357	0.7389	0.7422	0.7454	0.7486	0.7517	0.7549
0.7	0.7580	0.7611	0.7642	0.7673	0.7704	0.7734	0.7764	0.7794	0.7823	0.7852
0.8	0.7881	0.7910	0.7939	0.7967	0.7995	0.8023	0.8051	0.8078	0.8106	0.8133
0.9	0.8159	0.8186	0.8212	0.8238	0.8264	0.8289	0.8315	0.8340	0.8365	0.8389
1.0	0.8413	0.8438	0.8461	0.8485	0.8508	0.8531	0.8554	0.8577	0.8599	0.8621
1.1	0.8643	0.8665	0.8686	0.8708	0.8729	0.8749	0.8770	0.8790	0.8810	0.8830
1.2	0.8849	0.8869	0.8888	0.8907	0.8925	0.8944	0.8962	0.8980	0.8997	0.9015
1.3	0.9032	0.9049	0.9066	0.9082	0.9099	0.9115	0.9131	0.9147	0.9162	0.9177
1.4	0.9192	0.9207	0.9222	0.9236	0.9251	0.9265	0.9279	0.9292	0.9306	0.9319
1.5	0.9332	0.9345	0.9357	0.9370	0.9382	0.9394	0.9406	0.9418	0.9429	0.9441
1.6	0.9452	0.9463	0.9474	0.9484	0.9495	0.9505	0.9515	0.9525	0.9535	0.9545
1.7	0.9554	0.9564	0.9573	0.9582	0.9591	0.9599	0.9608	0.9616	0.9625	0.9633
1.8	0.9641	0.9649	0.9656	0.9664	0.9671	0.9678	0.9686	0.9693	0.9699	0.9706
1.9	0.9713	0.9719	0.9726	0.9732	0.9738	0.9744	0.9750	0.9756	0.9761	0.9767
2.0	0.9772	0.9778	0.9783	0.9788	0.9793	0.9798	0.9803	0.9808	0.9812	0.9817
2.1	0.9821	0.9826	0.9830	0.9834	0.9838	0.9842	0.9846	0.9850	0.9854	0.9857
2.2	0.9861	0.9864	0.9868	0.9871	0.9875	0.9878	0.9881	0.9884	0.9887	0.9890
2.3	0.9893	0.9896	0.9898	0.9901	0.9904	0.9906	0.9909	0.9911	0.9913	0.9916
2.4	0.9918	0.9920	0.9922	0.9925	0.9927	0.9929	0.9931	0.9932	0.9934	0.9936
2.5	0.9938	0.9940	0.9941	0.9943	0.9945	0.9946	0.9948	0.9949	0.9951	0.9952
2.6	0.9953	0.9955	0.9956	0.9957	0.9959	0.9960	0.9961	0.9962	0.9963	0.9964
2.7	0.9965	0.9966	0.9967	0.9968	0.9969	0.9970	0.9971	0.9972	0.9973	0.9974
2.8	0.9974	0.9975	0.9976	0.9977	0.9977	0.9978	0.9979	0.9979	0.9980	0.9981
2.9	0.9981	0.9982	0.9982	0.9983	0.9984	0.9984	0.9985	0.9985	0.9986	0.9986
3.0	0.9987	0.9987	0.9987	0.9988	0.9988	0.9989	0.9989	0.9989	0.9990	0.9990
3.1	0.9990	0.9991	0.9991	0.9991	0.9992	0.9992	0.9992	0.9992	0.9993	0.9993
3.2	0.9993	0.9993	0.9994	0.9994	0.9994	0.9994	0.9994	0.9995	0.9995	0.9995
3.3	0.9995	0.9995	0.9995	0.9996	0.9996	0.9996	0.9996	0.9996	0.9996	0.9997
3.4	0.9997	0.9997	0.9997	0.9997	0.9997	0.9997	0.9997	0.9997	0.9997	0.9998

Table D *t*-Distribution

		Confidence level				
		80%	90%	95%	98%	99%
		Area in one tail				
		0.10	0.05	0.025	0.01	0.005
		Area in two tails				
		0.20	0.10	0.05	0.02	0.01
df	1	3.078	6.314	12.706	31.821	63.657
	2	1.886	2.920	4.303	6.965	9.925
	3	1.638	2.353	3.182	4.541	5.841
	4	1.533	2.132	2.776	3.747	4.604
	5	1.476	2.015	2.571	3.365	4.032
	6	1.440	1.943	2.447	3.143	3.707
	7	1.415	1.895	2.365	2.998	3.499
	8	1.397	1.860	2.306	2.896	3.355
	9	1.383	1.833	2.262	2.821	3.250
	10	1.372	1.812	2.228	2.764	3.169
	11	1.363	1.796	2.201	2.718	3.106
	12	1.356	1.782	2.179	2.681	3.055
	13	1.350	1.771	2.160	2.650	3.012
	14	1.345	1.761	2.145	2.624	2.977
	15	1.341	1.753	2.131	2.602	2.947
	16	1.337	1.746	2.120	2.583	2.921
	17	1.333	1.740	2.110	2.567	2.898
	18	1.330	1.734	2.101	2.552	2.878
	19	1.328	1.729	2.093	2.539	2.861
	20	1.325	1.725	2.086	2.528	2.845
	21	1.323	1.721	2.080	2.518	2.831
	22	1.321	1.717	2.074	2.508	2.819
	23	1.319	1.714	2.069	2.500	2.807
	24	1.318	1.711	2.064	2.492	2.797
	25	1.316	1.708	2.060	2.485	2.787
	26	1.315	1.706	2.056	2.479	2.779
	27	1.314	1.703	2.052	2.473	2.771
	28	1.313	1.701	2.048	2.467	2.763
	29	1.311	1.699	2.045	2.462	2.756
	30	1.310	1.697	2.042	2.457	2.750
	31	1.309	1.696	2.040	2.453	2.744
	32	1.309	1.694	2.037	2.449	2.738
	33	1.308	1.692	2.035	2.445	2.733
	34	1.307	1.691	2.032	2.441	2.728
	35	1.306	1.690	2.030	2.438	2.724
	36	1.306	1.688	2.028	2.435	2.719
	37	1.305	1.687	2.026	2.431	2.715
	38	1.304	1.686	2.024	2.429	2.712
	39	1.304	1.685	2.023	2.426	2.708
	40	1.303	1.684	2.021	2.423	2.704
	50	1.299	1.676	2.009	2.403	2.678
	60	1.296	1.671	2.000	2.390	2.660
	70	1.294	1.667	1.994	2.381	2.648
	80	1.292	1.664	1.990	2.374	2.639
	90	1.291	1.662	1.987	2.368	2.632
	100	1.290	1.660	1.984	2.364	2.626
	1000	1.282	1.646	1.962	2.330	2.581
	z	1.282	1.645	1.960	2.326	2.576

Table E Chi-square (χ^2) distribution

Degrees of freedom	Area to the right of critical value									
	0.995	0.99	0.975	0.95	0.90	0.10	0.05	0.025	0.01	0.005
1	—	—	0.001	0.004	0.016	2.706	3.841	5.024	6.635	7.879
2	0.010	0.020	0.051	0.103	0.211	4.605	5.991	7.378	9.210	10.597
3	0.072	0.115	0.216	0.352	0.584	6.251	7.815	9.348	11.345	12.838
4	0.207	0.297	0.484	0.711	1.064	7.779	9.488	11.143	13.277	14.860
5	0.412	0.554	0.831	1.145	1.610	9.236	11.071	12.833	15.086	16.750
6	0.676	0.872	1.237	1.635	2.204	10.645	12.592	14.449	16.812	18.548
7	0.989	1.239	1.690	2.167	2.833	12.017	14.067	16.013	18.475	20.278
8	1.344	1.646	2.180	2.733	3.490	13.362	15.507	17.535	20.090	21.955
9	1.735	2.088	2.700	3.325	4.168	14.684	16.919	19.023	21.666	23.589
10	2.156	2.558	3.247	3.940	4.865	15.987	18.307	20.483	23.209	25.188
11	2.603	3.053	3.816	4.575	5.578	17.275	19.675	21.920	24.725	26.757
12	3.074	3.571	4.404	5.226	6.304	18.549	21.026	23.337	26.217	28.299
13	3.565	4.107	5.009	5.892	7.042	19.812	22.362	24.736	27.688	29.819
14	4.075	4.660	5.629	6.571	7.790	21.064	23.685	26.119	29.141	31.319
15	4.601	5.229	6.262	7.261	8.547	22.307	24.996	27.488	30.578	32.801
16	5.142	5.812	6.908	7.962	9.312	23.542	26.296	28.845	32.000	34.267
17	5.697	6.408	7.564	8.672	10.085	24.769	27.587	30.191	33.409	35.718
18	6.265	7.015	8.231	9.390	10.865	25.989	28.869	31.526	34.805	37.156
19	6.844	7.633	8.907	10.117	11.651	27.204	30.144	32.852	36.191	38.582
20	7.434	8.260	9.591	10.851	12.443	28.412	31.410	34.170	37.566	39.997
21	8.034	8.897	10.283	11.591	13.240	29.615	32.671	35.479	38.932	41.401
22	8.643	9.542	10.982	12.338	14.042	30.813	33.924	36.781	40.289	42.796
23	9.260	10.196	11.689	13.091	14.848	32.007	35.172	38.076	41.638	44.181
24	9.886	10.856	12.401	13.848	15.659	33.196	36.415	39.364	42.980	45.559
25	10.520	11.524	13.120	14.611	16.473	34.382	37.652	40.646	44.314	46.928
26	11.160	12.198	13.844	15.379	17.292	35.563	38.885	41.923	45.642	48.290
27	11.808	12.879	14.573	16.151	18.114	36.741	40.113	43.194	46.963	49.645
28	12.461	13.565	15.308	16.928	18.939	37.916	41.337	44.461	48.278	50.993
29	13.121	14.257	16.047	17.708	19.768	39.087	42.557	45.722	49.588	52.336
30	13.787	14.954	16.791	18.493	20.599	40.256	43.773	46.979	50.892	53.672
40	20.707	22.164	24.433	26.509	29.051	51.805	55.758	59.342	63.691	66.766
50	27.991	29.707	32.357	34.764	37.689	63.167	67.505	71.420	76.154	79.490
60	35.534	37.485	40.482	43.188	46.459	74.397	79.082	83.298	88.379	91.952
70	43.275	45.442	48.758	51.739	55.329	85.527	90.531	95.023	100.425	104.215
80	51.172	53.540	57.153	60.391	64.278	96.578	101.879	106.629	112.329	116.321
90	59.196	61.754	65.647	69.126	73.291	107.565	113.145	118.136	124.116	128.299
100	67.328	70.065	74.222	77.929	82.358	118.498	124.342	129.561	135.807	140.169

Right tail (used in Sections 9.6, 11.1, and 11.2)

α

χ^2_α

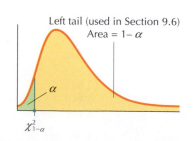

Left tail (used in Section 9.6)
Area = $1 - \alpha$

α

$\chi^2_{1-\alpha}$

Two tails (used in Sections 8.4 and 9.6)

Area = $\dfrac{\alpha}{2}$ Area = $\dfrac{\alpha}{2}$

$\chi^2_{1-\alpha/2}$ $\chi^2_{\alpha/2}$

The area to the right of $\chi^2_{1-\alpha/2}$ is $1 - \dfrac{\alpha}{2}$.

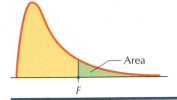

F

Area

Table F F-Distribution critical values

| | | | | | df_1 | | | | |
	Area in right tail	1	2	3	4	5	6	7	8
1	0.100	39.86	49.59	53.59	55.83	57.24	58.20	58.91	59.44
	0.050	161.45	199.50	215.71	224.58	230.16	233.99	236.77	238.88
	0.025	647.79	799.50	864.16	899.58	921.85	937.11	948.22	956.66
	0.010	4052.20	4999.50	5403.40	5624.60	5763.60	5859.00	5928.40	5981.10
	0.001	405284.00	500000.00	540379.00	562500.00	576405.00	585937.00	592873.00	598144.00
2	0.100	8.53	9.00	9.16	9.24	9.29	9.33	9.35	9.37
	0.050	18.51	19.00	19.16	19.25	19.30	19.33	19.35	19.37
	0.025	38.51	39.00	39.17	39.25	39.30	39.33	39.36	39.37
	0.010	98.50	99.00	99.17	99.25	99.30	99.33	99.36	99.37
	0.001	998.50	999.00	999.17	999.25	999.30	999.33	999.36	999.37
3	0.100	5.54	5.46	5.39	5.34	5.31	5.28	5.27	5.25
	0.050	10.13	9.55	9.28	9.12	9.01	8.94	8.89	8.85
	0.025	17.44	16.04	15.44	15.10	14.88	14.73	14.62	14.54
	0.010	34.12	30.82	29.46	28.71	28.24	27.91	27.67	27.49
	0.001	167.03	148.50	141.11	137.10	134.58	132.85	131.58	130.62
4	0.100	4.54	4.32	4.19	4.11	4.05	4.01	3.98	3.95
	0.050	7.71	6.94	6.59	6.39	6.26	6.16	6.09	6.04
	0.025	12.22	10.65	9.98	9.60	9.36	9.20	9.07	8.98
	0.010	21.20	18.00	16.69	15.98	15.52	15.21	14.98	14.80
	0.001	74.14	61.25	56.18	53.44	51.71	50.53	49.66	49.00
5	0.100	4.06	3.78	3.62	3.52	3.45	3.40	3.37	3.34
	0.050	6.61	5.79	5.41	5.19	5.05	4.95	4.88	4.82
	0.025	10.01	8.43	7.76	7.39	7.15	6.98	6.85	6.76
	0.010	16.26	13.27	12.06	11.39	10.97	10.67	10.46	10.29
	0.001	47.18	37.12	33.20	31.09	29.75	28.83	28.16	27.65
6	0.100	3.78	3.46	3.29	3.18	3.11	3.05	3.01	2.98
	0.050	5.99	5.14	4.76	4.53	4.39	4.28	4.21	4.15
	0.025	8.81	7.26	6.60	6.23	5.99	5.82	5.70	5.60
	0.010	13.75	10.92	9.78	9.15	8.75	8.47	8.26	8.10
	0.001	35.51	27.00	23.70	21.92	20.80	20.03	19.46	19.03
7	0.100	3.59	3.26	3.07	2.96	2.88	2.83	2.78	2.75
	0.050	5.59	4.74	4.35	4.12	3.97	3.87	3.79	3.73
	0.025	8.07	6.54	5.89	5.52	5.29	5.12	4.99	4.90
	0.010	12.25	9.55	8.45	7.85	7.46	7.19	6.99	6.84
	0.001	29.25	21.69	18.77	17.20	16.21	15.52	15.02	14.63
8	0.100	3.46	3.11	2.92	2.81	2.73	2.67	2.62	2.59
	0.050	5.32	4.46	4.07	3.84	3.69	3.58	3.50	3.44
	0.025	7.57	6.06	5.42	5.05	4.82	4.65	4.53	4.43
	0.010	11.26	8.65	7.59	7.01	6.63	6.37	6.18	6.03
	0.001	25.41	18.49	15.83	14.39	13.48	12.86	12.40	12.05

df_2 is shown on the left axis.

(Continued)

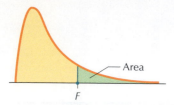

Table F F-Distribution critical values (continued)

	Area in right tail	9	10	15	20	30	60	120	1000
1	0.100	59.86	60.19	61.22	61.74	62.26	62.79	63.06	63.30
	0.050	240.54	241.88	245.95	248.01	250.10	252.20	253.25	254.19
	0.025	963.28	968.63	984.87	993.10	1001.4	1009.8	1014.0	1017.7
	0.010	6022.5	6055.8	6157.3	6208.7	6260.6	6313.0	6339.4	6362.7
	0.001	602284.0	605621.0	615764.0	620908.0	626099.0	631337.0	633972.0	636301.0
2	0.100	9.38	9.39	9.42	9.44	9.16	9.47	9.48	9.49
	0.050	19.38	19.40	19.43	19.45	19.46	19.48	19.49	19.49
	0.025	39.39	39.40	39.43	39.45	39.46	39.48	39.49	39.50
	0.010	99.39	99.40	99.43	99.45	99.47	99.48	99.49	99.50
	0.001	999.39	999.40	999.43	999.45	999.47	999.48	999.49	999.50
3	0.100	5.24	5.23	5.20	5.18	5.17	5.15	5.14	5.13
	0.050	8.81	8.79	8.70	8.66	8.62	8.57	8.55	8.53
	0.025	14.47	14.42	14.25	14.17	14.08	13.99	13.95	13.91
	0.010	27.35	27.23	26.87	26.69	26.50	26.32	26.22	26.14
	0.001	129.86	129.25	127.37	126.42	125.45	124.47	123.97	123.53
4	0.100	3.94	3.92	3.87	3.84	3.82	3.79	3.78	3.76
	0.050	6.00	5.96	5.86	5.80	5.75	5.69	5.66	5.63
	0.025	8.90	8.84	8.66	8.56	8.46	8.36	8.31	8.26
	0.010	14.66	14.55	14.20	14.02	13.84	13.65	13.56	13.47
	0.001	48.47	48.05	46.76	46.10	45.43	44.75	44.40	44.09
5	0.100	3.32	3.30	3.24	3.21	3.17	3.14	3.12	3.11
	0.050	4.77	4.74	4.62	4.56	4.50	4.43	4.40	4.37
	0.025	6.68	6.62	6.43	6.33	6.23	6.12	6.07	6.02
	0.010	10.16	10.05	9.72	9.55	9.38	9.20	9.11	9.03
	0.001	27.24	26.92	25.91	25.39	24.87	24.33	24.06	23.82
6	0.100	2.96	2.94	2.87	2.84	2.80	2.76	2.74	2.72
	0.050	4.10	4.06	3.94	3.87	3.81	3.74	3.70	3.67
	0.025	5.52	5.46	5.27	5.17	5.07	4.96	4.90	4.86
	0.010	7.98	7.87	7.56	7.40	7.23	7.06	6.97	6.89
	0.001	18.69	18.41	17.56	17.12	16.67	16.21	15.98	15.77
7	0.100	2.72	2.70	2.63	2.59	2.56	2.51	2.49	2.47
	0.050	3.68	3.64	3.51	3.44	3.38	3.30	3.27	3.23
	0.025	4.82	4.76	4.57	4.47	4.36	4.25	4.20	4.15
	0.010	6.72	6.62	6.31	6.16	5.99	5.82	5.74	5.66
	0.001	14.33	14.08	13.32	12.93	12.53	12.12	11.91	11.72
8	0.100	2.56	2.54	2.46	2.42	2.38	2.34	2.32	2.30
	0.050	3.39	3.35	3.22	3.15	3.08	3.01	2.97	2.93
	0.025	4.36	4.30	4.10	4.00	3.89	3.78	3.73	3.68
	0.010	5.91	5.81	5.52	5.36	5.20	5.03	4.95	4.87
	0.001	11.77	11.54	10.84	10.48	10.11	9.73	9.53	9.36

df_1

df_2

Table F *F*-Distribution critical values *(continued)*

df_2	Area in right tail	1	2	3	4	5	6	7	8	9	10
9	0.100	3.36	3.01	2.81	2.69	2.61	2.55	2.51	2.47	2.44	2.42
	0.050	5.12	4.26	3.86	3.63	3.48	3.37	3.29	3.23	3.18	3.14
	0.025	7.21	5.71	5.08	4.72	4.48	4.32	4.20	4.10	4.03	3.96
	0.010	10.56	8.02	6.99	6.42	6.06	5.80	5.61	5.47	5.35	5.26
	0.001	22.86	16.39	13.90	12.56	11.71	11.13	10.70	10.37	10.11	9.89
10	0.100	3.29	2.92	2.73	2.61	2.52	2.46	2.41	2.38	2.35	2.32
	0.050	4.96	4.10	3.71	3.48	3.33	3.22	3.14	3.07	3.02	2.98
	0.025	6.94	5.46	4.83	4.47	4.24	4.07	3.95	3.85	3.78	3.72
	0.010	10.04	7.56	6.55	5.99	5.64	5.39	5.20	5.06	4.94	4.85
	0.001	21.04	14.91	12.55	11.28	10.48	9.93	9.52	9.20	8.96	8.75
12	0.100	3.18	2.81	2.61	2.48	2.39	2.33	2.28	2.24	2.21	2.19
	0.050	4.75	3.89	3.49	3.26	3.11	3.00	2.91	2.85	2.80	2.75
	0.025	6.55	5.10	4.47	4.12	3.89	3.73	3.61	3.51	3.44	3.37
	0.010	9.33	6.93	5.95	5.41	5.06	4.82	4.64	4.50	4.39	4.30
	0.001	18.64	12.97	10.80	9.63	8.89	8.38	8.00	7.71	7.48	7.29
15	0.100	3.07	2.70	2.49	2.36	2.27	2.21	2.16	2.12	2.09	2.06
	0.050	4.54	3.68	3.29	3.06	2.90	2.79	2.71	2.64	2.59	2.54
	0.025	6.20	4.77	4.15	3.80	3.58	3.41	3.29	3.20	3.12	3.06
	0.010	8.68	6.36	5.42	4.89	4.56	4.32	4.14	4.00	3.89	3.80
	0.001	16.59	11.34	9.34	8.25	7.57	7.09	6.74	6.47	6.26	6.08
20	0.100	2.97	2.59	2.38	2.25	2.16	2.09	2.04	2.00	1.96	1.94
	0.050	4.35	3.49	3.10	2.87	2.71	2.60	2.51	2.45	2.39	2.35
	0.025	5.87	4.46	3.86	3.51	3.29	3.13	3.01	2.91	2.84	2.77
	0.010	8.10	5.85	4.94	4.43	4.10	3.87	3.70	3.56	3.46	3.37
	0.001	14.82	9.95	8.10	7.10	6.46	6.02	5.69	5.44	5.24	5.08
25	0.100	2.92	2.53	2.32	2.18	2.09	2.02	1.97	1.93	1.89	1.87
	0.050	4.24	3.39	2.99	2.76	2.60	2.49	2.40	2.34	2.28	2.24
	0.025	5.69	4.29	3.69	3.35	3.13	2.97	2.85	2.75	2.68	2.61
	0.010	7.77	5.57	4.68	4.18	3.85	3.63	3.46	3.32	3.22	3.13
	0.001	13.88	9.22	7.45	6.49	5.89	5.46	5.15	4.91	4.71	4.56
50	0.100	2.81	2.41	2.20	2.06	1.97	1.90	1.84	1.80	1.76	1.73
	0.050	4.03	3.18	2.79	2.56	2.40	2.29	2.20	2.13	2.07	2.03
	0.025	5.34	3.97	3.39	3.05	2.83	2.67	2.55	2.46	2.38	2.32
	0.010	7.17	5.06	4.20	3.72	3.41	3.19	3.02	2.89	2.78	2.70
	0.001	12.22	7.96	6.34	5.46	4.90	4.51	4.22	4.00	3.82	3.67
100	0.100	2.76	2.36	2.14	2.00	1.91	1.83	1.78	1.73	1.69	1.66
	0.050	3.94	3.09	2.70	2.46	2.31	2.19	2.10	2.03	1.97	1.93
	0.025	5.18	3.83	3.25	2.92	2.70	2.54	2.42	2.32	2.24	2.18
	0.010	6.90	4.82	3.98	3.51	3.21	2.99	2.82	2.69	2.59	2.50
	0.001	11.50	7.41	5.86	5.02	4.48	4.11	3.83	3.61	3.44	3.30
200	0.100	2.73	2.33	2.11	1.97	1.88	1.80	1.75	1.70	1.66	1.63
	0.050	3.89	3.04	2.65	2.42	2.26	2.14	2.06	1.98	1.93	1.88
	0.025	5.10	3.76	3.18	2.85	2.63	2.47	2.35	2.26	2.18	2.11
	0.010	6.76	4.71	3.88	3.41	3.11	2.89	2.73	2.60	2.50	2.41
	0.001	11.15	7.15	5.63	4.81	4.29	3.92	3.65	3.43	3.26	3.12
1000	0.100	2.71	2.31	2.09	1.95	1.85	1.78	1.72	1.68	1.64	1.61
	0.050	3.85	3.00	2.61	2.38	2.22	2.11	2.02	1.95	1.89	1.84
	0.025	5.04	3.70	3.13	2.80	2.58	2.42	2.30	2.20	2.13	2.06
	0.010	6.66	4.63	3.80	3.34	3.04	2.82	2.66	2.53	2.43	2.34
	0.001	10.89	6.96	5.46	4.65	4.14	3.78	3.51	3.30	3.13	2.99

(Continued)

Table F *F*-Distribution critical values (*continued*)

	Area in right tail	12	15	20	25	30	40	50	60	120	1000
						df_1					
9	0.100	2.38	2.34	2.30	2.27	2.25	2.23	2.22	2.21	2.18	2.16
	0.050	3.07	3.01	2.94	2.89	2.86	2.83	2.80	2.79	2.75	2.71
	0.025	3.87	3.77	3.67	3.60	3.56	3.51	3.47	3.45	3.39	3.34
	0.010	5.11	4.96	4.81	4.71	4.65	4.57	4.52	4.48	4.40	4.32
	0.001	9.57	9.24	8.90	8.69	8.55	8.37	8.26	8.19	8.00	7.84
10	0.100	2.28	2.24	2.20	2.17	2.16	2.13	2.12	2.11	2.08	2.06
	0.050	2.91	2.85	2.77	2.73	2.70	2.66	2.64	2.62	2.58	2.54
	0.025	3.62	3.52	3.42	3.35	3.31	3.26	3.22	3.20	3.14	3.09
	0.010	4.71	4.56	4.41	4.31	4.25	4.17	4.12	4.08	4.00	3.92
	0.001	8.45	8.13	7.80	7.60	7.47	7.30	7.19	7.12	6.94	6.78
12	0.100	2.15	2.10	2.06	2.03	2.01	1.99	1.97	1.96	1.93	1.91
	0.050	2.69	2.62	2.54	2.50	2.47	2.43	2.40	2.38	2.34	2.30
	0.025	3.28	3.18	3.07	3.01	2.96	2.91	2.87	2.85	2.79	2.73
	0.010	4.16	4.01	3.86	3.76	3.70	3.62	3.57	3.54	3.45	3.37
	0.001	7.00	6.71	6.40	6.22	6.09	5.93	5.83	5.76	5.59	5.44
15	0.100	2.02	1.97	1.92	1.89	1.87	1.85	1.83	1.82	1.79	1.76
	0.050	2.48	2.40	2.33	2.28	2.25	2.20	2.18	2.16	2.11	2.07
	0.025	2.96	2.86	2.76	2.69	2.64	2.59	2.55	2.52	2.46	2.40
	0.010	3.67	3.52	3.37	3.28	3.21	3.13	3.08	3.05	2.96	2.88
	0.001	5.81	5.54	5.25	5.07	4.95	4.80	4.70	4.64	4.47	4.33
20	0.100	1.89	1.84	1.79	1.76	1.74	1.71	1.69	1.68	1.64	1.61
	0.050	2.28	2.20	2.12	2.07	2.04	1.99	1.97	1.95	1.90	1.85
	0.025	2.68	2.57	2.46	2.40	2.35	2.29	2.25	2.22	2.16	2.09
	0.010	3.23	3.09	2.94	2.84	2.78	2.69	2.64	2.61	2.52	2.43
	0.001	4.82	4.56	4.29	4.12	4.00	3.86	3.77	3.70	3.54	3.40
25	0.100	1.82	1.77	1.72	1.68	1.66	1.63	1.61	1.59	1.56	1.52
	0.050	2.16	2.09	2.01	1.96	1.92	1.87	1.84	1.82	1.77	1.72
	0.025	2.51	2.41	2.30	2.23	2.18	2.12	2.08	2.05	1.98	1.91
	0.010	2.99	2.85	2.70	2.60	2.54	2.45	2.40	2.36	2.27	2.18
	0.001	4.31	4.06	3.79	3.63	3.52	3.37	3.28	3.22	3.06	2.91
50	0.100	1.68	1.63	1.57	1.53	1.50	1.46	1.44	1.42	1.38	1.33
	0.050	1.95	1.87	1.78	1.73	1.69	1.63	1.60	1.58	1.51	1.45
	0.025	2.22	2.11	1.99	1.92	1.87	1.80	1.75	1.72	1.64	1.56
	0.010	2.56	2.42	2.27	2.17	2.10	2.01	1.95	1.91	1.80	1.70
	0.001	3.44	3.20	2.95	2.79	2.68	2.53	2.44	2.38	2.21	2.05
100	0.100	1.61	1.56	1.49	1.45	1.42	1.38	1.35	1.34	1.28	1.22
	0.050	1.85	1.77	1.68	1.62	1.57	1.52	1.48	1.45	1.38	1.30
	0.025	2.08	1.97	1.85	1.77	1.71	1.64	1.59	1.56	1.46	1.36
	0.010	2.37	2.22	2.07	1.97	1.89	1.80	1.74	1.69	1.57	1.45
	0.001	3.07	2.84	2.59	2.43	2.32	2.17	2.08	2.01	1.83	1.64
200	0.100	1.58	1.52	1.46	1.41	1.38	1.34	1.31	1.29	1.23	1.16
	0.050	1.80	1.72	1.62	1.56	1.52	1.46	1.41	1.39	1.30	1.21
	0.025	2.01	1.90	1.78	1.70	1.64	1.56	1.51	1.47	1.37	1.25
	0.010	2.27	2.13	1.97	1.87	1.79	1.69	1.63	1.58	1.45	1.30
	0.001	2.90	2.67	2.42	2.26	2.15	2.00	1.90	1.83	1.64	1.43
1000	0.100	1.55	1.49	1.43	1.38	1.35	1.30	1.27	1.25	1.38	1.08
	0.050	1.76	1.68	1.58	1.52	1.47	1.41	1.36	1.31	1.24	1.11
	0.025	1.96	1.85	1.72	1.64	1.58	1.50	1.45	1.41	1.29	1.13
	0.010	2.20	2.06	1.90	1.79	1.72	1.61	1.54	1.50	1.35	1.16
	0.001	2.77	2.54	2.30	2.14	2.02	1.87	1.77	1.69	1.49	1.22

df_2

Notes and Data Sources

Chapter 1

1. As reported at **www.cnn.com/2001/ALLPOLITICS/03/11/palmbeach.recount/**.

2. T. J. Scanlon, R. N. Luben, F. L. Scanlon, and N. Singleton, "Is Friday the 13th bad for your health?" *British Medical Journal* 307 (December 1993).

3. U.S. Census Bureau, *The Population Profile of the United States: 2000*, **www.consensus.gov/population/www/pop-profile/profile2000**.

4. Pew Internet and American Life Project, "Cyberbullying and on-line teens," June 2007, **www.pewinternet.org**.

5. National Agricultural Statistics Service.

6. Iain McGregor and Wayne Hall, "MDMA (Ecstasy) neurotoxicity: assessing and communicating the risks," *Lancet* 355 (9217, May 20, 2000): 1818–21.

7. U.S. Department of Health and Human Services, *The Health Consequences of Involuntary Exposure to Tobacco Smoke: A Report of the Surgeon General—Executive Summary* (U.S. Department of Health and Human Services, Centers for Disease Control and Prevention, Coordinating Center for Health Promotion, National Center for Chronic Disease Prevention and Health Promotion, Office on Smoking and Health, 2006).

8. Michel de Lorgeril, Patricia Salen, Jean-Louis Martin, Isabelle Monjaud, Jacques Delaye, and Nicole Mamelle, "Mediterranean diet, traditional risk factors, and the rate of cardiovascular complications after myocardial infarction, final report of the Lyon Diet Heart Study," *Circulation: Journal of the American Heart Association* 99 (1999): 779–85. The American Heart Association (**www.americanheart.org**) identifies the following characteristics as common to most Mediterranean diets. There is a "high consumption of fruits, vegetables, bread and other cereals, potatoes, beans, nuts and seeds. Olive oil is an important monounsaturated fat source. Dairy products, fish and poultry are consumed in low to moderate amounts, and little red meat is eaten."

9. R. L. Bratton et al., "Effect of 'ionized' wrist bracelets on musculoskeletal pain: a randomized, double-blind, placebo-controlled trial," *Mayo Clinic Proceedings* 77 (2002):1164–68.

Chapter 2

1. National Health Interview Survey, U.S. Centers for Disease Control and Prevention, 2005.

2. M. A. Chase and G. M. Dummer, "The role of sports as a social determinant for children," *Research Quarterly for Exercise and Sport* 63 (1992): 418–24.

3. U.S. Bureau of Labor Statistics.

4. U.S. Centers for Disease Control and Prevention.

Chapter 3

1. U.S. Centers for Disease Control and Prevention, U.S. Department of Health and Human Services, **www.cdc.gov/flu/avian/index.htm**.

2. For more on clickstream analysis, see Zdravko Markov and Daniel Larose, *Data Mining the Web: Uncovering Patterns in Web Content, Structure, and Usage* (John Wiley and Sons, 2007).

3. U.S. Census Bureau.

4. P. A. Mackowiak, S. S. Wasserman, and M. M. Levine, "A critical appraisal of 98.6 degrees F, the upper limit of the normal body temperature, and other legacies of Carl Reinhold August Wunderlich," *Journal of the American Medical Association* 268 (1992): 1578–80.

5. Michael Brett and Charles Goldman, "A meta-analysis of the freshwater trophic cascade," *Proceedings of the National Academy of Sciences* 93 (July 1996).

6. Dr. Peter Nonacs, "Foraging habits of thatch ants," Department of Statistics, University of California at Los Angeles and the Sierra Nevada Aquatic Research Laboratory, **www.stat.ucla.edu/datasets/**.

7. Children's Bureau, Administration for Children and Families, U.S. Department of Health and Human Services.

8. National Water and Climate Center, U.S. Department of Agriculture.

9. B. S. Glenn et al., "Changes in systolic blood pressure associated with lead in blood and bone," *Epidemiology* 17 (September 2006).

10. National Center for Education Statistics, 2005.

11. M. Donald Thomas and William L. Bainbridge, "Grade inflation: the current fraud," *Effective School Research*, January 1997.

12. National Center for Health Statistics, *Health*, 2006.

Chapter 4

1. Roper Center, University of Connecticut.

2. T. Allison and D. V. Cicchetti, "Sleep in mammals: ecological and constitutional correlates," *Science* 194 (1976): 732–34. Web reference: Quantitative Environmental Learning Project, Joseph Hull and Greg Langkamp, Seattle Central Community College, **www.seattlecentral.org/qelp/sets/017/017.html**.

3. United Nations Educational, Scientific, and Cultural Organization (UNESCO).

4. *Crime in the United States, 2004*, **www.fbi.gov**.

Chapter 5

1. Amanda Lenhart et al., *Writing, Technology, and Teens*, Pew Internet and American Life Project, December 2007.

2. U.S. Census Bureau, 2004 American Community Survey.

3. Andrew Rocco Tresolini Fiore, "Romantic regressions: an analysis of behavior in online dating systems," master's thesis, Massachusetts Institute of Technology, 2004.

4. *Percentage Baseball* (MIT Press, 1966).

5. Washington Initiative (**greaterwashington.org**).

6. U.S. Bureau of Labor Statistics.

7. *Profile of Hired Farmworkers, A 2008 Update/ERR-60*, Economic Research Service/USDA.

Chapter 6

1. U.S. National Center for Education Statistics. The category "5 or more" has been changed to "5" for this exercise.

2. U.S. Census Bureau. The category "9 or more" has been changed to "9" for this exercise.

3. U.S. Census Bureau. The category "3 or more" has been changed to "3" for this exercise.

4. Gunter Hitsch, Ali Hortacsu, and Dan Ariely, "What makes you click: an empirical analysis of online dating;" available online at **www.aeaweb.org/annual_mtg_papers/2006/0106_0800_0502.pdf**.

5. D. L. Olds, C. R. Henderson Jr., R. Tatelbaum, et al., "Improving the delivery of prenatal care and outcomes of pregnancy: a randomized trial of nurse home visitation," *Pediatrics* 77 (1986): 16–28.

6. M. Donald Thomas and William L. Bainbridge, "Grade inflation: the current fraud," *Effective School Research*, January 1997.

7. Lynn Unruh and Myron Fottler, "Patient turnover and nursing staff adequacy," *Health Services Research*, April 2006.

8. See Note 6.

9. Harvard School of Public Health, survey of 5046 adults in hurricane high-risk areas, June–July 2007.

10. The Associated Press/Ipsos Poll actually contacted 1000 adults in June 2007.

11. Barbara Alving et al., "Trends in blood pressure among children and adolescents," *Journal of the American Medical Association* 291 (May 2004): 2107–13.

12. Phillida Bunkle and John Lepper, "Women's participation in gambling: whose reality? A public health issue," paper presented to the European Association for the Study of Gambling Conference, Barcelona, Spain, October 2002.

Chapter 7

1. George A. Millar "The magical number seven, plus or minus two: some limits on our capacity for processing information," *Psychological Review* 63 (1956): 81–97.

2. A small business is defined by the SBA as having fewer than 20 employees.

3. Murray Mittleman et al., "Determinants of myocardial onset study," *Circulation: Journal of the American Heart Association*, June 1999.

Chapter 8

1. Adapted from A. Johnson, "Results from analyzing metals in 1999 Spokane River fish and crayfish samples," Quantitative Environmental Learning Project, Washington State Department of Ecology report 00-03-017, **www.seattlecentral.edu/qelp/sets/021/021.html**.

2. Kevin Crowley et al., "Parents explain more often to boys than girls during shared scientific thinking," *Psychological Science* 12, no. 3 (May 2001): 258–61.

3. George Miller, "The magical number seven, plus or minus two: some limits on our capacity for processing information," *Psychological Review* 63 (1956): 81–97.

4. Mary H. Ward et al., "Proximity to crops and residential exposure to agricultural herbicides in Iowa," *Environmental Health Perspectives* 114, no. 6 (June 2006): 893–97.

5. Irene Yen et al., "Perceived neighborhood problems and quality of life, physical functioning, and depressive symptoms among adults with asthma," *American Journal of Public Health* 96, no. 5 (May 2006): 873–79.

6. Gary Bennett et al., "Television viewing and pedometer-determined physical activity among multiethnic residents of low-income housing," *American Journal of Public Health* 96, no. 9 (September 2006): 1681–85.

7. Mary C. Meyer, "Wider shoes for wider feet?" *Journal of Statistics Education* 14, no. 1 (2006), **www.amstat.org/publications/jse/v14n1/datasets.meyer.html**.

8. Clive Lipchin, "Report from the field: perceptions of water use in the Arava Valley of Israel and Jordan," *Journal of the International Institute, University of Michigan* (Summer 2004). A cubic meter of water is equivalent to 264 gallons.

9. Fact 2 in Section 7.1 stated that the standard deviation of the sampling distribution of the sample mean $\bar{x}$ is $\sigma_{\bar{x}} = \sigma/\sqrt{n}$. This quantity was used in Section 8.1 to calculate the Z interval for the population mean when σ is known. However, here in Section 8.2, we no longer assume that σ is known. When σ is not known, we use the following quantity to estimate the standard deviation of the sampling distribution of the sample mean: $s_{\bar{x}} = s/\sqrt{n}$. The term $s_{\bar{x}}$ is called the *standard error of the sample mean*. (Some textbooks use the term *standard error* to refer to the true value of the standard deviation of a statistic, for example, $\sigma/\sqrt{n}$ for the sample mean. In that case, the value of the standard error is unknown and not observed. However, the use of the term *standard error* as an observed value, deriving from a sample, is widespread, both in the methodological literature and in statistical software.)

10. Robert J. Pianta et al., "Teaching: opportunities to learn in America's elementary classroom," *Science* 315 (March 30, 2007): 1795–96.

11. Community College Survey of Student Engagement (CCSSE), 2007, **www.ccsse.org**. The survey reported that 178 of 307 (57.98045603%) students worked with classmates outside class to prepare a group assignment during the current academic year. The sample results in Example 8.15 (174 of 300, or 58%) were chosen for ease of calculation.

12. Christopher Reynolds, "Prey tell," *American Demographics* 25, no. 8 (October 2003): 48.

13. *MDMA (Ecstacy) Abuse*, National Institute on Drug Abuse Research Report Series, U.S. Department of Health and Human Services, February 2006.

14. Manchester (CT) *Journal-Inquirer*, June 9, 2008, p. 1.

15. Mildred Cho and Lisa Bero, "The quality of drug studies published in symposium proceedings," *Annals of Internal Medicine* 124, no. 5 (March 1996): 485–89.

16. See Note 7.

Chapter 9

1. Y. Kvach and C. A. Stepien, "The invasive round goby *Apollonia melanostoma* (Actinopterygii: Gobiidae)—a new intermediate host of the trematode *Neochasmus umbellus* (Trematoda: Cryptogonimidae) in Lake Erie, Ohio, USA," *Journal of Applied Ichthyology* 24, no. 1 (February 2008): 103–05.

2. NASA, "Climate data shows California has been heating up," *Science Daily* 31 (March 2007), **www.sciencedaily.com/releases/2007/03/070330221144.htm.** (Accessed June 5, 2008).

3. Press release, August 23, 2007: "Consumers report eating at home more in the wake of high gas prices," NPD Group, Inc., 900 West Shore Road, Port Washington, NY 11050.

4. "When it comes to height, Americans no longer stand tallest," *Research News*, The Ohio State University, **researchnews.osu.edu/**.

5. K. Marien, A. Conseur, and M. Sanderson, "The effect of fish consumption on DDT and DDE levels in breast milk among Hispanic immigrants," *Journal of Human Lactation* 14, no. 3 (1998): 237–42.

6. C. J. Earle, L. B. Brubaker, and G. Segura, International Tree Ring Data Base, NOAA/NGDC Paleoclimatology Program, Boulder, CO.

7. "Online shoppers will pay extra to protect privacy, Carnegie Mellon study shows," *ScienceDaily,* **www.sciencedaily.com/releases/2007/06/070606235316.htm**. (Accessed January 20, 2008).

8. *Digital Transactions News,* September 2007.

9. G. M. Darling, J. A. Johns, P. I. McCloud, and S. R. Davis, "Estrogen and progestin compared with simvastatin for hypercholesterolemia in postmenopausal women," *New England Journal of Medicine* 337, no. 9 (1997): 595–601.

10. **health.usnews.com/sections/health/west-hospitals**.

11. Caroline Davis, Elizabeth Blackmore, Deborah Katzman, and John Fox, "Anorexia nervosa case study," paper presented at Statistical Society of Canada Annual Conference, Montreal, 2004. I have reversed the research question from that of the original case study.

12. Courtesy American Heritage Center, University of Wyoming.

13. Data courtesy of OzDASL (Australian Data and Story Library) at **statsci.org**. The original source is Cara Dubois, ed., *Lowie's Selected Papers in Anthropology* (University of California Press, 1960).

14. Mary Madden and Amanda Lenhart, "Online dating," Pew Internet and American Life Project, 2006.

15. Barry Kosmin and Egon Mayer, principal investigators, American Religious Identification Survey, Graduate Center, City University of New York.

16. Brady Hamilton, Joyce Martin, and Stephanie Ventura, "Births: preliminary data for 2005," *National Vital Statistics Reports* 55 (11), U.S. Department of Health and Human Services.

17. "Patterns and trends in nonmedical prescription pain reliever use: 2002 to 2005," in *NSDUH Report,* Substance Abuse and Mental Health Services Administration, April 6, 2007.

18. Jeff Humphries, "The multicultural economy: minority buying power in the new century," Selig Center for Economic Growth, Terry College of Business, University of Georgia, 2006.

19. "Trends in the prevalence of alcohol use among eighth graders: Monitoring the Future Study, 1991–2003," NIAAA, National Institutes of Health.

20. "Fact sheet: National Survey on Environmental Management of Asthma and Children's Exposure to Environmental Tobacco Smoke," U.S. Environmental Protection Agency, May 17, 2005.

21. Energy Information Administration, "Annual electric generator report," Form EIA-906.

22. U.S. Census Bureau.

23. See Note 5.

24. See Note 6.

25. U.S. Bureau of Labor Statistics.

26. Mary C. Meyer, "Wider shoes for wider feet?" *Journal of Statistics Education* 14, No. 1 (2006), **www.amstat.org/publications/jse/v14n1/datasets.meyer.html**.

27. Steve Strand, Ian Deary, and Pauline Smith, "Sex differences in cognitive abilities test scores: a UK national picture," *British Journal of Educational Psychology* 76 (2006): 463–80.

28. Siobhan Banks and David Dinges, "Behavioral and physiological consequences of sleep restriction," *Journal of Clinical Sleep Medicine* 15 (2007): 519–28.

29. See the Infoplease Web site: **www.infoplease.com/ipa/A0934556.html**.

30. "A nation online: entering the broadband age," Economics and Statistics Administration, U.S. Department of Commerce.

31. U.S. Bureau of Justice Statistics.

32. Joyce A. Martin et al., "Births: final data for 2005," *National Vital Statistics Reports* 56, No. 6 (December 5, 2007).

Chapter 10

1. H. H. Kelley, "The warm-cold variable in first impression of persons," *Journal of Personality* 18 (1950): 431–39.

2. A. Towler and R. L. Dipboye, "The effect of instructor reputation and need for cognition on student behavior," poster presented at American Psychological Society conference, May 1998.

3. K. J. Thomas et al., "Randomized controlled trial of a short course of traditional acupuncture compared with usual care for persistent non-specific low back pain," *British Medical Journal* 23 (September 2006).

4. Karin Olson and John Hanson, "Using Reiki to manage pain," *Cancer Prevention and Control* 1, no. 2 (1997): 108–13.

5. P. A. Mackowiak, S. S. Wasserman, and M. M. Levine, "A critical appraisal of 98.6 degrees F, the upper limit of the normal body temperature, and other legacies of Carl Reinhold August Wunderlich," *Journal of the American Medical Association* 268 (1992):1578–80.

6. George W. Snedecor and William G. Cochran, *Statistical Methods,* 8th ed. (Iowa State University Press, 1989).

7. D. L. Olds, C. R. Henderson Jr., R. Tatelbaum, et al., "Improving the delivery of prenatal care and outcomes of pregnancy: a randomized trial of nurse home visitation," *Pediatrics* 77 (1986): 16–28.

8. Amanda Lenhart and Mary Madden, "Teens, privacy, and online social networks: how teens manage their online identities and personal information in the age of MySpace," Pew Internet and American Life Project, April 2007.

9. See Note 8.

10. Vijayakrishna K. Gadi et al., "Case-control study of fetal microchimerism and breast cancer," *PLoS one* 3 (March 5, 2008) (plos one, doi; 10:1371/journal.pone.0001706).

11. R. L. Bratton et al., "Effect of 'ionized' wrist bracelets on musculoskeletal pain: a randomized, double-blind, placebo-controlled trial," *Mayo Clinic Proceedings* 77 (2002): 1164–68.

Chapter 11

1. Joseph Maze, Richard Murphy, and Cheri Simonds, "I'll see you on Facebook: the effects of computer-mediated teacher self-disclosure on student motivation, affective learning, and classroom climate," *Communication Edition* 56 (2007): 1–17.

2. U.S. Department of Education, National Center for Education Statistics, Adult Education Survey of the 2005 National Household Education Surveys Program.

3. Derek M. Burnett et al., "Impact of minority status following traumatic spinal cord injury," *NeuroRehabilitation* 17 (2002): 187–94.

4. Pew Research Center for the People and the Press, *How Young People View Their Lives, Futures, and Politics: A Portrait of "Generation Next"* (Washington, D.C., 2007).

5. Mary Madden and Amanda Lenhart, *Online Dating,* Pew Internet and American Life Project, 2005.

6. M. A. Chase and G. M. Dummer, "The role of sports as a social determinant for children," *Research Quarterly for Exercise and Sport* 63 (1992): 418–24.

7. Peter Novacs, "Foraging Habits of Thatch Ants," **www.stat.ucla.edu/datasets/**.

8. See, for example, David S. Moore, George P. McCabe, and Bruce A. Craig, *Introduction to the Practice of Statistics,* 6th ed. (W. H. Freeman and Co., 2009), Section 2.2.

9. S. Blackman and D. Catalina, "The moon and the emergency room," *Perceptual and Motor Skills* 37 (1973): 624–26.

10. The data set is adapted from the **CEREALS** data set from the Data and Story Library, **Lib.stat.cmu.edu/DASL**.

11. William S. Cleveland, *Visualizing Data* (Hobart Press, 1993).

12. J. R. Knight, H. Wechsler, M. Kuo, M. Seibring, E. R. Weitzman, and M. Schuckit, "Alcohol abuse and dependence among U.S. college students," *Journal of Studies on Alcohol* 63, no. 3 (2002): 263–70.

13. Donald Garrow and Leonard Egede, "National patterns and correlates of complementary and alternative medicine use in adults with diabetes," *Journal of Alternative and Complementary Medicine* 12 (2006): 895–902.

14. J. E. Anderson and S. Sansom, "HIV testing in a national sample of pregnant US women: Who is not getting tested?" *AIDS Care* 19 (March 2007): 375–80.

15. National Agricultural Statistics Service, *Agricultural Statistics*, 2006, **www.usda.gov/nass**.

boldface indicates a definition   *italics* indicates a figure   *t* indicates a table

Table D *t*-Distribution

		Confidence level				
		80%	**90%**	**95%**	**98%**	**99%**
				Area in one tail		
		0.10	**0.05**	**0.025**	**0.01**	**0.005**
				Area in two tails		
		0.20	**0.10**	**0.05**	**0.02**	**0.01**
df	1	3.078	6.314	12.706	31.821	63.657
	2	1.886	2.920	4.303	6.965	9.925
	3	1.638	2.353	3.182	4.541	5.841
	4	1.533	2.132	2.776	3.747	4.604
	5	1.476	2.015	2.571	3.365	4.032
	6	1.440	1.943	2.447	3.143	3.707
	7	1.415	1.895	2.365	2.998	3.499
	8	1.397	1.860	2.306	2.896	3.355
	9	1.383	1.833	2.262	2.821	3.250
	10	1.372	1.812	2.228	2.764	3.169
	11	1.363	1.796	2.201	2.718	3.106
	12	1.356	1.782	2.179	2.681	3.055
	13	1.350	1.771	2.160	2.650	3.012
	14	1.345	1.761	2.145	2.624	2.977
	15	1.341	1.753	2.131	2.602	2.947
	16	1.337	1.746	2.120	2.583	2.921
	17	1.333	1.740	2.110	2.567	2.898
	18	1.330	1.734	2.101	2.552	2.878
	19	1.328	1.729	2.093	2.539	2.861
	20	1.325	1.725	2.086	2.528	2.845
	21	1.323	1.721	2.080	2.518	2.831
	22	1.321	1.717	2.074	2.508	2.819
	23	1.319	1.714	2.069	2.500	2.807
	24	1.318	1.711	2.064	2.492	2.797
	25	1.316	1.708	2.060	2.485	2.787
	26	1.315	1.706	2.056	2.479	2.779
	27	1.314	1.703	2.052	2.473	2.771
	28	1.313	1.701	2.048	2.467	2.763
	29	1.311	1.699	2.045	2.462	2.756
	30	1.310	1.697	2.042	2.457	2.750
	31	1.309	1.696	2.040	2.453	2.744
	32	1.309	1.694	2.037	2.449	2.738
	33	1.308	1.692	2.035	2.445	2.733
	34	1.307	1.691	2.032	2.441	2.728
	35	1.306	1.690	2.030	2.438	2.724
	36	1.306	1.688	2.028	2.435	2.719
	37	1.305	1.687	2.026	2.431	2.715
	38	1.304	1.686	2.024	2.429	2.712
	39	1.304	1.685	2.023	2.426	2.708
	40	1.303	1.684	2.021	2.423	2.704
	50	1.299	1.676	2.009	2.403	2.678
	60	1.296	1.671	2.000	2.390	2.660
	70	1.294	1.667	1.994	2.381	2.648
	80	1.292	1.664	1.990	2.374	2.639
	90	1.291	1.662	1.987	2.368	2.632
	100	1.290	1.660	1.984	2.364	2.626
	1000	1.282	1.646	1.962	2.330	2.581
	z	1.282	1.645	1.960	2.326	2.576

Table E Chi-square (χ^2) distribution

Degrees of freedom	Area to the right of critical value									
	0.995	0.99	0.975	0.95	0.90	0.10	0.05	0.025	0.01	0.005
1	—	—	0.001	0.004	0.016	2.706	3.841	5.024	6.635	7.879
2	0.010	0.020	0.051	0.103	0.211	4.605	5.991	7.378	9.210	10.597
3	0.072	0.115	0.216	0.352	0.584	6.251	7.815	9.348	11.345	12.838
4	0.207	0.297	0.484	0.711	1.064	7.779	9.488	11.143	13.277	14.860
5	0.412	0.554	0.831	1.145	1.610	9.236	11.071	12.833	15.086	16.750
6	0.676	0.872	1.237	1.635	2.204	10.645	12.592	14.449	16.812	18.548
7	0.989	1.239	1.690	2.167	2.833	12.017	14.067	16.013	18.475	20.278
8	1.344	1.646	2.180	2.733	3.490	13.362	15.507	17.535	20.090	21.955
9	1.735	2.088	2.700	3.325	4.168	14.684	16.919	19.023	21.666	23.589
10	2.156	2.558	3.247	3.940	4.865	15.987	18.307	20.483	23.209	25.188
11	2.603	3.053	3.816	4.575	5.578	17.275	19.675	21.920	24.725	26.757
12	3.074	3.571	4.404	5.226	6.304	18.549	21.026	23.337	26.217	28.299
13	3.565	4.107	5.009	5.892	7.042	19.812	22.362	24.736	27.688	29.819
14	4.075	4.660	5.629	6.571	7.790	21.064	23.685	26.119	29.141	31.319
15	4.601	5.229	6.262	7.261	8.547	22.307	24.996	27.488	30.578	32.801
16	5.142	5.812	6.908	7.962	9.312	23.542	26.296	28.845	32.000	34.267
17	5.697	6.408	7.564	8.672	10.085	24.769	27.587	30.191	33.409	35.718
18	6.265	7.015	8.231	9.390	10.865	25.989	28.869	31.526	34.805	37.156
19	6.844	7.633	8.907	10.117	11.651	27.204	30.144	32.852	36.191	38.582
20	7.434	8.260	9.591	10.851	12.443	28.412	31.410	34.170	37.566	39.997
21	8.034	8.897	10.283	11.591	13.240	29.615	32.671	35.479	38.932	41.401
22	8.643	9.542	10.982	12.338	14.042	30.813	33.924	36.781	40.289	42.796
23	9.260	10.196	11.689	13.091	14.848	32.007	35.172	38.076	41.638	44.181
24	9.886	10.856	12.401	13.848	15.659	33.196	36.415	39.364	42.980	45.559
25	10.520	11.524	13.120	14.611	16.473	34.382	37.652	40.646	44.314	46.928
26	11.160	12.198	13.844	15.379	17.292	35.563	38.885	41.923	45.642	48.290
27	11.808	12.879	14.573	16.151	18.114	36.741	40.113	43.194	46.963	49.645
28	12.461	13.565	15.308	16.928	18.939	37.916	41.337	44.461	48.278	50.993
29	13.121	14.257	16.047	17.708	19.768	39.087	42.557	45.722	49.588	52.336
30	13.787	14.954	16.791	18.493	20.599	40.256	43.773	46.979	50.892	53.672
40	20.707	22.164	24.433	26.509	29.051	51.805	55.758	59.342	63.691	66.766
50	27.991	29.707	32.357	34.764	37.689	63.167	67.505	71.420	76.154	79.490
60	35.534	37.485	40.482	43.188	46.459	74.397	79.082	83.298	88.379	91.952
70	43.275	45.442	48.758	51.739	55.329	85.527	90.531	95.023	100.425	104.215
80	51.172	53.540	57.153	60.391	64.278	96.578	101.879	106.629	112.329	116.321
90	59.196	61.754	65.647	69.126	73.291	107.565	113.145	118.136	124.116	128.299
100	67.328	70.065	74.222	77.929	82.358	118.498	124.342	129.561	135.807	140.169

Right tail (used in Sections 9.6, 11.1, and 11.2)

α

χ^2_α

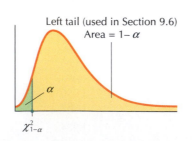

Left tail (used in Section 9.6)

Area = $1-\alpha$

α

$\chi^2_{1-\alpha}$

Two tails (used in Sections 8.4 and 9.6)

Area = $\dfrac{\alpha}{2}$

Area = $\dfrac{\alpha}{2}$

$\chi^2_{1-\alpha/2}$

$\chi^2_{\alpha/2}$

The area to the right of $\chi^2_{1-\alpha/2}$ is $1-\dfrac{\alpha}{2}$.